Strained-Layer Superlattices: Materials Science and Technology

SEMICONDUCTORS
AND SEMIMETALS
Volume 33

Semiconductors and Semimetals

A Treatise

Edited by R. K. Willardson
ENIMONT AMERICA, INC.
PHOENIX, ARIZONA

Albert C. Beer
BATTELLE COLUMBUS LABORATORIES
COLUMBUS, OHIO

Strained-Layer Superlattices: Materials Science and Technology

SEMICONDUCTORS
AND SEMIMETALS

Volume 33

Volume Editor

THOMAS P. PEARSALL

DEPARTMENT OF ELECTRICAL ENGINEERING
UNIVERSITY OF WASHINGTON
SEATTLE, WASHINGTON

ACADEMIC PRESS, INC.

Harcourt Brace Jovanovich, Publishers

*Boston San Diego New York
London Sydney Tokyo Toronto*

THIS BOOK IS PRINTED ON ACID-FREE PAPER.

COPYRIGHT © 1991 BY AT&T BELL LABORATORIES, INC.
ALL RIGHTS RESERVED.
NO PART OF THIS PUBLICATION MAY BE REPRODUCED OR
TRANSMITTED IN ANY FORM OR BY ANY MEANS, ELECTRONIC
OR MECHANICAL, INCLUDING PHOTOCOPY, RECORDING, OR
ANY INFORMATION STORAGE AND RETRIEVAL SYSTEM, WITHOUT
PERMISSION IN WRITING FROM THE PUBLISHER.

ACADEMIC PRESS, INC.
1250 Sixth Avenue, San Diego, CA 92101

United Kingdom Edition published by
ACADEMIC PRESS LIMITED
24–28 Oval Road, London NW1 7DX

The Library of Congress has cataloged this serial publication as follows:

Semiconductors and semimetals.—Vol. 1-—New York : Academic Press, 1966-

  v. : ill. ; 24 cm.

  Irregular.
  Each vol. has also a distinctive title.
  Edited by R. K. Willardson and Albert C. Beer.
  ISSN 0080-8784 = Semiconductors and semimetals

  1. Semiconductors—Collected works. 2. Semimetals—Collected works.
I. Willardson, Robert K. II. Beer, Albert C.
QC610.9.S48   621.3815′2—dc19   85-642319

ISBN 0-12-752133-X   (Vol. 33)

Printed in the United States of America
90 91 92 93   9 8 7 6 5 4 3 2 1

Contents

LIST OF CONTRIBUTORS vii
PREFACE ix

Chapter 1 Principles and Concepts of Strained-Layer Epitaxy   1
R. Hull and J. C. Bean

    I. Introduction 1
    II. Growth Techniques 3
    III. Importance of the Substrate Surface in Heteroepitaxial Growth 8
    IV. Nucleation and Growth Modes 16
    V. Strain Relief of Lattice-Mismatched Epitaxy 25
    VI. Atomic-Scale Structure of Epitaxial Layers 52
    VII. Summary 64
        References 67

Chapter 2 Device Applications of Strained-Layer Epitaxy   73
William J. Schaff, Paul J. Tasker, Mark C. Foisy, and Lester F. Eastman

    I. Introduction 73
    II. Properties of Strained Layers 77
    III. Electronic-Device Structures 100
    IV. Optoelectronic-Device Structures 120
    V. Optimized Lattice-Constant Epitaxy 127
    VI. Conclusions 131
        References 132

Chapter 3 Structure and Characterization of Strained-Layer Superlattices   139
S. T. Picraux, B. L. Doyle, and J. Y. Tsao

    I. Introduction 139
    II. Structure 143
    III. Ion-Scattering Characterization 147
    IV. X-ray Diffraction Characterization 170
    V. Other Characterization Techniques 189
    VI. Stability, Metastability, and Relaxation 198

VII.	Application to Single Strained Layers	205
VIII.	Application to Superlattices.	210
IX.	Conclusions.	219
	References	220

Chapter 4 Group-IV Compounds — 223
E. Kasper and F. Schäffler

I.	Introduction	223
II.	Material Properties	225
III.	Basic Principles	230
IV.	Stability of Strained-Layer Superlattices	240
V.	The $Si_{1-x}Ge_x/Si$ System	244
VI.	Outlook	303
	References	304

Chapter 5 Molecular-Beam Epitaxy of IV-VI Compound Heterojunctions and Superlattices — 311
Dale L. Partin

I.	Introduction	311
II.	Growth by Molecular-Beam Epitaxy	313
III.	Use of Lattice-Mismatched Substrates.	325
IV.	Strained-Layer Superlattices and Heterojunctions	329
V.	Conclusions.	333
	Acknowledgments	333
	References	333

Chapter 6 Molecular-Beam Epitaxy of II-VI Semiconductor Microstructures — 337
Robert L. Gunshor, Leslie A. Kolodziejski, Arto V. Nurmikko, and Nobuo Otsuka

I.	Introduction	338
II.	Growth by Molecular-Beam Epitaxy	340
III.	Use of Lattice-Mismatched Substrates.	363
IV.	Strained-Layer Superlattices and Heterojunctions	376
V.	Conclusions.	402
	Acknowledgments	403
	References	403

INDEX	411
CONTENTS OF PREVIOUS VOLUMES	423

Contributors

Numbers in parentheses indicate the pages on which the authors' contributions begin.

J. C. BEAN (1), *AT&T Bell Laboratories, 600 Mountain Avenue, Murray Hill, New Jersey 07974*

B. L. DOYLE (139), *Sandia National Laboratories, Division 1111, P.O. Box 5800, Albuquerque, New Mexico 87185-5800*

LESTER F. EASTMAN (73), *Department of Electrical Engineering, Phillips Hall, Cornell University, Ithaca, New York 14853*

MARK C. FOISY (73), *Department of Electrical Engineering, Phillips Hall, Cornell University, Ithaca, New York 14853*

ROBERT L. GUNSHOR (337), *School of Electrical Engineering, Purdue University, West Lafayette, Indiana 47907*

R. HULL (1), *AT&T Bell Laboratories, 600 Mountain Avenue, Murray Hill, New Jersey 07974*

E. KASPER (223), *Daimler-Benz Aktienqesellschaff, Research Center, Sedanstr. 10, D-7900 Ulm, Federal Republic of Germany*

LESLIE A. KOLODZIEJSKI (337), *Department of Electrical Engineering and Computer Science, Massachusetts Institute of Technology, Cambridge, Massachusetts 02139*

ARTO V. NURMIKKO (337), *Division of Engineering and Department of Physics, Brown University, Providence, Rhode Island 02912*

NOBUO OTSUKA (337), *School of Materials Engineering, Purdue University, West Lafayette, Indiana 47907*

DALE L. PARTIN (311), *Physics Department, General Motors Research Laboratory, Warren, Michigan, 48090-9055*

S. T. PICRAUX (139), *Sandia National Laboratories, P.O. Box 5800, Albuquerque, New Mexico 87185-5800*

WILLIAM J. SCHAFF (73), *Department of Electrical Engineering, Phillips Hall, Cornell University, Ithaca, New York 14853*

F. SCHÄFFLER (223), *Daimler-Benz AG, Research Center, Sedanstr. 10, D-7900 Ulm, Federal Republic of Germany*

PAUL J. TASKER (73), *Department of Electrical Engineering, Phillips Hall, Cornell University, Ithaca, New York 14853*

J. Y. TSAO (139), *Sandia National Laboratories, Organization 11-41, P.O. Box 5800, Albuquerque, New Mexico 87185-5800*

Preface

During the last decade, it has been acknowledged that the technology of silicon integrated circuits is approaching fundamental limits set by the atomic nature of matter. It is no longer possible to count on a doubling of chip capacity every few years. At the same time, the telecommunications and recording industries have driven the development of an economical and reliable optoelectronics technology. Optoelectronics offers *new functionality* to conventional silicon-based circuits. Development and integration of this new functionality is essential to the continued expansion of the information processing capacity of integrated circuits. Yet a vast gulf continues to separate the technologies for optoelectronics and silicon-based devices because they are based on dissimilar materials. These two volumes (Volumes 32 and 33) on strained-layer superlattices are dedicated to the idea that this gulf will be short-lived.

In 1982 Gordon Osbourn of Sandia Laboratories made the link between strained-layer structures and the need for new functionality in integrated circuit design.[1] Osbourn considered the conventional wisdom that all strain in semiconducting devices was bad and stood it on its head by proposing that the strain associated with the heteroepitaxy of dissimilar materials may itself offer *new functionality*, whose advantages may far outweigh the disadvantages of the presence of strain. In 1986, Temkin *et al.*[2] tested a device that illustrates the possibilities opened up by this breakthrough in thinking: a silicon-based photodiode with an absorption edge, strain-shifted to the 1.3–1.5 μm window for optical fiber communications.

The physics and technology of semiconductor strained-layer superlattices are surveyed in this two-volume set. Of course, the field of activity is wide and growing. The contents of this set should not be viewed as a review, but rather as a milestone in research and development that will play an important part in the evolution of semiconductor device technology.

Thomas P. Pearsall
March 28, 1990

[1] Osbourn, G. C. (1982), *J. Appl. Phys.* **53**, 1586.
[2] Temkin, H., Pearsall, T. P., Bean, J. C., Logan, R. A., and Luryi, S. (1986), *Appl. Phys. Lett.* **48**, 330.

CHAPTER 1

Principles and Concepts of Strained-Layer Epitaxy

R. Hull and J. C. Bean

AT&T BELL LABORATORIES
MURRAY HILL, NEW JERSEY

I.	Introduction	1
II.	Growth Techniques	3
	A. Introduction	3
	B. Molecular-Beam Epitaxy and Atomic-Layer Epitaxy	3
	C. Gas-Source Molecular-Beam Epitaxy	7
	D. Chemical Vapor Deposition	8
III.	Importance of the Substrate Surface in Heteroepitaxial Growth	8
IV.	Nucleation and Growth Modes	16
V.	Strain Relief in Lattice-Mismatched Epitaxy	25
	A. Introduction	25
	B. Review of General Dislocation Properties	27
	C. Critical Thickness	29
	D. Details of the Strain Relaxation Process	33
	E. Relaxation in Strained Clusters	43
	F. Critical-Thickness Phenomena in Multilayer Structures	44
	G. Techniques for Dislocation Reduction in Strained-Layer Epitaxy	47
VI.	Atomic-Scale Structure of Epitaxial Layers	52
	A. Introduction	52
	B. Theoretical Description of Isostructural Interface Structure	52
	C. Origins of Interface Roughness and Diffuseness	54
	D. Experimental Techniques	55
	E. Nonisostructural Interfaces	61
	F. Structure of Epitaxial Semiconductor Alloys	62
VII.	Summary	64
	Acknowledgments	66
	References	67

I. Introduction

Atomic-scale growth control of artificially modulated structures by advanced crystal-growth techniques such as molecular-beam epitaxy (MBE) and metal–organic chemical vapor deposition (MOCVD) has in the last decade extended the frontiers of materials science, physics, and semiconductor device performance. Historically, the initial focus has been on the growth of material combinations with essentially the same crystal lattice parameter and structure, e.g., $Al_xGa_{1-x}/GaAs$. Although the growth of ultrahigh electrical and structural quality material in these and similar systems has required both perseverance and inspiration, the fundamental physical constraints on successfully defining the epitaxial structure (i.e., point and line defect free, with planar surface morphology) are greatly eased when the equilibrium structure of the grown layers corresponds closely to that of the substrate.

Interest in extending the range of possible material combinations has, however, encouraged experimentation in lattice-mismatched epitaxial structures. In addition to the problems encountered (and to a large extent, solved) in lattice-matched epitaxy, the primary extra complication introduced by this extra degree of freedom is the introduction of extended defects that attempt to relieve elastic strain in the structure. Understanding and control of these defects appears to be the principal challenge faced in strained-layer epitaxy, and progress to date will be discussed in detail later in this chapter. Other phenomena present in lattice-matched epitaxy, e.g., surface diffusion and clustering phenomena, may be more significant in mismatched-epitaxy, due to the likelihood of a greater chemical dissimilarity between materials in the structure. Indeed, we shall aim to show in this chapter that each and every stage in the heteroepitaxial growth process, from substrate preparation to post-growth processing, is of critical importance in determining the final structure.

The field of strained-layer epitaxial growth is very much in its formative stages and is continually evolving. Care will be taken in this chapter to attemt to differentiate among those problems that appear presently to be understood, those in which an understanding is being developed, and those that to date remain intractable. The following fundamental stages of the heteroepitaxial growth process should be considered:

(i) Construction of a suitable growth chamber;
(ii) Preparation of the substrate surface;
(iii) Possible growth of a homoepitaxial buffer layer onto the as-cleaned substrate surface;
(iv) Nucleation of the heteroepitaxial layer: clustering or layer-by-layer growth;
(v) Introduction of extended defects (if critical-layer dimensions are exceeded) to relax the elastic strain introduced by lattice mismatch;
(vi) Evolution of the growth surface, i.e., coalescence of individual nuclei, or possibly roughening of a planar surface. Note that this stage could occur before or after step (v); and

(vii) Redistribution of defect populations within the epitaxial layer, such as by defect "filtering" via incorporation of strained-layer superlattices or termination of a dislocation at the edge of a structure.

Each of these processes will be discussed in succeeding sections of this chapter. Our aim will not be to review exhaustively all work done in the field (specific materials systems are reviewed in other chapters of this volume), but rather to outline the fundamental physical processes involved in strained-layer epitaxy and to illustrate them with specific examples.

II. Growth Techniques

A. INTRODUCTION

The fundamental tool of lattice-mismatched epitaxy is the crystal-growth chamber. The major requirements of a growth chamber for high-quality epitaxial growth include: a noncontaminating (ultrahigh vacuum or inert) environment; source purity; a source-substrate geometry that allows deposition uniformity; uniform substrate heating/cooling; absence of particulates and undesired impurity atoms; sufficient analytical techniques to allow in situ monitoring of growth quality and control of layer composition and thickness. In a research environment, throughput of wafers is less critical, therefore ultrahigh vacuum (UHV) techniques, particularly solid-evaporation-source MBE, are acceptable. MBE has two relevant advantages: (1) exceedingly good control of layer dimensions and composition; (2) growth at minimal temperatures (generally well below those required for significant bulk diffusion but not necessarily surface diffusion). Low-temperature growth is an absolute requirement if one is to attempt metastable strained-layer epitaxy (as in the germanium silicide system).

The following sections give a brief overview of the crystal growth techniques being used for strained-layer growth including molecular-beam epitaxy, atomic-layer epitaxy, gas-source molecular beam epitaxy and chemical vapor deposition (CVD). However, in subsequent sections, discussion will tend to focus on MBE. This is justified by the current dominance of MBE in strained-layer epitaxy and by the fact that, with MBE, it is particularly straightforward to define the role of substrate preparation, impurity effects, and nucleation. Nevertheless, the reader should bear in mind that these issues and discussions are generic to all crystal growth techniques.

B. MOLECULAR-BEAM EPITAXY AND ATOMIC-LAYER EPITAXY

Some important principles of the MBE growth technique are illustrated in Figs. 1a and 1b. The essential requirement for a MBE growth chamber is that

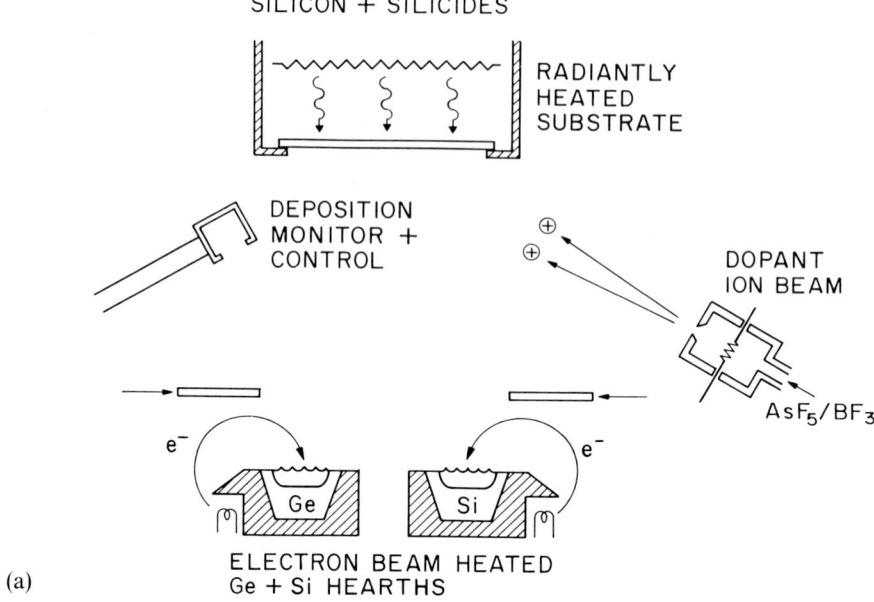

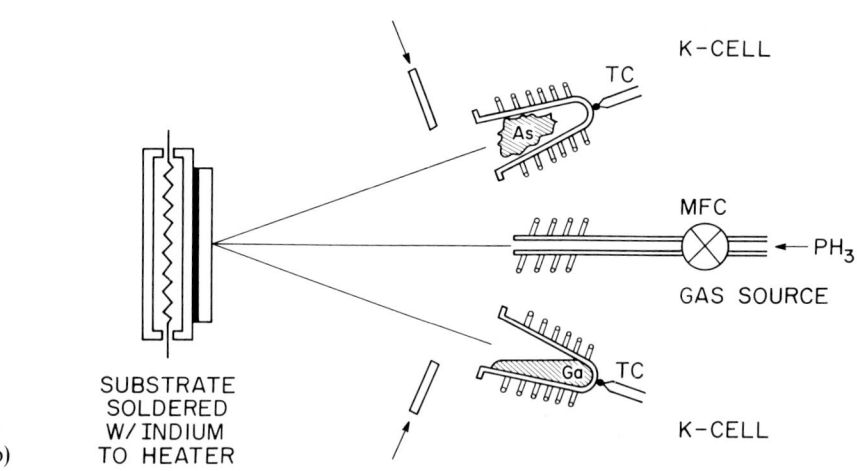

Fig. 1. Schematic diagram of a molecular-beam epitaxy growth chamber (a) for column-IV deposition, (b) for III–V deposition.

growth take place at a sufficiently high vacuum so that arrival and sticking of impurity atoms from the ambient occur at a negligible rate. This is particularly important in the growth of strained-layer heteroepitaxy. On growing surfaces, impurities can massively perturb growth by providing heterogeneous cluster or defect nucleation sites. Within the crystal, impurities may precipitate to form nucleation sites for strain-relieving dislocations.

To first order, impurity gas impingement varies inversely with partial pressure with a pressure of 10^{-6} torr producing a flux of 1 atomic monolayer per second. Thus, a total chamber pressure of 10^{-10} torr corresponds to an arrival rate of $\sim 10^{-4}$ atoms or molecules $\sec^{-1}$ cm^{-2} on the growth surface. For typical MBE growth rates of ~ 1 monolayer $\sec^{-1}$, this relatively low arrival rate would still produce an unacceptably high impurity concentration in the growing epilayer, were the sticking coefficient for the impurities unity.

Fortunately, at the elevated temperatures (of the order of 700–1200 K) generally employed during substrate cleaning and epilayer growth, sticking coefficients for most UHV constituents are many orders of magnitude less than one, permitting high purity growth. However, there are exceptions, and these exceptions strongly affect the design of the MBE growth chamber. In particular, growing AlAs and AlGaAs alloys have an extremely strong affinity for oxygen derived from the strength of the oxygen–aluminium chemical bond. Oxygen can affect layer morphology and electronic carrier transport. In a leak-free MBE system, this oxygen comes from decomposition of ambient water vapor held over from earlier venting of the vacuum chamber. This can be largely eliminated by the addition of extensive internal shrouds filled with liquid nitrogen on which the water is trapped. With the advent of plumbed-in liquid-nitrogen supply systems, these shrouds are frequently cooled not only during growth but continuously, over months of operation.

The strength of the carbon–silicon bond poses a similar problem in MBE of silicon and silicon alloys. Although carbon is relatively soluble in silicon, its presence at a growing crystal surface can lead to the formation of silicon carbide nuclei. These nuclei are extremely stable and, given their hexagonal crystal structure, they provide ready sites for defect nucleation in the cubic silicon lattice. Carbon can come from several sources. It is present in all but the most carefully controlled chemical cleaning solutions. Even if such control is maintained, an atomically clean surface will immediately react with carbon in ambient air. Ex situ cleaning is therefore generally terminated with the formation of a comparably inert chemical oxide that can be readily reduced and desorbed by heating in vacuum. While such surfaces are adequate for basic studies of silicon homoepitaxy, residual carbon can still adversely affect heteroepitaxial nucleation and strained-layer relief, as detailed later.

Carbon may also come from the decomposition of oils used in certain vacuum pumps. In the last decade or so, such pumps had been largely eliminated in favor of oil-free ion pumps and closed-cycle He cryogenic condensation pumps. Recently, however, manufacturability issues have stimulated renewed interest in gas-source MBE or MBE/CVD hybrid vacuum technologies. Ion pumps cannot handle significant gas loads, and the accumulation of typical gas-source chemicals within cryogenic pumps poses both toxic hazards and the possibility of ignition and/or explosion. Gas-source systems must therefore use nonaccumulating pumps such as diffusion pumps or turbo-molecular pumps. Carbon contamination from such pumps proved to be the bane of early attempts at silicon MBE (Joyce, 1974). The challenge is thus to develop hydrocarbon trapping techniques and chemically resistant oil-free fore-pumps. Whereas this might appear to be a problem unique to silicon MBE, the handling of fore-pump oils contaminated with III–V toxic materials is already a significant safety concern. At least one organization is already recommending the elimination of conventional oil fore-pumps on all gas-source systems.

Ironically, the reduction of oxygen and carbon contamination in solid-source MBE systems has introduced another defect-producing mechanism: particulates. In a solid-source MBE system (or a hot-wall CVD-like system), material will deposit in areas other than the targeted substrate. If vacuum is maintained for long periods, the buildup of material will produce strains leading to fracture and the release of very fine particulates. (Gross flaking may also occur, but it poses less of a problem.) Under the influence of applied electric fields or even typical MBE vapor fluxes, these particulates can actually be propelled upward onto a growing epitaxial surface (Matteson and Bowling, 1988).

Uncontrolled, these particulates can produce defect densities of 100–1000 per square centimeter (Bellevance, 1988). There are two emerging strategies to control this problem. This first is to tightly columnate the deposition fluxes to the substrate surface alone. The columnators are then replaced or cleaned of accumulated deposits at each vacuum break. The second approach is to grossly reduce thermal cycling within the MBE chamber, thereby maximizing deposit adhesion. In AlGaAs systems, this is accomplished by continuous cooling of liquid-nitrogen shrouds. In silicon-based MBE systems, the newest generation of equipment goes one step further. Because oxygen does not bond strongly to heated silicon, water vapor is not as critical a concern. Liquid-nitrogen shrouding is thus being removed from the growth area in favor of water cooling of either shrouds or chamber walls (e.g., Parker and Whall, 1988).

Assuming one can provide a suitably clean environment for molecular-beam epitaxy, the next problem is that of controlling layer composition. In an alloy system, such as Ge_xSi_{1-x}, significant errors may be tolerable. Growth of compounds, e.g. GaAs, however, requires control of stoichiometry to levels

1. PRINCIPLES AND CONCEPTS OF STRAINED-LAYER EPITAXY

better than one part per million. If such control is not maintained, second-phase inclusions will form and crystal growth will be massively disrupted. Figures 1a and 1b illustrate typical schemes for the growth of III-V and column-IV materials, respectively. Neither temperature-controlled Knudsen cells, mass-flows-controlled gas sources, nor sensor-controlled electron guns offer the requisite part-per-million regulation.

MBE thus depends on one of two mechanisms to maintain stoichiometry. For common III-V semiconductors, it was found that although the column-V species will bind tightly to a freshly deposited column-III layer, it will not bond well to another column-V layer. In terms of GaAs, this is to say that As will bind to Ga but tends to re-evaporate from another layer of As. Stoichiometry is thus maintained if one provides an excess flux of the column-V species. The crystal then adsorbs only the species it requires.

Atomic-layer epitaxy (ALE) simply takes this process one step further: For example, in growth of certain II-VI semiconductors, the complementary process is also active, and whereas column-II materials will bind to column VI, neither will bind to itself (at appropriate growth temperatures). Growth rate is then independent of the incoming flux and depends only on the number of times the substrate is exposed to alternate pulses of column-II and VI atoms (as each pulse produces a single, self-limiting atomic layer of deposition). This self-balancing process is essentially the same as that active in liquid-phase epitaxy or chemical vapor deposition.

The second self-balancing mechanism is operative in MBE growth of compounds such as $CoSi_2$ on Si. Neither Co nor Si will re-evaporate, but excess Co will readily diffuse through a thin $CoSi_2$ layer to react at the silicon substrate. In essence, stoichiometry is achieved by simultaneous vapor-phase crystal growth at the surface and solid-phase growth at the substrate. Excess metal fluxes may thus be used in thin layers, but as the epilayer thickness increases, diffusive transport of Co through the epilayer becomes increasingly difficult, and growth will ultimately break down.

C. Gas-Source Molecular-Beam Epitaxy

For the purposes of this chapter, gas-source MBE (GSMBE) can be considered a simple derivative of the solid-source process. Gas sources address three weaknesses of conventional MBE. First, Knudsen evaporation sources have a finite capacity (of the order of 50–100 cc) and may be depleted within as little as a month. This is particularly true for the column-V species, where excess fluxes must be maintained in order to assure stoichiometric growth of a III–V compound. Further, before a cell is fully depleted, there will be serious shifts in evaporation rates due to the reduction of charge size. To compensate for this, frequent, sacrificial calibration runs must be made, and for very critical structures, the shift of calibration within a single run may become unacceptable. The use of external, easily replaceable, gas cylinders eliminates this problem. Fluxes are continuously calibrated and regulated by mass-flow controllers or temperature baths. Moreover, the elimination of vacuum breaks not only increases chamber up-time but can result in a significant overall reduction of vacuum impurity levels.

Gas sources have a second advantage in that they overcome difficulties in handling pyrophoric species such as white phosphorus. Phosphine, although highly toxic, is not spontaneously flammable and with proper handling will not accumulate within the MBE system. This has opened the door to the highly successful growth of GaInAsP species by MBE.

Finally, with certain chemistries and growth temperatures, the species from gas sources may decompose at the heated substrate surface alone. There will be little or no wall deposition, and the elimination of such accumulations will largely eliminate particulate contamination due to flaking. Although not currently recognized as an issue in III-V MBE, it has been shown that solid-source wall deposits can limit silicon MBE materials to defect densities of above $100-1000/cm^2$, as discussed earlier.

D. CHEMICAL VAPOR DEPOSITION

Turning history around a bit, CVD can be viewed as a high-pressure variant of gas-source MBE (GSMBE). The chemistries are fundamentally similar; the differences depend primarily on what is done with the gaseous species. In CVD, gaseous species are delivered to the substrate wafer in their original form, that is, as column-V hydrides or column-III organics. In GSMBE, the hydride is often thermally "cracked" into a subhydride or pure column-V species by passing through a heated nozzle (often including a catalyzing metal). For the purposes of this chapter, the major possible advantage of this cracking process is that it can permit growth of epitaxy at lower substrate temperatures. This could be important if one is attempting to maintain strained-layer epitaxy to thicknesses above equilibrium limits. The reduction of temperature would then inhibit nucleation and growth of defects and thus reduce the relaxation of strain. However, to date, such metastable growth has only been well documented in the GeSi system. Thus, for the bulk of III-V growth, the absence of precracking places CVD at no significant disadvantage, and enhancements in wafer size and sample preparation may actually make it more desirable than MBE for the bulk of III-V growth systems.

III. Importance of the Substrate Surface in Heteroepitaxial Growth

The initial stage of the deposition process is to obtain a suitable surface on which to initiate growth. The primary concern is to obtain a surface that is atomically clean, i.e., stripped of its native oxide and free of any other surface or near-surface contaminants. To prevent reoxidation or contamination of the cleaned surface, at least the final stages of the cleaning process are usually done in situ under UHV conditions.

In general, the requirements for cleaning elemental column-IV substrates are more stringent than those for cleaning of III-V-compound semiconductor substrates. This is driven both by technological requirements and by the fact that native oxides desorb at significantly lower temperatures for the latter class of materials. By current standards, an adequate GaAs substrate can be prepared by ex situ wafer degreasing, followed by in situ oxide desorption at $\sim 600°C$ and growth of a homoepitaxial buffer layer. To produce state-of-the-art Si epitaxy, much more ingenuity and care are required. Tendencies toward islanded growth also make silicon substrate preparation a major issue in strained-layer growth of disparate materials such as GaAs on Si.

Conventional silicon substrate cleaning techniques fall into three general classes: (i) ex situ chemical cleaning followed by growth of a relatively volatile surface $SiO_x (x \sim 1)$ layer, followed by in situ desorption of the volatile oxide, this occuring at much lower temperatures than for the native SiO_2 oxide; (ii) in situ removal of the surface and near-surface region by sputtering with inert (generally argon) ions, followed by a thermal anneal to remove sputtering damage; (iii) in situ back-etching of the substrate under chlorine gas to remove several hundred angstroms of material. Only the later back-etching technique has demonstrated quality adequate for commercial production of complex integrated circuits. Unfortunately, its use is restricted to halogen CVD growth, and virtually all the work in this book is based on the first two approaches.

In the first class of cleaning techniques, the major requirements are first to chemically strip the Si-substrate surface in a manner that does not leave any detectable contaminants, especially carbon or metallic species, second, to produce a volatile oxide layer of uniform stoichiometry and thickness, and third, to successfully desorb the volatile oxide in situ in the growth chamber. None of these requirements are easy to satisfy. A variety of chemical cleaning techniques have been developed to satisfy the first two requirements above, e.g., the Henderson (1972) and Shiraki–Ishizaka (Ishizaka et al., 1983) cleans. The latter treatment is most widely used and consists of repeated oxidation and stripping of the surface in nitric and hydrofluoric acids, respectively, followed by growth of an approximately 10-Å-thick SiO layer in a hydrogen peroxide–hydrochloric acid mixture. In the original reference, it was suggested that desorption temperatures as low as 750°C would be sufficient to produce an oxygen-free surface; it has been our experience and that of others (e.g., Xie et al., 1986) that somewhat higher temperatures ($> \sim 900°C$) are required to produce truly clean surfaces using this technique. This is illustrated by secondary ion mass spectroscopy (SIMS) data showing the residual surface contamination following Shiraki oxide desorption at various temperatures in Fig. 2.

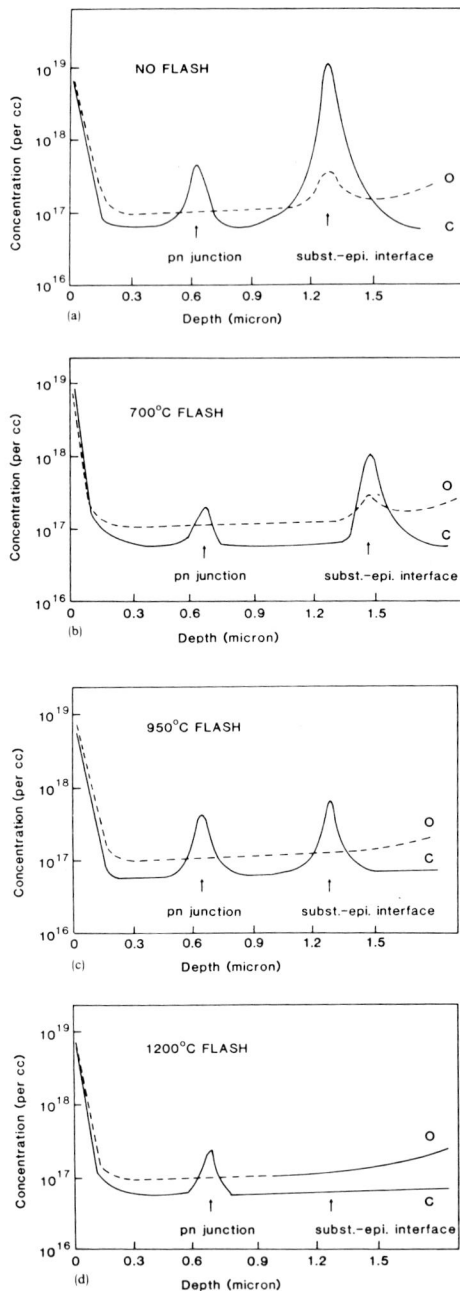

FIG. 2. Secondary ion mass spectroscopy of residual O and C surface contamination following Shiraki cleaning of Si(100) wafers at various SiO desorption temperatures (from Xie et al., 1986).

A number of variants of the basic Henderson and Shiraki methods have been developed. Oxide desorption has been shown to be more effective when a Si flux impinges upon the surface (Tabe et al., 1981; Hardeman et al., 1985), presumably because pockets of O-rich SiO_x can be converted to the correct monoxide stoichiometry. Another variant involves ozone cleaning under ultraviolet irradiation between the volatile oxide growth and desorption stages (Tabe, 1984); this technique has been shown to be extremely effective at reducing surface organic contaminants, reducing homoepitaxial defect densities by ~ 2 orders of magnitude on Si(100) and (111) surfaces for desorption temperatures of 800–900°C.

The second major class of cleaning techniques consists of in situ sputtering with inert gas ions to remove the near-surface region. In our laboratory, Si surfaces are cleaned using $\sim Ar^+$ ions accelerated to 0.2 keV, with the substrate held at room temperature. Approximately 100 Å of material is removed, leaving a clean but disordered surface. The substrate is then annealed at 750°C for five to fifteen minutes. The amorphous surface layer reorders by solid-phase epitaxy as the sample passes through 500–600°C, and residual point defects are annealed out at 750°C. Defect densities in homoepitaxial layers grown upon these cleaned surfaces are of the order of $10^2 - 10^3 \text{ cm}^{-2}$, comparable to or better than Shiraki-type cleaning techniques at 800°C.

A recently developed cleaning technique (Grunthaner et al., 1988) offers promise of clean Si surfaces at far lower temperatures. This process involves ex situ removal of a thin 10 Å chemical oxide by spinning in a N_2 dry box loadlocked to the MBE chamber, while rinsing/etching first in pure ethanol (added dropwise), then in 1:10 HF:ethanol, and finally in pure ethanol. This technique produces an atomically clean (to a level at least comparable to standard in situ volatile oxide desorption at temperatures of $\sim 800°C$) hydrogen-passivated 1×1 surface after heating to 150°C. Conversion to a 7×7 reconstruction on a (111) surface occurs at temperatures of $\sim 500°C$.

Although further improvements in Si cleaning techniques are both preferable and possible, it appears that careful control of these processes can produce Si surfaces that are sufficiently clean to allow high-quality heteroepitaxial growth upon them. An area that has hitherto received little attention, however, is the effect of the substrate cleaning on surface morphology and the resultant heteronucleation stage. For growth on Si substrates, we have observed that both major classes of cleaning techniques can affect the surface morphology and influence the initial stages of heteroepitaxial growth.

In Fig. 3, we show high-resolution cross-sectional transmission electron microscope (TEM) images of the early nucleation stages of GaAs grown on a Si substrate cleaned by the Ishizaka process (for exact details, see Koch et al., 1987). The substrate orientation is with the surface normal 4 degrees from

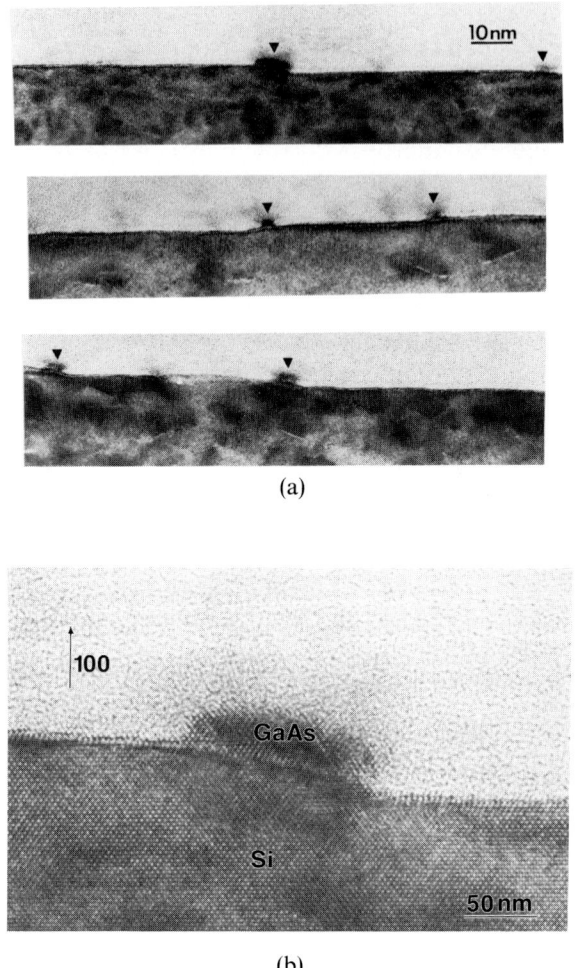

FIG. 3. (a) Cross-sectional TEM images of GaAs nucleation onto a vicinal (4 degrees toward $\langle 011 \rangle$)Si(100) wafer following SiO desorption at 880°C for 20 minutes. Note nucleation of GaAs on surface facets (arrowed) (from Hull et al., 1987b); (b) lattice structure image of one facet (from Hull et al., 1987a, b).

[100], the direction of the misorientation being towards an in-plane $\langle 011 \rangle$ azimuth. Growth of GaAs on such misoriented wafers historically believed to produce a uniform array of (200) steps on the Si surface, which prevents the formation of antiphase boundaries at the GaAs–Si interface (e.g., Kroemer, 1986).

From the images of Fig. 3, we note, however, that the Si substrate surface does not consist of a regular array of (200) steps and (100) terrace, but rather displays a distribution of "step groups" approximating low-angle facets many monolayers high. Note that nucleation of GaAs islands (arrowed) is associated with these facets. The effect of this facet–nucleation correlation is strongly demonstrated in Fig. 4, which is a "plan-view" TEM image where the electron beam is approximately (depending on the exact electron diffraction conditions used) perpendicular to the Si surface. The GaAs nuclei show as long strings of material nucleating along the [011] direction of the crystal, which equates to the direction of surface steps induced by the deliberate substrate misorientation. By comparing measured facet heights with the measured distances between "strings" of GaAs nuclei (Hull *et al.*, 1987a,b), we have been able to show statistically that the GaAs nucleation density is controlled by the substrate surface facet distribution.

We have ascertained by TEM that the interface between the as-grown volatile oxide and substrate surface (i.e., before the oxide desorption stage) is not similarly facetted. Thus the surface facetting appears to occur during the oxide desorption stage. In situ reflection high-energy electron diffraction (RHEED) measurements show that the initial surface structure of the misoriented Si(100) surface during the early stages of the oxide desorption process is a combination of the two orthogonal $(2 \times 1) + (1 \times 2)$ reconstructions possible on the (100) surface. The presence of these two domains indicates the existence of single monolayer (400) steps on the Si surface. As oxide desorption progresses, the RHEED pattern generally evolves into that expected from a single 2×1 surface domain, indicating reconstruction of pairs of (400) into (200) steps (see, e.g., Kaplan, 1980; Kroemer, 1986). Note that the efficiency of this process depends on the magnitude of the substrate misorientation toward [011], being most effective for tilts of 2–4 degrees (Koch *et al.*, 1987; Griffith *et al.*, 1988; Wierenga *et al.*, 1987). This is demonstrated beautifully by the scanning tunneling microscope images of

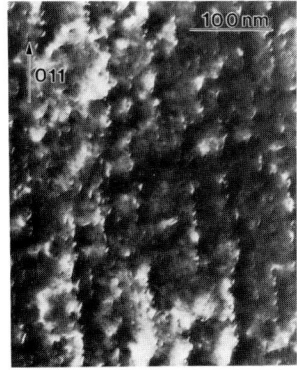

FIG. 4 Plan-view TEM image of the structure imaged in cross-section in Fig. 3. Note anisotropic nucleation of GaAs along ⟨011⟩ facets (Hull *et al.*, 1987a,b).

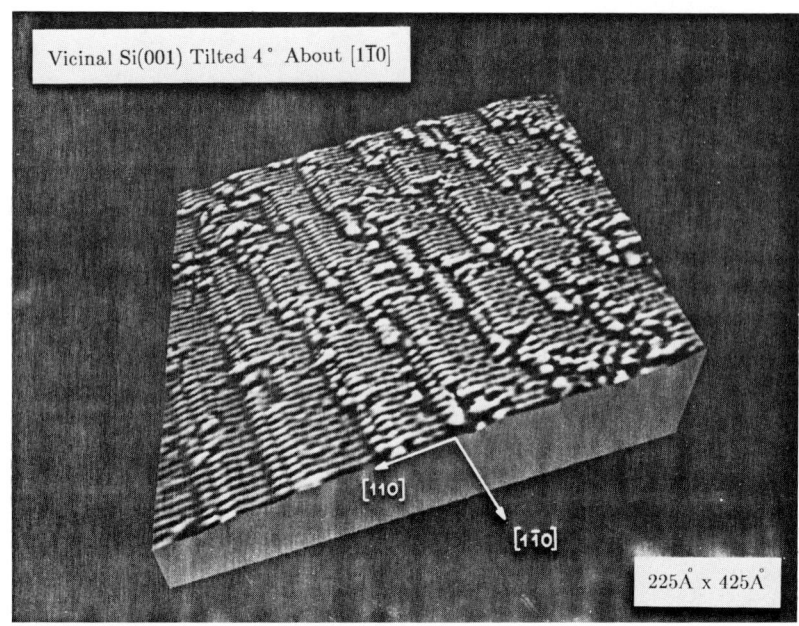

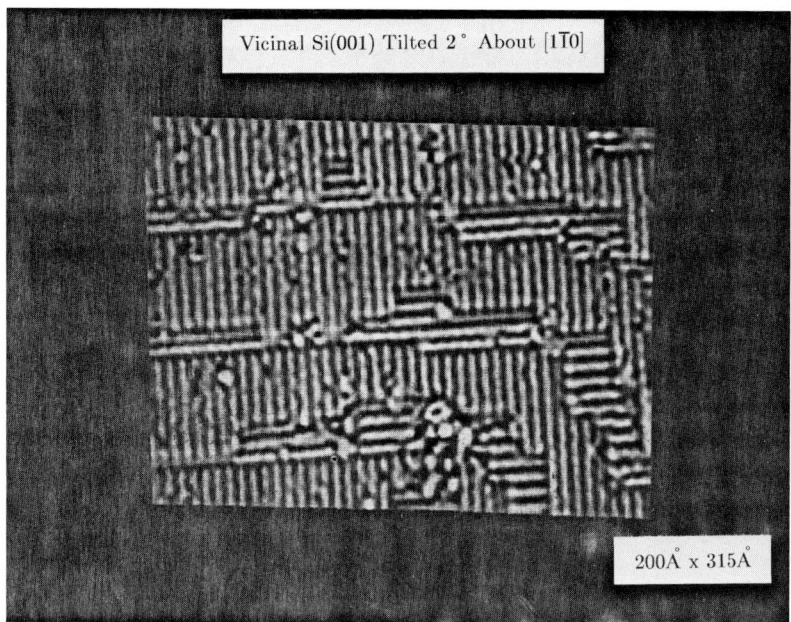

FIG. 5. Scanning tunneling microscope image of (a) a 4-degree and (b) a 2-degree misoriented (toward an [011] azimuth) silicon (100) surface, cleaned by Ar^+ annealing and sputtering. A higher proportion of double monolayer steps exists on the higher misorientation surface. (From Griffith, G. P. Kochanski, J. A. Kubby and P. E. Wierenga, J. Vac. Sci. Tech. **A1**, 1989).

Fig. 5, which demonstrate the difference in step structure for "low" and "high" substrate misorientations.

The oxide desorption temperature used in these experiments, $\sim 900°C$, is well below the expected roughening temperature on Si(100) surfaces. Measurements of the degree of surface facet height as a function of oxide desorption time (Hull *et al.*, 1987c) shows increasing facetting with longer time at the desorption temperature. A preliminary model for this facetting mechanism is shown in Fig. 6. Local nonstoichiometries (oxide rich, probably SiO_2) in the Shiraki oxide require excess Si to achieve the correct SiO composition for low temperature desorption. In the absence of an impinging Si flux, this excess Si may arrive only from the substrate. The easiest way to provide the required Si flux is via flow of surface steps (produced by the substrate misorientation) towards the O-rich region. In this manner, a facet is produced at this location. A similar mechanism of Si surface diffusion has been identified during decomposition of SiO_2 on Si(100) via reduction to SiO (Tromp *et al.*, 1985; Rubloff *et al.*, 1986). Motion of surface steps could also be impeded by surface impurities such as carbon, again producing step "groups." Nucleation studies of GaAs on vicinal Si(100) using UV-ozone substrate cleaning have demonstrated a much lower degree of facetting and significantly more isotropic nucleation of GaAs/Si (Biegelsen *et al.*, 1988), suggesting a more uniform Shiraki oxide and/or lower surface contamination levels by this technique. It has also been suggested that As adsorption onto a vicinal Si(100) surface can itself produce a facetting transition as the sample is cooled through a temperature of $\sim 800°C$ (Ohno and Williams, 1989).

The above results indicate that the heteronucleation process can be strongly influenced by the substrate cleaning technique and may in fact be controllable by suitable choice of a substrate surface "template." We note also that ion sputter cleaning of Si substrates may also influence surface structure and subsequent heteronucleation (Hull *et al.*, 1987c). Since cleaning of III-V compound semiconductor substrates is generally easier and more standardized, we are not aware of any similar reports in which the surface cleaning technique has been correlated to the nucleation mode in III-V epitaxy. Certainly, in the limit where the cleaning technique significantly affects surface stoichiometry and morphology, however, such effects would be very likely. These are most likely to arise where integrated growth of different

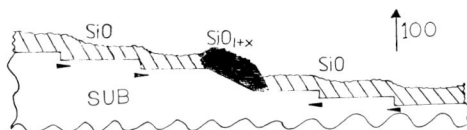

FIG. 6. Schematic illustration of possible model for surface facetting on Shiraki-cleaned vicinal Si(100) surfaces. (From Hull *et al.* 1987c. Paper originally presented at the Fall 1987 meeting of the Electrochemical Society held in Honolulu, Hawaii).

semiconductor "families" practically prevents growth of a homoepitaxial buffer layer onto the original cleaned surface. In the example given earlier, GaAs is usually grown on Si substrates in modified III-V MBE chambers or in MOCVD chambers without a silane source. Thus, GaAs is deposited directly onto the as-cleaned Si surface without intervention of a Si buffer layer. Such a buffer layer, if present, might be expected to modify the as-cleaned surface morphology. In compound semiconductor growth, before deposition of, say, a III-V heterostructure on a GaAs substrate (e.g., AlGaAs/GaAs or InGaAs/GaAs), growth of a homoepitaxial buffer layer on the as-cleaned substrate is standard. This might be expected to effectively bury any as-cleaned surface nonstoichiometries or other anomalies. In "mixed" deposition, such as deposition of II-VI materials onto a III-V substrate without intervention of a homoepitaxial buffer layer, the effect of the surface clean is more likely to be noticeable. An example of this is the fascinating phenomenon of "double positioning" observed for growth of CdTe/GaAs (Kolodziejski et al., 1986; Faurie et al., 1986a; Ponce et al., 1987), where CdTe deposited directly onto an as-cleaned GaAs substrate is observed to grow in two separate epitaxial orientations (100_{GaAs} parallel to 100_{CdTe} and 100_{GaAs} parallel to 111_{CdTe}), as shown in Fig. 7. This effect has been correlated to the extent of the presence of residual oxide on the cleaned GaAs surface. Development of linked "double-chamber" systems, with growth of III-V materials in one chamber and II-VI materials in the other will help overcome such difficulties. Note that it has also recently been reported (Reno et al., 1988) that deliberately induced substrate misorientation can improve the structural quality of growth of (111) CdTe on (100) GaAs, and that for growth of $Cd_{1-x}Zn_xTe/GaAs(100)$, the epitaxial orientational relationship between substrate and epilayer depends upon the composition x (Faurie et al., 1986b).

IV. Nucleation and Growth Modes

A critical factor in the practical applicability of strained-layer heteroepitaxial structures is the thickness uniformity of the grown layers. Three general heteroepitaxial growth modes have been observed, as illustrated in Fig. 8: (a) layer-by-layer or two-dimensional growth (Frank–Van der Merwe mode) of the epilayer; (b) clustered or three-dimensional growth (Vollmer–Weber mode); and (c) a hybrid growth mode known as Stranski–Krastanow (Stranski and Krastanow, 1939; Bauer and Poppa, 1972; Fisanick et al., 1988), in which the growth mode is initially layer-by-layer for a few monolayers and subsequently clusters. Unfortunately, the general growth mode of an epitaxial layer on a chemically dissimilar substrate consists of nucleation of clusters of the deposited material, as may be seen from the force balance equation for the

1. PRINCIPLES AND CONCEPTS OF STRAINED-LAYER EPITAXY

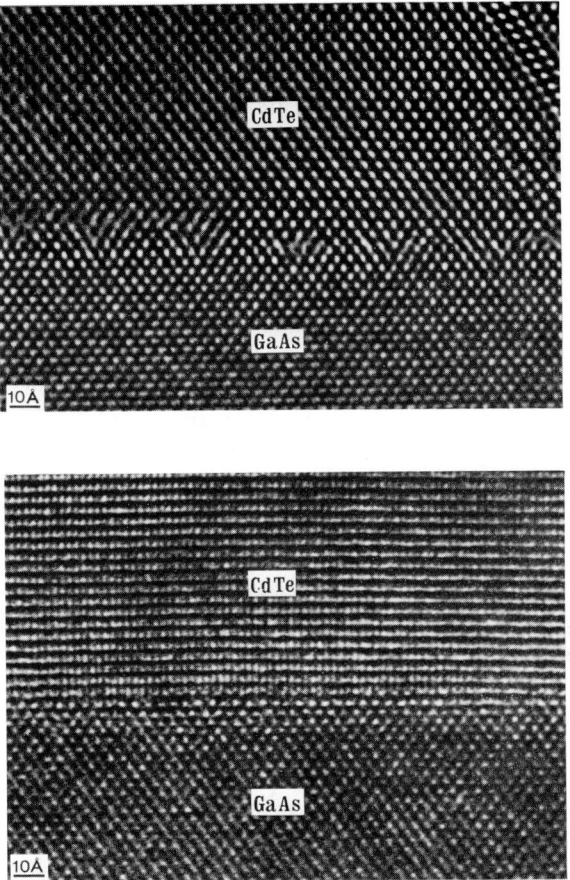

FIG. 7. Cross-sectional electron microscope images of double positioning of CdTe/GaAs(100): (a) [100] CdTe parallel to [100] GaAs; (b) [111] CdTe parallel to [100] GaAs (from Kolodziejski, et al., 1986. Micrographs supplied by N. Otsuka).

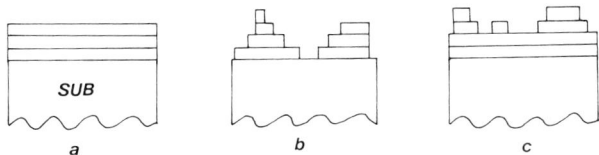

FIG. 8. Schematic illustration of the three primary heteronucleation modes: (a) layer-by-layer (Frank–Van der Merwe); (b) clustered (Vollmer–Weber) and (c) Stranski–Krastanow.

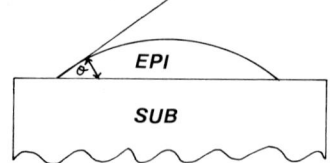

Fig. 9. Schematic illustration of the liquid-drop model for predictiong equilibrium cluster contact angles.

liquid-drop model of a nucleus,

$$\sigma_{sv} = \sigma_{es} + \sigma_{ev}\cos\theta, \tag{1}$$

where σ_{sv} is the energy per unit area of the substrate–vapor (for MBE, the substrate–vacuum) interface, σ_{ev} is the energy per unit area of the epilayer–vapor interface, σ_{es} is the energy per unit area of the substrate–epilayer interface, and θ is the contact angle between epilayer and substrate, as defined in Fig. 9. Unless there is a fortuitous combination of surface and interface energies, the above equation will only be identically true for layer-by-layer growth ($\theta = 0$) for homoepitaxy, where $\sigma_{sv} = \sigma_{ev}$ and $\sigma_{es} = 0$ or for the case where $(\sigma_{sv} - \sigma_{es})/\sigma_{ev} > 1$, i.e., where the substrate surface energy is high relative to the interface and epilayer surface energies. Thus, the general heteroepitaxial nucleation mode will be clustered, or three-dimensional, growth. Note that the diquid drop model does not take into account the effect of lattice-mismatch rain.

Equation (1), however, assumes that the structure can reach its equilibrium state and is thus effectively assuming infinite surface diffusion lengths to form the necessary clusters. In epitaxial growth, surface diffusion lengths may be controlled primarily by (i) substrate temperature, and (ii) by deposition rate. Lower substrate temperatures will mean lower, or perhaps even negligible, diffusion lengths and therefore promote layer-by-layer or two-dimensional growth, assuming uniform arrival of atomic/molecular flux at the growth surface. If the temperature is too low, however, this will be at the expense of crystalline perfection. In Fig. 10, we show the tendency for two-dimensional (2D) versus three-dimensional (3D) growth as functions of alloy composition and substrate temperature in the $Ge_xSi_{1-x}/Si(100)$ system (Bean et al., 1984). Note that the tendency for 3D growth is reduced both by lowering the substrate temperature and by decreasing the Ge concentration x in the alloy. In the limit of $x = 0$, we are regressing to the homoepitaxial case. Similar trends have been observed, for example, by RHEED in MBE growth of $In_xGa_{1-x}As$ on GaAs (Berger et al., 1988).

The effect of deposition rate on surface morphology has not been studied systematically to our knowledge and is not well understood. Presumably, the deposited species will generally diffuse most easily as single surface adatoms; as the deposition rate increases, the probability that further deposited atoms

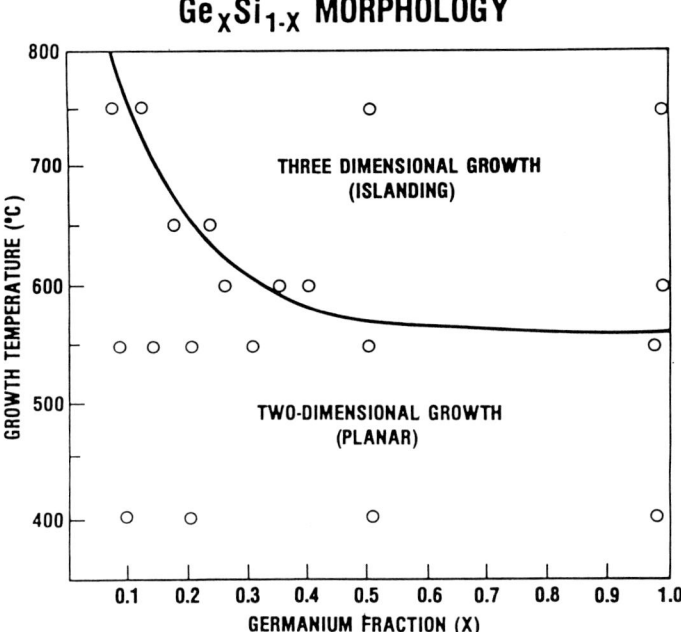

FIG. 10. Clustered versus layer-by-layer growth for MBE growth of $Ge_xSi_{1-x}/Si(100)$ (from Bean et al., 1984).

will impinge upon and bond with the diffusing adatom before it reaches a stable cluster will increase. It might therefore be expected that higher deposition rates would effectively inhibit surface diffusion and reduce epilayer surface roughness. Evidence contrary to this hypothesis, however, comes from analysis of low-temperature photoluminescence (PL) studies in AlGaAs/GaAs/AlGAs quantum well structures (e.g., Bimberg et al., 1986; Hayakawa et al., 1985). In these structures, it is commonly found that interrupting growth for a time of the order of seconds to minutes at the heteroepitaxial interfaces (this might be regarded as the limit of a vanishingly small growth rate) significantly reduces the photoluminescence line width, as illustrated in Fig. 11 (Note that in the 100 sec spectrum in this figure, the peak has split into a fine doublet structure, attributed to two discrete well widths differing by one monolayer). It is generally believed that this corresponds to a planarization of the growth surfaces during interruption, a significant factor in the PL peak broadening being well-width variation, and hence variations in exciton energy, arising from interface roughness. It would thus appear that in this system, reducing the growth rate allows surface diffusion to reduce surface roughness. This apparent dichotomy might be resolved by considering the specifics of the AlGaAs/GaAs system, where the constituents are

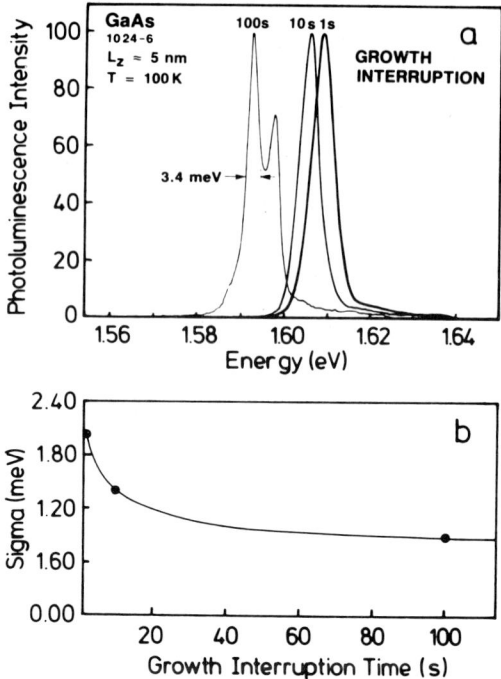

FIG. 11. (a) Comparison of $T = 100$ K photoluminescence spectra of 5 nm GaAs quantum wells grown between $Al_xGa_{1-x}As$ barriers, for growth interruptions at both sets of interfaces of 1, 10, and 100 seconds. Also shown (b) is the magnitude of line broadening as a function of growth interruption time (from Bimberg et al., 1986).

closely lattice-matched, isostructural, and chemically similar. Thus, the thermodynamic driving forces for cluster deposition are relatively small or non-existent. In these growth-interruption experiments, a continuous film of, say, AlGaAs is likely to have formed over the preceding GaAs layer before the interruption. At this stage, energy is lowered in the only possible way, i.e., by minimization, in this case by planarization, of the free growth surface. The liquid-drop model discussed earlier and defined in Eq. (1) might thus be expected to apply only when a significant fraction of free substrate surface remains.

Surface morphology may thus also be regarded as a function of the deposited epilayer thickness, in that depending on the nucleation density, at some mean deposit thickness h_m, the substrate will attain complete substrate surface coverage. Further deposition will thus approximate the homoepitaxial case, where the epitaxial material is growing onto itself. At this stage, surface diffusion would be expected to act such as to *planarize* the surface, because the sole free surface is now the growth surface, whose energy will be minimized for a minimum area, i.e., a single plane. Details of nucleation kinetics with temperature have been studied in many systems (e.g., Biegelsen *et al.*, 1987; Zinke-Almang *et al.*, 1987; Venables *et al.*, 1987); it is generally found that the nucleation density increases as substrate temperature and surface diffusion decrease, as illustrated in Fig. 12. Thus, at lower temperatures, the thickness h_m will be less than for higher temperatures, since individual nuclei require less lateral growth before coalescing with their neighbors. This is used to advantage in the growth of GaAs on Si, where the initial growth stages are highly three-dimensional (Biegelsen *et al.*, 1987; Hull and Fischer-Colbrie, 1987). In this system (e.g., Fischer *et al.*, 1985; Nishi *et al.*, 1985; Harris *et al.*, 1987), growth of GaAs normally proceeds with a low-temperature ($\sim 400°$C) "buffer" layer of thickness of $\sim 500-1000$ Å, prior to

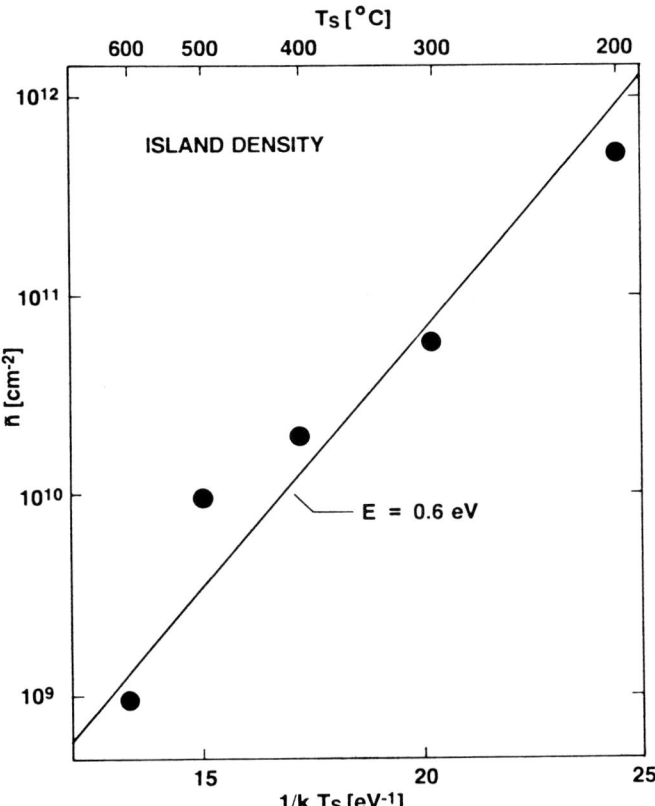

FIG. 12. Nucleation density for GaAs/vicinal Si(100) (2.5 degrees toward $\langle 011 \rangle$) as a function of substrate inverse temperature (Reprinted with permission from the Materials Research Society, from Biegelsen *et al.*, 1988.)

higher-temperature ($\sim 600°C$) growth under conditions more appropriate to homoepitaxial GaAs. This buffer layer produces better surface morphology than uniform growth at the higher temperature, because the higher nucleation density produces more rapid substrate coverage and an earlier simulation of homoepitaxial conditions. In summary, it is indeed fortunate that the very conditions that produce clustered growth (significant surface diffusion) can act so as to repair the surface morphology if the deposited epilayer is significantly thick. Note, however, that in the example of GaAs/Si, individual nuclei are typically relaxed before they coalesce (e.g., Hull and Fischer-Colbrie, 1987) and thus strain is not playing a significant role in the coalescence process. It has been suggested (e.g., Grabow and Gillmer (1987) and Bruinsma and Zargwill (1987)) that island formation on a previously planar-strained epilayer can be coincident with the strain relaxation process, as in the Stranski–Krastanow transition. It appears that kinetics may defeat this process, however, under typical growth conditions of semiconductors.

The above "liquid-drop" model should, in general, be modified to include the effect of variation of surface energy over different surfaces in the epitaxial cluster. As the contact angle θ increases, so will the range of surface normal directions over the cluster. If the cluster contains only a very small number of atoms, it is unlikely that surface atoms will experience an environment that closely approximates to uniform surfaces of atoms in low index planes, because the orientations of tangential planes to the cluster surface will vary so rapidly over the cluster. As the clusters grow, however, larger and larger numbers of surface atoms will approximate low-index planes. At some critical size, the cluster might thus be expected to facet into low-index planes so as to minimize its surface energy; this process will be encouraged for high substrate temperatures and for systems, such as III-V compounds, where different low-index planes may have very different energies.

The above arguments are illustrated with reference to the $InAs_{0.2}Sb_{0.8}$/GaAs system in Figs 13–15. Figures 13 and 14 are respectively cross-sectional and plan-view TEM images of different stages of the $InAs_{0.2}Sb_{0.8}$/GaAs deposition process (Yen et al., 1987; Hull et al., 1988a). It can be seen from Fig. 13 that in the earliest stages of growth (average nominal epilayer thickness of ~ 50 Å), approximately hemispherical $InAs_{0.2}Sb_{0.8}$ clusters nucleate on the GaAs surface. As deposition continues, individual nuclei grow, and at a critical dimension, corresponding to a nucleus radius of ~ 200 Å, starts to exhibit strong surface facetting behavior. As can be seen from Fig. 13, the facetted nuclei correspond to trapezoids, with large surface areas exposed on the top (100) surfaces, and with the trapezoid edges corresponding to $\{111\}$ faces. For compound semiconductors such as GaAs, two types of (111) surface exist: the column-III (or 111A) face and the column-V (or 111B) face. The (111)A and B faces are generally of different energies,

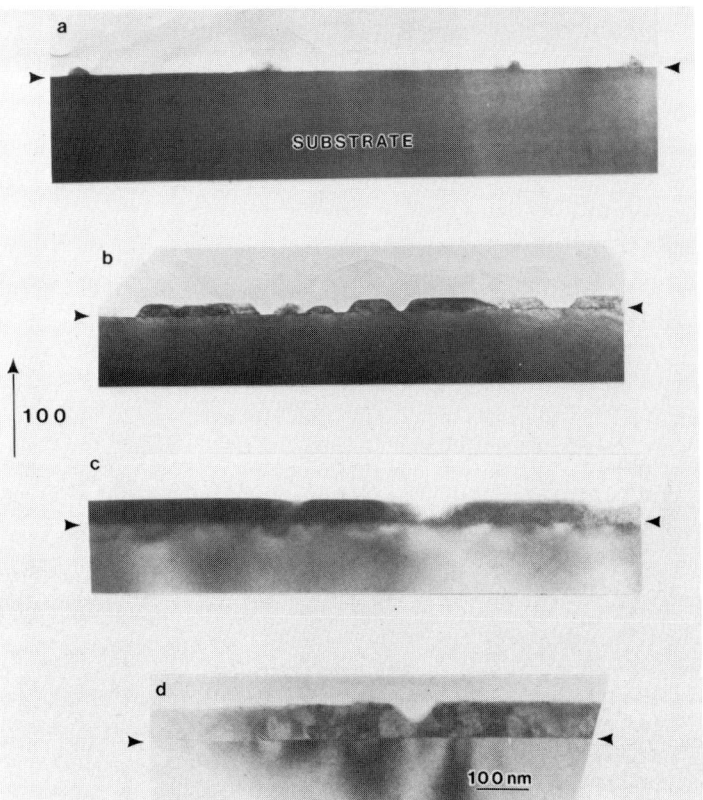

FIG. 13. Cross-sectional ⟨011⟩ TEM images of InAsSb alloys deposited on GaAs substrates. Nominal mean alloy thicknesses are: (a) 50 Å; (b) 100 Å; (c) 300 Å; (d) 600 Å. The substrate surface in each image is arrowed.

which often leads to anisotropically dimensioned facetted nuclei in epitaxial growth of III-V compound semiconductors (see, e.g., Pirouz et al., 1988).

The extreme sensitivity of the nuclei dimensions to surface facetting is illustrated in Fig. 15, which plots the height, h, and the width, w, of individual nuclei versus their cross-sectional area, A. It can be seen that in the earliest stages of growth, the nuclei are approximately hemispherical, i.e., $w \sim 2h$. As the nuclei pass through a critical value of $A \sim 100,000 \text{ Å}^2$, growth along the substrate normal almost arrests, and further growth is almost entirely lateral. In this regime, $dw/dA \sim 10(dh/dA)$. The enhanced lateral growth corresponds to the extension of the (100) trapezoid top and the long trapezoid edges of lower energy $\{111\}$ with respect to the higher energy $\{111\}$ short trapezoid edges. When lateral growth has been sufficient so that neighboring nuclei

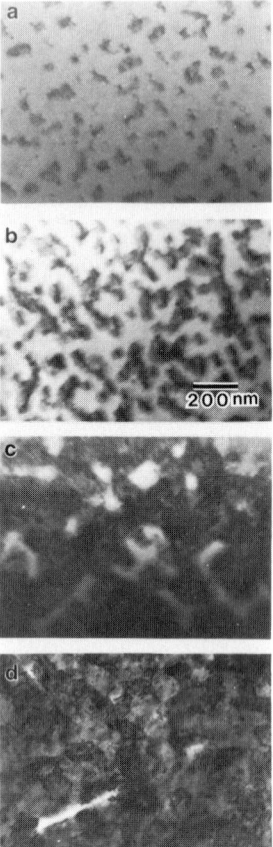

FIG. 14. Plan-view TEM images of InAsSb alloys deposited on GaAs substrates. Nominal mean alloy thicknesses are: (a) 50Å; (b) 100 Å; (c) 300 Å; (d) 600 Å. The lighter areas in each image correspont to uncovered substrate.

have coalesced and the substrate surface has been approximately entirely covered, further lateral growth of the deposit is not possible, and further growth has to be along the surface normal. This happens at a mean substrate deposit thickness of ~ 300 Å.

The above experiments were carried out for substrate temperatures of 500°C, which are very near the InSb melting point of 550°C. Thus, we would expect very high surface diffusivities and almost equilibrium surface cluster conditions to apply. It is interesting, and conceivably of practical importance, that the very conditions that promote surface clustering in this system also generate a self-planarizing mechanism. The very different surface energies of a compound semiconductor such as InAsSb, in this case, have produced dramatic surface facetting and enhanced lateral growth. Thus, by a mean deposit thickness of only 300 Å, a system that was initially highly three-dimensional has converted to a relatively planar surface morphology. These

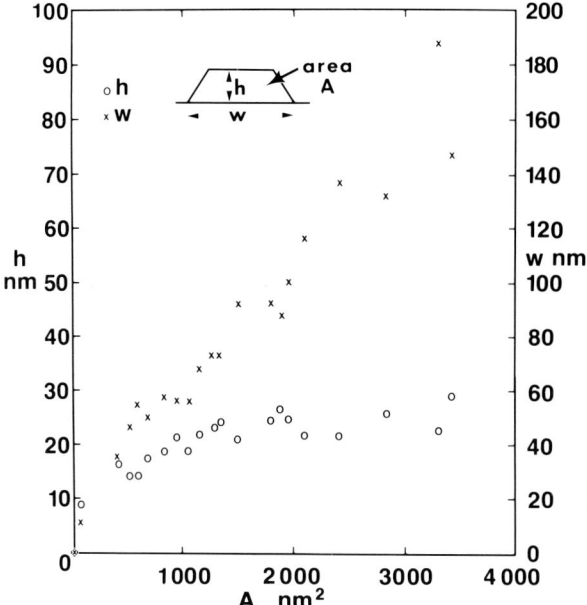

FIG. 15. Graphs of measured island height, h, and island width, w, versus island cross-sectional area, A. All measurements are from cross-sectional TEM images of nominal 100 Å and 300 Å InAsSb thickness samples. Error bars are not shown in the figure for clarity; the estimated errors on h and l are $\sim \pm 20$ Å. Experimental points for island height are denoted by dots, and for island width by crosses.

trends are in qualitative agreement with RHEED patterns in this thickness regime, which show the corresponding transition from sharp to streaked diffraction peaks (Yen et al., 1987).

V. Strain Relief in Lattice-Mismatched Epitaxy

A. INTRODUCTION

It has long been recognized that, in principle, it is possible to grow coherent lattice-mismatched epitaxial structures, where the lattice parameter of the deposit is different from that of the substrate (Nabarro, 1940; Mott and Nabarro, 1940; Frank and Van der Merwe, 1949a,b,c). This concept is illustrated in Fig. 16. In lattice-matched heteroepitaxy (Fig. 16a), the deposit and the substrate have the same lattice parameter, and deposition of the epilayer atoms onto the substrate surface allows them to easily locate the potential minima corresponding to the substrate lattice sites, assuming they have sufficient thermal energy (i.e., if the growth temperature is high enough)

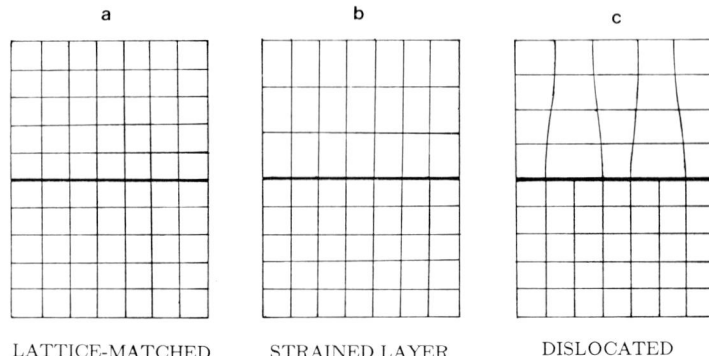

LATTICE-MATCHED   STRAINED LAYER   DISLOCATED

FIG. 16. Schematic illustration of (a) lattice-matched heteroepitaxy; (b) coherently strained lattice-mismatched heteroepitaxy; (c) relaxed lattice-mismatched heteroepitaxy. (Reprinted with permission from Plenum Pub. Corp., from Hull et al., 1989d.)

to move to the nearest minimum. In strained-layer epitaxy (Fig. 16b), despite the difference in substrate and deposit lattice parameters, deposit atoms are constrained to the substrate interatomic spacings in the plane of the interface. We designate such structures commensurate or coherent. Significant elastic strain energy is stored in the structure (accommodation of a lattice mismatch of just 1% in this fashion produces a stress field equivalent of 2 GPa, assuming a shear modulus of 5×10^{10} Pa and a Poisson's ratio of 0.33). A tetragonal distortion of the unit cell of the deposit is also produced, since elasticity theory shows that the planar stress parallel to the interface, σ_i, will produce a normal strain ε_n given by:

$$\varepsilon_n = \frac{1+v}{1-v}\varepsilon_i. \qquad (2)$$

Here v is Poisson's ratio, and ε_i is the interfacial strain produced by accommodation of the lattice mismatch, equal to $\sim(a_e - a_s)/a_s$, where a_s and a_e are the substrate and epilayer bulk (relaxed) lattice parameters, respectively. These relationships assume effectively that the substrate is of infinite thickness, such that all the elastic strain energy is stored in the deposit. In practice, the thickness of the substrate (typically $\sim 5 \times 10^{-4}$ m) is very much greater than the epilayer thickness (Å to microns), so this approximation is reasonably valid. In deposition onto thin film substrates, corrections have to be made for the finite substrate thickness.

For a given lattice mismatch, the elastic strain energy in the coherent deposit will increase approximately linearly with the substrate thickness. When the strain energy is sufficiently large, it will start to be relieved by deformation of the hitherto coherent structure. This process occurs via the introduction of slipped regions into the crystal, bounded by line defects known as "misfit" dislocations. As shown in Fig. 16c, these act so as to effectively remove planes of atoms from the deposit if the epilayer is of larger lattice parameter than the substrate (if the epilayer has the smaller lattice parameter, then extra planes will be effectively introduced into the deposit). The removal of these atomic planes increases the average spacings between deposit atoms, allowing the epilayer to relax toward its bulk structure. We designate such structures semicoherent or discommensurate.

B. REVIEW OF GENERAL DISLOCATION PROPERTIES

Before giving a more detailed description of present understanding of the strain relaxation process via the introduction of misfit dislocations, we will now briefly review the known salient properties of dislocations. Much classic work in this field has been done (see, e.g., Hirth and Lothe, 1968, and Nabarro, 1967), particularly in metals. The primary technique for experimental study of dislocation microstructure is TEM, as pioneered by Hirsch and co-workers (Hirsch *et al.*, 1977).

A perfect or total dislocation may be viewed as the boundary surrounding a slipped region of a crystal. As such, it is a line defect. It is a geometrical property of a perfect dislocation that it cannot simply terminate in the bulk of a crystal, but must rather terminate at a free surface, or upon itself by forming a continuous loop, or at a node with other defects. The structure of a dislocation is determined by its line direction, **L**, and by its Burgers vector **b**. The Burgers vector is defined by the direction and magnitude of the closure failure of a rectilinear loop drawn around the dislocation line. For a total or perfect dislocation, the Burgers vector links two atomic sites in the unit cell. The Burgers vectors of total dislocations in the cubic semiconductors discussed in this chapter are almost always $\frac{1}{2}\langle 011 \rangle$. The Burgers vector of a given dislocation is constant anywhere along that dislocation apart from a possible change in sign, according to convention. If **b** is parallel to **L**, then the dislocation is said to be of the screw type, whereas if **b** and **L** are orthogonal, the dislocation is of the edge type. Any intermediate configuration is said to be mixed screw and edge. Since a dislocation is not constrained to be straight, its screw/edge character (but not its Burgers vector) can change along its length.

Dislocations move on a given set of planes known as slip planes; in the cubic semiconductors considered here, these are $\{111\}$ planes. For a (100) interface, these planes make [011] or [01$\bar{1}$] intersections with the interface; thus, misfit dislocations typically lie in a square mesh along these directions. If the dislocation Burgers vector lies within its slip plane, it may move by a low-energy cooperative process known as glide. This process involves no mass transport of atoms and is thus relatively rapid. If a dislocation has its Burgers vector out of the slip plane, then it must move by a process known as climb, which involves mass transport of either vacancies or interstitials. Thus, this is generally a much slower process than glide.

The specific configurations of misfit dislocations to be considered here are shown in Fig. 17. For a (100) interface, consider for example the $(\bar{1}11)$ slip plane. This will intersect the interface along the $[01\bar{1}]$ direction. Total dislocations of $\mathbf{b} = \frac{1}{2}[101]$ or $\frac{1}{2}[110]$ will lie in this slip plane and thus be able to move by glide. Since $\mathbf{b}$ is neither parallel nor perpendicular to $\mathbf{L}$, these are mixed dislocations. Their Burgers vectors lie at an angle of 60 degrees to the interface and to their line directions (they are often referred to as 60-degree dislocations); thus, their component in the interfacial plane is only $\frac{b}{2}$. For this reason, these dislocations are only 50% effective (as a fraction of their total Burgers vector magnitude) in relieving lattice mismatch. Dislocations of the type $\frac{1}{2}[01\bar{1}]$, however, will have $\mathbf{b}$ perpendicular to $\mathbf{L}$ (edge or 90-degree dislocations), and their Burgers vectors will be 100% effective at relieving lattice mismatch. Since their Burgers vectors lie outside the slip plane, they must move by climb. Thus, although they may be regarded as more efficient than 60-degree dislocations, they will grow more slowly. The final possibility of screw-type $\frac{1}{2}[01\bar{1}]$ dislocations may be discounted, because they do not relieve any lattice mismatch in this configuration.

The deformation around the dislocation line is very well described by classic elastic theory, apart from the region very close to the dislocation center, known as its core. In this region, atomic displacements are sufficiently high so that Hooke's law is invalidated, and the total dislocation energy is given by the elastic energy outside of the core (which may be accurately

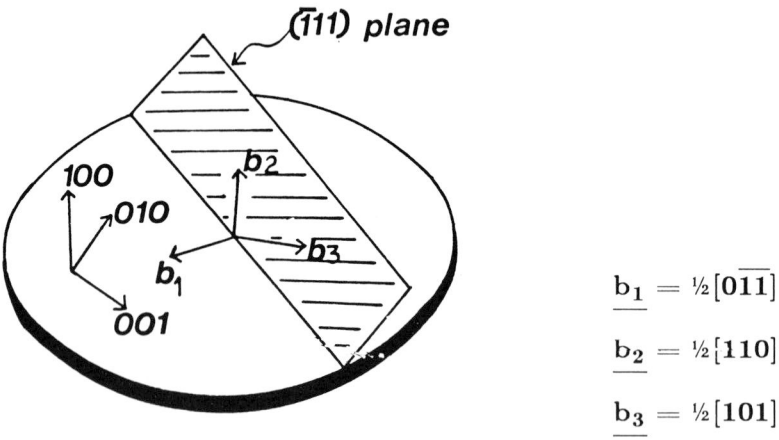

FIG. 17. Schematic illustration of the misfit-dislocation configurations encountered in cubic semiconductors at (100) interfaces.

calculated and is proportional to b^2), plus the energy within the core, which is not well known. The size of the core region is very small, being confined in semiconductors to a radial coordinate $< b$ from the deformation center.

The strain field around a dislocation produces a stress field. This field may act either upon the dislocation itself if it is curved (this gives rise to the concept of dislocation line tension), upon point defects, upon a free surface (giving rise to the concept of an image force), or upon other dislocations, causing interdislocation interactions. The force that two dislocations A and B exert on each other is proportional to the dot product of their Burgers vectors, $\mathbf{b}_A \cdot \mathbf{b}_B$, and is inversely proportional to the distance between them.

Finally, a total dislocation may dissociate into partial dislocations of lower magnitude Burgers vector. In the absence of external stress fields, this will be energetically favorable if the sum of the squares of the partial Burgers vectors is less than the square of the original total Burgers vector. The Burgers vectors of partial dislocations do not link lattice sites and thus produce stacking faults. For cubic semiconductors, $\frac{1}{6}\langle 112 \rangle$ and $\frac{1}{3}\langle 111 \rangle$ are the partial Burgers vectors most commonly observed. In a dissociation reaction, the partial dislocations will move apart producing a stacking fault between them. Their equilibrium separation will then be determined by the stacking fault energy. The total Burgers vector is always conserved in any such dissociation reaction. Note that other dislocation reactions are also possible involving total and/or partial dislocations, the criteria being that the total energy of the system is reduced and that the total Burgers vector is conserved.

C. CRITICAL THICKNESS

It might be expected that there will be a characteristic epilayer thickness (for a given lattice mismatch) at which the commensurate–discommensurate transition occurs, or at least dramatically accelerates. The magnitude of this "critical thickness," h_c, will be related to the balance between the relief of strain energy and the extra energy associated with the lattice distortions produced by the misfit dislocations. A number of equilibrium theories have been developed to attempt to describe h_c as functions of the elastic mismatch and the elastic constants of an epitaxial system. Early models by Frank and Van der Merwe and co-workers (Frank and Van der Merwe 1949a,b,c; Van der Merwe and Ball, 1975) attempted to model the commensurate–discommensurate transition by minimizing the energy of a misfit dislocation array at the interface. These formulations were originally developed for consideration of body-centered cubic metal systems. They are mathematically rigorous and have no analytical solutions. Different approximations are required for solution in the thin and thick epitaxial film limits (dimensions are defined here relative to the spacings of the misfit-dislocation arrays). Matthews and Blakeslee (Matthews and Blakeslee 1974a,b; Matthews, 1975) developed models for strain relaxation in epitaxial III-V compound semiconductor systems. In the most commonly quoted formulation of this model, they assumed that the source of misfit dislocations was the density of pre-existing "threading dislocations" in the compound semiconductor substrate. At the time of the original formulation of the Matthews and Blakeslee models, the threading-dislocation density in commercially available III-V substrates was of the order of 10^6cm^{-2}, producing a sufficiently high-density source for the misfit-dislocation process. Relaxation of the epilayer was predicted to occur via glide of the original threading dislocations akong the strained interface, as illustrated schematically in Fig. 18. To analyze the conditions required for this process to occur, Matthews and Blakeslee considered the forces acting on the laterally propagating threading arm. The force that drives extension of the misfit dislocation along the interfacial plane is the force due to the misfit stress in the structure, F_σ. Competing with this is the force due to the misfit-dislocation line tension, F_T. This force essentially represents the fact that as different segments of a curved dislocation exert a force on each other, extending the length of the dislocation requires work to be done against the stress field exerted by other dislocation segments (see Hirth and Lothe, 1968, for a complete discussion). The condition for misfit-dislocation propagation is then simply that $F_\sigma > F_T$; the epilayer thickness where $F_\sigma = F_T$ may be regarded as the critical thickness where misfit dislocation propagation begins to occur.

Simple elasticity theory yields

$$F_\sigma = S2Gbh \frac{(1 + v)}{(1 + v)} \varepsilon, \qquad (3)$$

where G, h, and v are the shear modulus, thickness, and Poisson ratio, respectively, of the epitaxial deposit, **b** is the dislocation Burgers vector, S is an angular factor resolving the misfit stress on the glide direction, and ε is the strain due to the mismatch between the lattice parameters of epilayer and substrate.

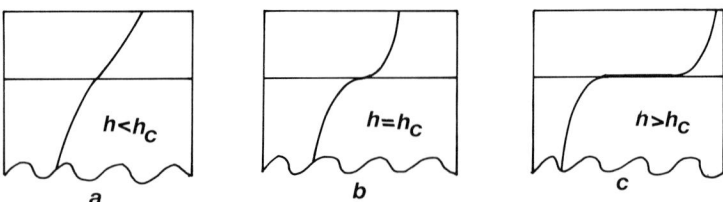

FIG. 18. Schematic illustration for the Matthews and Blakeslee model of misfit-dislocation propagation.

The line-tension force, F_T, is configuration dependent. An appropriate expression for the magnitude of this force for the threading-dislocation configuration is

$$F_T = \frac{Gb^2(1 - v\cos\theta^2)}{4\pi(1 - v)} \ln\left(\frac{\alpha h}{b}\right), \qquad (4)$$

where θ is the angle between the threading-dislocation line and its Burgers vector, and α is a constant representing the dislocation-core energy (α is generally taken to be 4 for covalent semiconductors; see Hirth and Lothe, 1968).

Equating (3) and (4) enables a solution for the critical thickness $h = h_c$. This original Matthews and Blakeslee formulation appears appropriate for the substrate threading-dislocation mechanism they propose, and has been experimentally verified for III-V compound semiconductor strained-layer epitaxy (see, e.g., Fritz et al., 1985), as illustrated in Fig. 19. Extension of this

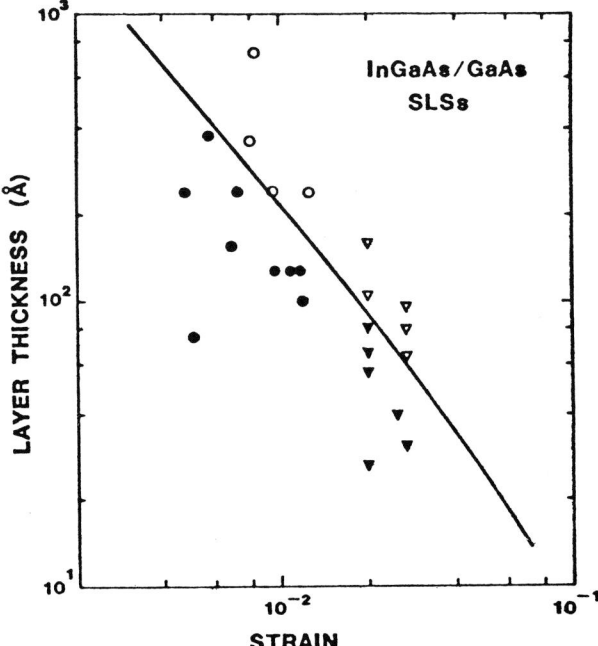

FIG. 19. Critical-thickness measurements as a function of composition in the $In_xGa_{1-x}As/GaAs(100)$ system (from Fritz et al., 1985). Solid points correspond to misfit-dislocation-free structures and open points to dislocated structures. Solid line is from theory of Matthews (1975).

model to systems such as those grown on Si substrates where there is not a high threading-dislocation density in the substrate, however, may not be appropriate as the activation barriers for dislocation nucleation and motion may prevent the essentially equilibrium configuration predicted by the Matthews–Blakeslee formulation from being attained (the nucleation barrier for the substrate threading-dislocation configuration will be zero). Such a discrepancy is illustrated in Fig. 20, where experimental data for strained-layer $Ge_xSi_{1-x}/Si(100)$ critical-thickness transitions are shown. Two different curves are shown for substrate temperatures of 550°C (Bean et al., 1984) and 750°C (Kasper et al., 1975), respectively. It can be seen that critical thicknesses are dramatically reduced at the higher substrate temperature. As discussed later in this section, we interpret this temperature dependence as being due to the activation barriers that have to be overcome for dislocation nucleation and motion. Also shown in Fig. 20 are the predictions of the Matthews and Blackeslee model for the "equilibrium" critical thickness. It should be noted that the exact form of this curve depends upon both the dislocation configuration and its structure (Burgers vector). Here we draw the curve appropriate to a $\frac{1}{2}\langle 011 \rangle$ dislocation of 60 degree character (i.e., with its Burgers vector lying within its glide plane) and in a hexagonal half-loop configuration, with one side of the semihexagon lying in the interfacial plane, and the other sides threading to the growth surface.

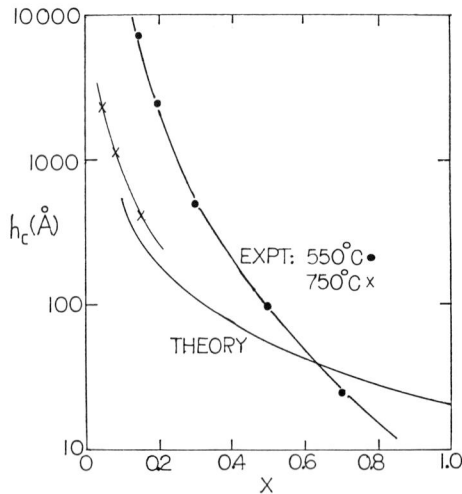

FIG. 20. Critical-thickness measurements in the GeSi/Si system (from Bean et al., 1984; 550°C; Kasper et al., 1975, 750°C), with equilibrium predictions of Matthews–Blakeslee theory. (Reprinted with permission from Elsevier Sequoia S. A., Hull, R., Bean, J. C., Eaglesham, D. J., Bonar, J. M., and Buescher, C. (1989), Thin Solid Films **183**, 117.)

The discrepancy between experiment and equilibrium theory is apparent. It should also be noted that, as has been pointed out by Fritz (1987), the exact positions of the experimental curves will depend on the sensitivity of the technique used to detect strain relaxation. The curves of Fig. 20 were determined using Rutherford backscattering (RBS) X-ray diffraction, and plan-view TEM, which are probably sensitive to strain relaxation of the order of one part in 10^3. More sensitive techniques would probably cause both experimental curves to fall at lower critical thicknesses. This effect has been experimentally verified in the InGaAs/GaAs system (Gourley et al., 1988), where it has been shown that application of a technique (in this case PL imaging) that is able to detect very low dislocation densities may reveal the onset of misfit-dislocation formation at epilayer thicknesses significantly less than those inferred by lower sensitivity techniques (e.g., RBS or TEM). Similar results have been demonstrated via X-ray topography in the GeSi/Si system (Eaglesham et al., 1988).

An elegant model that accounts for the temperature dependence of strain relaxation has recently been developed by Dodson and Tsao (Dodson and Tsao, 1987; Tsao et al., 1987). The model assumes the velocity of misfit dislocations, once nucleated, to be proportional to the excess stress in the strained epilayer. This excess stress is defined as the actual stress within the epilayer minus the residual stress that would be expected in the equilibrium state of the system. This excess stress may be derived straightforwardly in the framework of the Matthews and Blakeslee model as the stress due to the lattice mismatch, as obtained from Eq. (3), minus the self stress produced by the line tension in Eq. (4). If a dislocation source term (not explicitly identified in this model) is assumed, the solution of the relevant dynamic equations allows the strain state of the system to be predicted as functions of layer dimensions and compositions, and the growth temperature and time. Using bulk activation energies for dislocation glide in Si and Ge (Alexander and Haasen, 1968; Patel and Chaudhuri, 1966), this model has successfully predicted a wide range of experimental data in the $Ge_xSi_{1-x}/Si(100)$ system (Dodson and Tsao, 1987), where equilibrium theories have not described experimental results well. If the relevant experimental data for dislocation glide energies and elastic constants exist, then this model should be equally applicable to any other strained-layer system.

D. DETAILS OF THE STRAIN RELAXATION PROCESS

1. Introductory Remarks

Precise application of a kinetic theory such as the Dodson–Tsao model, however, requires detailed knowledge of dislocation nucleation mechanisms, activation energies for nucleation and propagation, knowledge of the precise dislocation configuration (Burgers vector, line direction, and dissociation state), and the details of defect interactions. Experimental and theoretical understanding of these various processes is still incomplete.

The main stages in relaxation of elastic strain via introduction of misfit dislocations are shown in Fig. 21. In Fig. 21a, dislocations are nucleating. In the absence of significant dislocations pre-existing in the substrate, this nucleation stage is perhaps the least understood event in the relaxation process. Given the requirement that a dislocation terminate upon itself at a node with another dislocation or at a free surface, the three generic possibilities for nucleation within the epitaxial layer of the first dislocations are shown: nucleation of a complete loop at a point within the epitaxial layer or at the substrate–epilayer interface, or nucleation of a half-loop at the free growth surface. In Fig. 21b, these initial loops are shown expanding such that the length of misfit dislocation in the substrate–epilayer interfacial plane is growing. Finally, the dislocation population will become high enough so that adjacent defects will interact, modifying their energy balance (Fig. 21c).

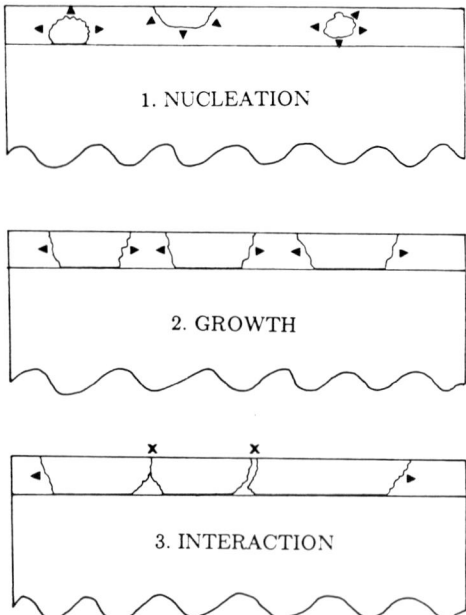

FIG. 21. Schematic illustration of (1) nucleation, (2) growth, and (3) interaction of misfit dislocations. (Reprinted with permission from Plenum Pub. Corp., from Hull *et al.*, 1989d.)

2. Nucleation of Misfit Dislocations

To illustrate the difficulties associated with the nucleation of misfit dislocations, consider the activation energy required to nucleate a half-loop at the growth surface. As a function of the loop radius, the total loop energy, $E_{tot}(r)$, will be given generically by the relationship

$$E_{tot}(r) = E_{self}(r) - E_\sigma(r) \pm E_{step}(r) - E_{dis}(r). \tag{5}$$

Here $E_{self}(r)$ is the self energy of the dislocation loop, $E_\sigma(r)$ is the strain energy relaxed by the dislocation loop, $E_{step}(r)$ is the energy associated with possible creation or removal of a surface step, and $E_{dis}(r)$ is the energy associated with possible dissociation of the total dislocation into partials. The exact forms of these individual components depend on the specific dislocation microstructure but have been evaluated for different systems and configurations by several authors (e.g., Matthews et al., 1976; Fitzgerald et al., 1989; Eaglesham et al., 1989; Hull and Bean, 1989a). The calculated loop energy generally passes through a maximum or activation barrier, δE, at a critical loop radius, R_c. The magnitude of R_c is typically of the order of 10–100 Å for common lattice mismatches of, say, 0.3–3%. The activation barriers generally become of the order of a few eV or less at strains $> \sim 4$–5%, making this nucleation process energetically feasible, but are extremely high (tens or even

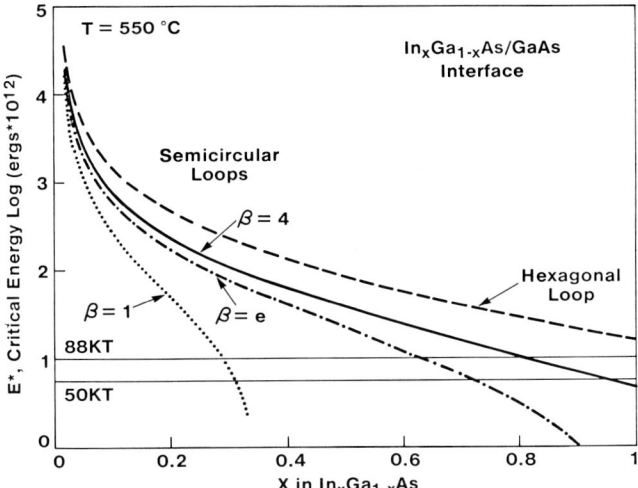

FIG. 22. Activation barriers for the nucleation of a surface half-loop in the $In_xGa_{1-x}As/GaAs$ system (from Fitzgerald et al., 1989). Different values of β correspond to different estimates of the dislocation core energy (see Hirth and Lothe, 1968; Eaglesham et al., 1989).

hundreds of eV) for more moderate stresses of $< \sim 1\%$, as illustrated for $In_xGa_{1-x}As/GaAs$ in Fig. 22. Similar numbers are obtained for the GeSi/Si system (Eaglesham et al., 1989; Hull and Bean, 1989a). It thus appears that in the relatively low mismatch regime, alternative nucleation sources have to be sought. Note that nucleation of full loops within the epilayer or at the epilayer–substrate interface would be expected to produce higher activation barriers, due to the necessity to nucleate a complete loop, as opposed to a half-loop at the surface.

Little work exists to date in establishing alternative nucleation paths. In a detailed analysis of dislocation densities in finite InGaAs pads grown on GaAs substrates, Fitzgerald et al. (1989) deduced that dislocation nucleation is due to "fixed" sources, i.e., sources inherent to the substrate or growth-chamber geometry such as substrate threading dislocations, oval defects, or particulates. Hagen and Strunk (1978) have proposed that intersection of orthogonal dislocations with equal Burgers vectors can produce a regenerative multiplication mechanism, a process that has been identified in several systems (Hagen and Strunk, 1978; Chang et al., 1988; Rajan and Denhoff, 1987; Kvam et al., 1988). This process, however, still cannot explain the generation of the initial defect density necessary to produce the observed intersection events. Eaglesham et al. (1989) have detected a novel generation mechanism in low-mismatch $(< 1\%)$ GeSi/Si systems, consisting of the dissociation of $\frac{1}{6}\langle 114 \rangle$ diamond-shaped loops. In the same system, Hull et al. (1989a) have measured a relatively low activation energy of 0.3 ± 0.2 eV for $Ge_{0.25}Si_{0.75}/Si(100)$ and have suggested that this may be associated with clustering of atoms in the alloy epilayer (Hull and Bean, 1989a). A widespread and detailed understanding of this nucleation process, however, is still clearly lacking and may be unique to particular systems and growth conditions, and even to specific growth chambers.

3. Direct Observations of Dynamic Misfit-Dislocation Events

Detailed understanding of the strain relaxation process has been hampered by the absence of direct techniques for observing dynamic misfit-dislocation phenomena. We have recently developed techniques for in situ relaxation experiments in a TEM that allow real-time observation of misfit-dislocation nucleation, propagation, and interaction phenomena.

The experimental configuration for these experiments is shown in Fig. 23. A suitable strained $Ge_xSi_{1-x}/Si(100)$ structure is grown in the MBE chamber at a substrate temperature of 550°C. The epilayer thickness is designed to lie between the equilibrium value of h_c (Matthews, 1975) and the experimentally measured value (Bean et al., 1984) for this substrate temperature. Thin plan-view TEM samples are then fabricated, and the thinned structure is subsequently annealed inside the electron microscope. While the annealing temperature is raised to the growth temperature and above, the essentially

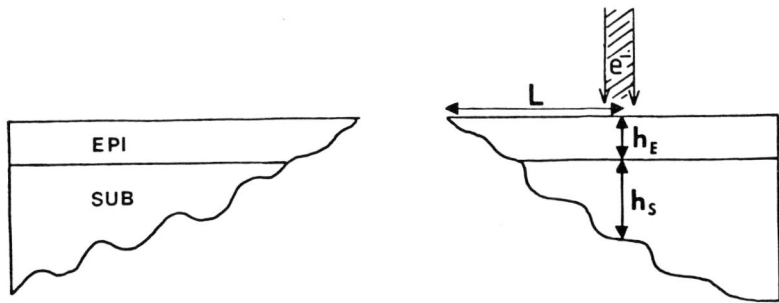

FIG. 23. Schematic illustration of the experimental geometry for TEM in situ strained-layer relaxation experiments. (Reprinted with permission from Plenum Pub. Corp., from Hull et al., 1989d.)

metastable 550°C structure relaxes as dislocations nucleate and then grow. Both static (photographic negative) and dynamic (video) images of the relaxation process may be recorded. Thus, the various strain relaxation processes (dislocation nucleation, growth, and interaction) may be observed directly and in real-time. As described in detail in Hull et al. (1988b), the crucial experimental complication is careful consideration of thin-foil relaxation effects in the TEM sample. Consideration of such effects leads to the conclusion that films of Ge composition $x \sim 0.25-0.30$ a few hundred Å thick represented the optimum geometry because (i) layer thicknesses are much less than the maximum thinned substrate thickness penetrable by imaging electrons, and (ii) surface roughness is negligible. Samples are always prepared in the plan-view geometry for these experiments, because annealing in the cross-sectional geometry would leave open the question of surface diffusion across heteroepitaxial interfaces at the top and bottom thin-foil surfaces. In the plan-view geometry, the Ge_xSi_{1-x}–Si interface is buried away from any free surfaces.

Plan-view TEM images of a typical annealing sequence are shown in Fig. 24. The dislocation density is observed to increase with annealing temperature. In Fig. 25, we show the average measured spacing between dislocations as a function of annealing temperature for two different sample thicknesses. It is observed that for $x = 0.25$, relaxation of $Ge_xSi_{1-x}/Si(100)$ heterostructures is relatively slow and continuous over the temperature range 550–900°C. As expected, the thinner epilayer relaxes more slowly, because its effective stress is less.

4. Misfit-Dislocation Propagation

By direct observation of dislocation motion via a video camera, image intensifier, and video recorder, we are able to directly measure dislocation velocities as a function of temperature. The time resolution of these

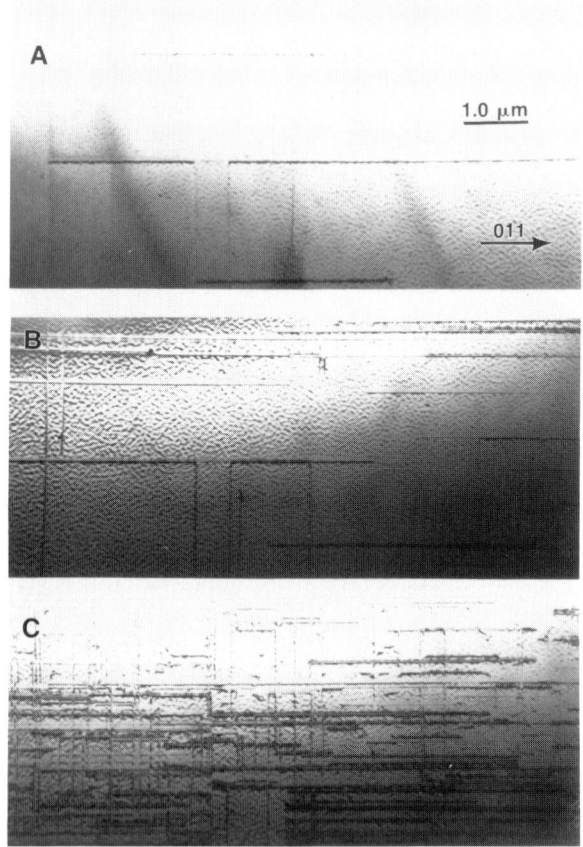

FIG. 24. Plan-view TEM images of a 350 Å $Ge_{0.25}Si_{0.75}/Si(100)$ structure (a) as-grown at 550°C; (b) annealed to 700°C; and (c) annealed to 900°C (from Hull et al., 1988b).

experiments is an individual TV frame, i.e., $\frac{1}{30}$sec. The minimum magnification we obtain on the viewing screen with the microscope objective lens excited is ∼ 45,000 × (1500 × on microscope screen and 30 × on video viewing screen via camera). Thus, we may measure dislocation velocities from ≪ 1 micron sec^{-1} to ∼ 100 micron sec^{-1}. The dislocation velocities we observe are typically of the order 1000 Å − 1 micron sec^{-1} at the growth temperature of 550°C, rising dramatically with annealing temperature until they are > 100 micron sec^{-1} at 800°C. The measured dislocation velocity at the growth temperature immediately tells us something of great importance about metastability in this system: Even ignoring questions of misfit-dislocation nucleation and interactions, the low growth-temperature velocity severely limits strain relaxation. Thus, for the structures studied here, a

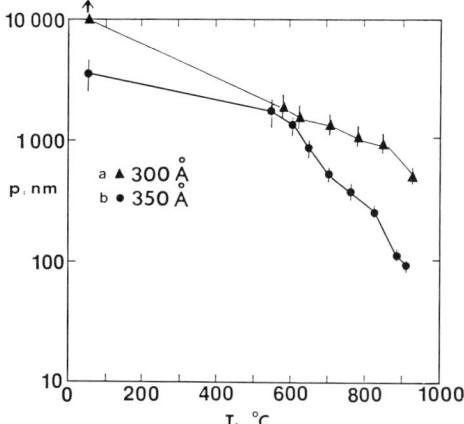

FIG. 25. Variation of average distance between misfit dislocations, p, versus annealing temperature, T, during relaxation of $Ge_{0.25}Si_{0.75}/Si(100)$ structures (Hull *et al.*, 1988c. Reprinted with permission from "Dislocations and Interfaces in Semiconductors," eds. K. Rajan, J. Narayan, and D. Ast, 1988, Metallurgical Society, 420 Commonwealth Drive, Warrendale, PA 15086.)

sample thickness of 350 Å and a growth rate of ~ 3 Å sec^{-1} implies a maximum misfit dislocation length of ~ 0.1 mm. Since this is very much less than the wafer diameter, the ends of the misfit dislocation have to terminate at the nearest free surface, which in general is the growth surface. These dislocation ends thus constitute so-called threading dislocations, which are so deleterious to transport/optical properties and device performance in the epilayer. In the $Ge_xSi_{1-x}/Si(100)$ system, they arise naturally as a result of the finite velocities of propagating dislocations.

Note that the above estimate of maximum dislocation length is very much an upper limit. In practice, measured dislocation lengths will be much lower, because (i) only for part of the growth time will the epitaxial film be obove the equilibrium critical thickness that will produce an excess stress to drive dislocation growth, (ii) misfit-dislocation kinetics are known to be partially nucleation limited in this system (Eaglesham *et al.*, 1989; Hull and Bean, 1989a), and (iii) as will be discussed later, dislocation interactions largely limit growth. These combined factors actually limit the average length of misfit dislocations in the $Ge_{0.25}Si_{0.75}/Si(100)$ structures discussed above to ~ 10 microns (Hull *et al.*, 1989a).

Temperature-dependent in situ measurements of the glide velocities of non-interacting 60° dislocations suggest activation energies that are generally of the same order as those expected from interpolation of measurements in bulk Ge and Si (1.6 and 2.2 eV, respectively, e.g., Alexander and Haasen (1968) and Patel and Chaudhuri (1966)). However, we have measured

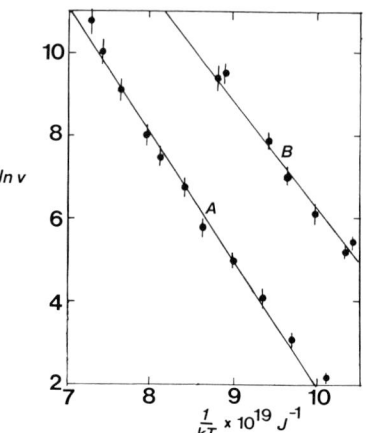

FIG. 26. Arrhenius plots of natural logarithm of dislocation glide velocity (velocities in Å sec^{-1}) vs. inverse energy for (A) 1200 Å Ge$_{0.15}$Si$_{0.85}$ buried beneath a 3000 Å Si(100) cap and (B) a surface layer of 600 Å Ge$_{0.28}$Si$_{0.72}$ on Si(100) (from Hull et al., 1989a).

dislocation velocities in very thin and highly stressed Ge$_x$Si$_{1-x}$/Si(100) structures where activation energies are a few tenths of an eV less than expected from bulk measurements, as illustrated in the Arrhenius plots of Fig. 26. Possible models for this apparent activation energy lowering are: (i) a stress dependence of the "Peierls" (Peierls, 1940) or glide activation energy barrier (Dodson, 1988a) and (ii) propagation of the dislocation by nucleation of single "kinks" (see Hirth and Lothe, 1968) at the free epilayer surface as opposed to the classic bulk double kink nucleation mechanism. We also note that in more relaxed films where dislocation interactions become more significant, apparent glide activation energies may be further reduced (Hull et al., 1989a). Also shown in Fig. 26 are in situ glide velocity measurements from a Ge$_x$Si$_{1-x}$ layer buried beneath a Si cap. In these structures, we generally observe activation energies equivalent to the bulk values and velocities that appear to be substantially lower than in uncapped Ge$_x$Si$_{1-x}$ layers, assuming a linear dependence of dislocation velocity on excess stress (Dodson and Tsao, 1987). Further work will be necessary to determine the precise defect propagation mechanisms.

We are not aware of any similar measurements of misfit-dislocation velocities in strained-layer compound semiconductor structures. In principle, the experiments we have described earlier could be applied to such materials in a relatively narrow temperature range between the original sample growth temperature and the temperature at which surface desorption (of, e.g., P or As) starts to significantly affect the surface stoichiometry.

5. Dislocation Interactions

It has been shown by Dodson (1988b) that interaction of high densities of misfit dislocations can significantly slow the rate at which a strained structure will relax. By experimentally measuring and theoretically modelling the rate at which a strained InAsSb/InSb with different types of compositional grading relaxed, it was shown that a simple "work-hardening" approach to misfit-dislocation interactions was reasonably successful at describing experimental data.

The importance of dislocation interactions has also been demonstrated by in situ GeSi/Si relaxation experiments (Hull et al., 1989a; Hull and Bean, 1989b). Shown in Fig. 27 are the variation of average dislocation spacing, p, average density of misfit dislocations, N, and average misfit-dislocation length, L (calculated from a from a simple geometrical relationship linking p, N, and L as described in Hull et al., 1989a), versus temperature during thermal relaxation of a $Ge_{0.25}Si_{0.75}/Si(100)$ film. It can be seen that although N increases and p decreases with temperature as expected, L does not increase

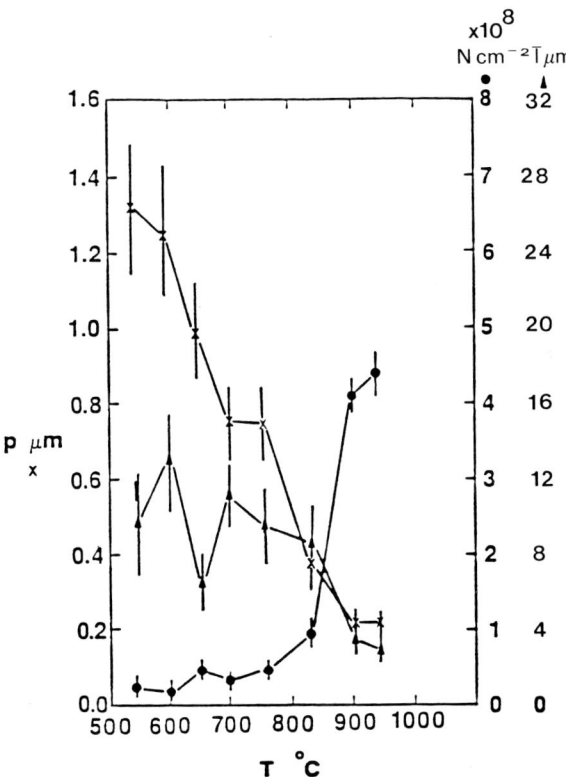

FIG. 27. Variation of average distance between dislocations, p, number density of dislocations, N, and average dislocation length, $\bar{l}$, during thermal relaxation of a 350 Å $Ge_{0.25}Si_{0.75}/Si(100)$ structure (from Hull et al., 1989a).

but remains relatively constant at ~ 10 microns and actually decreases at higher temperatures. This demonstrates that dislocation nucleation plays a pivotal role in the relaxation process. The data of Fig. 27 show clearly that relaxation does not occur solely via growth of existing defects. The fact that the average dislocation length is not increasing shows also that defects are being inhibited from continual growth. Inspection of TEM images (Hull and Bean, 1989b) has indeed revealed that intersection of orthogonal dislocations frequently pins propagating defects. This "pinning probability" is strongly dependent on epilayer thickness, as is suggested by Fig. 28, which shows representative relaxation data for 350 Å and 3000 Å $Ge_xSi_{1-x}/Si(100)$ films. It is observed that the higher thickness film is considerably more thermally unstable. Inspection of dynamic and still TEM images of the relaxation process (Hull and Bean, 1989b) has shown that this is due to the far greater ease of orthogonal dislocations crossing each other in the thicker films. This may be understood in terms of the relative misfit stress, line tension, interaction and image forces acting on the dislocations at the two different thicknesses (Hull and Bean, 1989b).

In summary, it is clear that dislocation nucleation, propagation, and interaction phenomena all play a role in determining strained-layer relaxation. The relative significance of these events in a given structure will determine its stability with respect to variations in temperature and epilayer thickness and composition.

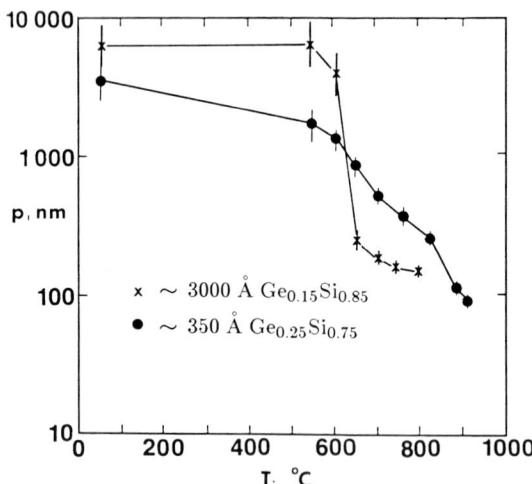

FIG. 28. Variation of average distance between misfit dislocations during thermal relaxation of (a) 350 Å $Ge_{0.25}Si_{0.75}/Si(100)$ and (b) 3000 Å $Ge_{0.15}Si_{0.85}/Si(100)$ (from Hull and Bean, 1989b).

E. RELAXATION IN STRAINED CLUSTERS

It should be stressed that the two-dimensional epitaxial structures discussed so far in this section actually represent a special case in heteroepitaxial growth. In the more general case of three-dimensional or clustered growth, we should discard the concept of critical "thickness" and consider all dimensions of the strained island. The limiting case of a strained rectilinear slab has been treated theoretically by Luryi and Suhir (1986), who related the strained–dislocated transition of a slab of height h and edge length $2L$ to the configuration of a two-dimensional epilayer of thickness h via the relation

$$h_c^L = \phi(L/h) h_c^{L=\infty}. \tag{6}$$

Here h_c^L is the critical slab dimension and $h_c^{L=\infty}$ is the two-dimensional critical thickness. The reduction factor $\phi(L/h)$ is obtained from the equation

$$\phi(L/h) = \left[1 - \text{sech}\left(\frac{\varepsilon L}{h\phi^2(L/h)}\right)\right]\left[1 - e^{-\pi h/L}\right]^{1/2}\left[\frac{L}{\pi h}\right]^{1/2}, \tag{7}$$

where

$$\varepsilon = \left[1.5\frac{(1-v)}{(1+v)}\right]^{1/2}.$$

These equations predict that as the slab width $2L$ decreases, the slab height at which the strained–dislocated transition occurs will increase. In fact, the Luryi–Suhir equations predict a critical edge half-length, L_c, below which the slab may be grown infinitely thick without misfit-dislocation introduction. This may be understood physically on the basis that if the slab width is less than approximately half the equilibrium average misfit-dislocation spacing for a planar structure, introduction of a single misfit dislocation into the slab–substrate interface will increase the magnitude of the lattice mismatch by making it larger but of the opposite sign to the undislocated structure.

The Luryi–Suhir model has been experimentally investigated in the GaAs–Si system (Hull and Fischer-Colbrie 1987). Cross-sectional TEM is used to probe the strain state of individual GaAs nuclei grown at 400°C on vicinal Si(100) substrates. Results are plotted in Fig. 29, which shows each island height, h, versus the ratio of island height to width (h/L). Dislocated islands are marked by crosses, whereas undislocated islands are marked by dots (note that in cross-sectional TEM, we will only observe island widths and components of dislocation Burgers vectors perpendicular to the electron beam. These experiments may thus be thought of as a two-dimensional version of the Luryi–Suhir model). A clear dependence of the undislocated–dislocated transition upon h/l is observed. Also shown are predictions of the Luryi–Suhir model. This model requires knowledge of the two-dimensional

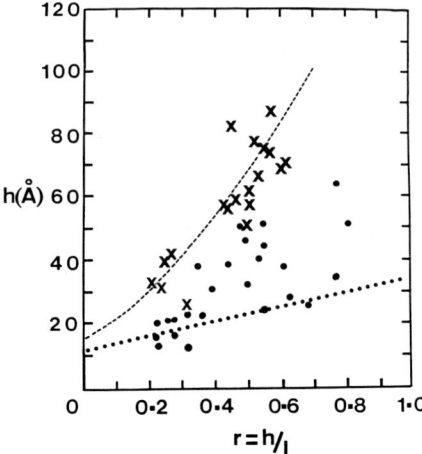

FIG. 29. Experimentally observed (via cross-sectional TEM) transitions between commensurate (●) and dislocated (×) GaAs nuclei grown on 4-degree misoriented Si(100) surfaces (from Hull and Fischer-Colbrie, 1987). The island height h is plotted on the ordinate, whereas the ratio of island height to width, (h/l) is plotted on the abscissa. The drawn curves represent the predictions of the Luryi–Suhir model using the two-dimensional critical-thickness models of People and Bean (dashed curve) and Matthews and Blakeslee (dotted line).

critical thickness as a function of varying strain in the system. Since this is not experimentally possible in the GaAs–Si system, where only one elastic mismatch is possible in the undislocated state, $h_c(\varepsilon)$ is obtained from theoretical models, namely the Matthews–Blakeslee model (1974a,b), and a more recent model by People and Bean (1985, 1986). The results of these models inputted into the Luryi–Suhir model are displayed in Fig. 29 by dotted and dashed lines, respectively. It is observed that these two theoretical curves straddle the experimental transition and are of similar form to the experimental results.

F. Critical-Thickness Phenomena in Multilayer Structures

Extension of critical-thickness concepts to multilayer structures is highly complex due to the greater degrees of freedom involved (individual layer strains and dimensions, total number of layers, total multilayer thickness, and so on). However, a simple energetic model has been developed (Hull et al., 1986a) based on the reduction of the multilayer to an equivalent single strained layer.

Consider a multilayer structure consisting of n periods of the bilayers A and B, of thickness d_A and d_B and lattice-mismatch strains with respect to the substrate of ε_A and ε_B, respectively. This structure is shown schematically in Fig. 30. The total thickness, T, of the multilayer is $n(d_A + d_B)$.

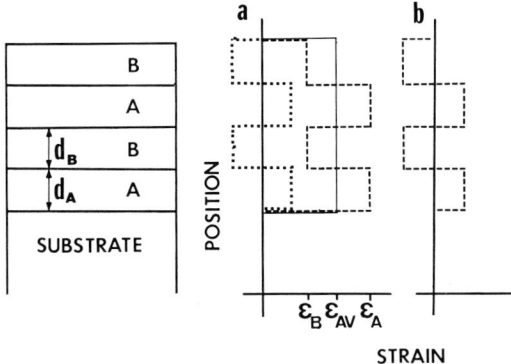

FIG. 30. Schematic illustration of criteria for structural stability of a strained-layer superlattice. (a) shows the unrelaxed superlattice and (b) its energy ground state (from Hull et al., 1986a). In (a) the total strain field is shown by the dashed line and may be regarded as the supposition of a uniform strain (solid line) and an oscillating strain (dotted line).

Experimentally, it is observed that provided that each of the individual multilayers is thinner than the critical thickness for that particular layer grown directly onto the substrate (if this condition is not met, then individual interfaces within the superlattice will relax), the great majority of the strain relaxation occurs via a misfit-dislocation network at the interface between the substrate and the first strained multilayer constituent. This is demonstrated by cross-sectional TEM images of GeSi/Si and InGaAs/GaAs multilayer structures in Fig. 31. Thus, the relaxation may be regarded as occurring between the substrate and the multilayer structure as a whole.

For an undislocated structure, the elastic strain energy of the configuration in Fig. 30 would be

$$E_{un} = n(k_A d_A \varepsilon_A^2 + k_B d_B \varepsilon_B^2), \qquad (8)$$

where k_A and k_B contain the relevant elastic constants for layers A and B.

Introduction of a misfit-dislocation array at the substrate–multilayer interface will change the lattice constant throughout the multilayer structure by an amount α, and the strain by an amount $\beta \sim \alpha/a_0$, where a_0 is the average bulk lattice parameter of the multilayer, such that

$$E_{dis} = n(k_A d_A (\varepsilon_A - \beta)^2 + k_B d_B (\varepsilon_B - \beta)^2). \qquad (9)$$

Energy is minimized by setting $dE/d\beta = 0$, yielding

$$E_{min} = \frac{n d_A k_A d_B k_B (\varepsilon_A - \varepsilon_B)^2}{d_A k_A + d_B k_B}. \qquad (10)$$

This may be regarded as the ground energy state of the superlattice with respect to dislocation formation at the substrate–multilayer interface. The

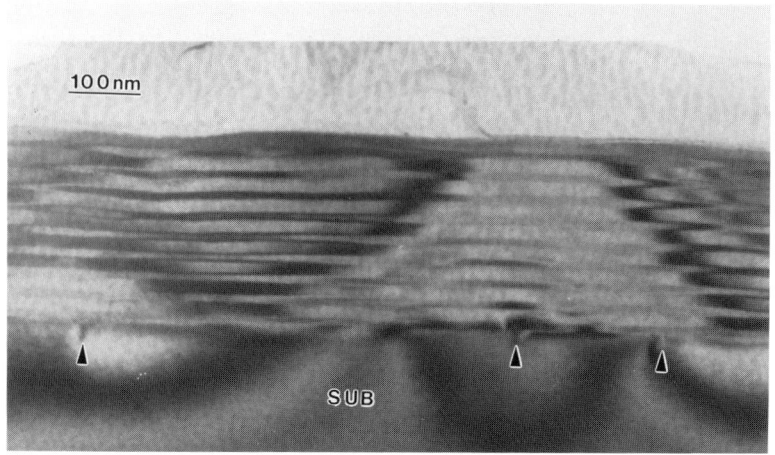

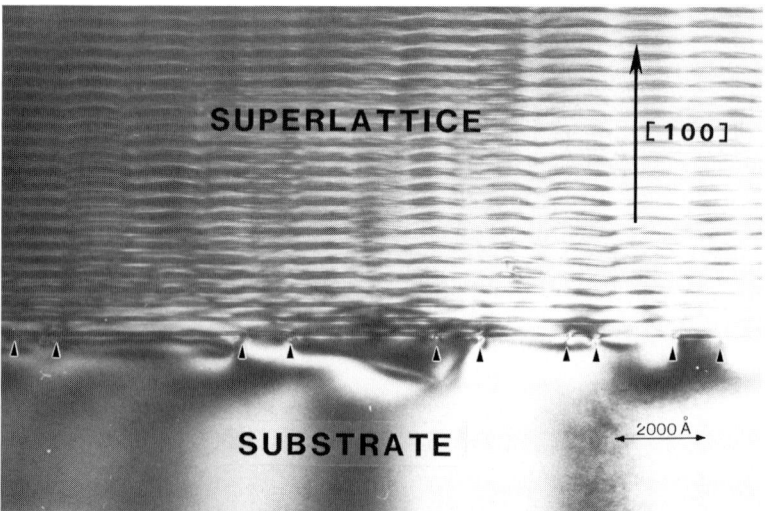

FIG. 31. Cross-sectional TEM images of (a) partially relaxed In GaAs/GaAs on GaAs (from Hull et al., 1987e) and (b) GeSi/Si on Si multilayer structures (from Hull et al., 1986a). Note that, as indicated by arrows, the vast majority of misfit dislocations are at the substrate–superlattice interface.

amount of energy available for relaxation by the dislocations is thus given by the energy of the undislocated state minus the energy of the ground state, (8)–(10), yielding

$$E_{\rm rel} = \frac{n(k_A d_A \varepsilon_A + k_B d_B \varepsilon_B)^2}{d_A k_A + d_B k_B}. \tag{11}$$

The form of this relaxable energy is equivalent to the average strain energy in the superlattice, weighted over the appropriate elastic constants and layer thicknesses. Consider a system with continuously varying strain as a function of varying composition, e.g., In_xGa_{1-x}/GaAs or Ge_xSi_{1-x}/Si, where one of the superlattice constituents (say B) is the same as the substrate. If we assume that $k_A \sim k_B$ and that the elastic strain varies linearly with x, then the relaxable strain energy simplifies to that in a uniform layer of the same total thickness as the superlattice equals $n(d_A + d_B)$, and the average superlattice composition, $x_{av} = xd_A/(d_A + d_B)$. This relationship may be used as a guide for predicting the superlattice stability, based upon the experimentally determined critical thickness of single uniform strained layers of Ge_xSi_{1-x} on Si(100). In Hull et al. (1986a), it is shown that this model is accurate to within 20% for varying superlattice layer thicknesses of $Ge_{0.4}Si_{0.6}$ on Si(100), although as pointed out by Miles et al. (1988), dislocation motion through a multilayer structure is more difficult than at a single strained interface, so that at moderate temperatures, kinetic effects will limit the accuracy of the above relationship. Thus, it should be stressed that the above approach to superlattice critical thickness, while energetically correct, does not take into account atomistic details of dislocation motion and interaction. Nevertheless, it represents a reasonable first-order approximation of relating the stresses in multilayer structures to those in equivalent single layers.

G. Techniques for Dislocation Reduction in Strained-Layer Epitaxy

1. The Fundamental Problem

It is apparent that for lattice mismatches of greater than $\sim 2\%$, the critical layer thickness (of the order 100 Å from theoretical modelling and experimental measurement in different systems) becomes prohibitively small for most practical applications. For some materials combinations of high potential importance (GaAs/Si, Ge/Si, InAsSb/GaAs), the critical thickness (assuming two-dimensional growth is possible) is predicted to be ~ 10 Å. Thus, in many areas of strained-layer growth, generation of enormous quantities of interfacial misfit dislocations is unavoidable. The challenge then becomes to control the misfit-dislocation structures and distributions after they are generated.

If an epitaxial layer is of sufficient thickness, dislocations constrained at or near the interface with the substrate need not be deleterious to the materials properties of the layer sufficiently far away from the interface. Thus, electronic device fabrication in the near-surface region of a 1-micron layer need not be impeded by misfit dislocations confined to the original interface.

Unfortunately, as described earlier, a simple geometrical property of dislocations precludes the possibility that all misfit dislocations will be confined to the interface region: the need for a dislocation to terminate upon itself, at a node with another defect, or at a free surface. A segment of interfacial misfit dislocation must then be associated with two free ends that must terminate in one of the above fashions. In general, this will mean termination at the nearest free surface, which will usually be the epilayer growth surface, because this will be at most microns from the interface, as opposed to the wafer edge, which will on average be centimeters away. Thus, each misfit-dislocation segment would be associated with two threading-dislocation ends.

The amount of strain relaxation in a lattice-mismatched system will be defined by (i) the total length of misfit dislocation and (ii) the dislocation Burgers vector. Considering the case of a constant Burgers vector, the total dislocation line length L will be given by Nl_{av}, where N and l_{av} are the number of misfit-dislocation segments and their average length, respectively. Assuming first that dislocations are noninteracting, i.e., that they must terminate at a free surface, the number of threading dislocations, N_{th}, will be given by $\sim 2N$ for $l_{av} \ll D$, the wafer diameter. We then obtain the relationship

$$N_{th} \sim \frac{2L}{l_{av}}. \tag{12}$$

Thus, as might intuitively be expected, the number of threading dislocations is inversely proportional to the average dislocation length for a given amount of strain relaxation. The ideal case, indicated schematically in Fig. 32a), is where the dislocations are essentially infinitely long, terminated only at the boundary of the wafer. The number of threading dislocations is then zero. In reality, misfit dislocations are generally of finite length (limited by dislocation nucleation, growth, and interaction rates) with $l_{av} \ll D$, as indicated schematically in Fig. 30b. High threading dislocation densities then result.

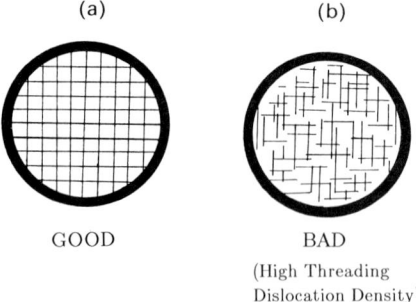

(a)   (b)

GOOD   BAD
(High Threading
Dislocation Density)

FIG. 32. Schematic illustration of the difference between (a) infinitely long misfit dislocations and (b) misfit dislocations of finite length. (from Hull et al., 1989d.)

Two approaches may be adopted in reducing threading dislocation densities: (i) increase l_{av} during growth or (ii) "filter" threading dislocations after they have formed. The former possibility requires detailed knowledge of the activation barriers for dislocation nucleation and motion, as well as defect interaction mechanisms, as we have outlined in previous sections of this chapter. The latter possibility generally involves three major concepts: thermal annealing, strained-layer superlattice incorporation, and patterned growth.

2. Effect of Thermal Annealing

It is widely recognized (e.g., Chand et al., 1986; Choi et al., 1987a; Lee et al., 1987) that thermal annealing either during or after the growth cycle may reduce threading-defect densities. In general, extra thermal energy will produce threading-dislocation motion, and if this process results in interaction of dislocations, defect reduction may result as indicated in Fig. 33. Interdislocation forces may either encourage or suppress defect interactions, depending upon the relative orientations of the defect Burgers vectors.

This process, although possibly producing significant improvement in material quality at relatively high defect densities (say, greater than $\sim 10^7 - 10^8 \, \text{cm}^{-2}$), is unlikely to be effective at lower defect densities, $< \sim 10^6 - 10^7 \, \text{cm}^{-2}$. This is because as defect densities decrease, the average spacing between threading defects necessarily increases, and this will in turn decrease the probability of defect interactions. If the original mismatched interface is fully relaxed, then the motion of the threading defects will, to a first approximation, be random during thermal annealing (local strain variations due to nonuniform defect distributions would be the major correction to this approximation), and there will be no systematic driving force for defect interactions at lower defect densities. In addition, thermal stresses arising from differential thermal expansion coefficients might act as a source for *further* defect generation in some systems.

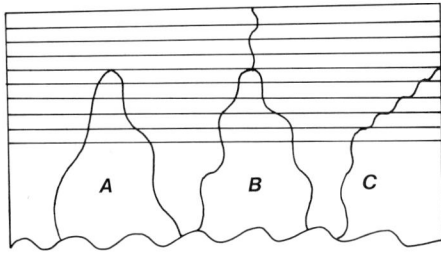

FIG. 33. Schematic illustrations of possible mechanisms for dislocation–threading arm interaction/annihilation events.

3. "Filtering" of Threading Dislocations by Strained-Layers Superlattices

A systematic driving force encouraging threading-defect motion in particular directions can be provided by incorporation of strained-layer superlattices into the epitaxial structure. It has long been recognized (e.g., Olsen et al., 1975; Fischer et al., 1986; Liliental-Weber et al., 1987; Dupuis et al., 1986) that strained interfaces in such superlattices may deflect threading dislocations into the interfacial planes between layers. The driving force for this process is essentially that described by the Matthews and Blakeslee mechanical equilibrium theory of lattice-mismatch relief, as outlined in an earlier section. The threading dislocations are bent into the interfacial planes to relieve mismatch at the interfaces. If care is taken with layer thicknesses, strains, and the structure growth rate, it may be possible to promote this process without generation of new mismatch-relieving dislocations.

For high threading-defect densities, this process is likely to be extremely effective at producing dislocation interactions and encouraging combination/annihilation events. For a threading-defect density n cm^{-2}, the average distance between the threading dislocations will be $1/\sqrt{n}$. Thus, at threading-defect densities of $\sim 10^{10}$ cm^{-2}, typical of near-interface regions in highly lattice-mismatched systems such as GaAs/Si and Ge/Si, the average dislocation separation will be only ~ 1000 Å. Since typical dislocation velocities during growth will be of the order of microns sec^{-1} at usual growth temperatures, very little lateral deflection of threading defects will be required to produce dislocation interactions at these densities. Indeed, it was pointed out by Gourley et al. (1986) that even elastic moduli (with lattice parameter matching) mismatches can be sufficient to produce the necessary deflections. Thus, for very high initial defect densities, crystal quality may be expected to improve dramatically during the first few thousand Å of epilayer growth.

At moderate (for highly lattice-mismatched materials!) defect densities, however, strained-layer superlattices will be expected to become much less efficient at defect filtering. At defect densities of 10^8 cm^2, the average threading-dislocation spacing becomes 1 micron. Although, lateral motion at strained interfaces of tens or even hundreds of microns may be possible under optimum growth conditions, one has also to consider the requirement that defects interact as they laterally propagate. Since there are only four possible directions for interfacial defect propagation at (100) interfaces for $\frac{1}{2}\langle 011\rangle$ Burgers vectors dislocations moving on $\{111\}$ planes in cubic semiconductors (the 011, $0\bar{1}\bar{1}$, $01\bar{1}$, and $0\bar{1}1$ directions as shown in Fig. 34), there is a relatively narrow range of defect positions that will allow interaction. It can easily be shown geometrically (Hull et al., 1989b) that the most favorable conditions for defect interaction correspond to dislocations moving towards each other on inclined $\{111\}$ glide planes with parallel intersections with the (100)

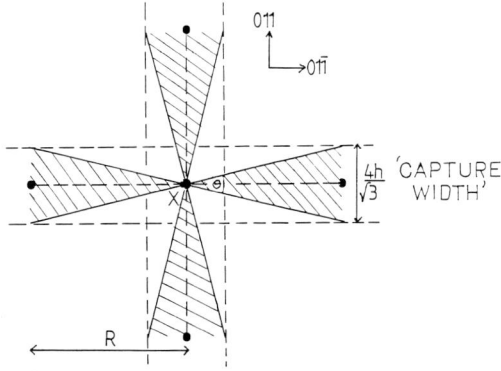

FIG. 34. Schematic diagram indicating probability of dislocation interactions (from Hull et al., 1989b). The dislocation X may move in one of four [011] directions. For an average interdislocation distance R, adjacent dislocations must lie within an angle θ of these directions, where $\theta \sim 4h/R\sqrt{3}$.

interface. Under these conditions, threading-arm intersection is possible if the corresponding misfit-dislocation segments in the interfacial plane are no further apart than $2h/\sqrt{3}$, where h is the epitaxial layer thickness. This leads to the concept of a defect "capture angle," $\theta = 4h/R\sqrt{3}$, where R is the average distance between defects in the dislocation propagation direction, as illustrated in Fig. 34. Four capture angles exist, corresponding to each of the in-plane $\langle 011 \rangle$ directions. Consideration of relative defect geometries and interaction probabilities in this theoretical framework (Hull et al., 1989b) leads to the result that the average distance each dislocation has to propagate before a threading-arm intersection event is $l_{av} = \sqrt{3/2hn}$. If we assume $h = 1000$ Å, we have for $n = 10^6$, 10^8, and 10^{10}, $L \sim 1000$ microns, 10 microns, and 0.1 micron, respectively. The actual distance a threading dislocation will be expected to propagate in the interfacial plane will depend upon a number of factors including the number of strained interfaces, the layer thicknesses and lattice mismatches, the growth rate, and the growth temperature. It is apparent, however, that the probability of defect interactions decreases dramatically with decreasing threading-dislocation densities. Of the figures quoted above, a 10-micron lateral propagation seems attainable, but 1000 microns seem unlikely. This suggests that threading-dislocation filtering via strained-layer superlattices, at least by the generally assumed mechanism of defect annihilation via interaction, is likely to become ineffective at threading-defect densities of the order of 10^6–10^8 cm^{-2} or below.

4. Growth on Patterned Substrates

An extremely promising technique for relaxing many of the constraints discussed previously is patterned growth onto bulk substrates. If instead of a uniform coverage of an epilayer onto a bulk substrate, selective growth (or etching) of features with dimensions much less than wafer diameter is successfully achieved, the probability of a threading defect terminating on a free edge of the structure (as opposed to the growth surface) is greatly enhanced. In the discussions above, we have ignored the possibility of threading-defect termination at the wafer edges, because in general, this would require lateral propagation of the order of centimeters (in structures that are at most microns thick). Patterned growth of structures of the order of microns or tens of microns wide will significantly relax this constraint so that annealing or strained-superlattice techniques could make these pads essentially defect free. In the ultimate extension of this technique, one could envisage pads of the order of hundreds of Å in dimension that would inhibit the *formation* of any misfit defects (Luryi and Suhir, 1986). Such pads would presently be very difficult to define by lithographic techniques, but natural seeding using porous Si substrates has been proposed (Luryi and Suhir, 1986; Lin *et al.*, 1987).

Recent work in growth of $In_xGa_{1-x}As$ on circular GaAs mesas (Fitzgerald *et al.*, 1988, 1989) has clearly shown that dramatic reductions in threading and threading-defect density may be achieved as the mesa diameter is reduced. This is believed to be both a function of reduced density of nucleation sites (reducing the number of misfit-dislocation segments) and of dislocation termination at the mesa edges, thus reducing the number of threading dislocations per misfit-dislocation segment. Representative results from this work are illustrated in Fig. 35. Recent work in epitaxial GaAs growth on patterned Si substrates has also been reported (e.g., Matyi *et al.*, 1988; Lee *et al.*, 1988).

VI. Atomic-Scale Structure of Epitaxial Layers

A. INTRODUCTION

For many devices (e.g., high-mobility modulation-doped structures and quantum-well-confined optoelectronic devices) and physical measurements (electron/hole mobilities, excitonic recombination energies, etc.), the detailed atomic atructure of epitaxial layers and their interfaces is of paramount importance. Many high-resolution experimental techniques have been developed to study this structure.

B. THEORETICAL DESCRIPTION OF ISOSTRUCTURAL INTERFACE STRUCTURE

Perhaps the most important class of semiconductor heteroepitaxial interfaces (including almost all III-V and II-VI structures, GeSi/Si, etc.) consists of

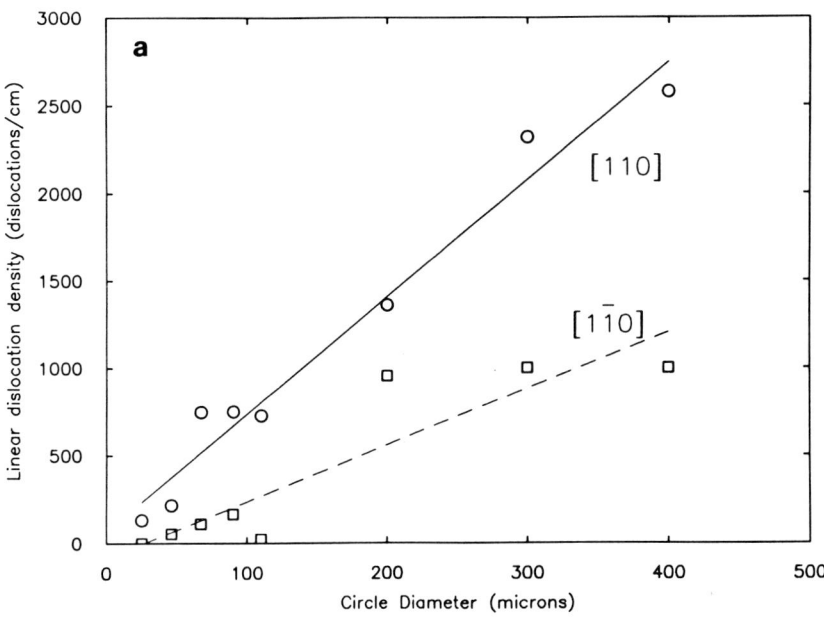

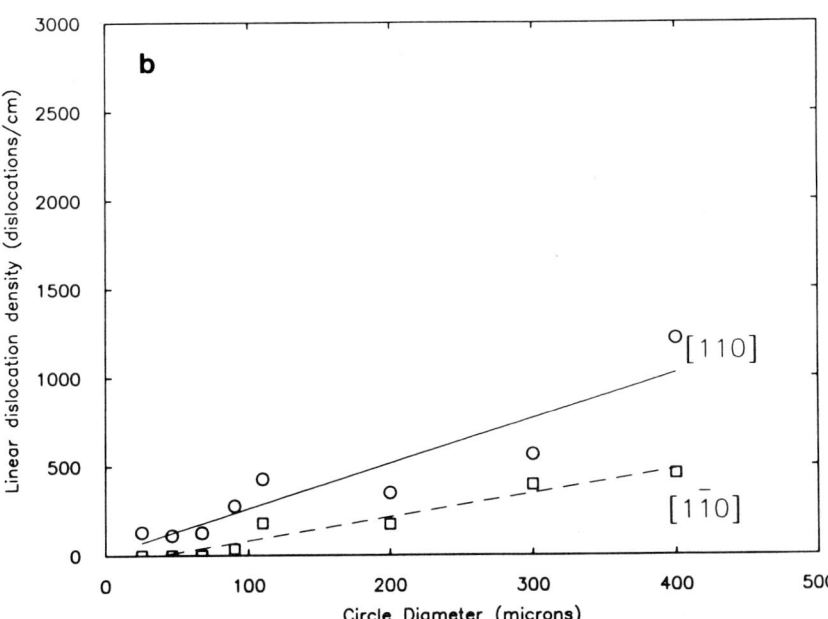

FIG. 35. Linear interface dislocation density versus circular mesa diameter for samples with 3500 Å of $In_xGa_{1-x}As$ ($x = 0.05$) grown onto GaAs with (a) $1.5 \times 10^5 \, cm^{-2}$ and (b) $10^4 \, cm^{-2}$ pre-existing dislocations in the substrate (from Fitzgerald et al., 1989.)

isostructural interfaces, i.e., those interfaces where the crystal unit cell is of identical symmetry on either side of the interface, and the only difference is in the atomic species across the interface. Two structural parameters are generally of importance: the interface diffuseness (variation in interface structure perpendicular to the interface) and the interface roughness (variation in structure parallel to the interface). In general, an ideal interface will have zero diffuseness and roughness.

In this section, we will discuss interface roughness and diffuseness by using the model illustrated in Fig. 36 and described in more detail in Hull et al., (1987d). An interface between two materials, A and B, is defined arbitrarily as lying at the boundary of 100% material A. The mean interface position is taken to be the plane $z = 0$, such that the interface lies in the x–y plane. The interface roughness is thus defined as the deviation of the interface at an arbitrary point (x_i, y_i) from the plane $z = 0$, z_i, where z_i is given by the "interface roughness function," $f(x, y)$. An interface diffuseness function, $g(z)$, then defines the perpendicular variation in composition from the interface coordinate (x_i, y_i, z_i), such that the composition at a point (x_i, y_i, z') is given by the function $g[z' - f(x_i, y_i)]$. A complete knowledge of the two functions $g(z)$ and $f(x, y)$ thus allows an exact description of the interface.

C. Origins of Interface Roughness and Diffuseness

A description of a perfect interface using the nomenclature of the previous section would be for $g(z)$ and $f(x, y)$ to be zero. Satisfaction of the condition that $f(x, y) = 0$ is essentially a thermodynamic requirement at least for oriented substrates. The primary necessity is that the growth be two-dimensional, i.e., either Frank–Van der Merwe or the earliest stages of Stranski–Krastinov, and not three-dimensional (Vollmer–Weber). Thus, as discussed in Section IV of this chapter, low growth temperatures should encourage planar heteroepitaxial interfaces. Growth of sufficiently thick layers in a system where nucleation is by the Vollmer–Weber mode may also reduce interface roughness, as individual nuclei coalesce, and further growth of that particular layer is essentially homoepitaxial, causing growth surface planarization.

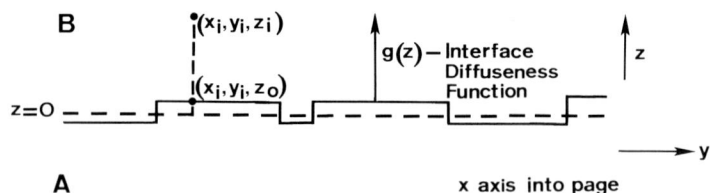

Fig. 36. Schematic illustration of the concepts of interface roughness and diffuseness, as measured by the functions $g(z)$ and $f(x, y)$, respectively (from Hull et al., 1987d).

To a higher degree of perfection, for an absolutely abrupt interface, it will be necessary to grow integral numbers of monolayers, otherwise monolayer interface roughness will necessarily result. In this case, it has been shown that growth interruptions at each individual interface in compound semiconductor growth can cause significant interface smoothing (e.g., Hayakawa et al., 1985; Bimberg et al., 1987) by allowing surface monolayer islands to coalesce.

Interface diffuseness will generally be caused primarily by (i) source switching conditions in the growth chamber, (ii) interdiffusion of the layers at elevated growth temperatures, or (iii) interfacial exchange reactions to improve local electroneutrality (see e.g., Kroemer, 1986). Thus, for reason (ii), lower growth temperatures again encourage sharper interfaces.

The effect of source switching conditions on interface abruptness is much more difficult to ascertain. Solid sources that may be simply shuttered, as in conventional MBE, might be expected to give the sharpest interface transition. Even in this case, however, slight interface transients might be expected due to both finite speed of shutter motion and redistribution of the temperature distribution around the source as the shutter is opened and closed (this effect will be minimized if the shutter is as far away from the source as possible). In gas-source growth, such as MOCVD or GSMBE, switching conditions are generally much less ideal, and optimization of this process for a given chamber is a science in itself. Characterization of switching effects using TEM, x-ray diffraction, and photoluminescence has been attempted (e.g., Carey et al., 1987; Vandenberg et al., 1986).

D. EXPERIMENTAL TECHNIQUES

In this section, we will briefly discuss experimental techniques for determining the atomic-scale structure of interfaces. For a more complete description of experimental techniques for studying epitaxial growth in general, see Chapter 3 by S. T. Picraux et al. in this volume.

1. High-Resolution Transmission Electron Microscopy

Many experimental techniques have been developed to probe the structure of buried interfaces. The highest spatial-resolution (better than 2 Å) imaging technique is high-resolution transmission electron microscopy (HRTEM). Under narrowly defined experimental conditions (Spence, 1981), an HRTEM image represents a two-dimensional map of the crystal potential projected through the specimen thickness onto a plane perpendicular to the electron beam (note that, in general, HRTEM images are interpreted on the basis of numerical simulations of dynamical electron scattering, e.g., the multislice method; Goodman and Moodie, 1974). If the electron beam is aligned parallel to the interface, atomic-scale information about interface structure may be obtained subject to three major limitations: (i) Structure variations parallel to the beam direction are averaged out. This necessitates either imaging of the same specimen area along different crystallographic zone axes (Hull et al., 1986b; Batstone et al., 1987) or assumptions about structural isotropy along crystallographically equivalent axes. (ii) Image noise effects tend to obscure fine image detail such as atomic steps (Gibson and McDonald, 1987). (iii) Thin-foil specimen preparation or the incident electron beam itself may modify interface structure. Despite these limitations, HRTEM has played a valuable role in elucidating interface structure in a wide variety of isostructural epitaxial systems including GeSi/Si (Hull et al., 1984), GaAs/Si (Hull et al., 1986b; Otsuka et al., 1986), InGaAs/InP (Ourmazd et al., 1987), AlGaAs/GaAs (Ourmazd et al., 1989), and InGaAs/InAlAs (Bimberg et al., 1989). Most notably, in some instances, it has been possible to correlate atomic structure with other interface or quantum confinement properties, e.g., with electronic mobilities at $Si/\alpha-SiO_2$ interfaces (Liliental et al., 1985; Goodnick et al., 1985) with interface state densities in CaF_2/Si (Batstone et al., 1989) and photoluminescent line broadening in InAlAs/InGaAs/InAlAs quantum-well structures (Bimberg et al., 1989). The experimental complications outlined earlier, however, particularly the projection and noise effects, severely limit the accuracy to which $g(z)$ and $f(x, y)$ may be determined. It has been shown (Hull et al., 1987d) that detailed analysis of intensity changes across the interface as a function of interface sampling area can yield information about the magnitude of $g(z)$ and the integral of spatial frequencies of $f(x, y)$ greater than approximately the inverse of the sampling area dimensions. An example of this method is shown in Fig. 37. Line scans perpendicular to the interface in a lattice image of a $Ge_{0.4}Si_{0.6}/Si(100)$ structure are recorded with varying "slit" lengths parallel to the interface and with lattice resolution perpendicular to the interface. Many scans are recorded for each slit length. The interface "width" is plotted as a function of sampling length and approximate specimen thickness, with the product of length and thickness giving the interface area sampled. The width is defined in this case as being the number of (200) monolayers in the interface region lying within the 20%–80% intensity interval, where the 0% and 100% levels correspond to the mean intensities of the materials on either side of the interface. The plot shown in Fig. 37 demonstrates a continuously increasing interface width with sampling area. In the limit of this area tending to zero (i.e., the ordinate intersects in Fig. 37), an upper limit to the spatial extent of $g(z)$ is obtained, in this case $< \sim 2 \times (200)$ monolayers. The roughness characteristic of a particular interface area may also be continuously read from the plot, increasing to several (200) monolayers for interface sampling areas $\sim 10,000 \text{ Å}^2$. These values essentially correspond to the integral of the Fourier spectrum of $f(x, y)$ up to spatial frequencies comparable to the inverse of the slit length and specimen thickness. The numbers thus obtained are of physical importance in determining, for

(a)

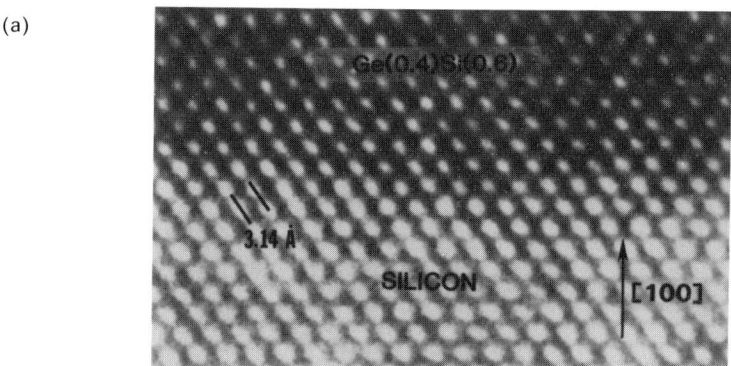

(b)

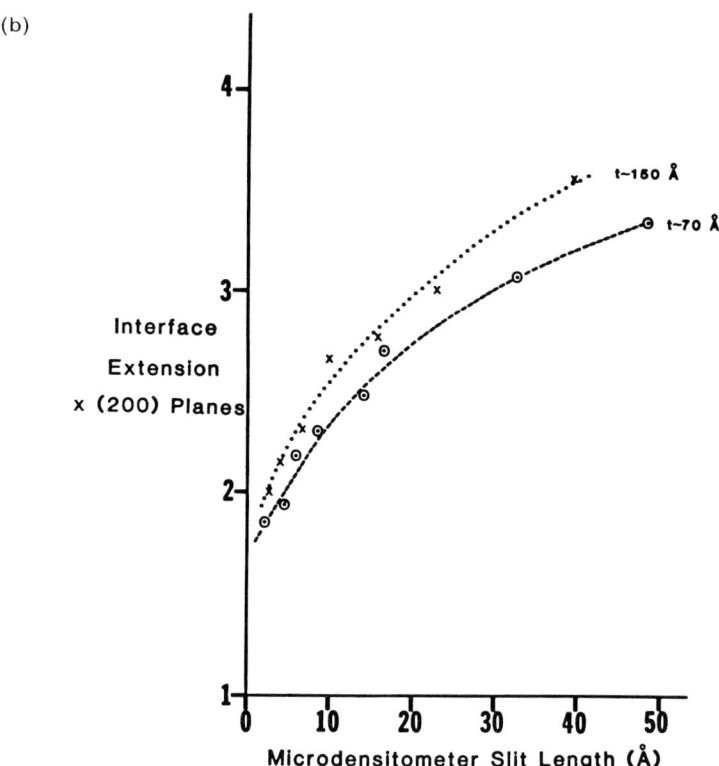

FIG. 37. (a) Cross-sectional TEM lattice structure image of a $Ge_{0.4}Si_{0.6}/Si(100)$ interface; (from Hull et al., 1987d); (b) measured interface "width," w, versus slit length, L, for microdensitometer scans across the interface of the structure imaged in (a). (From Hull et al., 1985. Paper originally presented at the Spring 1985 meeting of the Electrochemical Society, held in Toronto, Ontario, Canada.)

example, the characteristic interface roughness sampled within excitonic or electron wave-function diameters.

Other emerging electron microscopy techniques are proving to provide invaluable information about interface structure. Analysis of Fresnel fringe contrast at interfaces has been shown to be sensitive to interface diffuseness and is perhaps easier to interpret than lattice-structure images (Boothroyd et al., 1987). Measurement of extinction-contour positional variation across interfaces in 90° wedge-shaped samples has been demonstrated to be sensitive to absolute composition variations across the interface (Kakibayashi and Nagata, 1985, 1986), and high-angle annular detection in the scanning transmission electron microscope (STEM) can give very high signal-to-noise ratio interface images (Pennycook et al., 1989).

2. Reflection High-Energy Electron Diffraction

Perhaps the most prevalent in situ technique for determining growth-surface morphology (and thus, by inference, buried interface morphology) is RHEED. In its simplest form, it gives coarse information about surface morphology via the extent of reciprocal lattice reflections, a spotty pattern corresponding to islanded growth and a streaked pattern corresponding to two-dimensional growth (Note, however, that the observed streaks on "planar" surfaces generally arise from imperfections such as steps and terraces. For a perfectly planar surface, reciprocal lattice spots are expected—see Dobson, 1989 for a fuller discussion.) This technique, however, has recently been made dramatically more powerful and quantitative via the identification of oscillations in the RHEED intensity corresponding to completion of monolayers during two-dimensional growth (Harris et al., 1981; Neave et al., 1983). In principle, therefore, the smoothness of the growth surface can be determined to submonolayer precision. In addition, during a growth transition from one material to another (i.e., formation of an interface), where the amplitude of the RHEED oscillations may be expected to vary between the materials, information about the abruptness of the interface transition may, in principle, be obtained (Madhukar et al., 1985).

Although RHEED is an extremely powerful and now almost routine (particularly in compound semiconductor epitaxy) growth characterization tool, some complications remain in using this technique to deduce quantitative and detailed measurements of the interface structure. By their nature, RHEED oscillations characterize growth that is close to two-dimensional, and are therefore of limited use in the more general case of three-dimensional heteronucleation. The transition from a "spotty" to a "streaked" RHEED pattern can give information about this process, but in the absence of imaging capabilities, important parameters such as nucleation density cannot be directly deduced (unless the incident beam can be spatially collimated down to dimensions comparable to individual nuclei, as can be achieved in an electron microscope). The coherence length (approximately hundreds of Å) of the energetic electrons used in the RHEED process also naturally define a length scale on which this technique will be sensitive to surface roughness. Finally, the theory of electron scattering processes in RHEED is still at a formative stage, possibly rendering highly detailed interpretation unreliable. It should be stressed, however, that the very macroscopic nature of this technique, which lessens its impact in determining exact microscopic structure, is exactly what makes it such a powerful and crucial in situ tool for the crystal grower in determining epitaxial layer quality in real-time during growth.

Extension of the RHEED principle to imaging processes has led to the development of reflection electron microscopy (REM). For example, studies of atomic steps on clean semiconductor surfaces in ultrahigh vacuum electron microscopes have been reported (e.g., Osakabe et al., 1980; Cowley, 1984).

3. Photoluminescent Spectroscopy

A number of authors (e.g., Deveaud et al., 1984; Bimberg et al., 1986; Hayakawa et al., 1985; Tsang and Schubert, 1986; Ogale et al., 1987, 1988) have attempted to infer interface structure from photoluminescent measurements of buried quantum wells. Clearly, the quantum confinement energy of two-dimensional excitons will be highly sensitive to exact details of the well interfacial structure. By solving the one-dimensional Schrödinger equation for the particular well shape and dimensions, it should, in principle, be possible to determine the exact excitonic recombination energy. In addition, known variations in interface structure should produce calculable excitonic energy spreads. These calculations will clearly be most sensitive for narrow well widths where quantum confinement energies are a significant proportion of the bandgap discontinuity, and monolayer variations in the well structure produce appreciable energy shifts.

If the excitonic recombination spectrum is accurately measured by using photoluminescence and photoexcitation techniques, then information about the quantum-well structure can be inferred if certain simplifying assumptions are made. For example, if it is known that the interfaces are abrupt (from RHEED or HRTEM, for example), then the energy position of the maximum of the recombination peak allows the mean well width to be calculated from a single excitonic transition, subject to certain experimental and theoretical reservations that will be discussed in the following. In addition, it should be possible to relate the shape (most simply the full width at half maximum intensity (FWHM) of the excitonic peak to the variation in well width.

Many experimental and theoretical complications arise from such interpretation, however. In the simplest case, where the interface is totally abrupt, the following factors have to considered: (i) An accurate knowledge of electron/hole effective masses (particularly in the case of high confinement energies where the parabolic band approximation may not apply), two-dimensional excitonic binding energies, bandgaps and conduction-/valence-band discontinuities is required. Uncertainties in some of these parameters can significantly affect the accuracy of calculations at lower well widths. (ii) The material has to be low in doping and unintentional impurity levels to prevent excitonic screening effects. (iii) Other broadening effects such as band filling have to be taken into account. (iv) Barrier penetration by the wave function at low well widths becomes appreciable; thus, it is important to know the electronic properties of both the well and the barrier materials. (v) The excitonic spectrum is only sensitive to interface variations larger than a scale approximating the two-dimensional exciton diameter (itself not accurately known as a function of well width). Roughness on a scale smaller than this is "averaged out" in the spectrum.

Experimentally the simplest correlation is to measure the most intense recombination peak (i.e., the $n = 1$ to heavy-hole level) at low excitation intensities at liquid-helium temperature as a function of quantum-well width. With complementary experimental evidence regarding the interface diffuseness function, $g(z)$, this will allow the interface roughness function $f(x, y)$ to be determined for spatial periodicities greater than approximately the exciton diameter. In high-quality III-V epitaxial systems, it has been reported that single excitonic peaks can split into a doublet and triplet structure arising from monolayer interface roughness, producing relatively large regions of discrete well widths of varying sizes (Deveaud et al., 1984; Hayakawa et al., 1985; Bimberg et al., 1986); it has also been found that interrupting growth at GaAs/AlGaAs interfaces can provide the time scale necessary for surface diffusion of growth ledges to planarize the surface and remove the doublet/triplet structures.

Further refinement of these measurements is possible by monitoring the spectrum as a function of temperature; this allows assumptions about other spectral broadening mechanisms to be validated (Bimberg et al., 1986); and by varying bandgaps, it allows testing for self-consistent solutions. In systems where it is possible to simultaneously record several higher-order excitonic transitions (Tsang and Schubert, 1986), highly detailed interface interpretation may be possible, assuming theoretical modelling is sufficient. In this case, it may be possible to independently determine $g(z)$ and $f(x, y)$ from the excitonic spectrum.

As with RHEED, it should be pointed out that as a coarser tool for measuring interface quality, photoluminescent and photoexcitation techniques are powerful and convenient. In general, the recombination energy gives a relatively accurate measurement of well width, and as the width of the peak reduces, interface planarity will be increasing. More detailed conclusions, however, will require extensive experimental and theoretical interpretation.

4. Other Experimental Techniques

A number of other experimental tools have been developed to look either directly at the buried interface structure or indirectly via in situ monitoring of the growth surface. For example, photoemission spectroscopy has provided valuable information on the electronic structure of near-surface interfaces (e.g., Bringans et al., 1985; Himpsel et al., 1986; Olmstead et al., 1986), ion-scattering techniques have elucidated the structure of strained metal silicide–silicon interfaces (Van Loenen, 1986), and x-Ray diffraction is particularly sensitive at determining the average structure of many interfaces in a superlattice structure (e.g., Segmuller and Blakeslee, 1973; McWhan et al., 1983; Vandenberg et al., 1986). Recent time-resolved low-temperature cathodoluminescence images (Bimberg et al., 1987) have directly imaged monolayer high islands at GaAs/AlGaAs interfaces, and both low-energy electron diffraction (LEED) (see e.g., Henzler, 1983, for a review) and in situ electron microscopy (Gibson, 1988, 1989; Petroff, 1986; Takayanagi, 1978) have provided significant information during the growth process. Finally, Raman spectroscopy of interfacial atomic coordination in ultrafine-period (of the order of a few monolayers) structures has yielded information about interfacial planarity and diffuseness in Ge–Si multilayer structures (Tsang et al., 1987; Iyer et al., 1989).

E. Nonisostructural Interfaces

The above discussion has dealt primarily with isostructural interfaces, i.e., those in which the crystal structure and orientation is equivalent on either side of the interface such that the atomic positions are known but not the nature of the atomic species on each site. This turns out to be the most important class of interfaces in strained-semiconductor epitaxy because it embraces the majority of systems of technological and physical interest. A number of other interfaces are clearly of great interest, however. In some systems, the structure of the material on either side of the interface is similar or even equivalent but, generally due to a very large lattice mismatch between the materials or due to surface structure or contamination effects, the crystallographic zone axis along the growth direction is not constant across the interface. In this case, more than one orientational relationship across the interface may result. This concept is known as "double positioning" and occurs in epitaxial metal silicides on Si (Cherns et al., 1984; Tung and Gibson, 1985), epitaxial fluorides on Si (Phillips et al., 1988; Schowalter and Fathauer, 1986; Ishiwara et al., 1985), II-VI growth on III-V surfaces (Kolodziejski et al., 1986), and Al/GaAs (Liliental-Weber, 1987; Batstone et al., 1987), among other systems. The easiest method of determining the orientational relationships is using diffraction techniques, particularly electron diffraction when very thin epilayers and small grain sizes are involved. In many other nonisostructural systems, the arrangements of atoms at the interface are not self-evident (especially where the two materials have different forms of bonding), such as the epitaxial silicides and fluorides on Si discussed previously and GaAs/Si (Bringans et al., 1985; Kroemer, 1986; Patel et al., 1987). Arrangements of atoms at such interfaces can be studied using HRTEM (e.g., Batstone et al., 1988; Cherns et al., 1984; Tung and Gibson, 1985), photoemission (e.g., Bringans et al., 1985; Himpsel et al., 1986; Olmstead et al., 1986) or x-ray diffraction (e.g., Patel et al., 1987). Of these techniques, only HRTEM can analyze interface structure far away from the growth surface.

F. STRUCTURE OF EPITAXIAL SEMICONDUCTOR ALLOYS

1. Alloy Ordering

A great number of important epitaxial semiconductor systems, e.g., InAlAs/InGaAs, GeSi/Si, InGaAs/GaAs, InGaAs/InP, AlGaAs/GaAs, InGaAsP/InP, have one or both constituents in the form of an alloy. In the bulk form (where the bulk analogue has been studied), such alloys are generally assumed to be random structures, e.g., GeSi (Hansen, 1958). In thin-film form, it is often found that evidence for ordering of two or more of the alloy constituents exists, whether as a function of growth (kinetic) conditions or as a function of strain, such that the unit cell no longer has a random distribution of atoms within it. It is also possible that nonrandom distributions of alloy constituents may occur on a larger scale, either through clustering of like atoms, or through periodic decomposition of the alloy, such as spinodal decomposition.

The first direct evidence for long-range ordering of atoms within an epitaxial alloy unit cell was shown in the work of Kuan et al. (1985), who detected ordering within the unit cell of AlGaAs grown by MBE on (001) and, particularly, (110) GaAs substrates by electron diffraction, as illustrated in Fig. 38. The ordered unit-cell structure was equivalent to monolayers of AlAs and GaAs stacked along a $\langle 110 \rangle$ axis of the crystal, yielding a nominal ordered composition of $Al_{0.5}Ga_{0.5}As$. Since the work of Kuan and co-workers alloy ordering has been detected in a host of III-V ternary structures, with the ordered species lying on both the column-III and column-V sublattices (e.g., Jen et al., 1986; Gomyo et al., 1988; Shahid et al., 1987; Ihm et al., 1987). InGaAs, in particular, has been observed in several ordered states. Such semiconductor alloy ordering is analogous to ordering in metallic alloys, and virtually all ordered semiconductor states observed have their metallic analogues.

The presence of ordered domains within the alloy could strongly affect the optical and electronic properties of the material. The ordered state will have

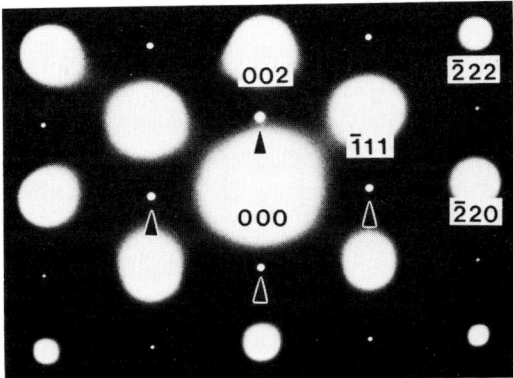

FIG. 38. Electron diffraction patterns illustrating the presence of column-III sublattice ordering in AlGaAs layers grown on GaAs. Reflections (arrowed) arise from ordering of Al and Ga atoms (from Kuan et al., 1985).

different band structures (and bandgaps) from the random alloy; thus, domains of ordered material in a disordered matrix are likely to produce spectral broadening in radiative recombination processes and extra electronic scattering mechanisms. Successful growth of a perfect uniform ordered alloy, however, could be extremely desirable because it raises the prospect of a material free of disorder scattering and free of random compositional variations that might produce spectral broadening in direct bandgap alloys. Such ordered alloys would offer advantages over artificially grown monolayer or short-period superlattices that have been grown to synthesize ordered alloys (Tamargo et al., 1985; Bevk et al., 1986; Petroff et al., 1978), since they would not be so subject to growth nonuniformities.

2. Alloy Clustering

Even if an alloy exhibits no signs of unit-cell ordering as discussed in the previous section nonrandom distributions of alloy species may still be present. The phenomenon of "clustering" of atoms of like species has been invoked (e.g., Holonyak et al., 1981; Singh and Bajaj, 1984, 1985) to explain anomalously high luminescence line widths from alloys, particularly those containing Al. Consider a three-dimensional exciton of radius R in an epitaxial alloy layer of thickness $\gg R$. For a unit cell containing N_c atoms on the clustering A_xB_{1-x} sublattice and with a volume V_c, the number of A atoms within the recombining exciton volume will be

$$N_A = \frac{4\pi R^3 \times N_c}{3V_c}.$$

Gaussian statistics using the binomial approximation for large N_A would predict a variance, $\sigma_A^2 = \sqrt{N_A x(1-x)}$. The spectral broadening $\delta E/E$, where δE represents the FWHM of the spectral line then becomes $2.35(\sqrt{x(1-x)}/\sqrt{N_A})$. Using the relationships given and $E \sim 1$ eV, $V_c = (5.6)^3$ Å^3, $N_c = 8$ (for one sublattice in the zinc-blende structures), $R = 150$ Å and $x = 0.5$, one obtains δE of the order of 2 meV. This represents the lower limit to the optical line width from excitonic recombination in semiconductor alloys (Schubert et al., 1984).

Experimental techniques for measurement of possible semiconductor clustering independent of optical measurements are relatively few and indirect. Experimental attempts have included those by Raman spectroscopy (Parayanthal and Pollak, 1984), infrared spectroscopy (Braunstein, 1963), and HRTEM (Hull et al., 1989c). More local environments (say, first- to third-nearest neighbors) have been studied by extended x-ray absorption fine structure (Mikkelsen and Boyce, 1982) and Raman spectroscopy (Renucci et al., 1971).

VII. Summary

In this chapter, we have attempted to outline present theoretical and experimental understanding of the physical processes involved in lattice-mismatched epitaxy. In this final section, we will summarize our understanding to date, emphasize those areas in which little understanding has yet emerged, and predict future directions for research.

In Section III of this chapter, we demonstrated that there can be a strong influence of the initial substrate cleaning process on subsequent hetero-epitaxial growth. This arises primarily from unexpected influences of the substrate cleaning on the resultant surface morphology. In addition, it is already well known that surface contaminants (e.g., Joyce, 1974) and surface reconstruction (e.g., Gossman, 1987) can significantly affect epitaxial growth.

Further knowledge of these processes will depend critically on development of experimental techniques to sensitively measure surface structure and contamination levels. Techniques such as LEED and RHEED give accurate information about surface reconstructions and some information about surface morphologies (terrace widths, etc.); and Auger spectroscopy can measure average surface contaminant levels to < 0.01 monolayers, but many significant effects can exist beyond the detectability level of these techniques. In situ TEM, for example (Gibson, 1989), has suggested that C contamination levels below the Auger detectability limit on a Si surface can still produce high silicon carbide particulate levels. The point here is not so much the absolute level of C contamination as its environment: a 0.01 monolayer that is evenly dispersed will be totally different from a 0.01 monolayer that has reacted to form a high density of SiC particulates. Such information is not readily available from Auger spectroscopy; and development of in situ imaging techniques such as TEM, scanning tunneling microscopy (if chemical identification can be more readily achieved), low-energy electron microscopy (Telieps, 1987), and reflection electron microscopy (see e.g., Cowley, 1984) may significantly extend our knowledge of surface processes in epitaxial growth.

The fundamentals of nucleation theory remain a much-studied topic, but the specific events that are not well understood during semiconductor epitaxial growth are interactions with surface structure and contamination, the role of nuclei coalescence boundaries in defect generation, cluster versus adatom mobility (see, e.g., Zinke-Allmang et al., 1987; Venables et al., 1987), and defect generation in clusters. Again, direct in situ imaging observations of these processes may help, and even nonimaging in situ techniques such as Rutherford backscattering (Zinke-Allmang et al., 1987) have proven invaluable.

The primary physical process inherent to strained-layer epitaxy is that of strain relaxation via misfit-dislocation generation. Although many details of this process remain poorly understood, much progress has been made in recent years. Earlier theories of Matthews and Blakeslee for misfit-defect propagation (Matthews and Blakeslee, 1974a,b; Matthews, 1975) have been developed and extended into a general kinetic model by Dodson and Tsao (1987). Preliminary measurements of activation energies necessary to describe these kinetic processes have been reported by Hull et al. (e.g., 1989a), and initial attempts to describe the role of defect interactions have also been published (Dodson, 1988b; Hull and Bean, 1989b). Attempts to critically examine dislocation nucleation modes in specific systems have been made (e.g., Matthews et al., 1976; Fitzgerald, 1989; Eaglesham et al., 1989; Hull and Bean, 1989a; Fitzgerald et al., 1989; Hagen and Strunk, 1978; Gibbings et al., 1989; Dodson, 1988c), and some new candidate sources have been identified. Molecular dynamics and other computer simulations of epitaxial growth are also providing significant insight (e.g., Grabow and Gilmer, 1987; Dodson, 1984; Thomsen and Madhukar, 1987a,b,c; Choi et al., 1987b). Refinement of our knowledge of these various processes should enable a complete predictive theory of strain relaxation to be developed, but much further work is required particularly in the areas of defect nucleation and interactions. Little work has been done to date on the extension of single strained-layer concepts to multilayer systems.

Experimental techniques for reducing threading-dislocation densities in thicker epitaxial layers exist, the primary tools being thermal annealing, filtering via strained-layer superlattices, and finite area growth. Of these techniques, it appears that the first two may be useful only down to moderate defect densities, of the order of $10^6 \, \text{cm}^{-2} - 10^7 \, \text{cm}^{-2}$, although recent reports suggest that electric field effects may extend the useful range of strained-layer superlattice filtering to lower defect densities (Liu et al., 1988; Shinohara, 1988). Techniques for patterned epitaxial growth are currently being developed with success in defect-density reduction being reported, for example, in InGaAs/GaAs (Fitzgerald et al., 1988; Fitzgerald, 1989; Fitzgerald et al., 1989) and GaAs/Si (Matyi et al., 1988; Lee et al., 1988).

Techniques for growing and measuring extremely abrupt and planar interfaces appear to be relatively advanced, as outlined in Section VI of this chapter. Molecular- and chemical-beam epitaxy techniques have advanced to the stage where interfaces that are planar and abrupt to small fractions of a monolayer may readily be synthesized (e.g., Petroff et al., 1978; Tsang et al., 1987; Bevk et al., 1986; Bimberg et al., 1986). Switching conditions in MOCVD growth, however, still appear to fall somewhat short of this. Developments of new crystal growth techniques such as ultrahigh-vacuum chemical vapor deposition (e.g., Meyerson, 1986; Meyerson et al., 1988), limited reaction processing (Gibbons et al., 1985, and atomic-layer epitaxy (e.g., Bedair et al., 1985; Nishizawa et al., 1985) may further extend epitaxial layer quality in some systems. Much fascinating science also remains to be done in understanding the exact nature of alloy materials, which are so important to many heteroepitaxial materials combinations.

In summary, much progress has been made in the last decade or so in understanding the limits of structural growth of lattice-mismatched epitaxial materials. In some systems, it is already possible to grow quasiperfect heteroepitaxial structures (misfit-defect free, with planar and abrupt interfaces), and the limits of growth of such structures are semi quantitatively known and understood. At this stage, however, it appears that in the field in general, more remains to be discovered than is already known.

Acknowledgments

We would like to acknowledge many of our colleagues who have been instrumental in their contributions to either experiments described in this chapter or our basic understanding of strained-layer epitaxy. At Bell Laboratories: D. Bahnck, J. M. Bonar, C. Buescher, N. Chand, A. Y. Cho, D. J. Eaglesham, L. C. Feldman, A. T. Fiory, G. Fitzgerald, J. M. Gibson, S. Luryi, D. M. Maher, A. Ourmazd, T. P. Pearsall, R. People, R. T. Lynch, Y. H. Xie and M. Yen; outside of Bell Laboratories: B. W. Dodson, I. J. Fritz and J. Y. Tsao (Sandia); C. M. Gronet, J. S. Harris, Jr., C. A. King, and S. M. Koch (Stanford University); K. W. Carey, A. Fischer-Colbrie, and S. J. Rosner

(Hewlett Packard Laboratories); E. P. Kvam and R. Pond (University of Liverpool); P. Pirouz (Case Western University); M. Tamargo (Bell Communications Research); D. K. Biegelsen and F. A. Ponce (Xerox PARC).

References

Alexander, H., and Haasen, P. (1968). *Solid State Phys.* **22**, p. 27.
Batstone, J. L., Gibson, J. M., Tung, R. T., Levi, A. F. J., and Outten, C. A. (1987). *Mat. Res. Soc. Symp.* **82**, R. W. Siegel, J. R. Weertman, and R. Sinclair, eds. Materials Research Society, Pittsburgh, Pennsylvania, p. 335.
Batstone, J. L., Phillips, J. M., and Hunke, E. C. (1988). *Phys. Rev. Lett.* **60**, 1394.
Bauer, E., and Poppa, H. (1972). *Thin Solid Films* **12**, 167.
Bean, J. C., Feldman, L. C., Fiory, A. T., Nakahara, S., and Robinson, I. K. (1984). *J. Vac. Sci. Technol.* **A2**, 436.
Bedair, S. M., Tischler, M. A., Katsuyama, T., and Al-Masry, N. A. (1985). *Appl. Phys. Lett.* **47**, 51.
Bellevance, D. (1988). In "Silicon Molecular Beam Epitaxy," Vol. II, Ch. 13, E. Kasper and J. C. Bean, eds. CRC Press, Boca Raton, California.
Berger, P. R., Chang, K., Bhattacharya, P., Singh, J., and Bajaj, K. K. (1988). *Appl. Phys. Lett.* **53**, 684.
Bevk, J., Mannearts, J. P., Feldman, L. C., Davidson, B. A., and Ourmazd, A. (1986). *Appl. Phys. Lett.* **45**, 286.
Biegelsen, D. K., Ponce, F. A., Smith, A. J., and Tramontana, J. C. (1987). *J. Appl. Phys.* **61**, 1856.
Biegelsen, D. K., Ponce, F. A., Krusor, B. S., Tramontana, J. C., Yingling, R. D., Bringans, R. D., and Fenner, D. B. (1988). *Proc. Mat. Res. Soc.* **116**, H. K. Choi, R. Hull, H. Ishiwara, and R. J. Nemanich, eds. Materials Research Society, Pittsburgh, Pennsylvania, p. 33.
Bimberg, D., Mars, D. E., Miller, J. N., Bauer, R., and Oertel, D. (1986). *J. Vac. Sci. Technol.* **B4**, 1014.
Bimberg, D., Christen, J., Fukunaga, T., Nakashima, H., Mars, D. E., and Miller, J. N. (1987). *J. Vac. Sci. Technol.* **B5**, 1191.
Bimberg, D., Oertel, D., Hull, R., Reid, G. A., and Carey, K. W. (1989). *J. Appl. Phys.*, **65**, 2688.
Boothroyd, C. B., Britton, E. G., Ross, F. M., Baxter, C. S., Alexander, K. B., and Stobbs, W. M. (1987). *Inst. Phys. Conf. Ser.* **87**. Institute of Physics, Bristol, England, p. 195.
Braunstein, R. (1963). *Phys. Rev.* **130**, p. 879.
Bringans, R. D., Uhrberg, R. I. G., Olmstead, M. A., Bachrach, R. Z., and Northrup, J. E. (1985). *Phys. Rev. Lett.* **55**, 533.
Bruinsma, R., and Zangwill, A. (1987), *Europhys. Lett.* **4**, 729.
Carey, K. W., Hull, R., Fouquet, J. E., Kellert, F. G., and Trott, G. (1987). *Appl. Phys. Lett.* **51**, 910.
Chand, N., People, R., Baiocchi, F. A., Wecht, K. W., and Cho, A. Y. (1986). *Appl. Phys. Lett.* **49**, 815.
Chang, K. H., Berger, P. R., Gibala, R., Bhattacharya, P. K., Singh, J., Mansfield, J. F., and Clarke, R. (1988). *Proc. of TMS/AIME Symposium* on "Defects and Interfaces in Semiconductors," K. K. Rajaj, J. Narayan, and D. Ast, eds. The Metallurgical Society, Warrendale, Pennsylvania, p. 157.
Cherns, D., Hetherington, C. J. D., and Humphreys, C. J. (1984). *Phil. Mag.* **49**, 165.
Choi, C., Otsuka, N., Munns, G., Houdre, R., Morkoc, H., Zhang, S. L., Levi, D., and Klein, M. V. (1987a). *Appl. Phys. Lett.* **50**, 992.

Choi, D. K., Halicioglu, T., and Tiller, W. A. (1987b). *Proc. Mat. Res. Soc.* **94**, R. Hull, J. M. Gibson, and D. A. Smith, eds. Materials Research Society, Pittsburgh, Pennsylvania, p. 91.
Cowley, J. M. (1984). *Proc. Mat. Res. Soc.* **31**, W. Krakow, D. A. Smith, and L. W. Hobbs, eds. Materials Research Society, Pittsburgh, Pennsylvania, p. 177.
Deveaud, B., Emery, J. Y., Chomette, A., Lambert, B., and Baudet, M. (1984). *Appl. Phys. Lett.* **45**, 1078.
Dobson, P. J. (1989). *NATO ASI Series B: Physics*, Vol. 203, D. Cherns ed., Plenum Press, New York, p. 267.
Dodson, B. W. (1984). *Phys. Rev.* **B30**, 3545.
Dodson, B. W. (1988a). *Phys. Rev.* **B38**, 12383.
Dodson, B. W. (1988b). *Appl. Phys. Lett.* **53**, 37.
Dodson, B. W. (1988c). *Appl. Phys. Lett.* **53**, 394.
Dodson, B. W., and Tsao, J. Y. (1987). *Appl. Phys. Lett.* **51**, 1325.
Dupuis, R. D., Bean, J. C., Brown, J. M., Macrander, A. T., Miller, R. C., and Hopkins, L. C. (1986). *J. Elect. Mat.* **16**, 69.
Eaglesham, D. J., Kvam, E. P., Maher, D. M., Humphreys, C. J., Green, C. S., Tanner, B. K., and Bean, J. C. (1988). *Appl. Phys. Lett.* **53**, 2083.
Eaglesham, D. J., Kvam, E. P., Maher, D. M., Humphreys, C. J., and Bean, J. C. (1989). *Phil. Mag.* **A59**, 1059.
Faurie, J. P., Hsu, C., Sivananthan, S. and Chu, X. (1986a). *Surf. Sci.* **168**, 473.
Faurie, J. P., Reno, J., Sivananthan, S., Sou, I. K., Chu, X., Boukerche, M., and Wijewarnasuriya, P. S. (1988b). *J. Vac. Sci. Technol.* **B4**, 585.
Fisanick, G. J., Gossman, H.-J., and Kuo, P. (1988). *Proc. Mat. Res. Soc.* **102**, R. Tung, L. R. Dawson, and R. L. Gunshor, eds. Materials Research Society, Pittsburgh, Pennsylvania, p. 25.
Fischer, R., Masselink, W. T., Klem, J., Henderson, T., McGlinn, T. C., Klein, M. V., Morkoc, H., Mazur, J. H., and Washburn, J. (1985). *J. Appl. Phys.* **58**, 374.
Fischer, R., Morkoc, H., Neumann, D. A., Zabel, H., Choi, C., Otsuka, N., Longerbone, M., and Erickson, L. P. (1986). *J. Appl. Phys.* **60**, 1640.
Fitzgerald, E. A. (1989). *J. Vac. Sci. Tech.* **B7**, 782.
Fitzgerald, E. A., Kirchner, P. D., Proano, R. E., Petit, G. D., Woodall, J. M., and Ast, D. G. (1988). *Appl. Phys. Lett.* **52**, 1496.
Fitzgerald, E. A., Watson, G. P., Proano, R. E., Ast, D. G., Kirchner, P. D., Pettit, G. D., and Woodall, J. M. (1989). *J. Appl. Phys.*, **65**, 2688.
Frank, F. C., and Van der Merwe, J. H. (1949a). *Proc. Roy. Soc.* **A198**, 205.
Frank, F. C., and Van der Merwe, J. H. (1949b). *Proc. Roy. Soc.* **A198**, 216.
Frank, F. C., and Van der Merwe, J. H. (1949c). *Proc. Roy. Soc.* **A200**, 125.
Fritz, I. J. (1987). *Appl. Phys. Lett.* **51**, 1080.
Fritz, I. J., Picraux, S. T., Dawson, L. R., Drummond, T. J., Laidig, W. D., and Anderson, N. G. (1985). *Appl. Phys. Lett.* **46**, 967.
Gibbings, C. J., Tuppen, C. G., and Hockley, M. (1989). *Appl. Phys. Lett.* **54**, 148.
Gibbons, J. F., Gronet, C. M., and Williams, K. E. (1985). *Appl. Phys. Lett.* **47**, 721.
Gibson, J. M. (1988). *Mat. Res. Soc. Proc.* **104**, M. Stavola, S. J. Pearton, and G. Davies, eds. Materials Research Society, Pittsburgh, Pennsylvania, p. 613.
Gibson, J. M. (1989). In "Surface and Interface Characterization by Electron Optical Methods" *NATO ASI Series B: Physics*, Vol. 191, A. Howie and U. Valdré, eds., Plenum Press, NY.
Gibson, J. M., and McDonald, M. L. (1987). In *Mat. Res. Proc. Soc.* **82**, R. W. Siegel, J. R. Weertman, and Sinclair, R. eds. Materials Research Society, Pittsburg, Pennsylvania, p. 109.
Gomyo, A., Suzuki, T., and Iijima, S. (1988). *Phys. Rev. Lett.* **60**, 2645.
Goodman, P., and Moodie, A. (1974). *Acta. Cryst. A* **30**, 280.
Goodnick, S. M., Ferry, D. K., Wilmsen, C. W., Lilliental, Z., Fathy, D., and Krivanek, O. L. (1985). *Phys. Rev.* **B32**, 8171.
Gossmann, H.-J. (1987). *Proc. Mat. Res. Soc.* **94**, R. Hull, J. M. Gibson, and D. A. Smith, eds

1. PRINCIPLES AND CONCEPTS OF STRAINED-LAYER EPITAXY

Materials Research Society, Pittsburgh, Pennsylvania, p. 53.
Gourley, P. L., Drummond, T. J., and Doyle, B. L. (1986). *Appl. Phys. Lett.* **49**, 1101.
Gourley, P. L., Fritz, I. J., and Dawson, L. R. (1988). *Appl. Phys. Lett.* **52**, 377.
Grabow, M., and Gilmer, G. (1987). *Proc. Mat. Res. Soc.* **94**, R. Hull, J. M. Gibson, and D. A. Smith, eds. Materials Research Society, Pittsburgh, Pennsylvania, p. 13.
Griffith, J. E., Kubby, J. A., Wierenga, P. E., and Kochanski, G. P. (1988). *Proc. Mat. Res.* **116**, H. K. Choi, R. Hull, H. Ishiwara, and R. J. Nemanich eds. Materials Research Society, Pittsburgh, Pennsylvania, p. 27.
Grunthaner, P. J., Grunthaner, F. J., Fathauer, R. W., Lin, T. L., Schowengerdt, F. D., Pate, B., and Mazur, J. H. (1988). *Proc. 2nd Int. Symp. on Si MBE*, J. C. Bean and L. J. Schowalter eds. Electrochemical Society, Pennington, New Jersey, p. 375.
Hagen, W., and Strunk, H. (1978). *Appl. Phys.* **17**, 85.
Hansen, M. (1958). In "Constitution of Binary Alloys," 2nd ed. McGraw-Hill, New York, p. 774.
Hardeman, R. W., Robbins, D. J., Gasson, D. B., and Daw, A. (1985). *Proc. 1st Int. Symp. on Si MBE*, J. C. Bean, ed. Electrochemical Society, Pennington, New Jersey, p. 16.
Harris, J. J., Joyce, B. A., and Dobson, P. J. (1981). *Surf. Sci.* **103**, L90.
Harris, J. S., Jr., Koch, S. M., and Rosner, S. J. (1987). In "Heteroepitaxy on Si II," *Proc. Mat. Res. Soc.* **91**, J. C. C. Fan, J. M. Phillips, and B.-Y. Tsaur eds. Materials Research Society, Pittsburg, Pennsylvania, p. 3.
Hayakawa, T., Suyama, T., Takahashi, K., Kondo, M., Yamamoto, S., Yano, S., and Hijikata, T. (1985). *Appl. Phys. Lett.* **47**, 952.
Henderson, R. C. (1972). *J. Electrochem. Soc.* **119**, 772.
Henzler, M. (1983). *Surf. Sci.* **132**, 82.
Himpsel, F. J., Karlsson, U. O., Morar, J. F., Rieger, D., and Yarmoff, J. A. (1986). *Phys. Rev. Lett.* **56**, 1497.
Hirsch, P. B., Howie, A., Nicholson, R. B., Pashley, D. W., and Whelan, M. J. (1977). "Electron Microscopy of Thin Crystals," 2nd ed. Robert E. Krieger, Malabar, Florida.
Hirth, J. P., and Lothe, J. (1968). "Theory of Dislocations." McGraw-Hill, New York.
Holonyak, N., Jr., Laidig, W. D., Camras, M. D., Morkoc, H., Drummond, T. J., Hess, K., and Burroughs, M. S. (1981). *J. Appl. Phys.* **52**, 7201.
Hull, R., and Bean, J. C. (1989). *J. Vac. Sci. Technol.* **A7**, 2580.
Hull, R., and Bean, J. C. (1989b). *Appl. Phys. Lett.* **54**, 925.
Hull, R., and Fischer-Colbrie, A. (1987). *Appl. Phys. Lett.* **50**, 851.
Hull, R., Gibson, J. M., and Bean, J. C. (1984). *Appl. Phys. Lett.* **46**, 179.
Hull, R., Bean, J. C., Fiory, A. T., Gibson, J. M., and Hartsough, N. E. (1985). *Proc. 1st Int. Symp. on Si MBE*, J. C. Bean, S. S. Iyer, E. Kasper, and Y. Shiraki, eds., Electrochemical Society, Pennington, NJ, p. 376.
Hull, R., Bean, J. C., Cerdeira, F., Fiory, A. T., and Gibson, J. M. (1986a). *Appl. Phys. Lett.* **48**, 56.
Hull, R., Rosner, S. J., Koch, S. M., and Harris, J. S., Jr. (1986b). *Appl. Phys. Lett.* **49**, 1714.
Hull, R., Fischer-Colbrie, A., Rosner, S. J., Koch, S. M., and Harris, J. S., Jr. (1987a). *Appl. Phys. Lett.* **51**, 1723.
Hull, R., Fischer-Colbrie, A., Rosner, S. J., Koch, S. M., Harris, J. S., Jr. (1987b). *Proc. Mat. Res. Soc.* **94**, R. Hull, J. M. Gibson, and D. A. Smith, eds. Materials Society, Pittsburgh, Pennsylvania, p. 23.
Hull, R., Bean, J. C., Leibenguth, R. E., Koch, S. M., and Harris, J. S., Jr. (1987c). *Proc. 2nd Int. Symp. on Si MBE*, J. C. Bean and L. J. Schowalter, eds. Electrochemical Society, Pennington, New Jersey, p. 293.
Hull, R., Reid, G. A., and Carey, K. W. (1987d). *Proc. Mat. Res. Soc.* **77**, J. D. Dow, I. K. Schuller, and J. Hilliard, eds. Materials Research Society, Pittsburgh, Pennsylvania, p. 261.
Hull, R., Turner, J. E., Fischer-Colbrie, A., White, A. E., Short, K. T., Pearton, S. J., and Tu, C. W. (1987e). *Proc. Mat. Res. Soc.* **93**, U. Gibson, A. E. White, and P. P. Pronko, eds. Materials Research Society, Pittsburgh, Pennsylvania, p. 153.
Hull, R., Yen, M. Y., Werder, D. J., Short, K. T., and Cho, A. Y. (1988a). Unpublished work.

Hull, R., Bean, J. C., Werder, D. J., and Leibenguth, R. E. (1988b). *Appl. Phys. Lett.* **52**, 1605.
Hull, R., Bean, J. C., Koch, S. M., and Harris, J. S., Jr. (1988c). In "Dislocations and Interfaces in Semiconductors," K. Rajan, J. Narayan, and D. G. Ast, eds. The Metallurgical Society, Warrendale, PA, p. 77.
Hull, R., Bean, J. C., Werder, D. J., and Leibenguth, R. E. (1989a). *Phys. Rev. B.* **40**, 1681.
Hull, R., Bean, J. C., Leibenguth, R. E., and Werder, D. J. (1989b). *J. Appl. Phys.* **65**, 4723.
Hull, R., Flores, J. R., and Bean, J. C. (1989c). Unpublished.
Hull, R., Bean, J. C., Bahek, D., Bonar, J. M., and Buescher, C. (1989d). *NATO ASI Series B: Physics*, Vol. 203, D. Cherns, ed., Plenum Press, New York, p. 381.
Hull, R., Bean, J. C., Eaglesham, D. J., Bonar, J. M., and Buescher, C. (1989e). *Thin Solids Films* **183**, 117.
Ihm, Y. E., Otsuka, N., Klem, J., and Morkoc, H. (1987). *Appl. Phys. Lett.* **51**, 2013.
Ishiwara, H., Asano, T., and Kanemura, S. (1985). *Proc. 1st Int. Symp. on Si MBE*, J. C. Bean, S. S. Iyer, E. Kasper, and Y. Shiraki, eds. Electrochemical Society, Pennington, New Jersey, p. 285, and references therein.
Ishizaka, A., Nakagawa, K., and Shiraki, Y. (1983). *Proc. of the 2nd Int. Conf. on Molecular Beam Epitaxy and Related Clean Surface Techniques*. Japan Society of Applied Physics, Tokyo, p. 183.
Iyer, S. S., Tsang, J. C., Copel, M. W., Pukite, P. R., and Tromp, R. M. (1989). *Appl. Phys. Lett.* **54**, 219.
Jen, H. R., Cherng, M. J., and Stringfellow, G. B. (1986). *Appl. Phys. Lett.* **48**, 1603.
Joyce, B. A. (1974). *Rep. Prog. Phys.* **37**, 363.
Kakibayashi, H., and Nagata, F. (1985). *Japan J. Appl. Phys.* **24**, L905.
Kakibayashi, H., and Nagata, F. (1986). *Japan J. Appl. Phys.* **25**, 1644.
Kaplan, R. (1980). *Surf. Sci.* **93**, 145.
Kasper, E., Herzog, H.-J., and Kibbel, H. (1975). *Appl. Phys.* **8**, 199.
Koch, S. M., Rosner, S. J., Hull, R., Yoffe, G. W., and Harris, J. S. Jr., (1987). *J. Crystal Growth* **81**, 205.
Kolodziejski, L. A., Gunshor, R. L., Otsuka, N., and Choi, C. (1986). *J. Vac. Sci. Technol.* **A4**, 2150.
Kroemer, H. (1986). *Proc. Mat. Res. Soc.* **67**, J. C. C. Fan and J. M. Poate, eds. Materials Research Society, Pittsburgh, Pennsylvania, p. 3.
Kuan, T. S., Kuech, T. F., Wang, W. I., and Wilkie, E. L. (1985). *Phys. Rev. Lett.* **54**, 201.
Kvam, E. P., Eaglesham, D. J., Maher, D. M., Humphreys, C. J., Bean, J. C., Green, G. S., and Tanner, B. K. (1988). *Proc. Mat. Res. Soc.* **104**, M. Stavola, S. J. Pearton, and G. Davies, eds. Materials Research, Pittsburgh, Pennsylvania, p. 623.
Lee, H. P., Huang, Y-H., Liu, X., Lin, H., Smith, J. S., Weber, E. R., Yu, P., Wang, S., and Lilliental-Weber, Z. (1988). *Proc. Mat. Res. Soc.* **116**, H. K. Choi, R. Hull, H. Ishiwara, and R. J. Nemanich, eds. Materials Research Society, Pittsburgh, Pennsylvania, p. 219.
Lee, J. W., Shichijo, H., Tsai, J. L., and Matyi, R. J. (1987). *Appl. Phys. Lett.* **50**, 31.
Lewis, B. F., Grunthaner, F. J., Madhukar, A., Lee, T. C., and Fernandez R. (1985). *J. Vac. Sci. Technol.* **B3**, 1317.
Liliental-Weber, Z. (1987). *J. Vac. Sci. Technol.* **B5**, 1007.
Liliental, Z., Krivanek, O. L., Goodnick, S. M., and Wilmsen, C. W. (1985). *Mat. Res. Proc.* **37**, J. M. Gibson and L. R. Dawson, eds. Materials Research Society, Pittsburgh, Pennsylvania, p. 193.
Liliental-Weber, Z., Weber, J., Washburn, J., Liu, T. Y., and Kroemer, H. (1987). *Proc. Mat. Res. Soc.* **91**, J. C. C. Fan, J. M. Phillips, and B.-Y. Tsaur, eds. Materials Research Society, Pittsburgh, Pensylvania, p. 91.
Lin, T. L., Sadwick, L., Wang, K. L., Kao, Y. C., Hull, R., Nieh, C. W., Jamieson, D. N., and Liu, J. K. (1987). *Appl. Phys. Lett.* **51**, 814.
Liu, T. Y., Petroff, P. M., and Kroemer, H. (1988). *J. Appl. Phys.* **52**, 543.
Luryi, S., and Suhir, E. (1988). *Appl. Phys. Lett.* **49**, 140.

Madhukar, A., Lee, T. C., Yen, M. Y., Chen, P., Kim, J. Y., Ghaisas, S. V., and Newman, P. G. (1985). *Appl. Phys. Lett.* **46**, 1148.
Matteson, S., and Bowling, R. A. (1988). *J. Vac. Sci. Technol.* **B6**, 2504.
Matthews, J. W. (1975). *J. Vac. Sci. Technol.* **12**, 126, and references therein.
Matthews, J. W., and Blakeslee, A. E. (1974a). *J. Crystal Growth* **27**, 118.
Matthews, J. W., and Blakeslee, A. E. (1974b). *J. Crystal Growth* **32**, 265.
Matthews, J. W., Blakeslee, A. E., and Mader, S. (1976). *Thin Solid Films* **33**, 253.
Matyi, R. J., Shichijo, H., and Tsai, H. L. (1988). *J. Vac. Sci. Technol.* **B6**, 699.
McWhan, D. B., Gurvitch, M., Rowell, J. M., and Walker, L. R. (1983). *J. Appl. Phys.* **54**, 3886.
Meyerson, B. S. (1986). *Appl. Phys. Lett.* **48**, 797.
Meyerson, B. S., Uram, K. J., and LeGoues, F. K. (1988). *Appl. Phys. Lett.* **53**, 2555.
Mikkelsen, J. C., and Boyce, J. B. (1982). *Phys. Rev. Lett.* **49**, 1412.
Miles, R. H., McGill, T. C., Chow, P. P., Johnson, D. C., Hauenstein, R. J., Nieh, C. W., and Strathman, M. D. (1988). *Appl. Phys. Lett.* **52**, 916.
Mott, N. F., and Nabarro, F. R. N. (1940). *Proc. Phys. Soc.* **52**, 86.
Nabarro, F. R. N. (1940). *Proc. Roy. Soc.* **A175**, 519.
Nabarro, F. R. N. (1967). "Theory of Crystal Dislocations." Clarendon Press, Oxford, England.
Neave, J. H., Joyce, B. A., Dobson, P. J., and Norton, N. (1983). *Appl. Phys.* **A31**, 1.
Nishi, S., Inomata, H., Akiyama, M., and Kaminishi, K. (1985). *Japan J. Appl. Phys.* **24**, L391.
Nishizawa, J., Abe, H., and Kurabayashi, T. (1985). *J. Electrochem. Soc.* **132**, 197.
Ogale, S. B., Madhukar, A., Voillot, F., Thomsen, M., Tang, W. C., Lee, T. C., Kim, J. Y., and Chen, P. (1987). *Phys. Rev.* **B36**, 1662.
Ogale, S. B., Madhukar, A., and Cho, N. M. (1988). *J. Appl. Phys.* **63**, 578.
Ohno, E. R., and Williams, E. D. (1989). *Jap. J. Appl. Phys.* **28**, L2061.
Olmstead, M. A., Uhrberg, R. I. G., Bringans, R. D., and Bachrach, R. Z. (1986). *J. Vac. Sci. Technol.* **B4**, 1123.
Olsen, G. H., Abrahams, M. S., Buiocchi, G. J., and Zamerowski, T. J. (1975). *J. Appl. Phys.* **46**, 1643.
Osakabe, N., Tanashiro, Y., Yagi, K., and Honjo, G. (1989). *Surf. Sci.* **97**, 393.
Otsuka, N., Choi, C., Nakamura, Y., Nagakuva, S., Fischer, R. F., Peng, C. K., and Morkoc, H. (1986). *Proc. Mat. Res. Soc.* **67**, J. C. C. Fan and J. M. Poate, eds. Materials Research Society, Pittsburgh, Pennsylvania, p. 85.
Ourmazd, A., Tsang, W. T., Rentschler, J. A., and Taylor, D. W. (1987). *Appl. Phys. Lett.* **50**, 1417.
Ourmazd, A., Taylor, D. W., Cunningham, J., and Tu, C. W. (1989). *Phys. Rev. Lett.* **62**, 933.
Parayanthal, P., and Pollak, F. H. (1984). *Phys. Rev. Lett.* **52**, 1822.
Parker, E. H. C., and Whall, T. E. (1988). *Proc. 2nd Int. Symp. on Si MBE*, J. C. Bean and L. J. Schowalter, eds. The Electrochemical Society, Pennington, New Jersey, p. 347.
Patel, J. R., and Chaudhuri A. R. (1966). *Phys. Rev.* **143**, 601.
Patel, J. R., Freeland, P. E., Hybertsen, M. S., Jacobson, D. C., and Golvechenko, J. A. (1987). *Phys. Rev. Lett.* **59**, 2180.
Peierls, R. E. (1940). *Proc. Phys. Soc.* **52**, 23.
People, R., and Bean, J. C. (1985). *Appl. Phys. Lett.* **47**, 322.
People, R., and Bean, J. C. (1986). *Appl. Phys. Lett.* **49**, 229.
Pennycook, S. J., Jesson, D. E., and Chisholm, M. J. (1989). *Inst. Phys. Conf. Ser. 100*, Institute of Physics, Bristol, England, p. 195.
Petroff, P. M., Gossard, A. C., Wiegmann, W., and Savage, A. (1978). *J. Crystal Growth* **44**, 5.
Petroff, P. M. (1986). *J. Vac. Sci. Technol.* **B4**, 874.
Phillips, J. M., Batstone, J. L., and Hensel, J. C. (1988). *Proc. Mat. Res. Soc.* **116**, H. K. Choi, R. Hull, H. Ishiwara, and R. J. Nemanich, eds. Materials Research Society, Pittsburgh, Pennsylvania, p. 403, and references therein.
Pirouz, P., Ernst, F., and Cheng, T. T. (1988). *Proc. Mat. Res. Soc.* **116**, H. K. Choi, R. Hull, H. Ishiwara, and R. J. Nemanich, eds. Materials Research Society, Pittsburgh, Pennsylvania, p. 57.

Ponce, F. A., Anderson, G. B., and Ballingal, J. M. (1967). *Proc. Mat. Res. Soc.* **90**, R. F. C. Farrow, J. F. Schetzina, and J. T. Cheung, eds. Materials Research Society, Pittsburgh, Pennsylvania, p. 199.
Rajan, K., and Denhoff, M. (1987). *J. Appl. Phys.* **62**, 1710.
Reno, J. L., Gourley, P. L., Monfroy, G., and Faurie, J. P. (1988). *Appl. Phys. Lett.* **53**, 1747.
Renucci, M. A., Renucci, J. B., and Cardona, M. (1971). In "Light Scattering in Solids." M. Balkanski, ed., Flammarion, Paris, France, p. 326.
Rubloff, G. W., Tromp, R. M., van Loenen, E. J., Balk, P., and LeGoues, F. (1986). *J. Vac. Sci. Technol.* **A4**, 1024.
Sakamoto, T., Funabashi, H., Ohta, K., Nakagawa, T., Kawai, N. J., Kojima, T., and Bando, Y. (1985). *Superlattices and Microstructure* **1**, 347.
Schowalter, L. J., and Fathauer, R. W. (1986). *J. Vac. Sci. Technol.* **A4**, 1026, and references therein.
Schubert, E. F., Gobel, E. O., Horikoshi, J., Ploog, K., and Queisser, H. J. (1984). *Phys. Rev.* **B30**, 813.
Segmuller, A., and Blakeslee, A. E. (1973). *J. Appl. Cryst.* **6**, 19.
Shahid, M. A., Mahajan, S., and Laughlin, D. E. (1987). *Phys. Rev. Lett.* **58**, 2567.
Shinohara, M. (1988). *Appl. Phys. Lett.* **52**, 543.
Singh, J., and Bajaj, K. K. (1984). *Appl. Phys. Lett.* **44**, 1075.
Singh, J., and Bajaj, K. K. (1985). *J. Appl. Phys.* **57**, 5444.
Spence, J. C. H., (1981). "Experimental High Resolution Electron Microscopy." Clarendon Press, Oxford.
Stranski, I. N., and Krastanow, L. (1939). *Akad. Wiss. Mainz L. Math.-Nat.* KlIIb **146**, 797.
Tabe, M. (1984). *Appl. Phys. Lett.* **45**, 1073.
Tabe, M., Arai, K., and Nakamura, H. (1981). *Japan J. Appl. Phys.* **20**, 703.
Takayanagi, K., Yagi, K., Kobayashi, K., and Honjo, G. (1978). *J. Phys.* **E11**, 441.
Tamargo, M. C., Hull, R., Greene, L. H., Hayes, J. R., and Cho, A. Y. (1985). *Appl. Phys. Lett.* **46**, 569.
Telieps, W. (1987). *Appl. Phys.* **A44**, 55.
Thomsen, M., and Madhukar, A. (1987a). *J. Crystal Growth* **80**, 275.
Thomsen, M., and Madhukar, A. (1987b). *J. Crystal Growth* **84**, 79.
Thomsen, M., and Madhukar, A. (1987c). *J. Crystal Growth* **84**, 98.
Tromp, R. M., Rubloff, G. W., Balk, P., LeGoues, F., and van Loenen, E. J. (1985). *Phys. Rev. Lett.* **55**, 2322.
Tsang, J. C., Iyer, S. S., and Delage, S. L. (1987). *Appl. Phys. Lett.* **51**, 1732.
Tsao, J. Y., Dodson, B. W., Picraux, S. T., and Cornelison, D. M. (1987). *Phys. Rev. Lett.* **59**, 2455.
Tsang, W. T., and Schubert, E. F. (1986). *Appl. Phys. Lett.* **49**, 220.
Tung, R. T., and Gibson, J. M. (1985). *J. Vac. Sci. Technol.* **A3**, 987.
Vandenberg, J. M., Chu, S. N. G., Hamm, R. A., Panish, M. B., and Temkin, H. (1986). *Appl. Phys. Lett.* **49**, 1302.
Van der Merwe, J. H., and Ball, C. A. B. (1975). In Epitaxial Growth," Part B, J. W. Matthews, ed. Academic Press, New York, pp. 193–528.
Van Loenen, E. J. (1986). *J. Vac. Sci. Technol.* **A4**, 939.
Venables, J. A., Doust, T., and Kariotis, R. (1987). *Proc. Mat. Res. Soc.* **94**, R. Hull, J. M. Gibson, and D. A. Smith, eds. Materials Research Society, Pittsburgh, Pennsylvania, p. 3, for a general review of nucleation theory.
Wierenga, P. E., Kubby, J. A., and Griffith, J. E. (1987). *Phys. Rev. Lett.* **59**, 2169.
Xie, Y.-H, Wu, Y. Y., and Wang, K. L. (1986). *Appl. Phys. Lett.* **48**, 287.
Yen, M. Y., Levine, B. F., Bethea, C. G., Choi, K. K., and Cho, A. Y. (1987). *Appl. Phys. Lett.* **50**, 927.
Zinke-Allmang, M., Feldman, L. C., and Nakahara, S. (1987). *Appl. Phys. Lett.* **51**, 975.

CHAPTER 2

Device Applications of Strained-Layer Epitaxy

William J. Schaff, Paul J. Tasker, Mark C. Foisy, and Lester F. Eastman

DEPARTMENT OF ELECTRICAL ENGINEERING
CORNELL UNIVERSITY
ITHACA, NEW YORK

I.	Introduction	73
	A. Strained- (Pseudomorphic) Layer Structures	74
	B. Strain-Relieved (Mismatched) Structures	76
	C. Summary and Overview	77
II.	Properties of Strained Layers	77
	A. Limits to Strained-Layer Critical Thickness	78
	B. Optical and Electrical Properties of GaInAs on GaAs and InP	84
	C. Strained-Layer Growth of GaInAs/GaAs Layers.	89
III.	Electronic-Device Structures	100
	A. Modulation-Doped Field-Effect Transistors	100
	B. Other Electronic-Device Structures	117
IV.	Optoelectronic-Device Structures	120
	A. Quantum-Well GRINSCH Lasers	120
	B. Detectors	127
	C. Modulators.	127
V.	Optimized Lattice-Constant Epitaxy	127
	A. MODFETs.	128
	B. MESFETs	128
	C. Detectors	130
VI.	Conclusions.	131
	References	132

I. Introduction

The epitaxial growth of elemental semiconductors, compound semiconductors, and compound semiconductor alloys had, until very recently, concentrated on the use of lattice-matched materials. This approach limited considerably the associated development of electronic and optoelectronic semiconductor device structures, restricting their performance and/or applications. Wavelengths of laser emission, frequency response, and output power of transistors all suffered limitations in lattice-matched systems. Recently, the availability of device-quality strained-layer epitaxy has provided additional flexibility to the design of device structures, allowing the rapid development of improved electronic and optoelectronic devices, thus expanding the applications that electronic and optoelectronic semiconductor devices will be able to address in the near future.

The application of strained-layer epitaxy can generally be divided into two very different categories. Structures using materials having different lattice constants can be either strained elastically (pseudomorphic) without the presence of unwanted defects, or can be strained beyond the limits of elastic deformation to become strain relieved (mismatched). The major interest in the device applications up to this point in time has been in strained- (pseudomorphic) layer structures that avoid the presence of defects, since defects degrade the performance of most devices. The stability of different layer designs to prevent formation of defects becomes an important factor in layer design.

A. Strained- (Pseudomorphic) Layer Structures

Strained (pseudomorphic) layers offer the ability to select an alternative energy band structure for application to semiconductor devices. This approach requires electrically and/or optically active strained layers.

1. *Bandgaps*

The ability to modify the energy bandgap using a strained layer and strained-layer heterojunctions has been exploited to great advantage in photodetectors (e.g., GaInAs strained layers on GaAs, InAsSb strained layers on InSb, Ge strained layers on Si, etc.). In this case, the strained layers modify the bandgap of the photodetector and thus provide for operation at the desired optical frequency or bandwidth by appropriate strained-layer design.

The emission wavelength of a quantum-well laser structure can also be modified by using a strained-layer active region and by modifying the bandgap of the active material. For example, the use of a strained layer of GaInAs in the quantum well of a laser structure can reduce the photon energy (increase the wavelength) of the laser emission by decreasing the bandgap of the quantum well.

2. *Band Offsets*

Heterojunctions are the key element in the realization of new high-performance semiconductor device structures. Strained layers provide modified heterojunction band offsets. The ability of heterojunctions to modify charge confinement parallel to the heterojunction has been used to great advantage, for example, in the modulation-doped field-effect transistor

(MODFET).[1] The availability of device-quality strained layers has led to the development of AlGaAs/Ga$_{1-y}$In$_y$As and GaInP/Ga$_{1-y}$In$_y$As strained-layer heterojunctions grown on GaAs and the AlInAs/Ga$_{.47-y}$In$_{.53+y}$As strained-layer heterojunction grown on InP. The application of these strained-layer heterojunctions to MODFET structures has resulted in significantly improved transistor performance because of the ability to increase the heterojunction conduction-band offset in an n-type (p-type) transistor by using the narrow bandgap GaInAs strained layer as the active channel. The increased conduction- (valence-) band offset allows more electrons (holes) to be accumulated in the two-dimensional electron (hole) gas, thus increasing both the speed and the current drive of the transistor.

The ability of heterojunctions to modify charge transport across the interface has been used to great advantage in the development of heterojunction bipolar transistors (HBTs) and resonant tunneling structures. The application of strained-layer heterojunctions has also led to improved performance from these device structures. The introduction of uniform and graded-composition GaInAs strained layers into the base of the HBT provides for both increased valence- and conduction-band offsets at the emitter–base heterojunction p–n diode and for a built-in field in the base region. The increased valence-band offset reduces the injection of holes into the emitter, hence increasing the gain of the transistor. A built-in field in the base is capable of forcing the electron minority carriers across the base in a shorter time at a nearly saturated velocity. This reduces the probability of recombination, therefore further increasing the gain, as well as increasing the speed of the transistor. The Si/Ge strained-layer heterojunction has allowed the realization of HBT structures in elemental semiconductors. The performance of tunnelling-device structures can be improved using increased band offsets, which can be achieved using strained-layer tunnelling barriers. The use of strained AlAs barriers, for example, in resonant tunnelling structures grown on InP provide improved performance, increased peak-to-valley current ratios over lattice-matched AlInAs barriers. An additional advantage of strained-layer binary material barriers is that they are not expected to suffer from random barrier-height variations over the wafer due to a random alloy distribution, as would be expected from ternary alloys, e.g., AlInAs.

3. *Band Structure*

In addition to the improvements predicted from the first-order effects already discussed, the transport properties of the semiconductor devices with active strained layers are also modified from that of unstrained material. The most

---

[1]Also referred to as the high-electron mobility transistor (HEMT), the selectively doped heterojunction transistor (SDHT), the two-dimensional electron gas field-effect transistor (TEGFET) and the heterojunction field-effect transistor (HFET).

significant effect that could be exploited in device applications is the splitting of the light- and heavy-hole valence bands in GaAs–GaInAs–GaAs quantum wells as a result of the biaxially compressive strain. The improved hole-transport properties including lower-mass and higher-mobility transport, which should result from this valence-band splitting, are expected to be reflected in (i) improved high-speed performance in strained-layer p-channel quantum-well MODFET structures and HBT structures and (ii) in improved high-speed performance and reduced threshold currents in strained-layer quantum-well graded-index separate-confinement heterostructure (GRINSCH) laser diodes.

Strained-layer heterojunctions also provide the opportunity to design multiquantum-well structures that have very high internal electric fields, as a result of strain, for application to optoelectronic devices. Modulators that use the strain-generated electric fields in (111)-orientated GaInAs/GaAs interfaces have demonstrated improved optical transmission selectivity compared to nonstrained structures.

B. Strain-Relieved (Mismatched) Structures

The second general use of strained-layer materials employs those that have been strained beyond the elastic limit of strain and have produced defects to accomodate stress relief, thus providing the ability to create an alternative substrate material. In most of these applications, the strain-relieving layers near the interface with the substrate are not generally electrically and/or optically active; thick buffer layers are usually epitaxially grown by molecular-beam epitaxy (MBE) or organometallic vapor-phase epitaxy (OMVPE) and involve various approaches, e.g., superlattice buffers, strained-superlattice buffers, etc., to define a new lattice constant and to limit the migration of dislocations into the active layers. The electrically active device layers are grown on top of these buffers largely unstrained. For example, the epitaxial growth of GaAs on a Si substrate produces GaAs with the same nominal lattice constant as bulk GaAs. The discussion presented here will be limited to GaInAs structures on GaAs substrates as an alternative to strained-layer GaInAs/GaAs.

1. Establishment of a New Constant

The two categories of strained-layer epitaxy can been combined in order to realize strained buffer layers (effectively a new substrate material) with new optimized lattice constants, referred to as optimized-lattice-constant epitaxy (OLE), on which new device structures with optimized electronic properties can be grown (strained or unstrained). For example, heterojunctions grown on these optimized buffer layers can be selected in order to maximize the conduction-band offset, avoiding the previous limitations of having (i) to select a specific substrate lattice constant, e.g., that of Si, Ge, GaAs, InP, etc., which gives compositions with small conduction-band offsets, or (ii) to grow strained active layers to increase the conduction-band offset. For example, the optimized $Al_{0.6}In_{0.4}As/Ga_{0.4}In_{0.6}As$ heterojunction using $Ga_{0.4}In_{0.6}As$ as a new lattice constant is interesting. These two compositions are strained in order to maximize the conduction-band offset achievable between AlInAs and GaInAs but use a substrate lattice constant different than any presently available substrate material.

C. SUMMARY AND OVERVIEW

The most significant improvements in device performance are expected in all the applications where the strained layer is electrically and/or optically active. In these applications, the constraints on the epitaxial growth techniques and layer design are the most demanding, since dislocations must be avoided. These applications will therefore be considered in detail in this chapter. First, the improvements that can be achieved by using *strained layers to modify heterojunction-band offsets* will be discussed in detail. Specifically the MBE growth, transistor design, and performance of the AlGaAs/GaInAs strained-layer MODFET will be covered, since this device structure most clearly demonstrates the advantages and potential for the use of strained layers in charge-confinement electronic semiconductor device structures. Other device structures, such as HBTs and resonant tunneling device structures will also be briefly discussed.

The improvements that can be achieved by using *strained layers to modify the bandgap* will be discussed in detail. The MBE growth and laser performance of the strained-layer GaInAs quantum-well GRINSCH laser will be covered, since this device structure most clearly demonstrates the advantages and potential for the use of strained layers in optoelectronic semiconductor device structures. Other device structures, such as photodetectors, will be briefly discussed.

Finally, we shall briefly discuss the application of strained-layer epitaxy to the development of OLE for optimized substrate material and associated device structures.

II. Properties of Strained Layers

Strained-layer structures open the door on a wide variety of new properties for device applications that cannot be obtained with nonstrained structures. Strain serves to modify the bandgap and offers a new range of band offsets. Some of these properties are sufficiently important, so that it would be desirable to grow thick layers of this material for device applications. Unfortunately, it is not possible to grow strained layers beyond a critical layer thickness (CLT) because large densities of misfit dislocations are formed to accomodate the strain. The effects of strain on forming dislocations, altering band structures, modifying transport properties, and introducing new restrictions on conditions of epitaxial growth are discussed in this section. Because of the importance of strained GaInAs on GaAs or InP, and due to the data available, discussion will focus on these materials to serve as a guide to typical strained-layer phenomena. While some of the conclusions reached in these studies may not be directly applicable to other materials systems, they can serve as a starting point when considering new materials.

A. LIMITS TO STRAINED-LAYER CRITICAL THICKNESS

The early work that served as a guide to the rapid increase in interest in strained-layer systems was performed by Matthews and Blakeslee during the early 1970s (Matthews and Blakeslee 1974). They developed expressions for limits to strain in epitaxial layers prior to formation of dislocations due to lattice mismatch. The value of critical layer thickness h_c is given as:

$$h_c = \frac{b}{[1 + v(x)]4\pi f(x)} \left[\ln\left(\frac{h_c}{b}\right) + 1 \right], \tag{1}$$

where v is Poisson's ratio, f is the lattice mismatch, and b is the magnitude of the Burgers vector; v, f, and b are functions of the indium mole fraction x. A plot of CLT for GaInAs grown in between layers of GaAs (Okamoto et al., 1987) is shown in Figs 1 and 2. It can be seen that increases in the mole fraction of In require that strained layers of GaInAs on GaAs be decreased in thickness to avoid the onset of dislocations.

Critical layer thickness has been examined extensively in the

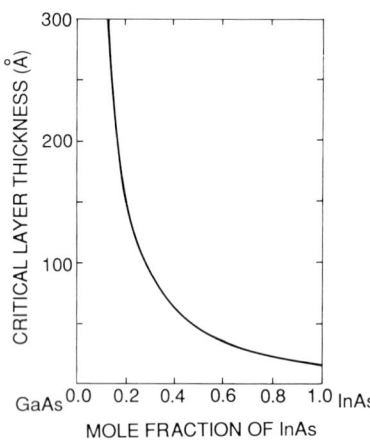

Fig. 1. CLT as a function of InAs mole fraction for single GaInAs quantum wells grown on GaAs.

GaInAs/GaAs system. The Matthews–Blakeslee CLT provides good guidance for epitaxial layer design, despite small discrepancies between the theoretical and experimental measurement of CLT. As discussed in Section III, the thicknesses of GaInAs on GaAs, or the mole fractions of In for a particular thickness, which are limited by CLT, establish the dislocation-free limits to layer design. Under some conditions to be discussed, it is not possible to reach CLT before the onset of dislocations. Much effort has gone into understanding the nature of defect formation in hopes of controlling dislocations to within tolerable densities. Further efforts have used the lessons of these studies to make strained-layer growth possible to thicknesses well beyond Matthews and Blakeslee CLT. Growth on patterned areas to greater CLT as one advancement is discussed below.

A variety of techniques have been used to observe defects in strained-layer structures to determine critical layer thickness. Physical characterization of strained layers and their interfaces has been performed by techniques that include transmission electron microscopy (TEM) and scanning tunneling microscopy (STM) (Kavanagh et al., 1988), cathodoluminescence (CL) (Fitzgerald et al., 1989), X-ray rocking curve (XRC) (Wie, 1989), and photoluminescence microscopy (PLM) (Gourley et al., 1988; Joyce, 1989). CL and PLM are most sensitive and therefore the best tools for observing the onset of dislocations at very low densities.

CL has been used to view GaInAs/GaAs layers from the surface looking down toward the interface (Woodall et al., 1983), as seen in Fig. 3. The crosshatched pattern results from the formation of dislocations that form orthogonally. Most, if not all, of the dislocations formed when strained layers of GaInAs with small amounts of strain ($< 2\%$) are grown on GaAs seem to

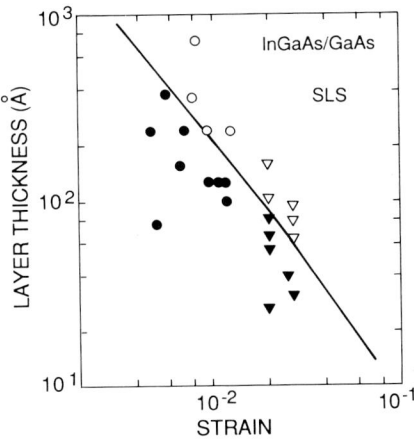

Fig. 2. Plot of GaInAs layer thickness versus strain for strained-layer superlattices (from Fritz et al., 1985). Solid symbols represent high-quality material as determined by low-temperature Hall mobility or photoluminescence measurements. Open symbols represent samples with structural defects. The solid line is the theoretical expression for critical layer thickness proposed by Matthews and Blakeslee.

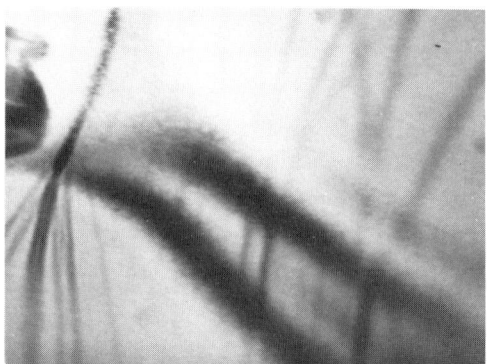

$Ga_{0.93}In_{0.07}As$ LAYER 100 nm THICK (PSEUDOMORPHIC)
0.5% MISFIT - (Mag. —— = 100 nm)

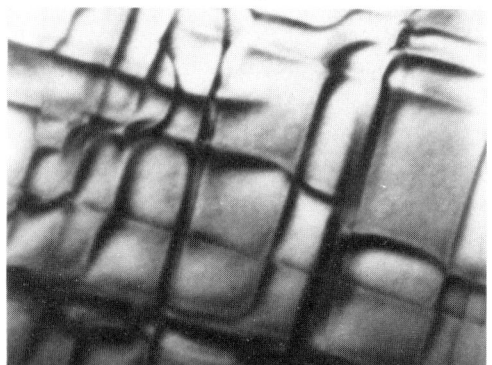

$Ga_{0.93}In_{0.07}As$ LAYER 1000 nm THICK
0.5% MISFIT - AVERAGE DISLOCATION SPACING = 140 nm

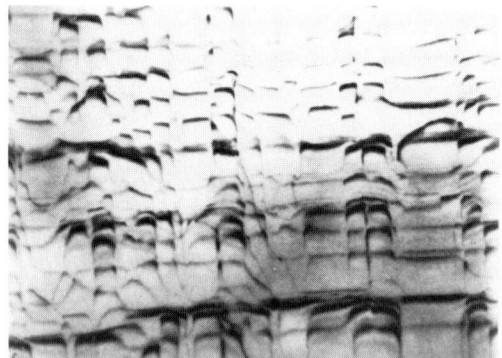

$Ga_{0.80}In_{0.20}As$ LAYER 1000 nm THICK
1.5% MISFIT - AVERAGE DISLOCATION SPACING = 50 nm

FIG. 3. Cathodoluminescence of the pseudomorphic interface and misfitted interface.

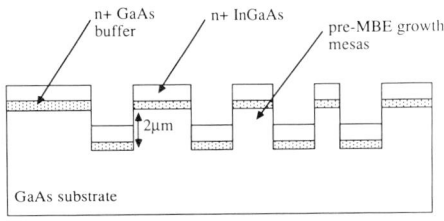

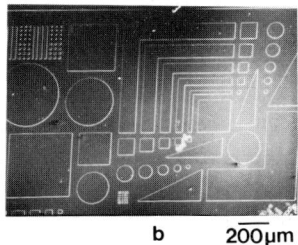

FIG. 4. (a) A schematic cross-sectional view of the patterns and structures used to investigate critical layer thicknesses for single layers of GaInAs grown on GaAs. (b) A plan-view scanning electron microscopy image of the mesa structures. (From Fitzgerald et al., 1989).

lie in a plane at the interface. Thus, the CL images are views of the interface buried below the GaInAs layers.

Cathodo-luminescence (CL) is the tool used to analyze growth of structures using patterned substrates. When critical thickness is exceeded, it is found that dislocation densities are related to the density of defects that existed prior to layer growth (Fitzgerald et al., 1989). These defects include substrate dislocations and other imperfections on the wafer surface prior to growth. In a study of growth of single GaInAs layers strained beyond the CLT on patterned area substrates, it has been found that it is possible to grow as much as 10 times thicker layers before the onset of dislocations. The structures studied are shown in Figs. 4a and 4b. CL images from growths on some of these areas are seen in Fig. 5. Dislocation densities are dramatically reduced through patterned growth. The smallest areas show no evidence of dislocations despite growth beyond CLT for large areas. The amount of improvement in patterned CLT is directly related to pattern dimensions; smaller-area structures permit greater thicknesses without the presence of dislocations. Devices have not yet been fabricated using this technology.

The strained-layer device structures to be discussed in Section III require layer dimensions that cannot be deduced from the simple Matthews–Blakeslee approximation of a single strained layer bounded by infinite cladding layers. Modifications to the original theory have been performed that take into account the different mechanisms of strain relief in order to examine the effects of cladding layers of arbitrary dimension (Tsao and Dodson, 1988). For single strained layers on a thick substrate, strain relief

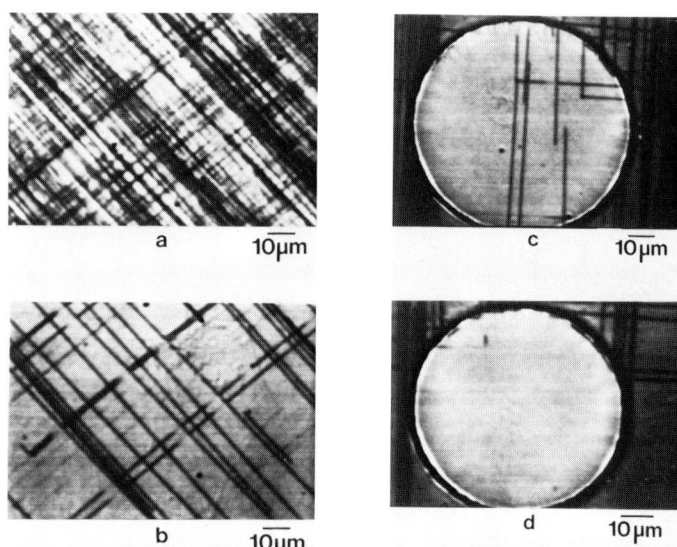

FIG. 5. CL images of a 3500 Å $Ga_{0.95}In_{0.05}As$ layer on varying area sizes: (a) Large-area control sample; (b) 200 μm circular mesa; (c) 90 μm circular mesa; (d) 67 μm circular mesa. (From Fitzgerald et al., 1989).

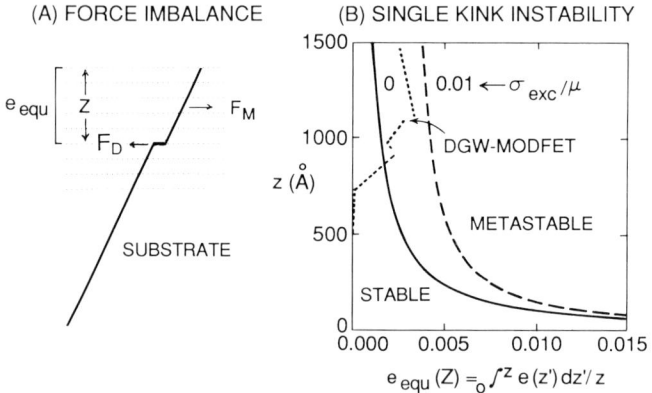

FIG. 6. (a) Strain relief by single kinking of a threading dislocation. (b) Constant single-kink excess stress contours on a depth/equivalent-strain plot. Dashed line shows depth-dependent single-kink equivalent strains for a double quantum-well MODFET.

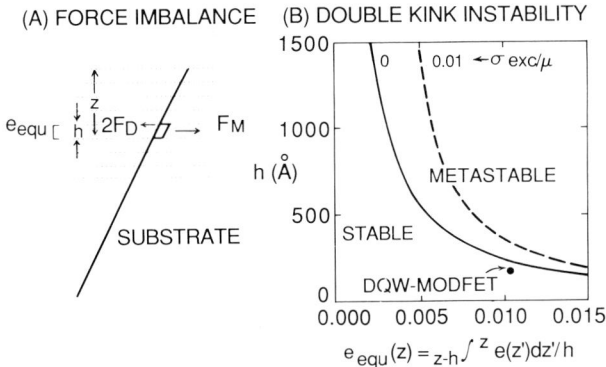

FIG. 7. (a) Strain relief by double kinking of a threading dislocation. (b) Constant double-kink excess stress contours on a thickness/equivalent-strain plot. Data point shows the double-kink equivalent strains for 175 Å layers of strained $Ga_{0.85}In_{0.15}As$ buried in GaAs and AlGaAs.

takes place by a "single-kink" mechanism; a single threading dislocation relieves the strain due to mismatch between the two materials of different lattice-constant. If a thin GaAs layer is then grown on the single GaInAs strained layer, the GaAs layer will be strained due to the presence of the GaInAs layer underneath it. The pair of layers can be viewed also as a single strained layer with an equivalent strain that exists all the way to the surface. This equivalent strained layer will also be subject to single-kink strain relief. In the case of MODFET structures, the equivalent strain is significant enough to limit channel thickness to less than conventional CLT before the onset of dislocations. The single-kink mechanism is schematically represented in Fig. 6a. Figure 6b shows single-kink equivalent strains for a double-channel strained-layer MODFET that is not stable to single-kink strain relief. Dislocations result from the equivalent thickness being beyond CLT.

In contrast to single-kink relief, strained layers bounded by more conventional thick cladding layers typically used to study strained-layer structures will relieve stress only within the strained layer by double kinking of a threading dislocation. This mechanism is shown schematically in Fig. 7a. Figure 7b shows that the strained MODFET will be stable to relief by double kinking. The GaAs/AlGaAs supply layers for MODFET designs unfortunately do not provide sufficiently thick cladding layers to be stable to single kinking despite being thought to be stable if only double kinking is considered. Thus, it is seen that MODFETs designed to follow "thick" cladding-layer designs will be metastable to defects.

The thermal stability of strained-layer structures has been evaluated by photoluminescence (PL) (Zipperian *et al.*, 1989), where thin surface cladding layers, such as those used in MODFETs, were analyzed. These structures are metastable to the formation of dislocations. If sufficient energy is supplied to these metastable structures through high-temperature annealing, dislocations will form. The design of MODFET structures must then pay attention to additional criteria beyond the simple CLT limitations. It is necessary to consider the post-growth processing temperatures to which the structure will be subjected when evaluating thermal stability of strained-layer design.

B. Optical and Electrical Properties of GaInAs on GaAs and InP

The presence of strain modifies the electrical and optical properties of semiconductor materials (Kuo et al., 1985; Massies and Sauvage-Simkin, 1983; Olsen et al., 1978). There is generally a splitting of the valence-band degeneracy due to anisotropic deformation of the strained-layer crystal lattice. A summary of all materials systems is beyond the scope of this discussion. GaInAs on GaAs or InP will be used as an example of one such system.

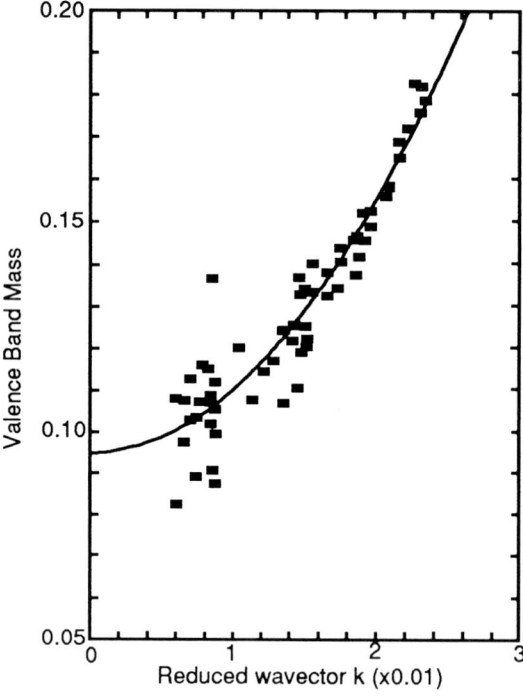

Fig. 8. The valence-band effective-mass dependence on the reduced wave-vector k. The data are obtained from magnetoluminescence with $m_c = 0.068$. The curve drawn through the data is the result of a fit of the form $m_v = m(0)(1 + Ak^2)$.

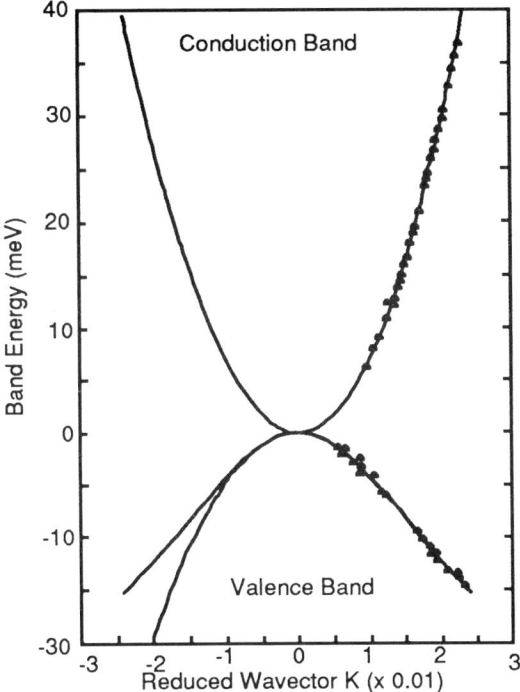

FIG. 9. Derived band energies for the conduction and valence bands as a function of the reduced wave vector k. The bandgap energy E_g has been substracted from the conduction-band energy. The upper (positive) portion is for the conduction band and the lower (negative) portion is for the valence band.

1. Modification of Bandgap, Effective Masses, and Band Splitting

Strain in GaInAs on GaAs modifies both the conduction- and valence-band structures. The valence band undergoes the most dramatic change in character. Strain induces a splitting between the heavy-hole and light-hole energy bands. The uppermost valence band (heavy-hole band) will have a light mass in the plane of conduction as shown in Fig. 8. The conduction band remains parabolic up to 40 meV above the bandgap and has an effective mass of 0.071 m_0 for InAs mole fractions of 0.15 to 0.25 (Jones et al., 1989a). The conduction-band energy dependence on the reduced wave vector k is shown in Fig. 9. The valence band is highly nonparabolic (Jones et al., 1989b) and has hole masses of 0.084_0 to 0.094 m_0 for the same composition range.

The light-hole in-plane conduction results in improved p-type mobility in GaInAs/GaAs MODFETs as a function of increasing InAs mole fraction in the channel as seen in Fig. 10. Some improvement would also be expected in

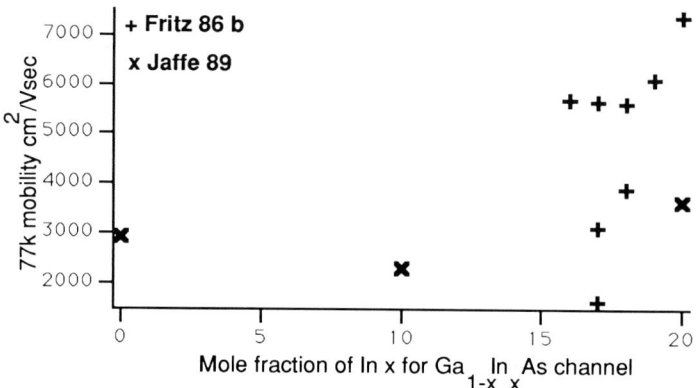

FIG. 10. 77 K p-MODFET mobilities as a function of InAs mole fraction for GaInAs strained-layer channels.

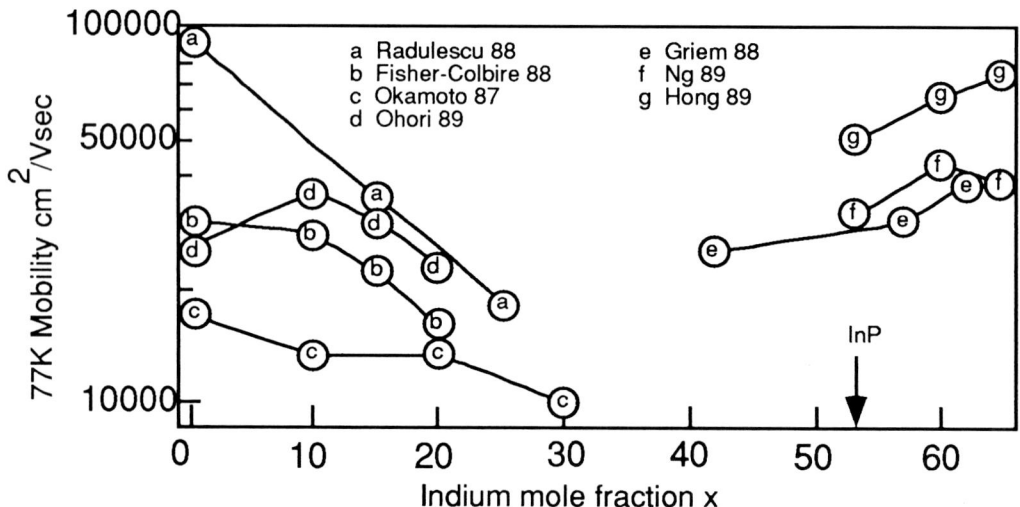

FIG. 11. Hall measurements of 2DEG mobility measured at 77 K for strained layer MODFETs grown on GaAs and InP substrates. In the GaAs substrate MODFETs (a)–(d), the structures vary between the studies, but are only different in InAs mole fraction within each study. Structures in (d) use GaInP electron supply layers, all others use AlGaAs supply layers.

the mobilities of *n*-type MODFETs, because effective masses also decrease with In mole fraction. When GaInAs is grown on InP, the 300 K electron mobilities for high-purity GaInAs are (at 77 K) as high as 11,000 cm^2/Vs due to the effective-mass reduction. This mobility is further improved through strained-layer AlInAs/GaInAs/InP structures with increasing amounts of InAs mole as seen in Fig. 11. In contrast, mobilities unexpectedly decrease with increased InAs mole fraction in *n*-type GaInAs/GaAs MODFETs. This problem is possibly due to ternary clustering and is discussed in Section II.C.

When the modifications to both conduction and valence bands are accounted for, the bandgap of strained-layer GaInAs grown on GaAs can be calculated. The bandgap as a function of In mole fraction at 300 K is seen in Fig. 12 (Mandeville, 1989). It can be readily seen that the overall effect of straining GaInAs is to increase its bandgap compared with the strain-relieved condition.

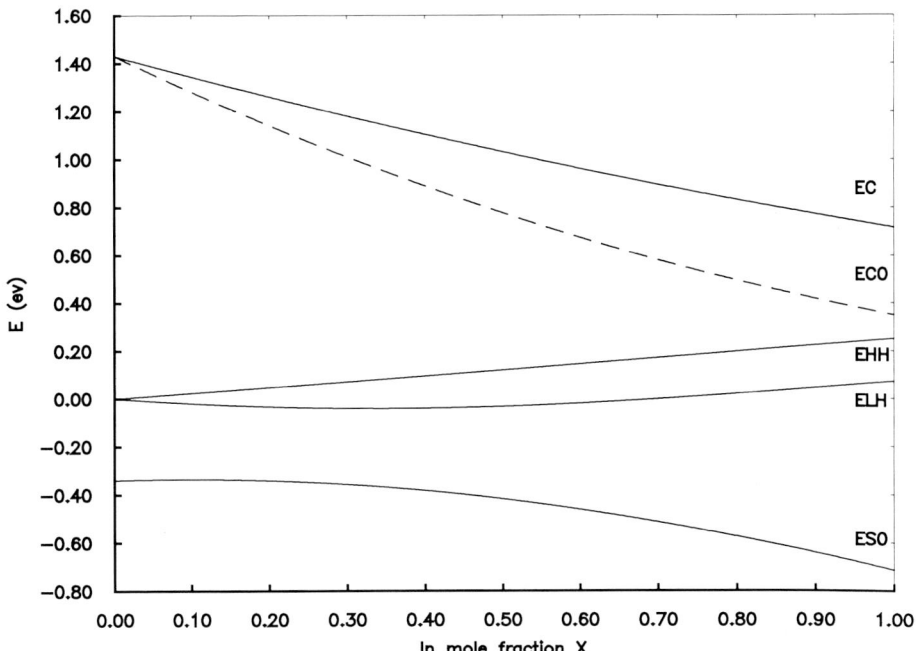

FIG. 12. Relative shifts of the band-edge extrema as a function of the InAs mole fraction *x* in GaInAs as 300 K. ECO is the variation of the unstrained conduction band with the valence band depth at 0 eV. EC is the variation of the conduction band due to the hydrostatic strain component. HH, LH, and SO are the variation of the heavy-hole, light-hole, and splitt-off band due to the shear-strain component (for convenience, the hydrostatic shift is attributed entirely to the conduction band).

2. Modification of Heterojunction Band Offset

The conduction- and valence-band offsets between GaInAs and GaAs need to be known accurately in order to design strained-layer device structures. Measurement of the offsets can be complicated by the restrictions of layer thicknesses by the values of CLT. One common technique for determining conduction- or valence-band discontinuities is the use of capacitance-voltage (C-V) profiling across a n-N heterojunction (Kroemer, 1985). This technique

TABLE I

CONDUCTION AND VALENCE BAND OFFSETS FOR GaInAs/GaAs HETEROJUNCTIONS. BOLD NUMBERS ARE REPORTED DATA, OTHERS ARE CALCULATED.

Reference	Technique	In fraction	Eg(eV)	ΔEc(eV)	ΔEv(eV)	Qc	Qv
Schaff, 1987	CV	4.0	1.386	**0.038**	0.016	0.628	0.372
Schaff, 1987	CV	5.0	1.375	**0.043**	0.023	0.574	0.426
Schaff, 1987	CV	7.0	1.353	**0.055**	0.037	0.522	0.478
Schaff, 1987	CV	7.5	1.348	**0.058**	0.040	0.517	0.483
Ramberg, 1987	IV	2.0	1.408	0.003	**0.019**	0.120	0.880
Ramberg, 1987	IV	5.0	1.375	0.006	**0.049**	0.117	0.883
Ramberg, 1987	IV	8.0	1.343	0.010	**0.078**	0.114	0.886
Ji, 1987	PL	13.0	1.289	0.099	0.043	0.698	**0.302**
Ji, 1987	PL	15.0	1.267	0.114	0.048	0.703	**0.297**
Ji, 1987	PL	15.0	1.267	0.115	0.048	0.704	**0.296**
Ji, 1987	PL	15.0	1.267	0.115	0.048	0.704	**0.296**
Ji, 1987	PL	19.3	1.222	0.146	0.063	0.699	**0.301**
Menendez, 1987	PL	5.0	1.375	**0.028**	0.027	0.503	0.497
Menendez, 1987	PL	5.0	1.375	**0.024**	0.031	0.437	0.563
Anderson, 1988	PL, PC	3.6	1.390	0.033	0.007	**0.830**	0.170
Anderson, 1988	PL, PC	7.3	1.350	0.066	0.014	**0.830**	0.170
Anderson, 1988	PL, PC	17.5	1.241	0.157	0.032	**0.830**	0.170
Anderson, 1988	PL, PC	18.7	1.228	0.168	0.034	**0.830**	0.170
Anderson, 1988	PL, PC	21.0	1.204	0.188	0.038	**0.830**	0.170
Dahl, 1987	PL	13.0	1.289	0.085	0.057	**0.600**	0.400
Joyce, 1988	PL	12.0	1.299	0.081	0.050	**0.620**	0.380
Joyce, 1988	PL	12.0	1.299	0.081	0.050	**0.620**	0.380
Huang, 1989	PL	9.0	1.332	0.069	0.030	**0.700**	0.300
Huang, 1989	PL	11.0	1.310	0.084	0.036	**0.700**	0.300
Huang, 1989	PL	13.0	1.289	0.099	0.042	**0.700**	0.300
Huang, 1989	PL	17.0	1.246	0.129	0.055	**0.700**	0.300
Huang, 1989	PL	20.0	1.214	0.151	0.065	**0.700**	0.300
Yu, 1988	PL	9.3	1.328	0.061	0.041	**0.600**	0.400
Yu, 1988	PL	13.8	1.280	0.045	0.105	**0.300**	0.700

was applied (Schaff, 1986; Schaff, 1987) to GaInAs/GaAs heterojunctions. C-V profiles through single layers of *n*-type GaInAs grown on *n* GaAs are shown in Fig. 13. There is depletion of electrons from the high bandgap GaAs layer that are accumulated in the low bandgap (GaInAs) layer. Analysis of the profiles using the technique described by Kroemer (Kroemer, 1985) results in conduction-band discontinuities of 0.043 eV and 0.058 eV for 5% and 7.5% InAs mole fractions, respectively. These and other results of experiments to determine conduction- and valence-band discontinuities are summarized in Table I and Fig. 14. In this figure, the conduction-band offsets ($Q_c = \Delta E_c/\Delta E_g$) and valence-band offsets ($Q_v = \Delta E_v/\Delta E_g$) are plotted along with a curve of offsets predicted by Schottky barrier theory. The offsets shown are for strained-layer structures, which means that the bandgap of GaInAs is modified for strain. This is the most useful way of representing offset data, as opposed to using nonstrained bandgap references for Q_c and Q_v, since actual devices will be subject to strain. The considerable scatter in the values reported in various works indicates that there still is no universal agreement on band offsets for GaInAs/GaAs strained layers.

C. STRAINED-LAYER GROWTH OF GaInAs/GaAs LAYERS

The most prevalent use of strained-layer growth technology is the strained-layer pseudomorphic MODFET (SMODFET): almost all of the SMODFET growth reported has been by MBE.

1. Substrate Temperature Effects on GaInAs Growth

The growth conditions that are employed for successful strained-layer growth

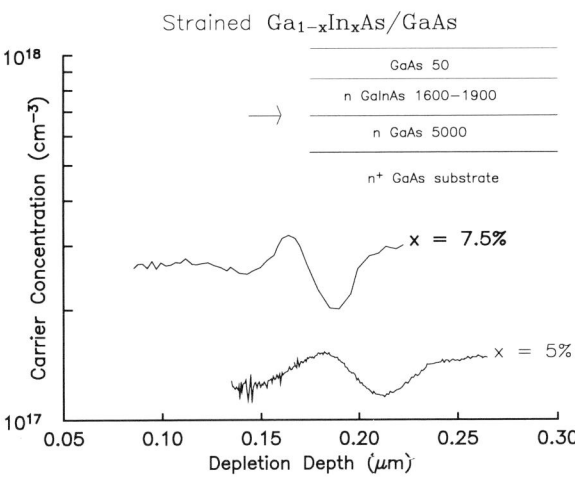

FIG. 13. Capacitance-voltage-free carrier profiles of GaInAs layers on GaAs for 5% and 7.5% InAs mole fraction.

of GaInAs on GaAs most closely resemble those used for lattice-matched GaInAs on InP rather than those of GaAs on GaAs. Substrate temperatures are chosen to be close to 500°C, and growth takes place under excess arsenic over pressure with a corresponding surface reconstruction. The range of substrate temperatures that has been reported for growth of device-quality strained GaInAs on GaAs is from 480°C to 580°C. Most reports are for growth close to 500°C. Deviations from laboratory to laboratory are possibly the result of differences in the difficult task of precise substrate-temperature measurement in different MBE machines. Direct measurement is by either a thermocouple or pyrometer. Thermocouple readings are often inaccurate due to poor thermal contact to the substrate, or are affected by radiation from nearby heater elements. Pyrometer measurements are troublesome because popular wavelengths commonly used in pyrometers limit the available measurement temperature range, or are subject to substantial corrections when changing from the growth of a material that is transparent to these wavelengths to a material that strongly absorbs in the wavelength region.

There are three reasons for the narrow range of substrate temperatures that are commonly employed. First, growth at temperatures much higher than 520°C may be plagued by difficulties in composition control. The temperature at which congruent sublimation (T_{cs}) occurs has been shown to be a function of InAs mole fraction as seen in Fig. 15 (Wood et al., 1982). Desorption of In from the substrate surface becomes significant for mole fractions typical for device applications at temperatures far below that of T_{cs} for GaAs. Small discrepancies in substrate-temperature reproducibility

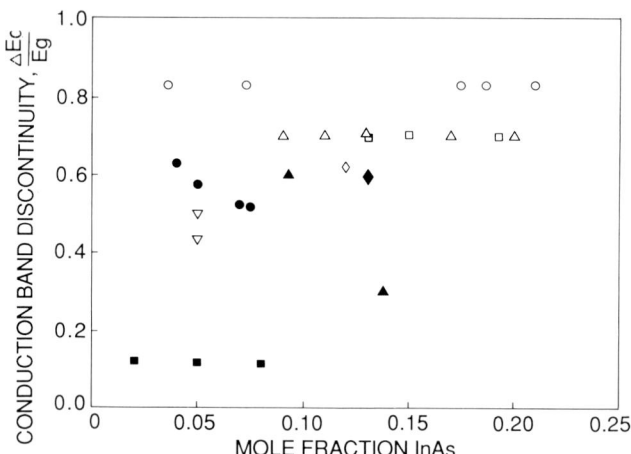

FIG. 14. Conduction-band discontinuities as a function of InAs mole fraction from various reported measurements.

would lead to significant irreproducibility of composition at substrate temperatures close to 600°C.

The second reason to avoid high-substrate-temperature growth is the difficulty in obtaining good surface morphology for layer thickness close to the critical thickness for the onset of dislocations. One of the most powerful tools for understanding the onset of conditions leading to poor surface morphology is the use of in situ reflection high-energy electron diffraction (RHEED). RHEED is used to observe the nature of the growing surface. When atomically flat (two-dimensional) Frank–Van der Merwe growth takes place, a streaked RHEED pattern is observed. A rough surface (three-dimensional) due to Stransky–Krastanov growth gives a RHEED pattern that is spotty, or decorated by chevrons. RHEED can further be used to observe the onset of critical thickness through observation of the change in

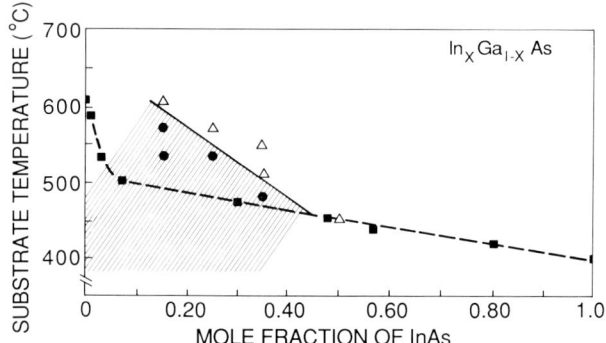

FIG. 15. Congruent sublimation temperature as a function of InAs mole fraction for GaInAs.

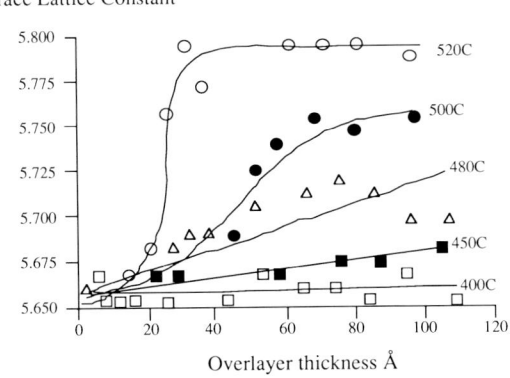

FIG. 16. CLT as a function of substrate temperature as determined from RHEED observations.

lattice constant (Whaley and Cohen, 1988; Berger et al., 1988; Price, 1988). In Fig. 16, this change is seen to be a strong function of substrate temperature. For moderate growth temperatures of GaInAs on GaAs, RHEED can be used for accurate measurement of small changes in lattice constant.

The transition from two-dimensional (2D) to three-dimensional (3D) growth of GaInAs on GaAs at higher temperatures is seen to be a strong function of substrate temperature (Fig. 17). The explanation for the formation of 3D surface is the segregation of In on the growing surface (Radulescu et al., 1989, Whaley and Cohen, 1988). At higher temperatures, In atoms cluster and can even ride along the growth front rather than incorporate where they are deposited. This behavior is seen in the Auger electron spectroscopy profiles taken from SMODFET structures and shown in Fig. 18. The In incorporation is well behaved at substrate temperatures near 500°C. All of it is found to be located in the quantum well where it was placed. At higher temperatures, In accumulates on the growing surface, results in the 3D surface observed by RHEED, and becomes incorporated closer to the surface of the epitaxial layer. Small improvements can be seen in high-temperature growth through the use of extremely high arsenic fluxes.

Substrate temperatures have not been selected much below 480°C. No work has been done to establish how low the substrate temperature can be reduced before the quality of strained GaInAs begins to suffer. Low-temperature growth of GaAs or AlGaAs produces inferior-quality material for carrier transport. Although this is undesirable for device applications requiring high-velocity or high-mobility carriers, low-temperature-grown material can be made to be of high resistivity, which has an application as a

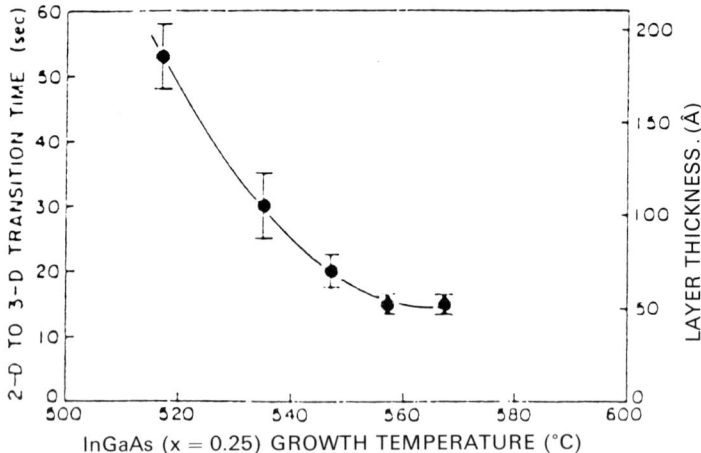

FIG. 17. Transition from 2D to 3D growth of GaInAs on GaAs as determined from RHEED transition from streaky to spotty reconstruction.

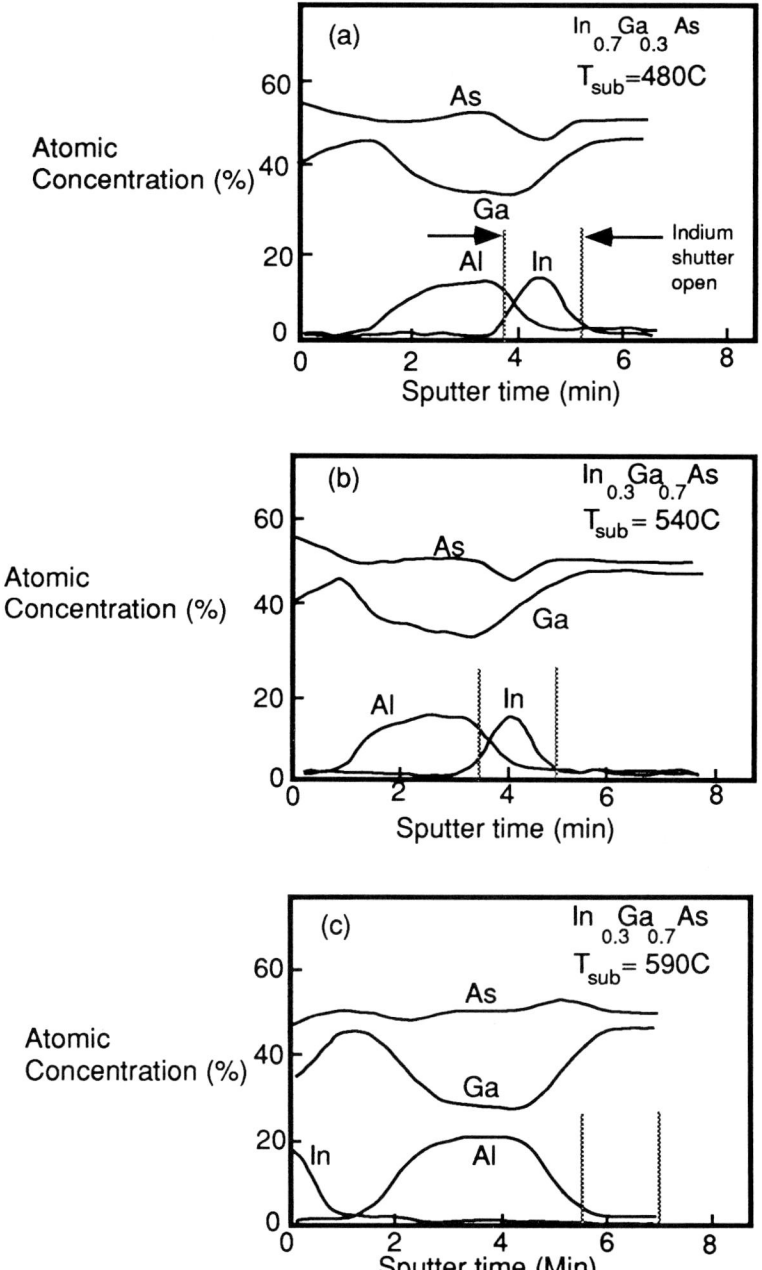

FIG. 18. Sputter Auger depth composition profiles showing In segregation at high substrate temperatures.

buffer layer (Smith, 1987). This technique has not yet been applied to the SMODFET. Conventional MODFET growth takes place at temperatures at or above 600°C. A compromise typically employed in the growth of SMODFETs is to grow GaInAs near 500°C, after which the temperature is ramped to near 600°C while growth may or may not be interrupted.

2. *Electrical Quality of MBE-Grown GaInAs/GaAs Interfaces*

The quality of the interface between the high- and low-bandgap materials is critical to the performance of the MODFET and SMODFET. If electrically active defects or impurities exist at these interfaces, they can act as scattering centers, which will reduce mobility, velocity, or quasi-two-dimensional electron-gas (2DEG) density in the channel. The electrical properties of this interface can be investigated through electrical measurement techniques. C-V profiling and deep-level transient spectroscopy (DLTS) across the interface, current-voltage (I-V), and Hall measurements of the 2DEG parallel to the interface provide clues to the nature of defects at the interface.

The C-V profiles of Fig. 13 show the clear accumulation and depletion of carriers, which suggests very good interface quality. When critical thickness is exceeded, C-V profiles show evidence of severe electrical degradation of GaInAs at the interface with GaAs (Fig. 19). In Fig. 19, there is no electron accumulation at the GaInAs at interface. Rather, there is severe electron loss to the dislocations at the interface between GaInAs and GaAs, and those that extend into the GaInAs. It is interesting to note that there is not much

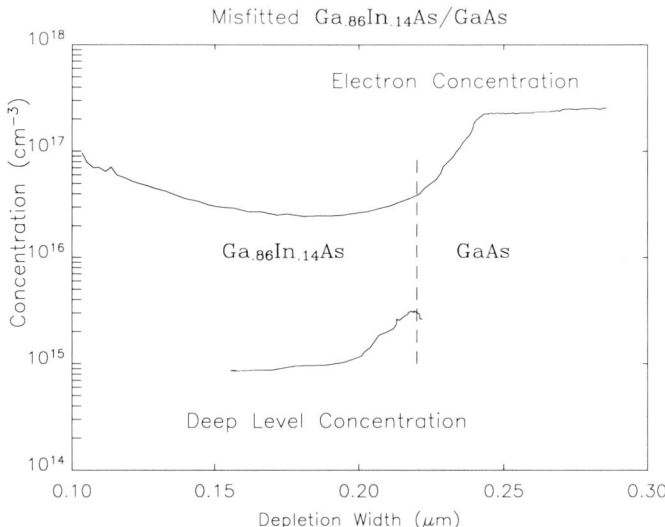

FIG. 19. Capacitance-voltage-free carrier profile of GaInAs on GaAs that has exceeded CLT.

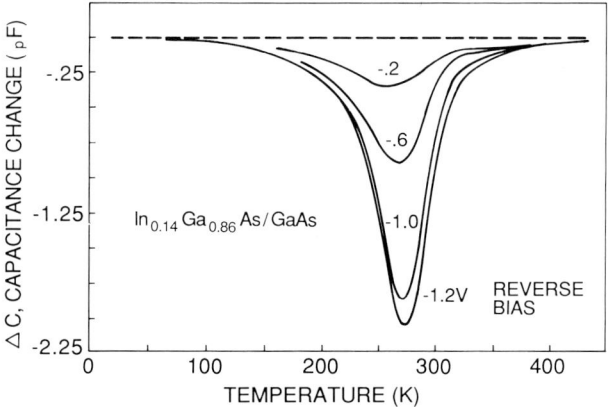

FIG. 20. DLTS boxcar plot of GaInAs on GaAs that has exceeded CLT.

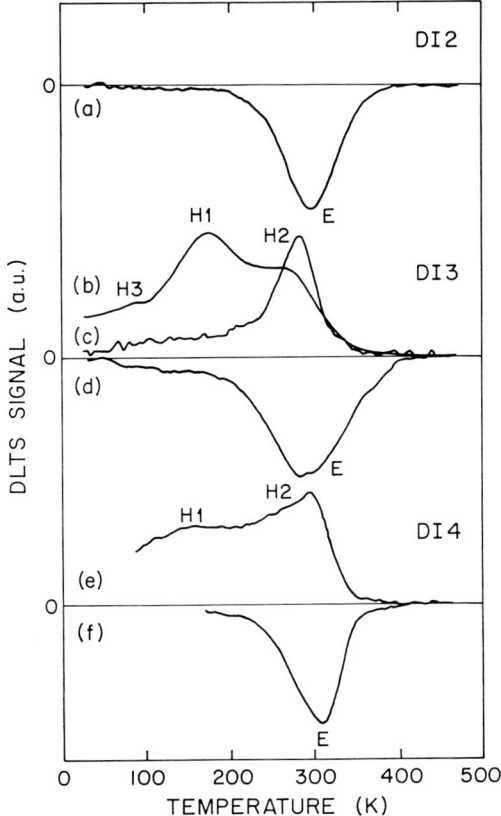

FIG. 21. DLTS profiles of defects and impurities at the GaInAs–GaAs interface.

electron loss in the GaAs layer, which may indicate that defects do not extend into this material. Physical characterization such as TEM supports this observation.

DLTS (Lang, 1974) is a technique used to observe defects or impurities that change the charge state under transient conditions. DLTS measurements of both N–n ($x = 0.14$) (Schaff, 1987) and P–n GaInAs/GaAs ($x = 0.05$) (Ashizawa et al., 1988) layers beyond critical thickness find similar electrical activity. These results show that a single electron trap is spatially located at the interface between the materials with an activation energy of approximately 0.5 eV. DLTS profiles are seen in (designated H2) Figs. 20 and 21. A hole trap that is thought to be due to Fe is also seen. The Arrhenius plots for the hole and electron traps are seen in Fig. 22. Fe is known to be seen in MBE-grown GaAs and originates from the substrate (Schaff, 1984).

Further insight into the electrical properties of the defects that occur due to strained-layer growth beyond critical thickness comes from I-V behavior across this junction. A single layer of n-type GaInAs on n-GaAs grown beyond critical thickness shows significant asymmetric rectification when biased (Fig. 23) (Woodall et al., 1983). The explanation for the rectification is Fermi level pinning at the interface by the electron trap level seen by DLTS.

Measurement of reverse-bias I-V characteristics in GaInAs/GaAs n–N

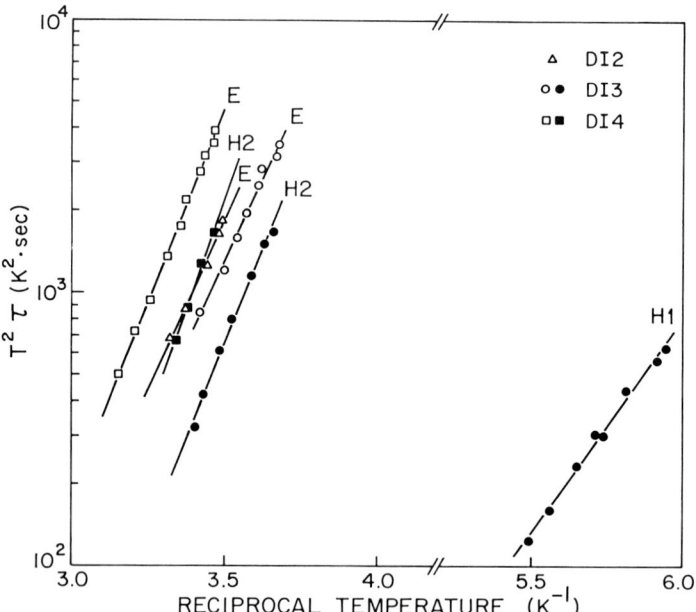

FIG. 22. Arrhenius plots for the hole and electron traps.

2. DEVICE APPLICATIONS OF STRAINED-LAYER EPITAXY

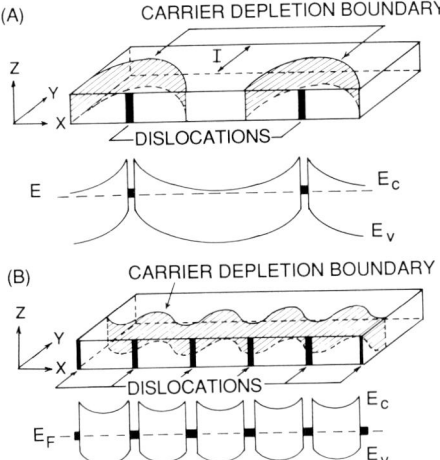

FIG. 23. I-V characteristics of the misfitted heterojunction showing rectification.

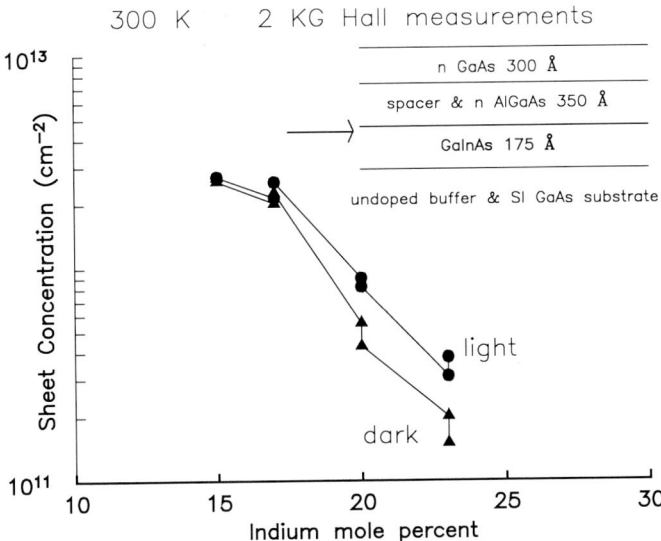

FIG. 24. Hall measurements of 2DEG density in strained-layer MODFETs with channels below and beyond CLT.

heterojunctions located just beyond the base and into the collector of HBTs showed little effect from the influence of the interface defects (Ashizawa *et al.*, 1988). HBT current gains were virtually unaffected by the presence of these dislocations contained entirely within the collector. Poor collector-emitter current saturation was noted in comparison with control samples without regions of GaInAs bases, but these effects were only present in very-broad-area test devices. Devices with dimensions closer to logic-integrated-circuit or microwave-device scale showed no differences between devices with or without the presence of GaInAs–GaAs interface defects.

The characterization of the electrical properties for transport perpendicular to the GaInAs–GaAs junction does not reveal interface effects on MODFET performance. The best electrical probe of the interface for device applications consists of Hall effect measurements of the 2DEG. Hall measurements have been performed on SMODFET structures where the GaInAs channel composition has been chosen to span the range from below to beyond CLT (Fig. 24). The lowest mole-fraction channels show typical 2DEG sheet densities, whereas mole fractions that result in channels beyond critical thickness exhibit severe carrier loss. The loss is due to capture of the channel electrons by the interface traps. This magnitude of carrier loss is unacceptably large for device operation. The data indicate that critical thickness limits must be strictly observed for the design of SMODFETs with

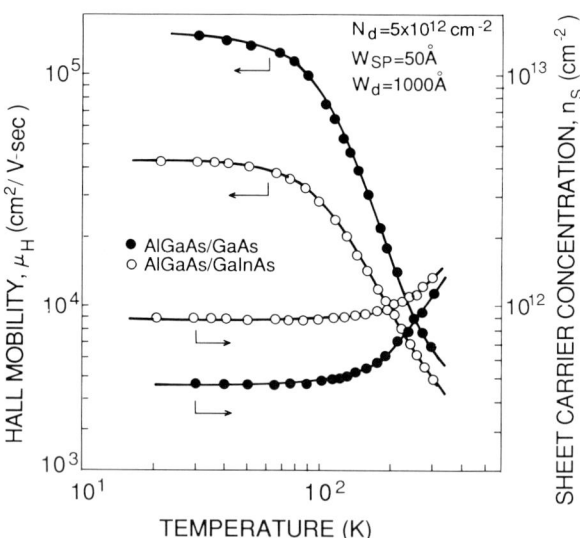

FIG. 25. Temperature-variable Hall-effect characteristics for strained- and unstrained-layer MODFETs.

best performance. It is not known whether there are significant interface states that might still exist for structures below critical thickness. The question of whether very small interface-state densities may exist in sufficient quantity to affect such SMODFET characteristics as noise figure has not been carefully examined yet.

A potentially more serious limitation to SMODFET performance is also investigated through Hall effect measurement studies. It was seen in Fig. 11 that 2DEG mobility in the GaInAs channels below CLT falls with increased InAs mole fraction. The cause of this drop in mobility has been suggested by Ohno et al. (1989) to be an additional scattering due to alloy clustering. The temperature-variable Hall effect characteristics for AlGaAs/GaInAs SMODFETs have been compared to AlGaAs/GaAs MODFETs in Fig. 25.

Interface roughness is ruled out as a source of mobility limitation by noting that mobilities were insensitive to channel widths from 50 to 200 Å. Interface roughness would be expected to increase with decreasing channel thickness. Alloy scattering is further dismissed by analysis of the temperature-variable Hall behavior and by the simple observation that GaInAs matched to InP exhibits very high mobility despite being a ternary material. The dominant scattering mechanism proposed is due to alloy clustering or nonuniformity. A comparison of the competing scattering mechanisms is seen in Fig. 26. Nonuniformities in strained GaInAs growth on GaAs might not be surprising when viewed in light of the results of growth of thin InAs layers on GaAs (Muneketa, 1987), where it was reported that InAs grows very nonuniformly strained to GaAs. Large area nonuniform growth of strained GaInAs/GaAs is also reported over 2-inch diameter wafers by Grider et al. (1989).

Clustering in GaInAs grown on GaAs might not be too surprising due to the kinetics of strained growth. Clustering may adequately explain the fall in mobility as a function of InAs mole fraction in n-type GaInAs/GaAs SMODFETs. Contrary to growth on GaAs, n-type GaInAs/InP SMODFETs exhibit a rise in mobility as a function of InAs mole fraction.

3. Growth and Characterization Summary

Very-high-quality strained layers of GaInAs on GaAs have been demonstrated through test structures and in SMODFETs. It is critical to avoid the onset of misfit-dislocation generation for SMODFET applications. Techniques using structures with dimensions that can minimize the generation of misfit dislocations need to be investigated. Perhaps patterned growth techniques to extend CLT will be successful (Fitzgerald et al., 1989). The problem of degraded mobility with increased In mole fraction needs to be looked at in greater depth.

III. Electronic-Device Structures

The application of strained-layer epitaxy for the modification of heterojunction band offsets in high-speed and high-frequency electronic-compound semiconductor-device structures is very important for the development of new optimized device structures.

A. MODULATION-DOPED FIELD-EFFECT TRANSISTORS

To date, the application of strained, dislocation-free (pseudomorphic) layers to modulation-doped transistor structures has produced significant advances in device performance, with further advances expected in the future as a result of the increased heterojunction conduction-band offset.

1. *Modification of Heterojunction Band Offset*

The key ingredient of the MODFET transistor is a high-quality abrupt heterojunction—a junction formed between two different bandgap-compound semiconductors such as $Al_xGa_{1-x}As$ and GaAs (or a strained

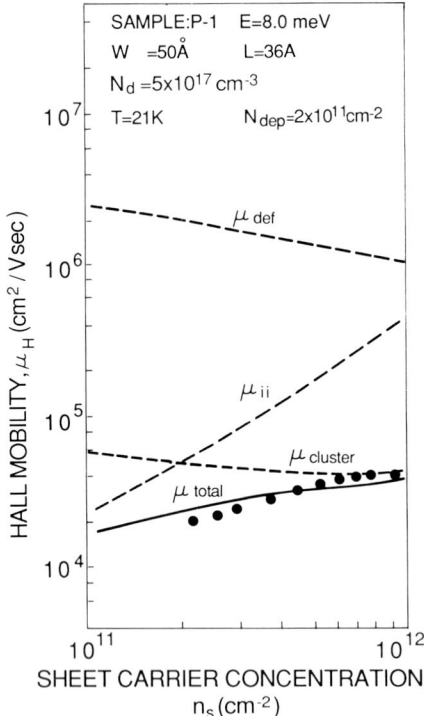

FIG. 26. A comparison of the competing scattering mechanisms at different sheet concentrations.

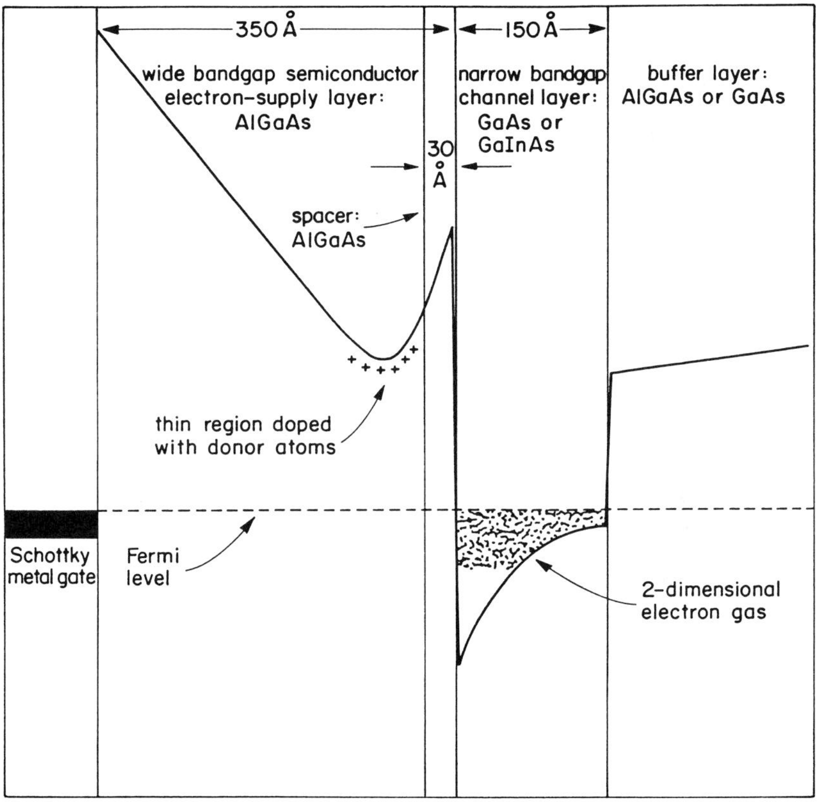

FIG. 27. Diagram of the epitaxial layer structure and energy-band diagram of a typical Schottky-gate-controlled modulation-doped structure.

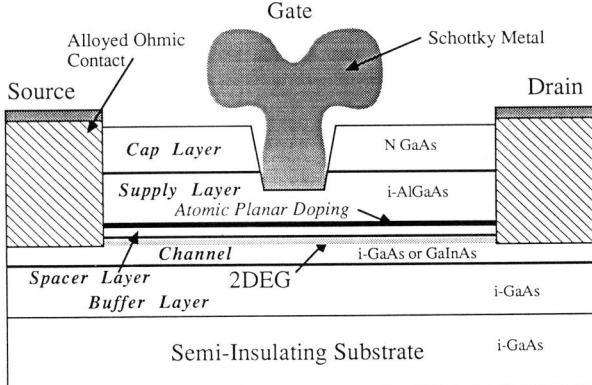

FIG. 28. A schematic diagram of a typical microwave MODFET transistor design, with a gate-recess and a "T"-shaped cross-section gate metalization.

$Ga_{1-y}In_yAs$)—and a unique way to introduce electrons (or holes) into the conducting channel called *modulation doping*. The donor atoms are placed in the higher bandgap semiconductor, e.g., $Al_xGa_{1-x}As$ at a small distance (commonly referred to as the spacer layer) away from the heterojunction, while the electrons transfer to the undoped narrow bandgap material, e.g., GaAs or $Ga_{1-y}In_yAs$. Because quantum mechanical and space-charge effects cause these electrons to accumulate in a thin region (< 200 Å), they are frequently referred to as a quasi-2DEG. Diagrams of the epitaxial layer structure and energy-band diagram of a typical modulation-doped structure are shown in Fig. 27. A typical microwave MODFET design is shown schematically in Fig. 28. Its mode of operation is similar to that of other FETs (e.g., MOSFET, metal-semiconductor field-effect transistor (MESFET)) in that a potential applied to the gate electrode is used to control the number of electrons that conduct current between source and drain contacts. It is unique, however, in that electrons in the low field contact regions travel primarily in an undoped and hence higher mobility channel. Improved high-frequency and high-speed performance is expected to result from the reduction in parasitic resistance in the source and drain contact regions.

Initial research in MODFETs focused on lattice-matched heterojunctions, mainly the AlGaAs/GaAs heterojunction lattice matched to GaAs. This structure has great promise for both high-speed and high-frequency transistor applications, with improved performance over the doped-channel GaAs MESFET. High transconductance (> 450 mS/mm) (Camnitz, 1984), high current-gain cut-off frequency (> 55 GHz at $L_g = 0.25 \mu m$) (Camnitz et al., 1984; Mishra et al., 1985), fast switching speeds 12.2 ps/gate at 77 K (Lee et al., 1983), 10.1 GHz frequency divider operation at 77 K (Hendel, 1984), a very low room-temperature noise figure of 2.1 dB with an associated gain of 7.0 dB at 40 GHz (Chao et al., 1985a), a 10.5 K noise temperature at 8.3 GHz at cryogenic temperatures (12.5 K lattice temperature) (Camnitz et al., 1985a), and excellent power-added efficiencies of 33% at 30 GHz, 28% at 40 GHz, and high maximum-output power density 0.44 W/mm at 30 GHz (Smith et al., 1985) had all been achieved by 1985. These impressive results along with the development of low-resistance submicron ($< 0.25 \mu m$) gate technology (Chao et al., 1985b) had allowed the development by 1987 of both high-speed logic circuits; 1.5 K HEMT gate array (Watanabe et al., 1987) and millimeter-wave systems; two-stage Ka-band amplifier with a 4.0 dB noise figure and 16.5 dB gain at 37.5 GHz (Upton et al., 1987), and four-stage V-band amplifier with a 5.65 dB noise figure and 16.57 dB gain at 60.4 GHz (Yau et al., 1987) incorporating AlGaAs/GaAs MODFETs.

In spite of these impressive results, detailed analysis and modelling performed as early as 1985 indicated that serious limitations resulted from using the AlGaAs/GaAs system. The presence of a deep-donor DX-center in

AlGaAs caused persistent photoconductivity and other I-V anomalies (Drummond et al., 1983, Fischer et al., 1984). Whereas the small conduction-band offset (< 0.23 eV) limited the maximum useable 2DEG sheet density ($< 1.0 \times 10^{12}$ cm^{-2}) (Camnitz et al., 1985b), alternative MODFET structures were required to overcome these problems and satisfy the ever-increasing system demands for high-speed logic and millimeter-wave transistors.

Remaining within the constraints of lattice-matched materials, the $Al_{0.48}In_{0.52}As/Ga_{0.47}In_{0.53}As$ heterojunction lattice matched to InP is an obvious alternative. This MODFET structure has a larger conduction-band offset (> 0.5 eV), hence higher 2DEG sheet densities ($> 2.5 \times 10^{12}$ cm^{-2}), and utilizes GaInAs channel that has improved transport properties over GaAs. Despite considerable problems in the MBE growth of these materials (Pearsall et al., 1983) and anomalous I-V characteristics (Kuang et al., 1988a,b), AlInAs/GaInAs MODFETs have recently demonstrated state-of-the art performance (Palmateer et al., 1987; Mishra et al., 1988a,b): high transconductance (> 1000 mS/mm), high current-gain cut-off frequency (> 120 GHz), and fast switching speed (< 6.0 ps/gate).

In 1985, based on the earlier work of Zipperian et al. (1983), where a MODFET was realized using a modulation-doped strained-layer super-lattice, Rosenberg et al. (1985) demonstrated that a high-performance MODFET structure incorporating a single thin (200 Å) strained but dislocation-free (pseudomorphic) $Ga_{0.85}In_{0.15}As$ channel could be obtained. This success stimulated considerable interest, since the combination of a strained (pseudomorphic) $Ga_{0.85}In_{0.15}As$ channel with an AlGaAs supply layer grown on a GaAs substrate provided a solution to almost all of the AlGaAs/GaAs MODFET problems: larger conduction-band offset (> 0.35 eV) and a reduced DX-center problem (which allows the use of low $< 23\%$ aluminum compositions), without the need to resort to the use of an InP substrate. By 1988, optimized pseudomorphic AlGaAs/GaInAs strained-layer SMODFET structures incorporating the conservative InAs mole fractions between 15% and 20% had demonstrated superior performance over lattice-matched AlGaAs/GaAs MODFETs and GaAs MESFETs at room temperature, for all the important figure of merit: higher transconductance (> 540 mS/mm) (Chao et al., 1987), higher current-gain cut-off frequency (> 46 GHz for $L_g = 0.35$ μm) (Nguyen et al., 1988a), lower noise figure (2.4 dB at 60 GHz) (Chao et al., 1987), and higher power-added efficiencies (22% at 60 GHz) and output power (9 mW at 94 GHz) (Smith et al., 1988). Strained (pseudomorphic) GaInAs channels have also demonstrated superior performance over GaAs channels when used in double-doped SMODFETs (Nishii et al., 1988), doped-channel SMODFETs (Saunier et al., 1988) and doped-channel MISFETs (Kim et al., 1988), where very high current densities (> 900 mA/mm) and high power densities at millimeter-wave frequencies (> 0.76 W/mm at 60 GHz) have been achieved.

Most of the research has focused on the millimeter-wave applications of strained-layer (pseudomorphic) SMODFETs, because of the possibility of improved performance, of complementary logic circuits. The compressive strain in GaInAs channels produces an energy splitting of the light-hole and heavy-hole valence bands (Zipperian et al., 1987). This may provide for improved p-channel SMODFET performance. Initial results have indicated improved hole mobility (860 cm^2/Vs at 300 K and 2815 cm^2/Vs at 77 K) resulting in improved transconductance (113 mS/mm at 300 K and 800 mS/mm at 77 K) and increased current drive (94 mA/mm at 300 K and 180 mA/mm at 77 K) in p-channel strained-layer (pseudomorphic) $Ga_{0.2}In_{0.8}As$ SMODFETs (Daniels et al., 1988) compared with p-channel lattice-matched GaAs MODFETs (Daniels et al., 1986); however, complementary circuits using strained layers have not yet been demonstrated.

Zipperian and coworkers (Zipperian et al., 1989) have indicated a possible limitation to the use of strained layers in logic circuits which generally require a high-temperature annealing cycle during fabrication. They reported that thick strained-layer structures that have no dislocations (pseudomorphic layers) suitable for device applications are unstable during the high-temperature annealing cycle. This limitation can be overcome by either decreasing the strained-layer thickness or providing a thick stabilizing cap layer, either of which may not be practical for some integrated device/circuit applications. The previously reported p-channel SMODFET transistors were made with a high-temperature annealing cycle, indicating that strained layers can survive such a high-temperature processing step. In addition, microwave and millimeter-wave transistor structures generally do not involve high-temperature processing cycles and so may not be subject to this problem.

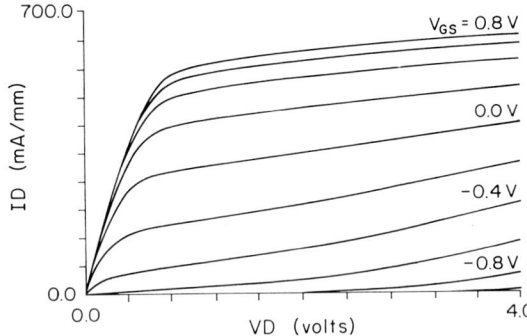

FIG. 29. The I-V characteristics of the high InAs mole fraction 120 Å-thick single-doped quantum-well $Ga_{0.75}In_{0.25}As$ pseudomorphic MODFET grown on GaAs with a 0.15 μm gate length and 150 μm gate width (Nguyen 1988b; Nguyen, 1988c).

This work, however, does probe the question of the long-term reliability of strained layers, which is still to be determined.

Further improvement in SMODFET performance would result if the InAs mole fraction of these strained pseudomorphic channels could be further increased. The growth of sufficiently thick (≈ 100 Å) high InAs mole fraction SMODFET channels is difficult (see Section II.C.), hence the concentration of effort on 15%–20% mole fraction strained channels. 1987 researchers from Hewlett Packard (Moll et al., 1988) reported high-performance pseudomorphic transistors with an InAs mole fraction as high as 25%. This SMODFET

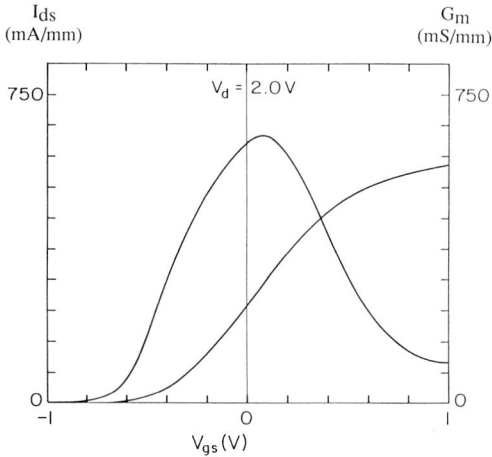

FIG. 30. The transfer characteristic and gate bias dependance of the transconductance g_m of the high InAs mole fraction 120-Å-thick single-doped quantum-well $Ga_{0.75}In_{0.25}As$ pseudomorphic MODFET grown on GaAs with a 0.15 μm gate length and 150 μm gate width (Nguyen, 1988b; Nguyen, 1988c).

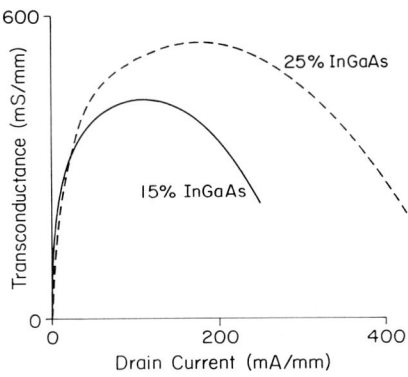

FIG. 31. The transconductance g_m plotted as a function of drain current for the high InAs mole fraction, 120 Å-thick single-doped quantum-well $Ga_{0.75}In_{0.25}As$ pseudomorphic MODFET grown on GaAs with a 0.15 μm gate length and 150 μm gate width (Nguyen, 1988b; Nguyen, 1988c), compared directly with that obtained for a similar transistor structure using a lower InAs mole fraction $Ga_{0.85}In_{0.15}As$ pseudomorphic channel (Nguyen, 1988a). It clearly illustrates the improvements obtained with the high InAs mole fraction channel.

structure provides for a large conduction-band offset (>0.45 eV), hence a high single-channel 2DEG sheet density (>2.5 × 10^{12} cm^{-2}). In December 1988, the first, very-high-performance short-gate-length strained-layer SMODFET structures on both GaAs and InP were reported. The I-V characteristics of the high InAs mole fraction, 120-Å-thick single doped quantum-well $Ga_{0.75}In_{0.25}As$ pseudomorphic SMODFET grown on GaAs with a 0.15 μm gate length is shown in Fig. 29 (Nguyen et al., 1988b, c). These characteristics show the superior performance and ultimate potential of the strained-layer transistor: a high current drive (>600 mA/mm), sharp saturation at a low knee voltage (<0.75 volts), and a high transconductance (>650 mS/mm), all of which result directly from a high 2DEG sheet density (2.5 × 10^{12} cm^{-2}).

The transfer characteristic and gate-bias dependance of the transconductance g_m, shown in Fig. 30, indicate a Gaussian characteristic, g_m, with a maximum value of 650 mS/mm. In order to better observe the improvements produced by the high InAs mole fraction pseudomorphic channel and for easy comparison with other strained and unstrained MODFET structures, it is more informative to plot the g_m as a function of drain current instead of gate voltage. Presenting the measured data in this form (see Fig. 31) indicates (i) the high current density at maximum transconductance ($\approx$ 180 mA/mm) and (ii) the large current swing over which this high value of transconductance is maintained (from 120 mA/mm to 300 mA/mm). These features result directly from the increased conductance-band offset. The shape of the transfer characteristic of this SMODFET transistor is more "box like," i.e., ideal, as required for high performance and predicted theoretically (see Section III.A.2) for modulation-doped structures, than the transfer characteristic previously obtained (Gaussian-like) with AlGaAs/GaAs MODFET transistors. Figure 32 shows the on-wafer measured S-parameters and the calculated current-gain variation with frequency for this high InAs mole fraction, 150 μm wide, pseudomorphic MODFET biased at the optimum for maximum f_T. The calculated current gain demonstrates a well-behaved 20 dB/decade roll-off, which can therefore be accurately extrapolated (Tasker, 1987, Nguyen, 1987a) to give the unity current-gain cutoff frequency f_T = 152 GHz. This very high value demonstrates the high frequency potential of this transistor structure.

Transistor performance for the short-gate-length strained-layer AlInAs/GaInAs SMODFETs on InP is equally impressive: high transconductance (>1000 mS/mm), large current drive (>1000 mA/mm), and high current-gain cutoff frequency (>210 GHz) (Mishra et al., 1988a).

The superior performance of n-channel AlGaAs/GaInAs pseudomorphic SMODFETs has been clearly illustrated by Nguyen et al. (1988a) and Hikosaka et al. (1988) by comparing the unity current-gain cutoff frequencies,

f_T, for conventional and strained pseudomorphic $Ga_{0.85}In_{0.15}As$ and $Ga_{0.8}In_{0.2}As$ MODFETs with gate lengths between 0.25 and 1.0 μm. The value of f_T is compared, since it is a very important figure of merit that demonstrates the potential for high-frequency and high-speed operation and relates directly to the effective electron velocity:

$$f_T = \frac{v_{eff}}{2\pi(L_g + d)}, \qquad (2)$$

where v_{eff} is the effective electron velocity, L_g is the gate length, and d is the extension of the gate length caused by parasitics and fringing effects. The value f_T is not a strong function of gate-to-channel separation, but the transconductance g_m, which is often used to compare performance and extract velocity, is a strong function of the gate-to-channel separation; hence, analysis of the f_T values gives more meaningful comparisons of the real

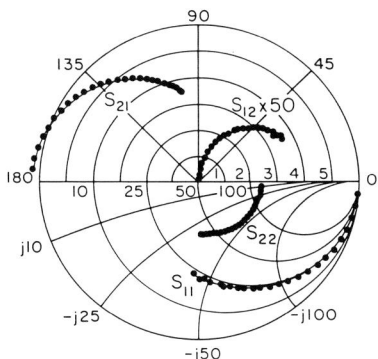

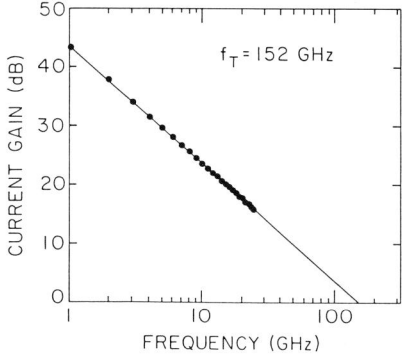

FIG. 32. (a) The on-wafer measured S-parameters and the calculated current-gain variation with frequency for the high InAs mole fraction 120-Å-thick single-doped quantum-well $Ga_{0.75}In_{0.25}As$ pseudomorphic MODFET grown on GaAs with a 0.15 μm gate length and 150 μm gate width, biased at the optimum for maximum f_T (Nguyen, 1988b; Nguyen, 1988c). (b) Modulation band width versus gain.

potential of the various structures. Figure 33 summarizes the results obtained for strained-layer pseudomorphic SMODFETs and compares them directly with those obtained by using lattice-matched MODFETs. The results show that for similar gate lengths, the strained-layer pseudomorphic devices have consistently higher f_T, with the value increasing with higher InAs mole fraction. By replotting this data as $1/(2\pi f_T)$ versus L_g (gate length), as shown in Fig. 34, Tasker and Eastman extracted effective velocities of 1.2×10^7 cm/s for conventional MODFETs, 1.5×10^7 cm/s for $Ga_{0.85}In_{0.15}As$ pseudomorphic SMODFETs, and 1.8×10^7 cm/s for $Ga_{0.75}In_{0.25}As$ pseudomorphic SMODFETs (Tasker, 1987). The strained-layer pseudomorphic SMODFET structure can provide for an improvement of over 50% in effective electron velocity, resulting from the improved modulation efficiency (ME), see Section III.A.2, obtained in these transistor structures because of the increased conduction-band offset. The measured ME of the 25% InAs mole fraction MODFET is greater than 90% at room temperature, indicating *almost ideal modulation-doped transistor behavior*, i.e., a very efficient thermal transfer of charge occurs because of the very large conduction-band offset. This accounts for the significant improvements measured in transistor performance. In lattice-matched MODFET structures, this desired performance is only obtained at cryogenic (< 77 K) temperatures.

The epitaxial growth of these high ($> 25\%$) InAs mole fraction strained-

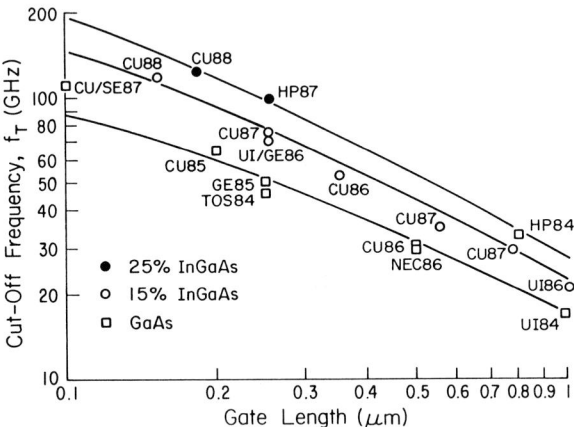

FIG. 33. Summary of the variation of f_T with gate length L_g for strained-layer pseudomorphic MODFETs and lattice-matched MODFETs, indicating the superior performance of the pseudomorphic transistors. The data points were obtained from papers published by, and from private communications with, researchers from several universities and companies: Cornell University (labeled CU), General Electric (GE), Hewlett-Packard (HP), Nippon Electric Corporation (NEC), Siemens (SE), Toshiba (TOS), and the University of Illinois (UI).

layer SMODFET structures is presently not fully understood and difficult to reproduce. As a consequence, whereas basic research transistors have been fabricated that demonstrate the significant intrinsic performance potential of these SMODFET structures as discussed, optimized high-performance millimeter-wave transistor structures for low-noise and high-power applications are still to be fabricated and tested. As the material growth is further understood and improved, it is predicted that such optimized high ($> 25\%$) InAs mole fraction transistor structures will provide for further improvement in circuit performance over that presently obtained (as discussed) by using moderate InAs composition $Ga_{0.85}In_{0.15}As$ and $Ga_{0.80}In_{0.20}As$ strained-layer channels, higher current (> 700 mA/mm for single channel, > 1400 mA/mm for double channel), higher gain (> 15 dB) at 94 GHz, higher power density (> 1.0 W/mm), and increased efficiency ($> 40\%$) and lower noise figures (< 1.8 dB at 60 GHz). Further improvements are possible if the strained-layer growth technology can be developed to further increase the InAs mole fraction in SMODFET channels.

To explain the role of pseudomorphic layers in achieving this significant improvement in effective electron velocity, and hence improved MODFET transistor performance, the theory of MODFET operation will be briefly reviewed.

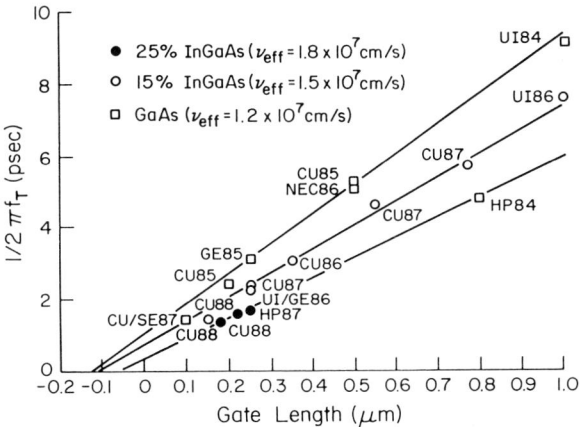

FIG. 34. $1/(2\pi f_T)$ versus L_g (gate length) for strained-layer pseudomorphic MODFETs and lattice-matched MODFETs, from which the effective electron velocity can be determined. The data points were obtained from papers published by, and from private communications with, researchers from several universities and companies: Cornell University (labeled CU), General Electric (GE), Hewlett-Packard (HP), Nippon Electric Corporation (NEC), Siemens (SE), Toshiba (TOS), and the University of Illinois (UI).

2. The Need for Higher Sheet Densities in MODFETs

The experimental data cited earlier clearly demonstrates the advantage of incorporating indium into the channel of a MODFET, even at the expense of introducing strain. Pioneers in this work, such as Zipperian *et al.* (1983), Rosenberg *et al.* (1985), and Ketterson *et al.* (1985), typically cited two reasons for the advantage of strained-layer MODFETs. The first is the establishment of a significant bandgap discontinuity without using the higher aluminum mole fraction AlGaAs leading to elevated levels of DX-centers and their undesirable effects. The second involves the improved transport properties of $Ga_{0.47}In_{0.53}As$ relative to GaAs. This latter reason is often presumed to be the primary cause for the higher f_T and the effective velocities observed in strained-layer MODFETs. Unfortunately, the frequently cited theoretical work of Cappy *et al.* (1980) and the experimental observation of Bandy *et al.* (1981) are strictly applicable to $Ga_{0.47}In_{0.53}As$ MESFETs and not to strained, lower InAs mole fraction GaInAs channels. Both works attribute the superior velocities of $Ga_{0.47}In_{0.53}As$ to the larger mobility and greater $\Gamma-L$ valley separation relative to GaAs. However, room-temperature Hall mobilities for strained GaInAs MODFET channels with 5–30% InAs are typically lower or comparable to those of GaAs MODFETs. This has been demonstrated experimentally by Okamoto *et al.* (1987), Fischer-Colbrie *et al.* (1988), and Radulescu (1988). Moreover, since the energy barrier to real-space transfer (the conduction-band discontinuity) is lower than that to k-space transfer (the $\Gamma-L$ valley separation) in both types of MODFET, the importance of the $\Gamma-L$ valley separation is diminished.

A clearer picture of the underlying physical mechanisms that account for the superior performance of strained GaInAs channel MODFETs follows from an examination of more recent experimental and theoretical work. The emerging consensus is that strained-layer MODFETs derive their advantage from the greater conduction-band discontinuity and hence from larger 2DEG sheet density. The seeds of this understanding were planted by Camnitz *et al.* (1985a, 1985b), who studied the role of charge control to explain experimentally observed differences between GaAs channel MODFETs with different undoped spacer thicknesses.[2] They observed that even for moderate sheet densities, the MODFET capacitance began to increase relative to the value predicted from the gate-to-2DEG separation. The increase in capacitance was attributed to the accumulation of electrons in the AlGaAs electron-supplying layer. By integrating the area under the measured C-V character-

---

[2]Because a thicker undoped spacer extends the conduction band in the electron-supplying layer closer to the Fermi level prior to its bending by ionized donors, it decreases the effective conduction-band discontinuity. The results of this work are therefore applicable to GaInAs-channel MODFETs.

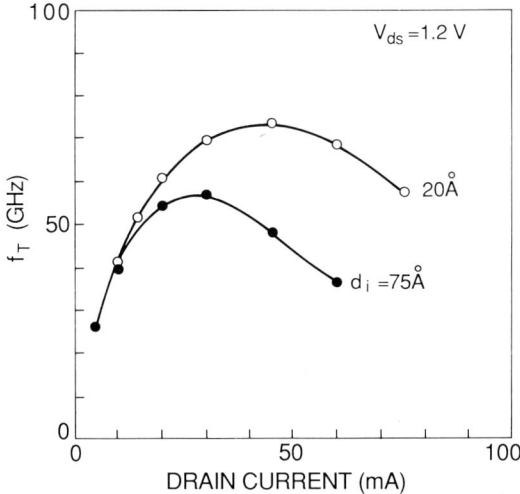

FIG. 35. Comparison of the bias dependence of the measured f_T for 0.2 μm AlGaAs/GaAs gate length transistors with 20 Å and 75 Å spacer-layer thickness.

istic up to the point of sharp increase in capacitance, they calculated n_{sm}, the maximum 2DEG sheet density achievable before the onset of excess charge modulation. They reported a 77 K n_{sm} value of 4.2×10^{11} cm^{-2} for a conventional GaAs/Al$_{0.3}$Ga$_{0.7}$As MODFET with 75 Å spacer-layer thickness.[3] This value increased to 7×10^{11} cm^{-2} when the spacer layer was decreased to 20 Å. Figure 35 (Camnitz, 1985) shows the bias dependence of the measured f_T for 0.2 μm gate length transistors fabricated from these layer structures. It can be clearly seen that the higher 2DEG sheet density produces a larger total current swing, higher maximum f_T, and greater current density at maximum f_T. These results are particularly significant because ionized impurity scattering is more severe for thinner spacer devices (the 300 K Hall mobilities were 6300 and 5100 cm^2/Vs, respectively). This work clearly placed charge control on equal ground with electron transport in influencing the performance of a MODFET.

Experimental work with strained GaInAs channel devices also supports the basic observation that higher n_{sm} leads to higher f_T. Nguyen performed Hall measurements on Ga$_{0.85}$In$_{0.15}$As/Al$_{0.2}$Ga$_{0.8}$As and Ga$_{0.75}$In$_{0.25}$/Al$_{0.3}$Ga$_{0.7}$As MODFETs with the GaAs cap partially removed and reported n_H values of 1.7×10^{12} cm^{-2} and 2.4×10^{12} cm^{-2}, respectively, at

---

[3]Although n_{sm} is significantly lower at 300 K, these calculations were performed using 77 K C-V data, because its generally sharper shape allows more accurate determination of the bias at which excess charge modulation begins.

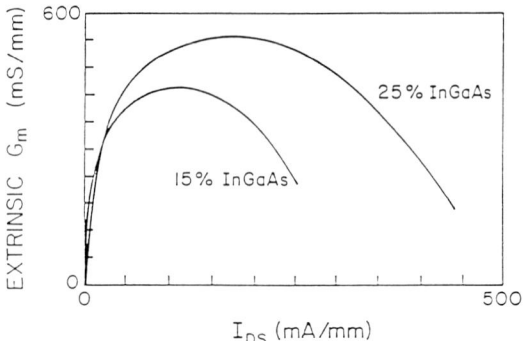

FIG. 36. Comparison of the bias dependence of the transconductance g_m for 0.3 μm gate length 15% and 25% in mole fraction channel transistors.

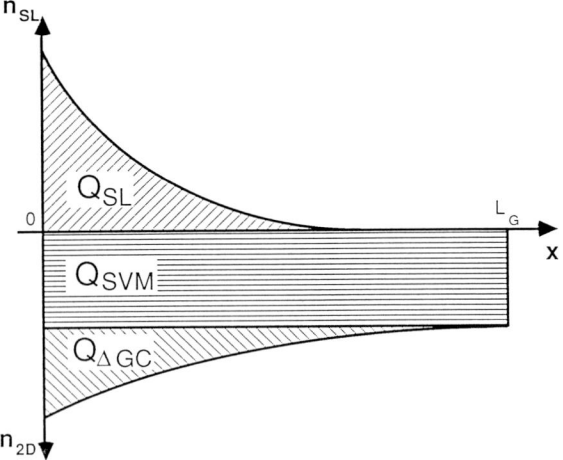

FIG. 37. A schematically depicted variation in sheet density with position along the gate length in a MODFET transistor.

300 K (Nguyen, 1989).[4] Figure 36 compares the bias dependence of the transconductance[5] for 0.3 μm gate-length transistors fabricated from these layer structures, indicating a dependence on InAs composition similar to the spacer-thickness dependence of Fig. 37. Moll et al. (1988) also successfully

---

[4]For comparison with the data of Camnitz, note that Nguyen calculated a 77 K n_{sm} of 1.4×10^{12} cm^{-2} on a similar $Ga_{0.85}In_{0.15}As/Al_{0.2}Ga_{0.8}As$ MODFET with n_H of 1.7×10^{12} cm^{-2}.

[5]In general, the transconductance is more sensitive to gate-to-channel separation than the f_T. However, because Nguyen recessed the gates identically and carefully chooses devices with similar thresholds, the comparison presented is a fair one.

fabricated MODFETs with $Ga_{0.75}In_{0.25}As$ channels. Their extensive analysis of the experimental data in terms of transit and charging delays led them to conclude that the superior microwave performance of these strained GaInAs channel devices arose not from a higher electron velocity in $Ga_{0.75}In_{0.25}As$ relative to GaAs but from a reduction of parasitic and drain delays. They suggest that maximizing the 2DEG sheet density without significantly degrading mobility is the key to enhancing the performance of MODFETs. These correlations between f_T and increased sheet density strongly indicate that excess charge modulation is a fundamental limiting mechanism in the high-frequency performance of the MODFET.

During this period of experimental advance, the theoretical understanding of parasitic modulation of electrons in the AlGaAs supply layer also progressed. Although early MODFET models such as those by Delagebeaudeuf and Linh (1982) assumed complete depletion of electrons in the electron-supplying layer, subsequent analyses such as those by Vinter (1984), Ponse et al. (1985), Gray and Lundstrom (1985), and Moloney et al. (1985) accounted for the presence of electrons in the AlGaAs layer. Hirakawa et al. (1984) studied the dependence of the 2DEG sheet density on the spacer-layer thickness both experimentally and numerically and found monotonically increasing sheet densities but decreasing mobilities with thinner spacers. After stressing the need to maximize the layer conductivity, they concluded that either the AlGaAs layer needed to be replaced by a higher bandgap material or that multilayered channels were necessary. Foisy et al. (1987) used numerical charge-control simulations and the saturated-velocity approximation to study the effect of the parasitic charge-modulation layer on f_T. They predicted improved performance with higher conduction-band discontinuities but also found that for the GaAs/AlGaAs system, increases in the donor ionization energy caused a saturation in this improvement at around a 30% aluminum composition. This was experimentally confirmed by Nguyen et al. (1987a). The technique of increasing the conduction-band discontinuity by adding indium to the conducting channel specifically addresses the issues raised by Hirakawa and Foisy.

Absent in the above-cited literature is a conceptual explanation as to why higher sheet densities lead to improved high-frequency performance. To more fully understand and quantify the effect of excess charge modulation, the concept of *modulation efficiency* was introduced by Foisy et al. (1988). This theory, which unifies the experimental data and models reviewed earlier, is outlined in the following.

In the field-effect transistor, the drain current, I_D, is modulated by varying the density of electrons in the conducting channel. Modulation of this charge produces a frequency-dependent displacement current through the gate terminal. The current-gain cutoff frequency, f_T, is defined as the frequency at which this current equals the ac drain current. Denoting the instantaneous source-drain voltage as V_{GS} and the total charge modulated by the gate as Q_{TOT}, this equality leads to an equation for f_T,

$$f_T = \frac{1}{2\pi} \frac{\partial I_D / \partial V_{GS}}{\partial Q_{TOT} / \partial V_{GS}} = \frac{1}{2\pi} \frac{\partial I_D}{\partial Q_{TOT}}. \quad (3)$$

This formulation shows that f_T can be increased by minimizing the charge that must be modulated for a given change in drain current. This requires higher electron velocities and more efficient utilization of charge. Since electrons in the AlGaAs layer are immobile or have low mobility, they increase the total charge modulated by the gate and while shielding the high-mobility charge in the 2DEG channel. To minimize this effect, the gate should not be forward biased past the point that results in the charge n_{sm}.

This second mechanism is one common to both MESFETs and MODFETs of all gate lengths and is responsible for the decrease in g_m seen at low currents. Electron velocities along the channel are nonuniform under all operating conditions. Current continuity demands that more electrons be present in the high-mobility conducting channel than would be necessary if all traveled at the maximally achievable velocity. Voltage drops occur along the channel causing the drain end to be more reverse biased than the source end. These self-consistent effects result in a variation in sheet density similar to that schematically depicted in Fig. 37. This figure shows three different charge components. Q_{SL} has already been discussed and is simply the low-velocity charge present in the electron-supplying layer. Q_{SVM} is the charge required to support the drain current if all electrons traveled at a uniform velocity v_{sat}. This situation is analogous to the saturated velocity model (Williams and Shaw, 1978). $Q_{\Delta GC}$ is simply the additional charge necessitated by nonuniform velocities and voltage drops along the channel. In the case of (Foisy et al., 1988) the *delta*, the in-channel charge was approximated by the gradual-channel model (Grebene and Ghandhi, 1969) using a saturated-velocity boundary condition (Turner and Wilson, 1969).

In the case of the saturated-velocity model, I_D and Q_{SVM} are simply related by

$$I_D = \frac{v_{SAT}}{L_G} Q_{SVM}, \quad (4)$$

where L_G is the gate length. Substitution of Eq. (4) into Eq. (3) and identification of the major components of Q_{TOT} gives a general equation for f_T

$$f_T = \left[\frac{v_{SAT}}{2\pi L_G} \right] \left[\frac{\partial Q_{SVM}}{\partial [Q_{SVM} + Q_{\Delta GC} + Q_{SL}]} \right]. \quad (5)$$

This formulation allows the isolation of the term dependent on the saturation velocity and the term dependent on charge utilization. The second term in Eq. (5) is the ME of the device:

$$\text{ME} = \left[\frac{\partial Q_{\text{SVM}}}{\partial [Q_{\text{SVM}} + Q_{\text{AGC}} + Q_{\text{SL}}]} \right] \quad (6)$$

It can now be seen that the presence of Q_{AGC} and Q_{SL} leads to degradation of the performance of the MODFET by reducing ME from the 100% efficiency assumed by the saturated-velocity model.

Figure 38 (Foisy, 1988) shows the comparison of a 1 μm gate-length conventional and pseudomorphic MODFET f_T behavior, simulated by using nearly identical epitaxial layer structures and identical transport properties. The GaInAs-channel MODFET has a higher cutoff frequency and improved current-driving capabilities. The frequency response has a peak f_T of 22 GHz at 140 mA/mm and remains within 80% of the peak f_T over a current swing of 190 mA/mm. In contrast, f_T of the conventional MODFET only reaches 18 GHz, with the peak occurring at the significantly lower current of 70 mA/mm. It maintains 80% or more of its peak f_T over a 110 mA/mm range. The 22% improvement in the peak f_T is consistent with the values shown in Fig. 34 and with the data assembled by Hikosaka et al. (1988).

The improvement of f_T and its bias dependence can be better understood

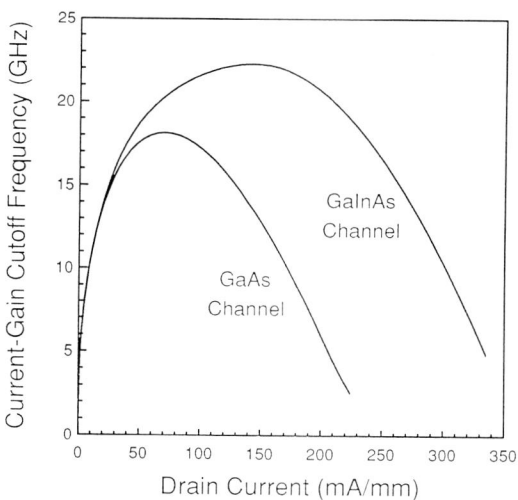

FIG. 38. Comparison of a 1 μm gate length conventional and pseudomorphic MODFET f_T behavior, simulated using nearly identical epitaxial layer structures and identical transport properties.

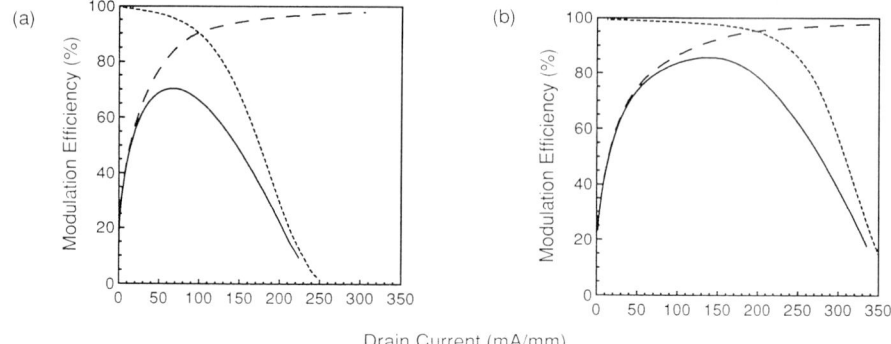

FIG. 39. The bias dependence of ME, ME_{AGC}, and ME_{SL} for (a) conventional and (b) pseudomorphic devices. ME = solid curve, ME_{AGC} = dashed curve, ME_{SL} = dotted curve.

by generating two modulation-efficiency envelopes, each describing the effects of one limiting mechanism. ME_{AGC} presents the effect of deviations from the saturated-velocity model while excluding the effects of parasitic charge in the electron supply layer. ME_{SL} presents the effects of parasitic charge in the electron-supplying layer while using the saturated-velocity model. Figures 39a and 39b show the bias dependence of ME, ME_{AGC}, and ME_{SL} for the conventional and pseudomorphic devices. From these figures, it can be seen that the shape of both ME and f_T follows from the overlap of ME_{AGC} and ME_{SL}. Whereas these envelopes reach 100% at high and low currents, respectively, their overlap and interaction limit the peak ME to 70% and 86%, respectively, for the two devices simulated. The apparent increase in electron velocity observed experimentally is accounted for by a higher modulation efficiency. Hence a plot of $1/(2\pi f_T)$ versus L_G would have a slope $1/(ME \cdot v_{sat})$. If we were to neglect the effect of inefficient charge modulation and simply equate the slope with $1/v_{eff}$, the velocity calculated would be a factor of ME less than the saturation velocity. Hence, poor modulation efficiency reduces the apparent electron velocity. ME is clearly capable of accounting for the low velocities observed experimentally (Tasker and Eastman, 1987a).

The ability to increase n_{sm} via a larger conduction-band discontinuity is illustrated in Fig. 40. This figure overlays the conduction-band diagrams for the conventional and pseudomorphic devices compared earlier. In each case, the gate bias has been adjusted to accumulate exactly $8.0 \times 10^{11}\,\text{cm}^{-2}$ electrons in the high-mobility channel. Note that in order to achieve this 2DEG sheet density, the AlGaAs conduction band (and donor energy) in the AlGaAs/GaAs device must come much closer to the Fermi energy than in the AlGaAs/GaInAs device. Because significant parasitic supply-layer modulation is already occurring in the former case, the n_{sm} is only $4.4 \times 10^{11}\,\text{cm}^{-2}$.

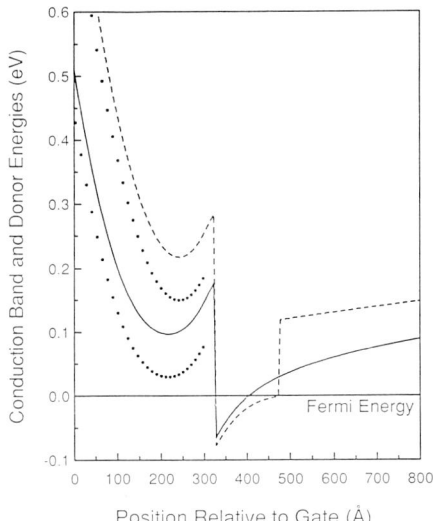

FIG. 40. Conduction band energy vs. position for GaAs and GaInAs channel MODFETS (solid and broken lines respectively). In both cases, the total sheet density below the upper heterojunction is $8 \times 10^{11} \, \text{cm}^{-2}$. The deep donor energies are marked by small circles.

In contrast, significant parasitic modulation has yet to occur in the strained-channel device. Its n_{sm} is $9.1 \times 10^{11} \, \text{cm}^{-2}$.

As shown in this section, both experimental and theoretical evidence for MODFETs show that achieving accumulated high 2DEG sheet densities prior to the accumulation of electrons in the electron-supplying layer is essential to achieving a f_T limited only by the maximum possible electron velocity (i.e., ME = 100%). The use of $\text{Ga}_{1-y}\text{In}_y\text{As}$ as the narrow bandgap material provides an elegant solution to this problem. Even though $\text{Ga}_{1-y}\text{In}_y\text{As}$ is not lattice matched to GaAs, thin layers may be kept below a critical thickness so that the strain is taken up coherently because elastic strain is required for dislocation free, pseudomorphic channels. These structures have large bandgap discontinuities ($\Delta E_c > 0.35 \, \text{eV}$) providing for larger 2DEG concentrations ($n_{sm} > 1.8 \times 10^{12} \, \text{cm}^{-2}$) and improved speed ($v_{\text{eff}} > 1.8 \times 10^7 \, \text{cm/s}$).

B. OTHER ELECTRONIC-DEVICE STRUCTURES

In addition to the use of heterojunctions to modify charge confinement parallel to the interface, they can also be used to control (modify) current transport. Modified current transport across the heterojunction interfaces has been exploited, for example, in the HBT and in the resonant tunneling diode. The introduction of strained-layer heterojunctions can also improve the performance of this class of heterojunction devices.

1. Heterojunction Bipolar Transistors

The heterojunction bipolar transistor has advantages over the homojunction version. Using a larger bandgap material for the emitter, as compared to the base, can yield higher injection efficiency at the emitter–base junction. Normally, a portion of the increased bandgap is exhibited as a conduction-band potential step, and the remainder is a valence-band potential step. If this heterojunction is abrupt in its composition change, these steps serve two purposes. First, the conduction-band step in an $n-p-n$ transistor gives the corresponding potential increment to the ballistically injected electrons. This potential step is $\sim .25$ V for the typical $Al_{0.30}Ga_{0.70}As/GaAs$ heterojunction. Second, the valence-band step is an added barrier for the holes in the base, reducing their thermionic emission from the base to the emitter. The injection ratio of electron current to total current at the emitter–base junction is thus improved. Thus, with an $\sim .15$ eV valence-band step, typical of the $Al_{0.30}Ga_{0.70}As/GaAs$ heterojunction, this hole emission reduces by a factor of ~ 400. It is thus possible to raise the acceptor concentration in the base to values of $\sim 1 \times 10^{20}/cm^3$, for low base resistance, while still maintaining an injection ratio near unity. Thus, the transistor has a very high overall current gain even when the base is heavily doped. Such transistors have yielded up to 218 GHz = f_{max} and 105 GHz = f_T (Asbeck, 1989) for this case of GaAs base.

Another heterojunction bipolar transistor uses an InP emitter and $In_{0.53}Ga_{0.47}As$ bases and collectors (Nottenberg et al., 1989). In this case, there is a ~ 0.40 V potential step in the valence band, and a 0.20 V potential step in the conduction band. With such a large potential step in the valence band, base acceptor concentration up to $5 \times 10^{20}/cm^3$ is useful for low base resistance. In early research, $f_T = 165$ GHz has been obtained with this device, and $f_{max} = 100$ GHz. With improved parasitic circuit elements, this device will eventually yield much higher f_{max}.

In both cases, even though the electrons are ballistically injected into the thin ($\sim 500–1000$ Å) bases of these devices, the electron forward momentum is not maintained. The energy distribution of injected carriers is cooled by inelastic collisions and recombination. Since recombination of minority carriers (electron) is predominantly radiative, electroluminescence can be used to help characterize the electron transport in these devices. It has been found that recombination probability of hot electrons with equilibrium holes decreases with increasing energy.

The use of a strained, graded bandgap base further improves the base transport factor, which is the ratio of the minority current density at the collector junction to that at the emitter junction. A built-in electric field, resulting from the graded reduction of the base bandgap, causes the electrons to keep moving rapidly across the base. Field strengths of 3–6 KV/cm are adequate to keep electrons moving fast in GaAs or GaInAs. By using a thin ($\leqslant 1000$ Å), graded ($0 < y < 0.06$) $Ga_{1-y}In_yAs$ base, in place of a GaAs base, gain has been improved (Enquist et al., 1987) by this built-in field. In order to apply the Matthews–Blakeslee limit to this case, an average indium fraction of ~ 0.03 would be used, yielding an $\sim 800\,\text{Å}$ thickness limit. It is possible to have even larger composition variations in the case of the approximately $In_{0.53}Ga_{0.47}As$ base, where the InAs mole fraction y can range from $y < 0.53$ at the emitter side (layer in tension) to $y > 0.53$ at the collector side (layer in compression). In this latter case, experiments have not yet been carried out.

2. Resonant Tunneling Oscillator Structures

When two, thin potential barriers are separated by one half of an electron wavelength, there is a resonant current rise when electrons are raised in energy to the energy of the resonant state in the region between the barriers. This device is called a resonant tunneling device. The electrons tunnel through both barriers, but only when their energy is at the resonant state determined by the barrier separation. Early versions used GaAs-active regions with AlGaAs barriers. The leading effort at MIT Lincoln Laboratory (Brown et al., 1988) has used very thin AlAs barriers (11.3 Å and 17 Å) in order to avoid nonuniform alloy composition. The GaAs between the barriers was $\sim 45\,\text{Å}$ thick. Above the resonant tunneling peak current, the current falls with increasing bias voltage, yielding negative conductance. The current minimizes in a "valley" before rising again with further voltage increases. The "valley" current in these devices is higher than expected from theory.

Scattering allows electrons to tunnel through this structure even when the electron energy is well above the resonance energy. For electrons with energy of more than 0.31 eV above the conduction band in GaAs, upper-valley scattering occurs. This upper-valley energy is $\sim 0.55\,\text{V}$ for the case of $In_{0.53}Ga_{0.47}As$, causing this scattering mechanism to be removed to higher bias voltages. Even elastic scattering from potentials caused by atomic steps in barrier thickness gives rise to such anomalous tunneling. The ratio of the peak-to-valley current must be high to yield high efficiency in oscillators using the negative conductance. These AlGaAs/GaAs resonant tunneling oscillators have operated at up to 420 GHz in the MIT Lincoln Laboratory experiments.

Using $In_{0.53}Ga_{0.47}As$ active regions, grown on InP substrates, and with strained-layer AlAs barriers, very high current peak-to-valley ratios ($> 30:1$) have been achieved (Inata et al. 1987). In this strained-layer structure, the thin AlAs is under tension. It yields an electron barrier potential height of $\sim 1.4\,\text{V}$ in the conduction band, sharply lowering the valley current. The speed of tunneling has also been lowered by this high barrier height. Another strained, binary compound barrier, that of GaAs, could also be used. Because this layer has a 0.3–0.4 V barrier, rather than $\sim 1.4\,\text{V}$ for AlAs, the tunneling speed is ~ 100 times faster. This would allow both high-frequency and high-efficiency operation.

IV. Optoelectronic-Device Structures

A. Quantum-Well GRINSCH Lasers

Semiconductor lasers are an important addition to the list of strained-layer devices fabricated to date. There are three reasons for moving to strained GaInAs quantum-well laser structures. First, specific longer wavelengths with photon energies below the bandgap of GaAs can be reached, allowing GaAs to be used as a transparent substrate for fabrication of surface-emitting lasers (Huang et al., 1989). Emission out to 1.03 μm (Offsey et al., 1989) and 1.07 μm (York et al., 1989) have been obtained in this way. Second, the strain-induced splitting of the valence bands predicted to lower the density of states in the highest energy valence band permits lasing to occur at lower threshold currents than for unstrained lasers (Adams, 1988; Yablonovitch, 1988). Third, the strain-induced splitting of the valence bands results in a reduction in the hole mass in the plane of the junction, which is predicted to result in higher speed modulation (Suemune, 1988).

Materials for lasers must be of very high quality for efficient lasing operation with low unintentional impurity concentrations and low defect densities. In order to utilize GaInAs as a strained-laser material, laser designs and conditions of growth must be determined such that the presence of the strained material does not result in the formation of undesirable defects that might limit performance or operational lifetime. Unfortunately, the different materials in the GRINSCH structure are grown optimally at very different temperatures. Typical growth temperatures are 710°C for AlGaAs (Offsey et

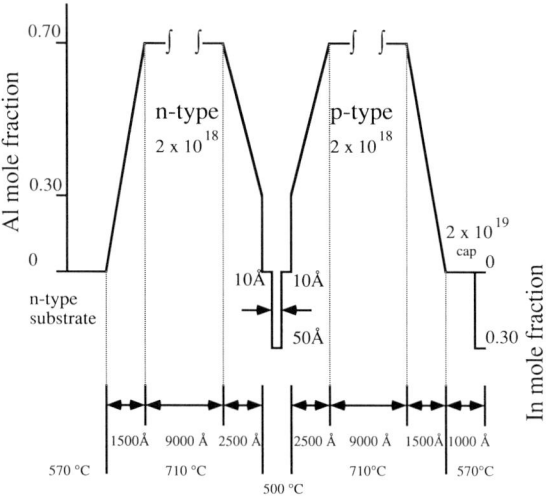

Fig. 41. A typical structure of a GRINSCH laser.

al., 1989) by MBE and up to 800°C (Gavrilovic *et al.*, 1988) by OMVPE. As mentioned in Section II.C.1, optimal GaInAs quantum-well growth temperatures are more typically 500°C. As a result, growth interruptions have been used in MBE growth to allow for the temperature change and to increase interface smoothness (Offsey *et al.*, 1989). A typical structure is shown in Fig. 41.

The lasing wavelength is determined by the bound quantum-well state energies. The strain-modified bandgap has been taken into consideration to produce the predicted emission energy versus quantum-well thickness for different In mole fractions seen in Fig. 42 (Kolbas *et al.*, 1988). For surface-emitting lasers, emission energies below the bandgap of GaAs are required for low loss due to absorption (Huang *et al.*, 1989). A summary of the laser emission energies reported to date are shown in Table II.

All strained-layer lasers fabricated on GaAs substrates have used some form of GaInAs quantum well (Laidig *et al.*, Fekete *et al.*, 1986; Fischer *et al.*, 1986; Kolbas *et al.*, 1988; Vawter *et al.*, 1989; Bour *et al.*, 1988; Offsey *et al.*, 1989). The highest-performance lasers of unstrained GaAs quantum wells are

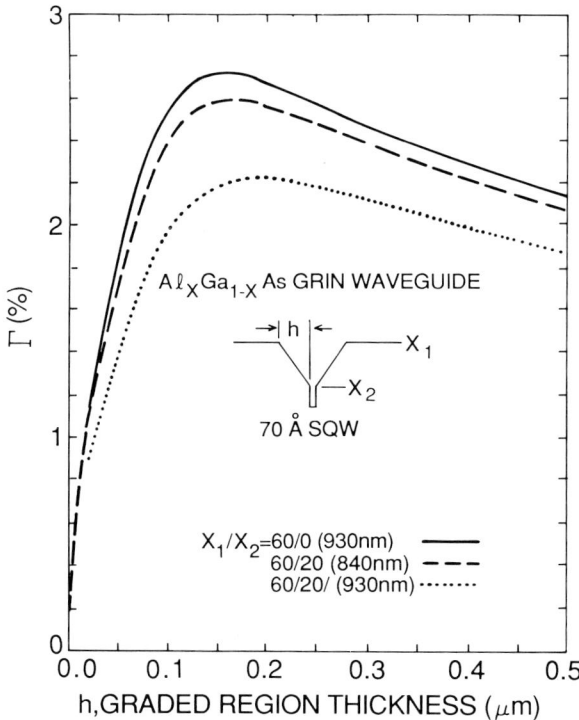

FIG. 42. The confinement factor calculated for three different GRIN waveguide structures.

those of the GRINSCH variety because of superior optical confinement. One disadvantage of GaInAs strained-layer laser is that their longer emission wavelength results in a reduction in optical confinement (Bour 1989). In Fig. 43, Bour shows the confinement factor calculated for three different graded-index (GRIN) waveguide structures. The confinement factors of strained-layer lasers are further reduced, since the quantum-well widths of strained-

TABLE II

Year	Growth technique	Length μm	Width μm	Wavelength nm	Threshold current density A/cm^2	Reference
1984	MBE	270	110	1000	1200	Laidig
1986	OMVPE	527	146	1002	152	Feketa
1988	OMVPE	600	90	930	200	Bour
1988	OMVPE	250	90	950	170–260	Stutius
1988	OMVPE	750	250	950	152	Gavrilovic
1989	OMVPE	400	90	930	194	Bour
1989	OMVPE	405	3.5	1074	493–634	York
1989	MBE	375	12		1333	Vawter
1989	MBE	800	150	1026	167	Offsey
1989	MBE	300	4	950	2100	Caldwell
1989	OMVPE		2	911		Major

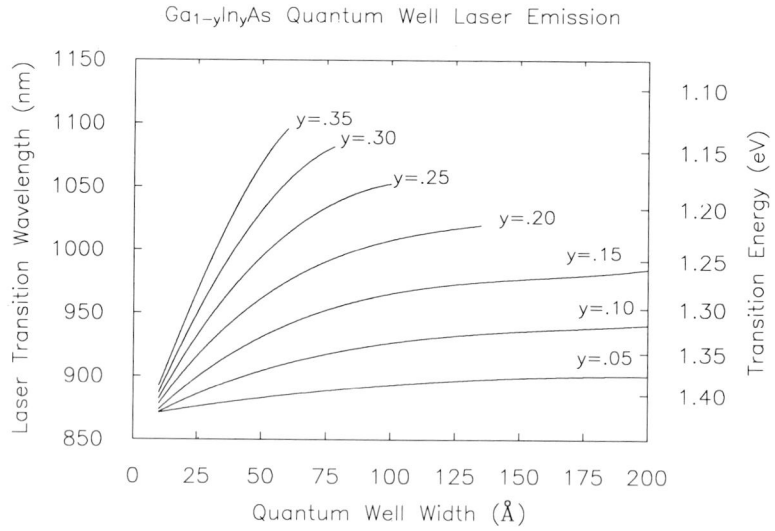

FIG. 43. The predicted emission energy versus quantum-well thickness for different In mole fractions.

layer lasers are limited to smaller dimensions than the quantum wells of unstrained lasers due to critical-thickness limitations.

Table II also reports on the threshold-current densities for strained-layer lasers. It is difficult to make comparisons of lasers of different sizes and different structures. Both the epitaxial layer design and the fabricated structure (simple mesa, ridge waveguide, buried heterostructure, or use of different mirror coatings) can be responsible for wide variations in performance. It is not sufficient to simply remove device dimensions from each reported performance by normalizing threshold currents to unit area, since performance is fundamentally tied to dimension. For broad-area devices, where performance is not as dramatically affected by size, the threshold-current density is the best figure of merit. Smaller devices show higher threshold-current densities, as seen in Table II, since the mirror loss

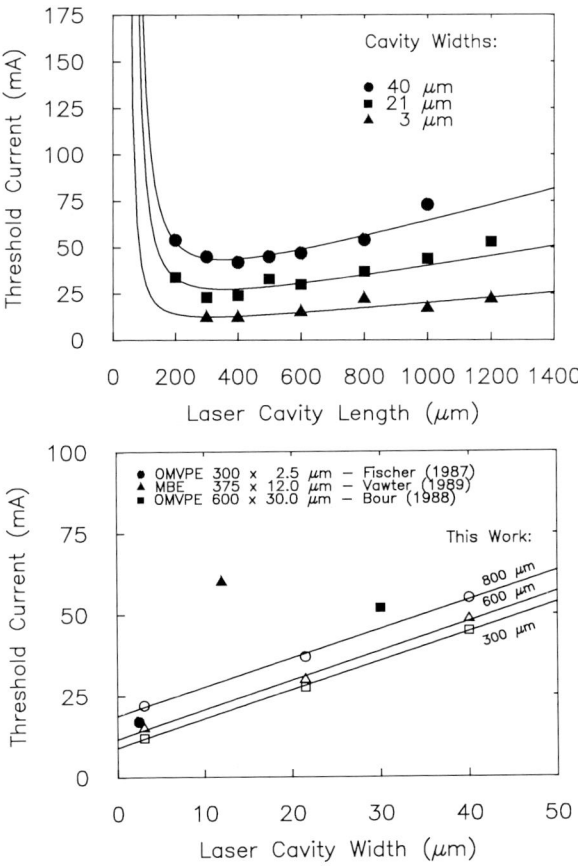

FIG. 44. Comparisons of threshold currents between strained GRINSCH lasers of different dimensions.

constitutes a higher fraction of the total loss. For smaller devices suitable for high-speed digital modulation, the threshold current is the figure of merit, since lower threshold currents produce smaller turn-on delays. One attempt to make comparisons between lasers of smaller dimensions is seen in Fig. 44 (Offsey *et al.*, 1989). Since the threshold current is a strong linear function of width, small devices can be more easily compared in this manner.

The threshold performance of the strained-layer lasers reported to date seem to be comparable, or possibly better than, lasers using GaAs quantum wells. Thus, the increase in gain due to the strain-modified valence band more than compensates for the decreases in optical confinement mentioned earlier. In addition, these results indicate that high-quality strained-quantum-well lasers can survive the growth process, which tends to occur with much higher temperatures and for much longer times than those used for SMODFETs.

Useful operating lifetimes for strained-layer devices has been a concern. There has been speculation about the stability of strained layers where the added stresses of elevated operating temperature and high carrier and photon-flux densities might contribute to device degradation. The first report of continuous-wave (CW) life-test results for a strained-layer laser (Stutius 88) extrapolated mean time to failure in excess of 20,000 hours. Life tests to 5,000 hours were reported in 1989 (Fischer *et al.*, 1989). The results are shown in Fig. 45. The current required to maintain 70 mW/facet CW output power is plotted as a function of time for 15 different devices. Little degradation and no sudden failures were seen. Additional data for a smaller sample of devices showed operation beyond 10,000 hours.

Strained-layer lasers have also been shown to be stable when subject to

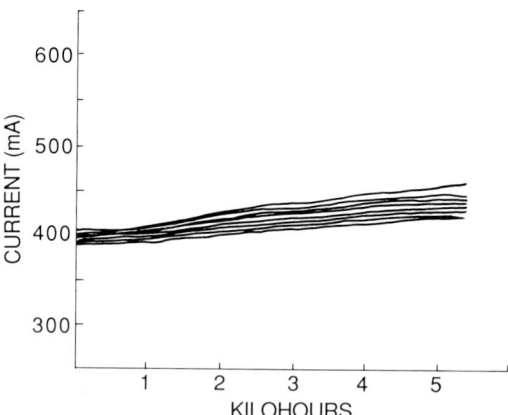

FIG. 45. The current required to maintain 70 mW/facet CW output power is plotted as a function of time for 15 different devices.

high-temperature disorder-defined buried heterostructure processing (Major et al., 1989). A scanning electron microscope image of a quantum-well laser following Si–O diffusion (at 825°C for 24 hours) is seen in Fig. 46. A TEM image of the quantum-well region of the same structure shows both clear interdiffusion and inhibition from interdiffusion where Si–O is, or is not, introduced, respectively (Fig. 47). It is important for integration and array fabrication that this stability has been demonstrated.

One of the promises of strained GaInAs quantum-well lasers has just recently been demonstrated. It is predicted that lower threshold currents can be obtained through the modification of the valence band (Suemune et al., 1988), which will result in faster digital switching and higher microwave-modulation rates. Direct comparisons of GaAs and GaInAs quantum-well lasers are complicated by such differences as the confinement factor for nearly identical structures operating at different wavelengths. Preliminary evidence suggests that improvements in microwave response are seen in GaInAs

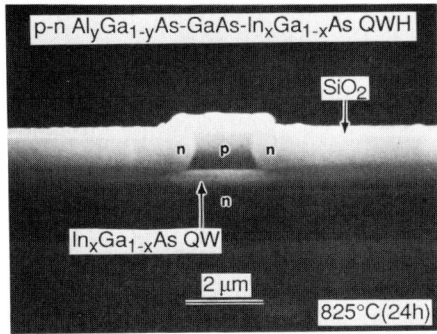

FIG. 46. A scanning electron microscope image of a quantum-well laser following Si–O diffusion (825C 24h). (From Major et al., 1989).

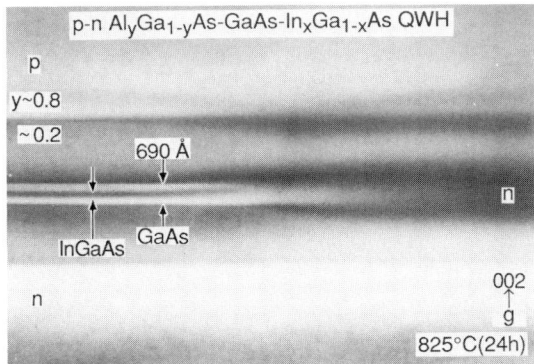

FIG. 47. A TEM image of the quantum-well region following Si–O diffusion (at 825°C for 24 hr) shows both clear interdiffusion and protection from interdiffusion where Si–O is present. (From Major et al., 1989).

quantum-well structures as compared to GaAs wells (Offsey, 1989(2)) as a result of strain. The microwave-frequency response as a function of drive current for a strained-layer laser is seen in Fig. 48. The 3 dB cutoff frequency is seen to vary as the square root of the drive current (and therefore optical power). The microwave-frequency response results are compared for lasers with GaAs or GaInAs quantum wells in Fig. 49. The GaInAs lasers clearly demonstrate superior high-frequency performance.

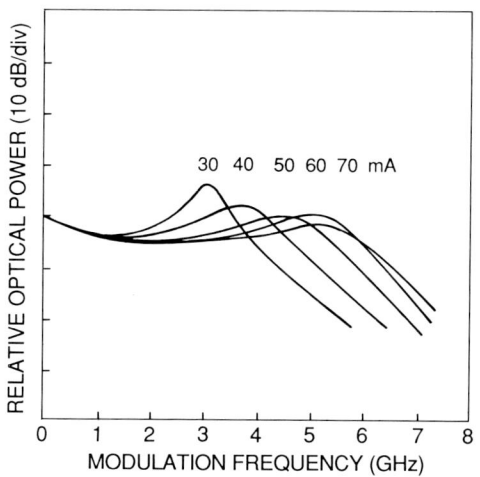

FIG. 48. Microwave performance of a strained GRINSCH laser as a function of drive current.

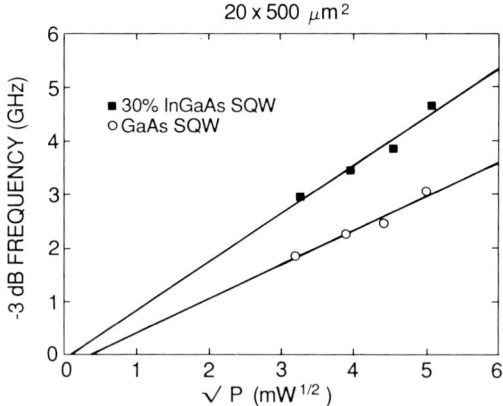

FIG. 49. Comparison of microwave cutoff frequencies of GaAs and GaInAs quantum-well GRINSCH lasers as a function of optical power.

The last advantage of GaInAs strained-layer lasers is the transparency of the GaAs substrate to the lasing wavelength, and low-loss surface-emitting lasers have been demonstrated. Bour (Bour 89(3)) reported on the fabrication and operation of strained-layer GaInAs/GaAs grating-surface emitting laser arrays. These arrays used a typical GRINSCH structure followed by fabrication of grating-surface-emitting distributed Bragg reflectors. The arrays consisted of groups of ten 2-μm wide ridges spaced on 4 μm centers and 150 μm length. Peak output powers of 2 W were obtained from a 10×10 section operating at 12 A. The highest output power was 200 W obtained by driving a larger number of elements to produce a 1.1 kW/cm^2 power density.

B. DETECTORS

The selection of new wavelengths of operation also extends to detectors. Unfortunately, most detectors require absorbing layer thicknesses that exceed CLT. The detectors that operate at longer wavelengths than possible and using intrinsic GaAs have all been fabricated from strain-relieved GaInAs growths on GaAs (Rogers, 1988; Rhazeghi, 1987). These devices are described later in the discussion on strain-relieved materials.

C. MODULATORS

High-speed modulation of light by quantum-well phenomena is expected to play an important role in integrated optoelectronic applications. One efficient class of high-speed modulators are the quantum-well modulators that use the Stark shift in absorption, a second-order nonlinear optical phenomena to provide either transmission or absorption of incident light (Miller, 1986; Ralston, 1989). With strained GaInAs/GaAs superlattices grown on the (111) orientation, large first-order electro-optic effects can be observed due to strain-generated internal piezoelectric fields (Mailhiot 1989). Devices of this class have been fabricated and found to exhibit superior selectivity (Caridi 1989). The internal field due to strain provides a blue shift to transmission under no applied bias, thus resulting in a wider selectivity under reverse bias, which red-shifts the passband. This improvement in dynamic range is an important figure of merit for integrated optoelectronic systems.

V. Optimized Lattice-Constant Epitaxy

One of the limitations of strained-layer technology for SMODFETs consists of the restraints on the value of In mole fraction that can be effectively used. Higher InAs mole fractions give rise to the desired increased conduction-band discontinuity; however, the maximum amount of InAs is strictly limited by critical-thickness boundaries. It is clear that superior carrier-confinement properties can be obtained through the use of mismatched lattice-constant epitaxy, and some technique must be found to permit the growth of arbitrary lattice-constant structures that have been selected for their superior electronic properties.

A. MODFETs

A great deal of attention has been focused on the growth of GaAs on Si despite a 4% lattice mismatch. This work has encouraged new efforts in other materials systems. One approach to demonstrate mismatched growth of high-performance MODFETs is the growth of AlInAs/GaInAs at the composition that would lattice-match InP and instead grow these materials on a GaAs substrate (Schaff, 1987; Wang et al., 1988). A structure used in this approach is shown in Fig. 50. This structure attempts to minimize the effect of lattice mismatch, and the accompanying defects, through the use of a superlattice buffer of AlInAs/GaInAs. The buffer is seen to be effective in halting the propagation of dislocations in Fig. 51. These TEM images show dramatic reduction in dislocation densities to concentrations in the area of $1 \times 10^{+8} \mathrm{cm}^{-2}$ near the surface where the MODFET active channel is located. It would be expected that large concentrations of defects would act to compensate the 2DEG concentration and prevent satisfactory device performance. Apparently, this residual concentration of defects is insufficient to significantly disrupt device operation as seen in the IV characteristics of Fig. 52.

B. MESFETs

Other techniques for obtaining single-crystal materials with a new lattice constant include use of graded lattice-constant layers, and use of step-graded lattice-constant layers as buffers between the substrate and active device.

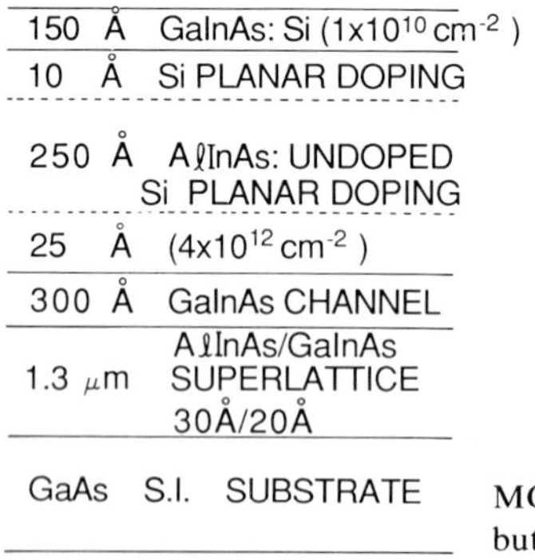

150 Å	GaInAs: Si (1×10^{10} cm^{-2})
10 Å	Si PLANAR DOPING
250 Å	AlInAs: UNDOPED Si PLANAR DOPING
25 Å	(4×10^{12} cm^{-2})
300 Å	GaInAs CHANNEL
1.3 µm	AlInAs/GaInAs SUPERLATTICE 30Å/20Å
GaAs S.I. SUBSTRATE	

FIG. 50. Layer structure of a AlInAs/GaInAs MODFET with compositions that would lattice-match InP but instead grown on a GaAs substrate.

2. DEVICE APPLICATIONS OF STRAINED-LAYER EPITAXY

FIG. 51. A TEM image of a superlattice buffer of AlInAs/GaInAs grown on GaAs showing the dramatic reduction in the propagation of dislocations.

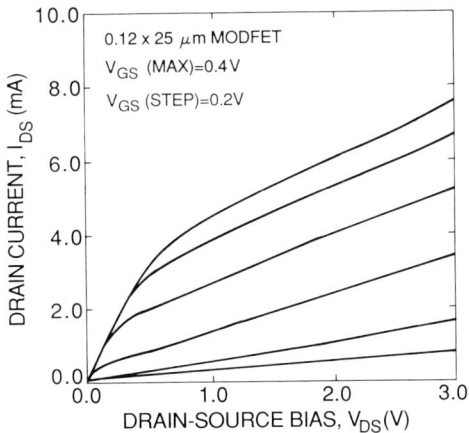

FIG. 52. I-V characteristics of a AlInAs/GaInAs MODFET with compositions that would lattice-match InP but instead grown on a GaAs substrate.

Another approach is to grow layers with slight mismatch that will cause all defects arising due to mismatch to be confined in the plane of the mismatched interface. With this technique, a new lattice constant is achieved through use of several homogenous layers, each with increasing mismatch to the substrate. Two device structures have been fabricated in this way. A photodector is described in the following. MESFETs with GaInAs channels and buffers have also been grown in this way (Shih 1989).

C. DETECTORS

The growth of arbitrary lattice-constant structures has further advantages beyond permitting optimized lattice-constant growth. It would be desirable to integrate devices of different lattice constants in a single wafer for a number of optoelectronic applications. One important integrated application is the growth of GaInAs for optical detectors on the same GaAs substrate used for MESFET or MODFET integration. This has been demonstrated by using material grown by OMVPE (Razeghi, 1986), as seen in Fig. 53.

Pseudomorphic layers cannot be used for detectors in the surface illumination configuration, since the critical thicknesses to which they are

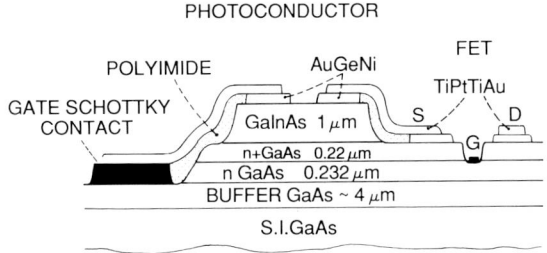

FIG. 53. An integrated photoreceiver has been performed using material grown by OMVPE.

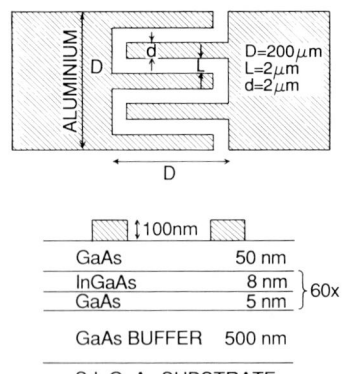

FIG. 54. A growth of GaInAs for detector applications on a GaAs substrate.

limited are far smaller than the absorption depth of the light they are meant to detect. For edge-illuminated detectors, this constraint does not apply. GaInAs detector grown on a GaAs substrate is seen in Fig. 54 (Rogers 1989). The problem of lattice mismatch was addressed by using step-graded layers in an attempt to confine dislocations to the plane of each interface. The goal of step grading is to turn threading dislocations, which are both electrically and optically active, into 60-degree misfit dislocations that are confined to the interfaces. Even though the 60-degree misfit dislocations are also electrically and optically active, they are deep enough below the absorption depth of the light that the electrical bias applied to the device overcomes the diffusion of photogenerated carriers into the dislocation array. Using this method, high-performance detector operation was obtained.

An additional technique for growth of optimized lattice-constant structures is the use of a strained-layer superlattice to control the propagation of defects that arise from lattice mismatch. This technique has been employed in the growth of GaAs on Si. To reach fiber-application wavelengths (1.3–1.55 μm), a lattice constant larger than GaAs is necessary. One study to demonstrate lasers used strained-layer GaInAs/GaAs layers that exceeded CLT to establish the lattice constant of $Ga_{0.92}In_{0.08}As$ as a buffer for subsequent laser growth (Caldwell, 1989). The multiple quantum-well structures yielded broad-area thresholds of 2.1 kA/cm^2. Detectors operating at fiber wavelength could also use this approach to obtaining low-dislocation-density films.

VI. Conclusions

The availability of *device-quality strained-layer epitaxy* has demonstrated its potential to provide additional flexibility to the design of device structures, allowing the development of new and/or improved electronic and optoelectronic device structures, and thus will play a key role in expanding the applications for electronic and optoelectronic semiconductor devices. Strained-layer epitaxy has successfully demonstrated the ability to select alternative substrate materials and modify the energy-band structure of active channels.

The use of strained GaInAs layers in AlGaAs/GaInAs and AlInAs /GaInAs SMODFET structures, epitaxially grown on GaAs and InP substrates, respectively, has provided for significant improvements in high-frequency performance, and these device structures are expected to have a considerable impact on future electronic applications at high-speed and millimeter-wave frequencies. The strained-layer structure can increase the heterojunction conduction-band offset between the strained-layer channel and the AlGaAs or AlInAs supply layer (0.23 eV to 0.45 eV in GaAs-based structures). This produces increased accumulation of charge in the active channel ($1.0 \times 10^{12}\,\text{cm}^{-2}$ to $2.5 \times 10^{12}\,\text{cm}^{-2}$ in GaAs-based structures), which is directly reflected in improved transistor performance, high current drive ($> 650\,\text{mA/mm}$ on GaAs, $> 1000\,\text{mA/mm}$ on InP), high transconductance ($> 650\,\text{mS/mm}$ on GaAs, $> 1000\,\text{mS/mm}$ on InP), and increased frequency response ($f_T > 152\,\text{GHz}$ on GaAs, $> 210\,\text{GHz}$ on InP). Further improvements are predicted as the growth of strained layers is further refined so that higher InAs mole fraction strained layers can be obtained for device applications. In addition, the use of strained layers in other high-speed device structures, e.g., HBTs, resonant tuneling diodes, are still in their infancy, and significant improvements are predicted via the use of strained-layer epitaxy.

Strained-layer epitaxy has considerably expanded the applications for semiconductor photodetectors, modulators, and lasers. The strained layers are used to modify the bandgap of the photodetector or laser and thus change their optical absorption or emission spectrum, providing for operation at the desired optical frequency or bandwidth. This is expected to have a considerable impact on the development of integrated optoelectronic circuit functions.

If the strained layer is electrically and/or optically active, their electrical properties are modified from those of bulk material. For example, the splitting of the light- and heavy-hole bands in GaInAs quantum wells as a result of the compressive strain can be an advantage. This should provide for improved hole-transport properties, which are expected to be reflected in improved high-speed performance in strained-layer p-channel quantum-well MODFET structures and improved high-speed performance and reduced threshold currents in strained-layer quantum-well GRINSCH laser diodes.

References

Adams, A. R. (1986). *Electron. Lett.* **22**, 250.
Asbeck, P. (1989). Private communication.
Ashizawa, Y., Akbar, S., Schaff, W. J., Eastman, L. F., Fitzgerald, E. A., and Ast, D. A. (1988). *J. Appl. Phys.* **46**, 4065.
Bandy, S., Nishimoto, C., Hyder, S., and Hooper, C. (1981). "Saturation velocity determination for $\text{In}_{0.53}\text{Ga}_{0.47}\text{As}$ field-effect transistors," *Appl. Phys. Lett.* **38**, 817–819.
Berger, P. R., Chang, K., Battacharya, P., and Singh, J. (1988) *J. Appl. Phys.* **53**, 684.
Bour, D. P., Gilbert, D. B., Elbaum, L., and Harvey, M. G., (1988). *Appl. Phys. Lett.* **53**, 2371.
Brown, E. R., Goodhue, W. D., and Sollner, T. C. L. G. (1988). *J. Appl. Phys.* **64**, 1519.
Caldwell, P. J., Leavitt, R. P., Whisnant, J. K., Horst, S. C., and Towner, F. J. (1989). "InGaAs/InAlGaAs Injection Lasers Grown on GaAs Substrates," *Conf. on Lasers and Electro-Optics* (*Cleo*), Baltimore, Maryland.
Camnitz, L. H. (1985). "Design Principles and Performance of Modulation-Doped Field-Effect Transistors for Low-Noise Microwave Amplification," Ph.D. Thesis, Cornell University.
Camnitz, L. H., Tasker, P. J., Lee, H., van der Merwe, D., and Eastman, L. F. (1984). "Microwave characterization of very high transconductance MODFET," in *IEDM Tech. Dig. 1984*, 360–363.

Camnitz, L. H., Maki, P. A., Tasker, P. J., and Eastman, L. F. (1985a). "Sub-micrometer quantum-well HEMT with $Al_{0.3}Ga_{0.7}As$ buffer layers," *Inst Phys. Conf. Ser.* **74**, 333–338.

Camnitz, L. H., Tasker, P. J., Maki, P. A., Lee, H., Huang, J. and Eastman, L. F. (1985b). "The role of charge control on drift mobility in AlGaAs/GaAs MODFETs," *Proc. IEEE Cornell Conference on Advanced Concepts in High-Speed Devices and Circuits*, 199–208.

Cappy, A., Carnez, B., Fauquembergues, R., Salmer, G., and Constant, E. (1980). "Comparative potential performance of Si, GaAs, GaInAs, InAs submicrometer-gate FETs," *IEEE Trans. Electron Devices* **ED-27**, 2158–2160.

Cardi, E. A., Chang, T. Y., Goossen, K. W., and Eastman, L. F. (1989). "Direct Demonstration of a Misfit Strain-Generated Electric Field in a [111] Growth Axis Zinc-Blende Heterostructure," *Appl. Phys. Lett.*, **56**, 659–661.

Chao, P. C., Palmateer, S. C., Smith, P. M., Mishra, U. K., Duh, K. H. G., and Hwang, J. C. M. (1985a). "Millimeter-wave low-noise high electron mobility transistors," *IEEE Electron Devices Lett.* **EDL-6**, 531–533.

Chao, P. C., Smith, P. M., Palmateer, S. C., and Hwang, J. C. M. (1985b). "Electron-beam fabrication of GaAs low-noise MESFETs using a new trilayer resist technique," *IEEE Trans. Electron Devices* **ED-32**, 1042–1046.

Chao, P. C., Tiberio, R. C., Duh, K. H. G., Smith, P. M., Ballingal, J. M., Lester, L. F., Lee, B. R., Jabra, A., and Gifford, G. G. (1987). "0.1-μm gate-length pseudomorphic HEMTs," *IEEE Electron Devices Lett.* **EDL-8**, 489–491.

Daniels, R. R., Mactaggart, R., Abrokwah, J. K., Tufte, O. N., Shur, Baek, J., and Jenkins, P. (1986). "Complementary heterostructure insulated gate FET circuits for high-speed, low power VLSI," *Tech. Dig. IEDM*, 448–451.

Daniels, R. R., Ruden, P. P., Shur, M., Grider, D., Nohava, T. E., and Arch, D. K. (1988). "Quantum-well p-channel AlGaAs/InGaAs/GaAs heterostructure insulated-gate field-effect transistor with very high transconductance," *IEEE Electron Devices Lett.* **EDL-9**, 355–357.

Delagebeaudeuf, D., and Linh, N. T. (1982). "Metal-(n) AlGaAs-GaAs two-dimensional electron gas FET," *IEEE Trans. Electron Devices* **ED-29**, 955–960.

Drummond, T. J., Fischer, R. J., Kopp, W. F., Morkoç, H., Lee, K., and Shur, M. S. (1983). "Bias dependence and light sensitivity of (Al,Ga)As/GaAs MODFETs at 77 K," *IEEE Trans. Electron Devices* **ED-30**, 1806–1811.

Enquist, P. M., Ramberg, L. P., Najjar, F. E., Schaff, W. J., Kavanagh, K. L., Wicks, G. W., and Eastman, L. F. (1987). *J. Crystal Growth* **18**, 378–382.

Fekete, D., Chan, K. T., Ballantyne, J. M., and Eastman, L. F. (1986). *Appl. Phys. Lett.* **49** 1659–1660.

Fischer, R., Drummond, T. J., Kliem, J., Kopp, W., Henderson, T. S., Perrachione, D., and Morkoç, H. (1984). "On the collapse of drain I-V characteristics in modulation-doped FETs at cryogenic temperatures," *IEEE Trans. Electron Devices* **ED-31**, 1028–1032.

Fischer, S. E., Waters, R. G., Fekete, D., Ballantyne, J. M., Chen, Y. C., and Soltz, B. A. (1989). *Appl. Phys. Lett.* **54**, 1861.

Fischer-Colbrie, A., Miller, J. N., Laderman, S. S., Rosner, S. J., and Hull, R., (1988). *J. Vac. Sci. Technol. B.* **6**, 620–624.

Fitzgerald, E. A., Watson, G. P., Proano, R. E., Ast, D. G., Kirchner, P. D., Pettit, G. D., and Woodall, J. M. (1989). *J. Appl. Phys.* **65**, 2220.

Foisy, M. C., Huang, J. C., Tasker, P. J., and Eastman, L. F. (1987). "Modulation efficiency limited high-frequency performance of the MODFET," in *Picosecond Electronics and Optoelectronics II* (F. J. Leonberger, C. H. Lee, F. Capasso, and H. Morkoç, eds.). Springer-Verlag, Berlin, pp. 181–183.

Foisy, M. C., Tasker, P. J., Hughes, B., and Eastman, L. F. (1988). "The role of inefficient charge modulation in limiting the current-gain cutoff frequency of the MODFET," *IEEE Trans. Electron Devices* **ED-35**, 871–878.

Fritz, I. J., Picraux, S. T., Dawson, L. R. Drummond, T. J., Laidig, W. D., and Anderson, N. G. (1985). *Appl. Phys. Lett.* **46**, 967.

Fritz, I. J., Drummond, T. J., Osbourn, G. C., Schirber, J. E., and Jones, E. D. (1986a). *Appl. Phys. Lett.* **48**, 1678.

Fritz, I. J., Doyle, B. L., Schirber, J. E., Jones, E. D., Dawson, L. R., and Drummond, T. J., (1986b), *Appl. Phys. Lett.* **49**, 581.

Gavrilovic, P., Meehan, K., Stutius, W., Williams, J. E., and Zarrabi, J. H. (1988). *11th IEEE Intl. Semiconductor Laser Conf.*, Boston, MA.

Gray, J. L. and Lundstrom, M. S. (1985). "Numerical solution of Poisson's equation with application of C-V analysis of III-V heterojunction capacitors," *IEEE Trans. Electron Devices* **ED-32**, 2102–2109.

Grebene, A. B., and Ghandhi, S. K. (1969). "General theory for pinched operation of the junction-gate FET," *Solid-State Electronics* **12**, 573–589.

Griem, H. T. (1988). "Study of DC and optical characteristics of $In_xGa_{1-x}As/In_yAl_{1-y}As$ (on InP) quantum-well MODFETs," Ph.D. Thesis Cornell University.

Gourley, P. L., Fritz, I. J., and Dawson, L. R. (1988). *Appl. Phys. Lett.* **52**, 377.

Hendel, R. H., Pei, S. S., Kiehl, R. A., Tu, C. W., Feuer, M. D., and Dingle, R. (1984). "A 10-GHz frequency divider using selectively doped heterostructure transistors," *IEEE Electron Devices Lett.* **EDL-5**, 406–408.

Hirakawa, K., Sakaki, H., and Yoshino, J. (1984). "Concentration of electrons in selectively doped GaAlAs/GaAs heterojunction and its dependence on spacer-layer thickness and gate electric field," *Appl. Phys. Lett.* **45**, 253–255.

Hikosaka, K., Sasa, S., Harada, N., and Kuroda, S. (1988). "Current-gain cutoff frequency comparison of InGaAs HEMTs," *IEEE Electron Devices Lett.* **EDL-9**, 241–2438.

Hong, W-P., Derosa, F., Bhat, R., Allen, S. J., Hayes, J. R. and Bhattacharya, P. (1989). *Inst. Phys. Conf. Ser.* **96**, 237–242.

Huang, K. F., Tai, K., Jewell, J. L., Fischer, R. J., McCall, S. L., and Cho, A. Y. (1989). *Appl. Phys. Lett.* **54**, 2192.

Inata, T., Muto, S., Nakata, Y., Sasa, S., Fujii, T., and Hiyamiza, S. (1987). *Japan J. Appl. Phys.* **26**, L1332–L1334.

Ishibashi, T. and Yamauchi, Y. (1988). "A possible near-ballistic collection in an AlGaAs/GaAs HBT with a Modified Collector Structure," *IEEE Trans. Electron Devices* **35**, 401–404.

Jaffe, M., Oh, J. E., Pamulapati, J., Singh, J. and Bhattacharya, P. (1989). *Appl. Phys. Lett.* **54**, 2345–2346.

Jones, E. D., Lyo, S. K., Fritz, I. J., Klem, J. E., Schriber, J. E., Tigges, C. P., and Drummond, T. J. (1989a). *Appl. Phys. Lett.* **54**, 2227.

Jones, E. D., Lyo, S. K., Klem, J. F., Schriber, J. E., and Tigges, C. P. (1989b). "Simultaneous measurement of the conduction and valence-band masses in strained-layer structures," *Inst Phys. Conf. Ser.* **96**, 243–247.

Joyce, M. J., Gal, M., and Tann, J. (1989). "Observation of Interface Defects in Strained InGaAs-GaAs by Photoluminescence Spectroscopy," *J. Appl. Phys.* **65**, 1377.

Kavanagh, K. L., Capano, M. A., Hobbs, L. W., Barbour, J. C., Maree, P. M. J., Schaff, W. J., Mayer, J. W., Pettit, D., Woodall, J. M., Stroscio, J. A., and Feenstra, R. M. (1988). *J. Appl. Phys.* **64**, 4843.

Ketterson, A., Moloney, M., Masselink, W. T., Peng, C. K. Klem, J., Fisher, R., Kopp, W., and Morkoç, H. (1985). "High transconductance in InGaAs/AlGaAs pseudomorphic modulation-doped field-effect transistors," *IEEE Electron Devices Lett.* **EDL-6**, 628–630.

Kim, B., Matyi, R. J., Wurtele, M., Bradshaw, K., and Tserng, H. Q. (1988). "Millimeter-wave AlGaAs/InGaAs/GaAs quantum-well power MISFET," *Tech. Dig. IEDM*, 168–171.

Kolbas, R. M., Anderson, N. G., Laidig, W. D., Sin, Y., Lo, Y. C., Hsieh, K. Y., and Yang, Y. J. (1988). *IEEE J. Quantum Electron.* **QE-24**, 1605.

Kroemer, H. (1985). *Appl. Phys. Lett.* **46**, 504.
Kuang, J. B., Tasker, P. J., Chen, Y. K., Wang, G. W., Eastman, L. F., Aina, O. A., Hier, H., and Fathimulla, A. (1988a). "I/V anomaly and device performance of submicrometre-gate $Ga_{0.47}In_{0.53}As/Al_{0.48}In_{0.52}As$ HEMT," *Electron. Lett.* 1571–1572.
Kuang, J. B., Tasker, P. J., Wang, G. W., Chen, Y. K., Eastman, L. F., Aina, A. O., Hier, H., and Fathimulla, A. (1988b). "Kink effect in submicrometer-gate MBE-grown InAlAs/InGaAs/InAlAs heterojunction MESFETs," *IEEE Electron Devices Lett.* **EDL-9**, 630–632.
Kuo, C. P., Vong, S. K., Cohen, R. M., and Stringfellow, G. B. (1985). *J. Appl. Phys.* **57**, 5428.
Lee, C. P., Hou, D., Lee, S. J., Miller, D. L., and Anderson, R. J. (1983). "Ultra high-speed digital integrated circuits using GaAs/GaAlAs high electron mobility transistors," *GaAsIC Symp. Tech. Dig.* 162–165.
Mailhiot, C., and Smith, D. L. (1989). "Modulation of Internal Piezoelectric Fields in Strained-Layer Superlattices Grown Along the [111] Orientation," *J. Vac. Sci. Technol.* **A7**, 609–615.
Major, J. S., Guido, L. J., Hsieh, K. C., Holonyak, N., Stutius, W., Gavrilovic, P., and Williams, J. E. (1989). *Appl. Phys. Lett.* **54**, 913.
Mandeville, P. (1989). Unpublished.
Massies, J., and Sauvage-Simkin, M. (1983). *Appl. Phys. Lett.* **43**, 8465.
Matthews, J. W., and Blakeslee, A. E. (1974). *J. Cryst. Growth* **27**, 118.
Matthews, J. W., and Blakeslee, A. E. (1976). *J. Crystal Growth* **32**, 265–273.
Miller, D. A. B., Weiner, J. S., and Chemla, D. S. (1986). *IEEE J. Quantum Electron.* **QE-22**, 1816.
Mishra, U. K., Palmateer, S. C., Chao, P. C., Smith, P. M., and Hwang, J. C. M. (1985). "Microwave performance of a 0.25-μm gate-length high electron mobility transistors," *IEEE Electron Devices Lett.* **EDL-6**, 142–145.
Mishra, U. K., Brown, A. S., and Rosenbaum, S. E. (1988a). "DC and RF performance of 0.1 μm gate length $Al_{0.48}In_{0.52}As$–$Ga_{0.38}In_{0.62}As$ pseudomorphic HEMTs," *Tech. Dig. IEDM* 180–183.
Mishra, U. K., Brown, A. S., Rosenbaum, S. E., Hooper, C. E., Pierce, M. W., Delaney, M. J., Vaughn, S., and White, K. (1988b). "Microwave performance in AlInAs-GaInAs HEMTs with 0.2- and 0.1 μm gate-length," *IEEE Electron Devices Lett.* **EDL-9**, 647–649.
Mishra, U. K., Jensen, J. F., Brown, A. S., Thompson, M. A., Jelloian, L. M., and Beaubien, R. S. (1988c). "Ultra-high-speed digital Circuit performance in 0.2-μm gate-length alInAs/GaInAs HEMT technology," *IEEE Electron Devices Lett.* **EDL-9**, 482–484.
Moll, N., Hueschen, M. R., and Fisher-Colbrie, A. (1988). "Pulse-doped AlGaAs/InGaAs pseudomorphic MODFETs," *IEEE Trans. Electron Devices* **ED-35**, 879–886.
Moloney, M. J., Ponse, F., and Morkoc, H. (1985). "Gate Capacitance-Voltage Characterization of MODFET's: Its Effect on Transconductance," *IEEE Trans. Electron Devices* **ED-32**, 1675–1684.
Muneketa, H., Segmuller, A., and Chang, L. L. (1987). "Inhomogeneous Lattice Distortion in the Heteroepitaxy of InAs on GaAs," *Appl. Phys. Lett.* **51**, 587–589.
Ng, G. I., Weiss, M., Pavlidis, D., Tutt, T., Bhattacharya, P. and Chen, C. Y. (1989). *Inst. Phys. Conf. Ser.* **96**, 465–470.
Nguyen, L. D. (1989). "Realization of Ultra–High Speed Modulation-Doped Field Effect Transistors," Ph.D. Dissertation, Cornell University.
Nguyen, L. D., Foisy, M. C., Tasker, P. J., Schaff, W. J., Lepore, A. N., and Eastman, L. F. (1987a). "Carrier deconfinement limited velocity in pseudomorphic AlGaAs/InGaAs modulation-doped field-effect transistors (MODFETs)," in *Proc. IEEE/Cornell Conf. on Advanced Concepts in High Speed Semicond. Dev. & Ckts.*, 60–69.
Nguyen, L. D., Tasker, P. J., and Schaff, W. J. (1987b). "Comments on 'A new low-noise AlGaAs/GaAs 2DEG FET with a surface undoped layer'," *IEEE Trans. Electron Devices* **ED-34**, 1187–1187.
Nguyen, L. D., Radulescu, D. C., Tasker, P. J., Schaff, W. J., and Eastman, L. F. (1988). "0.2-μm

gate-length atomic-planar doped pseudomorphic $Al_{0.3}Ga_{0.7}As/In_{0.25}Ga_{0.75}As$ MODFETs with f_T over 120 GHz," *IEEE Electron Devices Lett.* **EDL-9**, 374–370.

Nguyen, L. D., Schaff, W. J., Tasker, P. J., Lepore, A. N., Palmateer, L. F., Foisy, M. C., and Eastman, L. F. (1988a). "Charge control, DC, and RF performance of a 0.35-μm pseudomorphic AlGaAs/InGaAs modulation-doped field-effect transistor," *IEEE Trans. Electron Devices* **ED-35**, 139–143.

Nguyen, L. D., Tasker, P. J., Radulescu, D. C., and Eastman, L. F. (1988b). "Design, fabrication and characterization of ultra high speed AlGaAs/InGaAs MODFETs," *Tech. Dig. IEDM*, 176–178.

Nguyen, L. D., Tasker, P. J., Radulescu, D. C., and Eastman, L. F. (1989). "Characterization of ultra-high speed pseudomorphic," Submitted to *IEEE Trans. Electron Devices*.

Nishii, K., Matsuno, T., Ishikawa, O., Yagita, H., and Inoue, K. (1988). "Noval high-performance N-AlGaAs/InGaAs/N-AlGaAs pseudomorphic double-heterojunction modulation-doped FETs," *Japan J. of Appl. Phys.* **27**, 2216–2218.

Nottenburg, R. N., Chen, Y. K., Panish, M. B., Humphrey, D. A., and Hamm, R. (1989). "Hot-electron InGaAs/InP heterostructure bipolar transistors with f_T of 110 GHz," *IEEE Electron Device Lett.* **10**, 30–32.

Offsey, S. D., Schaff, W. J., Tasker, P. J., Braddock, W. D., and Eastman, L. F. (1989). "Microwave performance of GaAs-AlGaAs and strained-layer InGaAs-GaAs-AlGaAs graded-index separate confinement single quantum well lasers," *12th Biennial IEEE/Cornell Conference on Advanced Concepts in High Speed Semiconductor Devices and Circuits*, Ithaca, New York.

Ohari, T., Takechi, M., Takikawa, M. and Komeno, J. (1989). *Inst. Phys. Conf. Ser.* **96**, 131–134.

Ohno, H., Luo, J. K., Matsuzaki, K., and Hasegawa, H. M. (1989). *Appl. Phys. Lett.* **54**, 36.

Okamoto, A., Toyoshima, H., and Ohata, K. (1987). "Strained N-$Ga_{0.7}Al_{0.3}As/In_xGa_{1-x}As/GaAs$ modulation-doped structures," *Japan J. Appl. Phys.* **26**, 539–542.

Olsen, G. H., Neuse, C. J., and Smith, R. T. (1978). *J. Appl. Phys.* **49**, 5523.

Palmateer, L. F., Tasker, P. J., Itoh, T., Brown, A. S., Wicks, G. W., and Eastman, L. F. (1987). "Microwave characterisation of 1 μm gate $Al_{0.48}In_{0.52}As/Ga_{0.47}In_{0.53}As/InP$ MODFETs," *Electron. Lett.* **23**, 53–55.

Pearsall, T. P. (1984). "Two-dimensional electronic systems for high-speed device applications," *Surface Sci.* **142**, 529–544.

Pearsall, T. P., Hendel, R., Connor, P. O., Alavi, K., and Chao, A. Y. (1983). "Selectively-doped $Al_{0.48}In_{0.52}As/Ga_{0.47}In_{0.53}As$ heterojunction field effect transistor," *IEEE Electron Devices Lett.* **EDL-4**, 5–8.

Ponse, F., Masselink, W. T., and Morkoc, H. (1985). "The quasi-Fermi level bending in the MODFETs and its effect on the FET transfer characteristics," *IEEE Trans. Electron Devices* **ED-32**, 1017–1023.

Price, G. L. (1988). *Appl. Phys. Lett.* **53**, 1288.

Radulescu, D. C. (1988) "Molecular-Beam Epitaxial Growth and Characterization of Aluminum Gallium Arsenide/Indium Gallium Arsenide Single Quantum-Well Modulation-Doped Field-Effect Transistor Structures," Ph.D. Thesis, Cornell University.

Radulescu, D. C., Schaff, W. J., Eastman, L. F., Ballingall, J. M., Ramseyer, G. O., and Hersee, S. D. (1989). "Influence of substrate temperature and InAs mole fraction on the incorporation of indium during molecular beam epitaxial growth of InGaAs single quantum-wells on GaAs," *J. Vac. Sci. Technol.* B **7**, 111–115.

Ralston, J. D., Schaff, W. J., Bour, D. P., and Eastman, L. F. (1989). "Room-Temperature Exciton Electroabsorption in Partially Intermixed GaAs/AlGaAs Quantum Well Waveguides," *Appl. Phys. Lett.* **54**, 534–536.

Rogers, D. L., Woodall, J. M., Pettit, G. D., and McInturff, D. (1989). "High-Speed 1.3 μm GaInAs Detectors Fabricated on GaAs Substrates," *IEEE Electron Dev. Lett.* **9**, 515–517.

Rosenberg, J. J., Benlamri, M., Kirchner, P. D., Woodall, J. M., and Pettit, G. D. (1985). "An $In_{0.15}Ga_{0.85}As/GaAs$ pseudomorphic single quantum well HEMT," *IEEE Electron Devices Lett.* **EDL-6**, 491–493.

Saunier, P., Matyi, R. J., and Bradshaw, K. (1988). "A doubled-heterojunction doped-channel pseudomorphic power HEMT with a power density of 0.85 W/mm at 55 GHz," *IEEE Electron Devices Lett.* **EDL-9**, in *Proc IEEE/Cornell Conference on Advanced Concepts in High Speed Semiconductor Devices and Circuits*, 189–198.

Schaff, W. J., and Eastman, L. F. (1984). "Superlattice Buffers for GaAs Power MESFETs Grown by MBE," *J. Vac. Sci. Technol.* **B2**, 265–268.

Schaff, W. J., Kavanagh, K. L., Batson, P. E., Kirchner, P., Woodall, J. M., and Mayer, J. W. (1986). "Capacitance Characterization of GaInAs Grown on GaAs by MBE," *Electronic Materials Conference.* Univ. of Massachusetts, Amherst, Massachusetts.

Schaff, W. J., and Eastman, L. F. (1987). "MBE Growth and Materials and Device Characterization of Strained Layer GaInAs/GaAs for Application to MODFETs," *E-MRS Meeting* **XVI**, (Les Editions de Physique, Paris), 295–301.

Shih, H. D., Kim, B., Bradshaw, K., Tserng, H. Q., Wurtele, M., Duncan, W. M., and Moore, T. M. (1989). "Millimeter-Wave $In_{0.17}Ga_{0.83}As$ Power MESFETs on GaAs (100) Substrates," *Microwave and Optical Tech. Letts.* **2**, 153–155.

Smith, F. W., Calawa, A. R., Chen, C. L., Manfra, M. J., and Mahoney, L. J. (1988). *Electron Dev. Lett.* **9**, 77.

Smith, P. M., Chao, P. C., Mishra, U. K., Palmateer, S. C., Duh, K. H. G., and Hwang, J. C. M. (1985). "Millimeter wave power performance of 0.25 μm HEMTs and GaAs FETs," *Proc. IEEE/Cornell Conference on Advanced Concepts in High Speed Semiconductor Devices and Circuits*, 189–198.

Smith, P. M., Chao, P. C., Lester, L. F., Smith, R. P., Lee, B. R., Ferguson, D. W., Jabra, A. A., Ballingall, J. M., and Duh, K. H. G. (1988). "InGaAs pseudomorphic HEMTs for millimeter wave power applications," *Tech. Dig. MTT Symposium*, 927–930.

Strid, E. W., and Gleason, K. R. (1984). "Calibration methods for microwave wafer probing," *Tech. Dig. Monolithic Circuit Symposium*, 78–82.

Stutius, W., Gavrilovic, P., Williams, J. E., Meehan, K., and Zarrabi, J. H. (1988). *Electron. Lett.* **24**, 1493.

Suemune, I., Coldren, L. A., Yamanishi, M., and Kan, Y. (1978). *Appl. Phys. Lett.* **53**, 1378.

Tasker, P. J., and Eastman, L. F. (1987a). "The correct determination of intrinsic F_T of FET devices from measured microwave data," ††th *Workshop on Compound Semicond. Dev. and Integrated Ckts.*, Grainau, West Germany.

Tasker, P. J., and Eastman, L. F. (1987b). "Determination of intrinsic f_T of FET devices from measured microwave data," *WOCSEMMAD*, Hilton Head Island, South Carolina.

Tasker, P. J., and Hughes, B. (1989). "Bias dependence of the intrinsic model element values at microwave frequencies," *IEEE Trans. Electron Devices* **ED-36**, 2267–2273.

Tsao, J. Y., and Dodson, B. W. (1988). *Appl. Phys. Lett.* **53**, 848–850.

Turner, J., and Wilson, B. (1969). "Implications of Carrier Velocity Saturation in a Gallium Arsenide Field-Effect Transistor," *Proc. 1968 Symp. GaAs*, Inst. Phys. Conf. Series 7, 195–204.

Upton, M. A. G., Smith, P. M., and Chao, P. C. (1987). "HEMT low noise amplifier for Ka-band," *Tech. Dig. MTT Symposium*, 1007–1010.

Vawter, G. A., Myers, D. R., Brennan, T. M., Hammons, B. E., and Hohimer, J. P. (1990). *IEEE Photon. Tech. Lett.* **1**, 153.

Vinter, B. (1984). "Subbands and charge control in a two-dimensional electron gas field-effect transistor," *Appl. Phys. Lett.* **44**, 307–309.

Wang, G. W., Chen, Y. K., Eastman, L. F., and Whitehead, B. (1988). "Reduction of gate resistance in tenth-micron gate MODFETs for microwave applications," *Solid-State Electron.* **31**, 1247–1250.

Watanabe, Y., Kakjii, K., Asada, Y., Odani, K., Mimura, T., and Abe, M. (1987). "A high-speed HEMT 1.5 K gate array," *IEEE Trans. Electron Devices* **ED-34**, 1253–1258.

Whaley, G. J., and Cohen, P. I. (1988). *J. Vac. Sci. Tech. B* **6**, 625.

Wie, C. R. (1989). *J. Appl. Phys.* **65**, 2267.

Williams, R. E., and Shaw, D. W. (1978). "Graded channel FETs: improved linearity and noise figure," *IEEE Trans. Electron Devices* **ED-25**, 600.

Wood, C. E. C., Singer, K., Ohashi, T., Dawson, L. R., and Noreika, A. J. (1983). *J. Appl. Phys.* **54**, 2732.

Woodall, J. M., Pettit, G. D., Jackson, T. N., Lanza, C., Kavanaugh, K. L., and Meyer, J. W. (1983). *Phys. Rev. Lett.* **51**, 1783.

Yablonovitch, E., and Kane, E. O. (1986). *J. Lightwave Technol.* **LT-4**, 961.

Yau, E. T., Watkins, S. K., Wang, K., and Klatskin, B. (1987). "A four stage V-band MOCVD HEMT amplifier," *Tech. Dig. MTT Symposium June* ††††, 1015–1018.

York, P. K., Beernink, K. J., Fernandez, G. E., and Coleman, J. J. (1989). "InGaAs/GaAs Strained Layer Quantum Well Heterostructure Lasers and Laser Arrays by Metalogranic Chemical Vapor Phase Deposition," *Conf. on Lasers and Electro-Optics*, Baltimore, Maryland.

Zipperian, T. E. Private communication.

Zipperian, T. E., Dawson, L. R., Osbourn, C. E., and Fritz, I. J. (1983). "An $In_{0.2}Ga_{0.8}As$/GaAs, Modulation-doped, strained-layer surerlattice field-effect transistor," *Tech. Dig. IEDM*, 696–699.

Zipperian, T. E., Drummond, T. J., Fritz, I. J. (1987). "Operation of a *p*-channel GaAs/(In,Ga)As, strained quantum well field-effect transistor at 4 K," *Proc. IEEE Cornell Conference on Advanced Concepts in High Speed Devices and Circuits*, 80–88.

Zipperian, T. E., Jones, E. D., Dodson, B. W., Klem, J. F., Gourley, P. F., and Plut, T. A. (1989). "A study of the thermal stability of (In,Ga)As strained quantum well structures as a function of overlayer thickness," *Inst. Phys. Conf.* **96**, 365–370.

CHAPTER 3

Structure and Characterization of Strained-Layer Superlattices

S. T. Picraux, B. L. Doyle, and J. Y. Tsao

SANDIA NATIONAL LABORATORIES
ALBUQUERQUE, NEW MEXICO

I.	Introduction	139
II.	Structure	143
	A. Perfect Coherence	143
	B. Dislocations and Imperfect Coherence	146
III.	Ion-Scattering Characterization	147
	A. RBS: Layer Thickness and Composition	148
	B. Channeling: Layer Strain and Perfection	155
IV.	X-ray Diffraction Characterization	170
	A. Structure of the Reciprocal Lattice	171
	B. Mapping of the Reciprocal Lattice	178
V.	Other Characterization Techniques	189
	A. Lattice-Structure Methods (RHEED, Raman Scattering)	189
	B. Defect-Structure Methods (TEM, XRT, EBIC, PLT)	194
VI.	Stability, Metastability, and Relaxation	198
	A. Driving Force: Excess Stress	199
	B. Materials Response: Stability Diagrams	202
VII.	Application to Single Strained Layers	205
	A. The Coherent–Partially Relaxed Boundary	205
	B. Dislocation Nucleation and Propagation	208
	C. The Stable–Metastable Boundary	210
VIII.	Application to Superlattices	210
	A. Coherent Structures and Buffer Layers	212
	B. Nonperiodic Strain Profiles	214
	C. Imperfect Structures	216
IX.	Conclusions	219
	References	220

I. Introduction

Strained-layer superlattices and related strained-layer structures made from semiconducting materials have unique electronic and optoelectronic properties (Osbourn *et al.*, 1983). There is widespread interest in these structures, both for their solid-state properties and for their potential applications. A key element to the success of this new materials area is the ability to grow layered structures with slightly different lattice constants in a fully commensurate fashion. Under that condition, the lattice mismatch is taken up by biaxial strain in the plane of the layers. This feature is essential for semiconductor devices for two reasons. First, certain properties, such as the preferential population of the light-hole states due to valence-band splitting (see Fritz *et al.*, 1986), derive their effect directly from the presence of strain. Second, the alternative to a commensurate structure is one that is strain relieved by the introduction of defects (e.g., misfit dislocations), and these defects strongly degrade the electrical and optical properties of these structures.

For these reasons, quantitative understanding and characterization of the structure of strained-layer material is crucial. Such understanding must include the strains present, the strain-relief mechanisms, and the limits of strained-layer growth. One purpose of this chapter is to provide an understanding of techniques used to characterize strain and strain relief in these structures. Two techniques, Rutherford backscattering spectrometry (RBS)/ion channeling and x-ray diffraction, are discussed in detail, whereas numerous other techniques are discussed in more general terms and compared by way of example. In addition, a more detailed treatment of the transmission electron microscopy (TEM) technique for characterizing strained layers can be found in Chapter 1 by Hull and Bean. A second purpose of this chapter is to provide a framework for understanding stability, metastability, and relaxation of strained-layer structures. We do this in terms of the concepts of the driving force, as characterized by the excess stress, and the materials response, as characterized by stability diagrams. However, a detailed discussion of the dynamics of strain relaxation is not included in the present work. The characterization techniques and the stability/metastability/relaxation concepts are illustrated for both single strained layers and strained-layer superlattices.

The range of lattice constants for the common group IV, III–V, and II–VI semiconductors is shown in Fig. 1. Also shown is the percentage lattice mismatch relative to the substrates Si, GaAs, and CdTe. Typically, the layer mismatches used for unrelaxed strained-layer structures are in the range of 0.1 to 2%. When one considers that ternary and quarternary alloys with intermediate lattice constants can usually be grown within each group, it is apparent that an enormous range of material combinations are available for commensurate strained-layer structures.

The basic concept of the formation of a single strained layer is illustrated in Fig. 2. Under commensurate growth, the layer is biaxially strained, and the atom relaxation normal to the plane of the layer leads to a tetragonal distortion of the strained layer. As the layer thickness increases, the strain

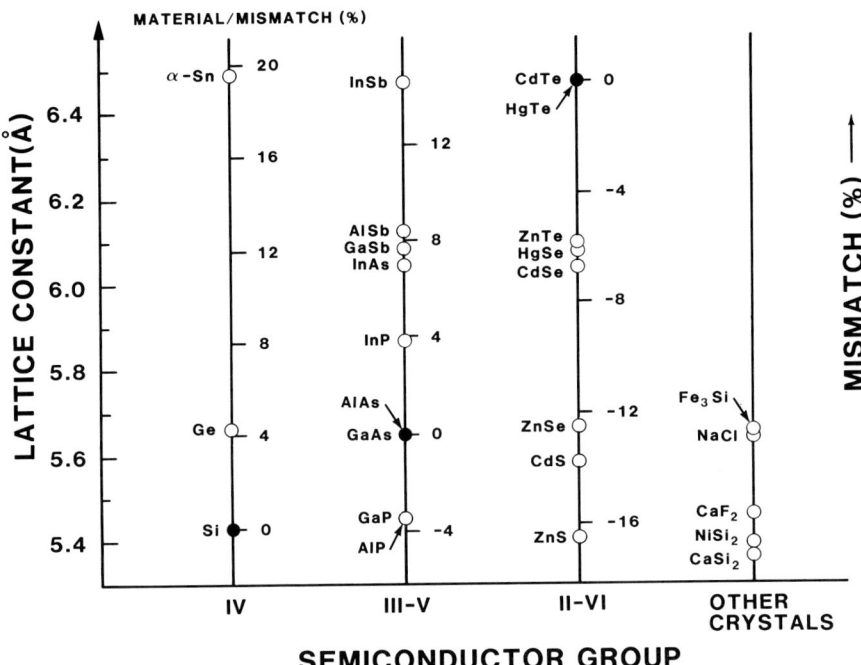

FIG. 1. Lattice constants and mismatch for the main group IV, III–V, and II–VI semiconductor materials. In addition, several insulator and silicide structures are shown that also exhibit epitaxial relationships with the (001) zinc-blende structures.

energy increases and the strain may become partially relieved by the introduction of misfit dislocations.

For a fully commensurate structure, the magnitude of the strain depends directly on the composition of the layer and the substrate. For a partly relieved structure, the remaining strain in the layer is determined by the areal density of strain-relieving defects. In the latter case, one might determine the residual strain in a partly relieved structure either directly, by measurement of the strain in the layers, or *indirectly*, by measurement of the composition and the density of misfit dislocations of known type. Thus, for perfect or imperfect strained layers, *we may view the characterization of the composition, strain, and defect densities as interrelated, complementary measurements.*

The structure of a commensurate strained-layer superlattice is shown schematically in Fig. 3. The single strained layer may be considered the essential building block of the strained-layer superlattice. Thus, a thorough understanding of the single-layer case is essential to understanding superlattices. In the ideal case, the superlattice is grown on a substrate of

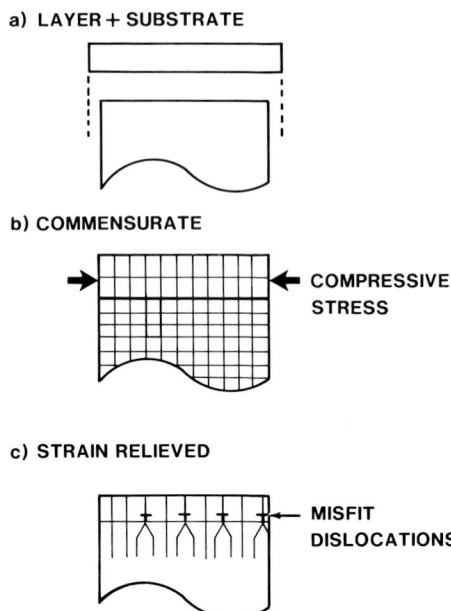

FIG. 2. Schematic diagram of a layer with slightly larger lattice constant than substrate for (a) isolated layer and substrate in equilibrium, (b) commensurately grown strained-layer on substrate, and (c) strain-relieved layer on substrate.

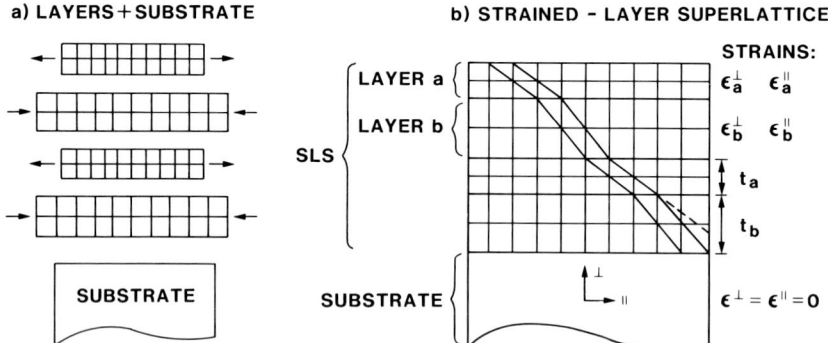

FIG. 3. Schematic diagram of (a) isolated (unstrained) layers and substrate in equilibrium, and (b) commensurate superlattice with alternating layers in compressive and tensile stress.

intermediate lattice constant, so that the layers are in alternating compressive the tensile strain, and the forces acting on all the layers just balance. However, in the non-ideal case the superlattices may not be lattice matched to the substrate so that a net force in the strained-layer structure remains.

In what follows, we first discuss the structure of strained and strain-relieved layers. We then discuss techniques that are useful for characterizing the structure of strained layers. Primary emphasis is given to ion-scattering and x-ray diffraction methods, in part because these are broadly applicable techniques that require no special specimen preparation, and in part because the authors have a greater degree of expertise in these areas. We then briefly discuss and give examples of other techniques for measuring strain (reflection high-energy electron diffraction (RHEED), Raman scattering), for imaging strain-relieving defects (TEM, electron-beam-induced current (EBIC), etc.), and for detecting low levels of disorder (photoluminescence (PL), etc.). An overview of the various characterization techniques and the key aspects of their structural sensitivities is given in Table I. The characterization sections are followed by a treatment of the stability, metastability, and relaxation of these structures. In that section, the concept of excess stress is used to provide insight into the driving force for strain relaxation, and stability diagrams are used to systematize, in a natural way, the materials response to this driving force under various conditions. The chapter concludes with sections giving examples of features encountered in the application of these concepts and techniques to single strained layers and superlattices.

II. Structure

A. PERFECT COHERENCE

If the interface between two layers is coherent, the lateral positions of the atoms do not change across the interface. Then, as illustrated in Fig. 4, the two layers must have the same in-plane lattice constant, and we refer to these as commensurate epitaxial layers. Furthermore, if, as is almost always the case, the substrate on which the heterostructure is built is much wider than it is thick, then the in-plane lattice constant may be considered a characteristic of the entire substrate–heterostructure combination. In other words, the in-plane lattice constant may be considered depth independent.

For the simple case of a single epitaxial layer having the same elastic constants as its substrate, that in-plane constant is

$$a^{\parallel} = \frac{a^0_{\text{epi}} h_{\text{epi}} + a^0_{\text{sub}} h_{\text{sub}}}{h_{\text{epi}} + h_{\text{sub}}}, \qquad (1)$$

TABLE I

SELECTED CHARACTERIZATION TECHNIQUES FOR THE ANALYSIS OF STRAINED-LAYER STRUCTURES AND THEIR APPROXIMATE SENSITIVITIES FOR THIN LAYERS, STRAIN, AND STRAIN-RELAXING DEFECTS

Technique	Depth sensitivity (nm)	Probing depth (nm)	Strain sensitivity (%)	Defect detection[aa] (dislocations/cm)	Lateral resolution (μm)	Comments
RBS/channeling	5[a]	1,000	0.1[b]	> 10^5	10^3	Depth-resolved measurement
X-ray diffraction	0.1[c]	> 10,000	0.01	> 10^4	10^3	Integrates over sampling depth
TEM	≤ 1[d]	100	0.5	> 10^3	10^{-3}	Images defect structure; sample preparation can modify structure
RHEED	0.1	1	0.1	—	10^2	In situ convenience
Raman	10–100	10–100	0.2	> 10^4	1	Nondestructive, no sample preparation
EBIC	~ 100[e,f]	1,000	—	< 10^4	1	Image electrically active defects
X-ray topography	~ 100[e]	> 10,000	—	< 10^4	1	Image defects with strain
Photoluminescence topography	~ 1,000	> 1,000	—	< 10^4	1	Image recombination-enhancing defects
Catholuminescence topography	~ 1,000	≥ 1,000	—	< 10^4	1	Combined structural and optical information in SEM
Etch-contrast	≥ 1,000	≥ 1,000	—	< 10^4	1	Careful wet chemistry required for defect specificity

[aa] Here we consider the total length of misfit dislocations (lying in a plane parallel to the surface) per unit area in the plane of the layer.

[a] Medium-energy ion scattering (e.g., 100 keV protons) with electrostatic energy analysis can extend depth sensitivities down to one to five monolayers near the surface.

[b] When the planar resonance channeling technique called catastrophic dechanneling can be used (thicker layers and small strains), strain sensitivities of 0.02% can be achieved.

[c] For superlattices, 0.1 nm resolution is achieved for highly perfect uniform structures, whereas for single layers, thickness is not determined and the minimum thickness for sensitive analysis is ~ 10 nm.

[d] For cross-section analysis.

[e] Determined by layer configurations.

[f] Determined by bias potentials.

3. STRUCTURE AND CHARACTERIZATION OF SUPERLATTICES

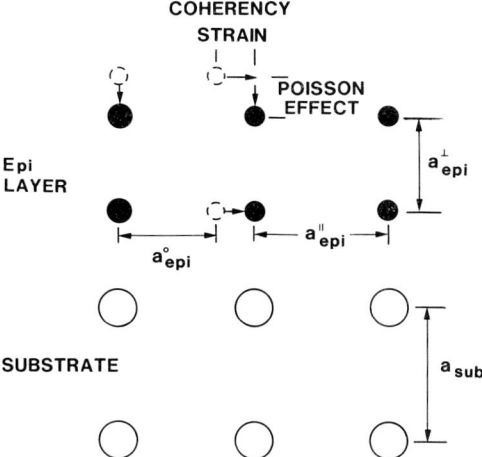

FIG. 4. Atom positions for a strained epitaxial layer on a thick substrate illustrating the coherency strain and Poisson effect for commensurately grown layer.

where a_{epi}^0 and a_{sub}^0 are the unstrained lattice constants, and h_{epi} and h_{sub} are the unstrained thicknesses of the epitaxial layer and substrate, respectively. Since the substrate is generally vastly thicker than the heterostructure grown on it, $a^\parallel$ is nearly always virtually the same as a_{sub}^0. The in-plane strain of the epitaxial layer is then referred to as the coherency strain (see Fig. 4).

If the in-plane lattice constant of the heterostructure is a constant, forced to be equal to that of its substrate, then the unit cell will distort "tetragonally" in order to (very approximately) preserve its volume per unit cell. For example, if the unit cell of the epitaxial layer is smaller than that of the substrate, it must be "stretched" in the in-plane direction, and its height will decrease (the Poisson effect as illustrated in Fig. 4).

In the usual notation of elasticity theory, the lattice strain is given by

$$e_x = \frac{1}{E}[\sigma_x - v(\sigma_y + \sigma_z)], \tag{2a}$$

$$e_y = \frac{1}{E}[\sigma_y - v(\sigma_x + \sigma_z)], \tag{2b}$$

$$e_z = \frac{1}{E}[\sigma_z - v(\sigma_x + \sigma_y)], \tag{2c}$$

where E is the modulus of elasticity and v is Poisson's ratio. If the layer is free to expand vertically, then $\sigma_z = 0$, and if the crystal is initially cubic, then

$e^{\|} = e_x = e_y$ and $\sigma^{\|} = \sigma_x = \sigma_y$. Either of the first two equations then determines the in-plane stress ($\sigma^{\|}$) in terms of the in-plane strain ($e^{\|}$):

$$\sigma^{\|} = \frac{Ee}{1-v} = 2\mu\left[\frac{1+v}{1-v}\right]e^{\|}, \qquad (3)$$

where we have introduced the shear modulus $\mu = \frac{E}{2(1+v)}$. The third equation then determines the perpendicular strain $e^{\perp} = e_z = \Delta_a^{\perp}/a^{\perp}$ in terms of the parallel strain

$$e^{\perp} = \frac{-2v}{1-v}e^{\|}. \qquad (4)$$

Note that v, Poisson's ratio, is approximately 0.25–0.30 for most materials, and that deviations of $2v/(1-v)$ from 2 represent deviations from incompressibility.

It is also common to write the stress–strain relations in a cubic crystal as

$$\begin{bmatrix}\sigma_x\\ \sigma_y\\ \sigma_z\end{bmatrix} = \begin{bmatrix}c_{11} & c_{12} & c_{12}\\ c_{12} & c_{11} & c_{12}\\ c_{12} & c_{12} & c_{11}\end{bmatrix}\begin{bmatrix}e_x\\ e_y\\ e_z\end{bmatrix}, \qquad (5)$$

where, by symmetry, all the off-diagonal terms are equal to c_{12}, and all the diagonal terms are equal to c_{11}. Setting $\sigma_z = 0$, $e^{\|} = e_x = e_y$, and $e^{\perp} = e_z$ in Eq. (5) then gives

$$e^{\perp} = \frac{-2c_{12}}{c_{11}}e^{\|}. \qquad (6)$$

Comparison with Eq. (4) implies the correspondence $v = c_{12}/(c_{11} + c_{12})$. Also, using Eq. (5) and comparison with Eq. (4) give the parallel stress

$$\sigma^{\|} = \left(c_{11} + c_{12} - \frac{2c_{12}^2}{c_{11}}\right)e^{\|}. \qquad (7)$$

Finally, comparison with Eq. (3) implies the correspondence

$$\mu = \frac{c_{11}^2 + c_{11}c_{12} - 2c_{12}^2}{2(c_{11} + 2c_{12})} \qquad (8)$$

B. Dislocations and Imperfect Coherence

If the interface between two layers is not fully coherent, then the two layers need not have the same in-plane lattice constant. In the limit where the deviation from perfection is small (analogous to low-angle grain boundaries in metallurgy), the interface may be considered *essentially* perfect but (periodically) interrupted by "misfit" dislocations lying in the plane of the interface. Loosely speaking, misfit dislocations are those dislocations that act to relieve strain, so that they must have some edge character, and their Burgers vectors must have some component in the plane of the interface.

In (001) epitaxy of diamond-cubic semiconductors, the most common type of misfit dislocation is the so-called 60° type, whose direction and Burgers vectors are both along inclined $\langle 110 \rangle$ directions, 60° from each other. At very highly strained ($> 2\%$) interfaces, however, pure edge dislocations tend to form. These misfit dislocations are illustrated in Fig. 5, the greatest reduction in strain energy being achieved by locating the dislocations at the layer–substrate interface. In compound semiconductors, such as GaAs, there are additional distinctions between types of dislocations, as discussed, e.g., by George and Rabier (1987).

In the extreme case of perfect "incoherence," misfit dislocations are so numerous as to relieve all the strain in the epitaxial layer. This implies a misfit dislocation spacing of $l_{\text{disl}} = b^{\|}/f$, where $b^{\|}$ is the magnitude of the component of the Burgers vector parallel to the plane of the interface, and f is the lattice mismatch. Note that although it is interesting to consider a case of perfect "incoherence," as discussed in Section VI, such a case can be difficult to achieve in practice, due to the finite kinetics of structural relaxation.

III. Ion-Scattering Characterization

Ion-beam analysis techniques are particularly useful for characterizing the composition and structure of layered materials. Typically, a monoenergetic beam of ions, such as 1 MeV He$^+$, is directed onto the sample, and the

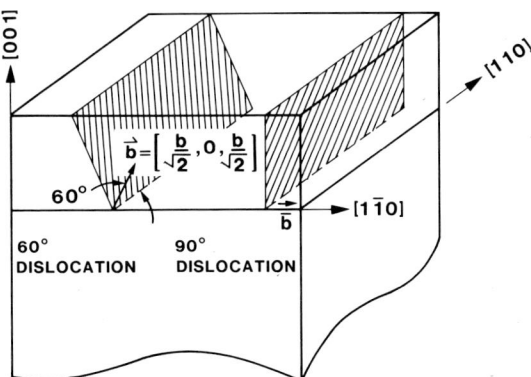

FIG. 5. Schematic diagram of the geometry of 60° and 90° dislocations in a partly strain-relieved epitaxial layer on a thick substrate.

scattered ions, recoiled atoms, ion-induced x-rays, or ion-induced nuclear reaction products are detected. The most useful of these techniques for the study of layered structures is RBS, which is based on the energy analysis of elastically scattered particles (Chu et al., 1978). The present discussion will be restricted to this case.

RBS analysis allows determination of layer composition, uniformity, and thickness and can be applied readily to superlattices (Mayer et al., 1973; Saris et al., 1980; Picraux et al., 1983a). When the incident ion beam is directed along a high-symmetry crystal direction, ion channeling occurs (Feldman et al., 1982). The use of the channeling effect in conjunction with RBS provides a measure of the crystalline quality as a function of depth and also often allows determination of the strain in epitaxial layered structures. In the remainder of this section, we shall first discuss the application of RBS for the determination of layer thickness and composition in strained-layer superlattices. We then consider the use of RBS in conjunction with ion channeling for the assessment of strain and crystal quality in these structures.

A. RBS: Layer Thickness and Composition

RBS analysis of the composition versus depth of layered structures (Chu et al., 1978) is based on the principle that monoenergetic ions incident on a target will lose their energy by two basic mechanisms: nuclear and electronic energy loss. The energy distribution of ions scattered by large angles contains the mass- and depth-scale information. When an energetic ion undergoes large-angle Rutherford scattering it will retain a fraction, k, of its energy, E, which increases with the mass of the atom struck. Thus, for scattering from the surface, the mass of the atoms at the surface directly determines the energy of the backscattered ions. In addition to this nuclear energy-loss mechanism, the ions lose energy by inelastic excitation of electrons as they penetrate the target. Therefore, for scattering from a given mass atom, the energy distribution of the backscattered ions will extend to lower energies due to scattering from deeper depths. The depth scale is constructed based on tabulated energy-loss rates for a given ion (E, Z_1) and target (Z_2), where Z is the atomic number.

With more than one element, as in layered structures, the resulting scattering is a superposition of the various elements present. The simplest superlattice case of layers "a" consisting only of element A and layers "b" with only element B is shown in Fig. 6. This figure illustrates computer simulation calculations with the code RUMP (Doolittle, 1986) in order to decompose the individual scattering contributions of the two elements involved for the case of scattering from alternating layers of pure Si and pure Ge. The resulting composite spectrum corresponds closely to that observed experimentally (Picraux et al., 1987). The scattering at the surface and interfaces is

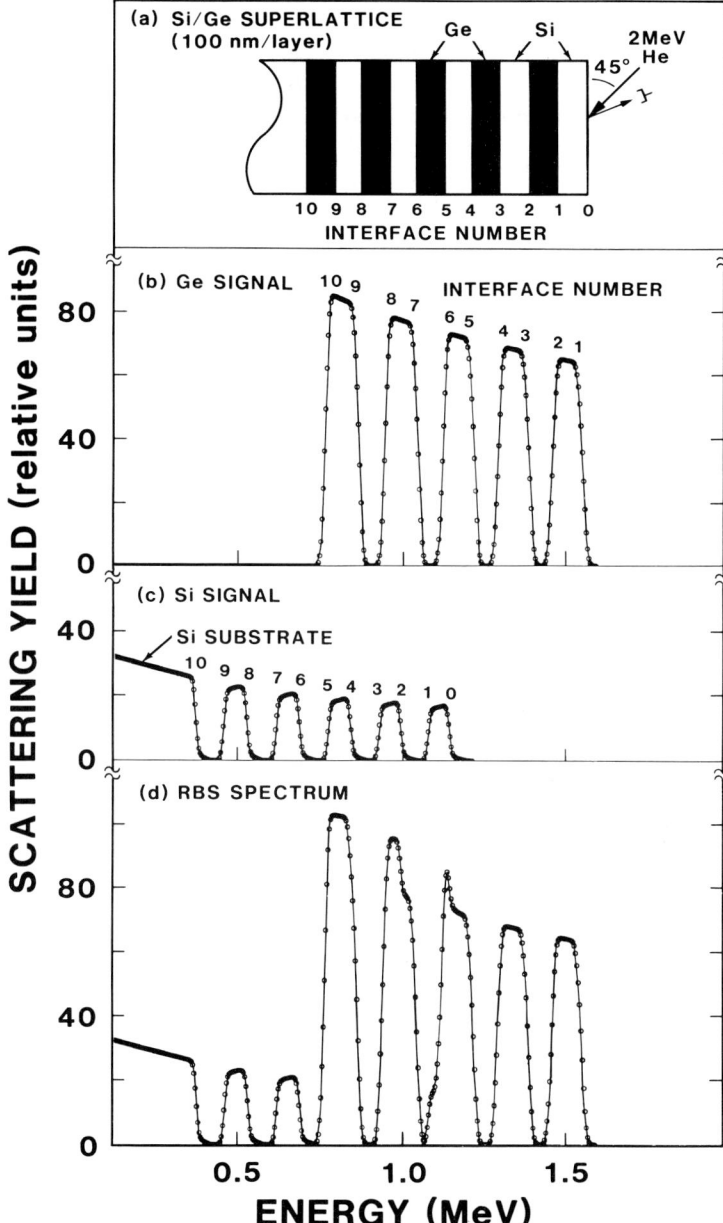

FIG. 6. Computer-simulated RBS spectrum for a Si/Ge 5-period superlattice (100 nm/layer) on a Si substrate and 2-MeV He incident at 45° relative to the surface: (a) superlattice configuration; (b) Ge signal; (c) Si signal; and (d) composite RBS spectrum.

easily identified, and the energy width for scattering from a given layer is referred to as ΔE. The thickness of the layers is determined to a good approximation (Picraux, 1974; Chu et al., 1978) by

$$\Delta x = \Delta E / n \sec \theta_1 [S]_A^{AB}, \tag{9}$$

where the energy loss factor, $[S]_A^{AB}$, in units of energy loss per unit areal thickness (number of target atoms per unit area) for the incoming ion at energy E_1, scattered by A atoms in target AB to an outgoing energy E_2 is given by

$$[S]_A^{AB} = k_A \varepsilon_{AB}(E_1, x) + c \varepsilon_{AB}(E_2, x). \tag{10}$$

Here k is the fractional energy retained upon scattering ($E_2 = kE_1$) and is

$$k = [\{M_1 \cos \theta + (M_2^2 - M_1^2 \sin^2 \theta)^{1/2}\}/(M_1 + M_2)]^2, \tag{11}$$

where n is the atomic density, θ is the scattering angle in laboratory coordinates, $c = \sec \theta_2 / \sec \theta_1$, θ_1 and θ_2 correspond to the angle of the beam on ingoing and outgoing paths relative to the surface normal, and M_1 and M_2 refer to the masses of the incident ion and target atom, respectively. The energy loss rate, ε, can be found in tables and depends on the projectile energy and atomic number as well as on the target composition (Chu et al., 1978). Usually, additivity of elemental energy-loss rates for pure elemental targets is assumed, so that a material AB with fractional atomic compositions N_A and N_B will have an energy loss rate given by $\varepsilon_{AB} = N_A \varepsilon_A + N_B \varepsilon_B$, where ε_A and ε_B are the individual elemental energy-loss rates.

In a standard RBS system, a surface barrier detector is used to detect backscattered He ions with an energy resolution of about 12 keV and a corresponding depth resolution full width at half maximum (FWHM) of the order of 20 nm. Since superlattices are often of this thickness or thinner, depth resolution by standard RBS is limited. Higher-precision thickness determinations can be achieved by using magnetic or electrostatic energy-loss spectrometers to analyze the scattered particles, or (for superlattices) by x-ray double-crystal diffraction (DCD) as discussed in the next section. RBS analysis has the particular advantage over x-rays of separately resolving the signals from individual layers. This is particularly useful for nonperiodic structures having variations in the thickness of layers where RBS analysis can be used to determine each layer's thickness separately. For flat surfaces, as is typical of strained-layer superlattices, the depth resolution can be increased to values of $\lesssim 5$ nm by tilting the target to glancing angles and thereby increasing the ion path length per layer. For typical RBS arrangements with the scattering angle $\theta = 160°$, tilting the target relative to the scattering plane to about $70°$ from normal incidence provides close to optimum resolution. For normal incidence and typical analyses with 2-MeV He beams, good depth resolution is retained to target depths of several hundred nanometers before energy straggling becomes significant; total probing depths at more moderate depth resolutions extend to several micrometers.

The composition of the superlattice is determined from the relative yields (or "counts") for scattering from different elements. The yield per unit energy width, Y, for scattering from A atoms in a uniform alloy AB is

$$Y_A^{AB} = N_A(x)\sigma(E_1, \theta)d\Omega\phi/[S]_A^{AB}, \tag{12}$$

where N_A is the atomic fraction of element A, $d\Omega$ the solid angle of the detector, ϕ is the total number of incident ions, and the cross section for scattering from element A is given by

$$\sigma^{RBS} \simeq \frac{d\sigma}{d\Omega} = (Z_1 Z_2 e^2/4E_1)^2 (4/\sin^4\theta) \frac{\{[1-(M_1/M_2)\sin\theta)^2]^{1/2} + \cos\theta\}^2}{[1-((M_1/M_2)\sin\theta)^2]^{1/2}} \tag{13}$$

in the laboratory frame of reference for Rutherford scattering conditions. A convenient form for evaluating layer compositions is to reference the yield for scattering from A atoms in alloy layer AB to that found for scattering from a layer of pure A,

$$\frac{Y_A^{AB}}{Y_A^A} = N_A [S]_A^A / [S]_A^{AB}. \tag{14}$$

Alternatively, the ratio of the scattering from elements A and B in the alloy layer AB is given by

$$\frac{Y_A^{AB}}{Y_B^{AB}} = \frac{N_A \sigma_A}{N_B \sigma_B} \frac{[S]_B^{AB}}{[S]_A^{AB}}, \tag{15}$$

where again N_A and N_B are the fractional atomic compositions in the alloy layer. More complex cases involving ternary or quarternary layers are straightforward to derive based on Eq. (12), although identification of the components of an actual spectrum at a given energy are not always simple, as will be illustrated in the following.

It can be seen from the previous equations that the sensitivity for scattering varies as the square of the atomic number of the elements in the layers, so that RBS analysis is quite sensitive to the high Z elements of a superlattice. Correspondingly, it is less sensitive to low Z elements, for example for the case of P in GaP/GaAsP or Si in SiGe/Ge superlattices. The net yield for a given element within a layer is then determined (Eq. (12)) by the atomic fraction of that element in the layer, by the cross section for scattering by that element, and by the reciprocal of the energy-loss factor (which depends on the energy-loss rate of *all* the elements within the layer).

An example of an RBS spectrum for an $In_xGa_{1-x}As/GaAs$ superlattice with relatively thick layers of 31 nm each is shown in Fig. 7. The components of the scattering from Ga, As, and In, as well as the composite spectrum, are illustrated in the upper panels by computer simulation; experimental results for the same system are given in the lowest panel. The In fraction ($x = 0.19$) in alternating layers gives rise to the major oscillations observed and is still resolvable after 16 layers (a depth of ~ 500 nm). Less apparent in the spectrum is the reduction in the Ga signal in the $In_xGa_{1-x}As$ layers due to (i) the reduced amount of Ga and (ii) the small increase in the energy-loss factor(s) (note Eq. (12)) due to the presence of the higher atomic number In. The contribution of the energy-loss factor is reflected directly by the As signal, which is slightly reduced in the $In_xGa_{1-x}As$ layers due to the increased energy-loss rate, even though the As fraction remains at 50% throughout all the layers. The relatively low In fraction of $x = 0.19$ gives rise to appreciable oscillations in the signal height, since the In-scattering cross section is $2\frac{1}{2}$ times larger than that for Ga. The layers are therefore clearly distinguished, and In fractions of $\sim 5\%$ or less are easily resolved for such a superlattice combination. In contrast, small variations in lower Z components relative to the other elements present, such as for As in $InAs_xSb_{1-x}/InSb$ superlattices, are more difficult to resolve. For example, to determine the As fraction in such superlattices in detailed RBS studies, we have found that for reasonably high counting statistics and detailed computer-simulation analysis, As fractions at the $x = 8 \pm 2\%$ level could be just resolved from the reduction in the In and Sb yields in those layers containing the As (Picraux and Lee, 1988).

Scattering interferences between two or more elements in the different layers of a superlattice can make determination of the interface positions difficult under certain conditions. For example, in the schematic of Fig. 6 corresponding to a superlattice of pure layers with elements A and B, the positions of the interfaces are still clearly identifiable after superposition of the two scattering contributions. However, if the layers of pure B are replaced by AB alloy layers, it is possible for the scattering kinematics to locate the scattering from B exactly in the reduced-scattering regions of element A in the alloy layers. In this case, interface identification and composition determinations are less straightforward. In general, to confirm the interpretation of RBS spectra and to avoid such interferences, the scattering conditions can be changed simply by tilting the sample to change the path length per layer. In this regard, computer simulation of the spectra (such as with the code RUMP by Doolittle, 1986) is useful in optimizing experiments, particularly for cases of more complex spectra with overlapping signals from three or more elements.

These "scattering interference" considerations, as well as the increased depth resolution with tilting, are illustrated in the computer simulation

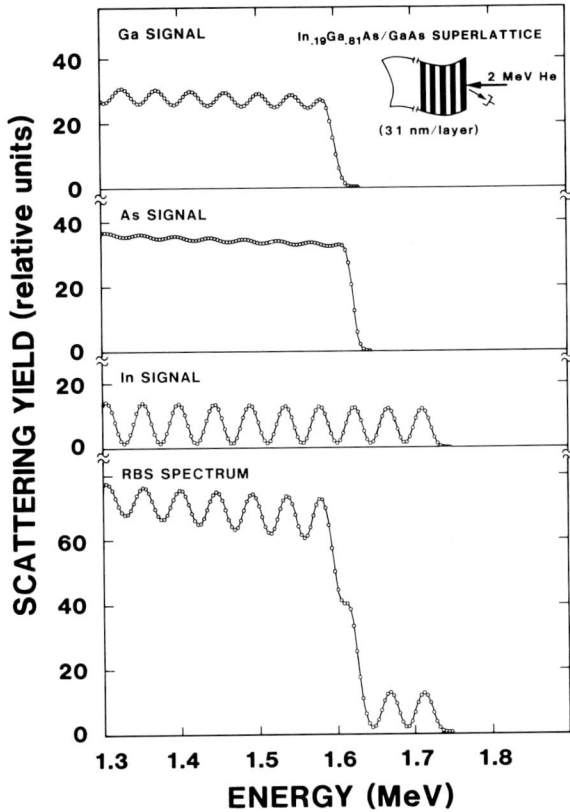

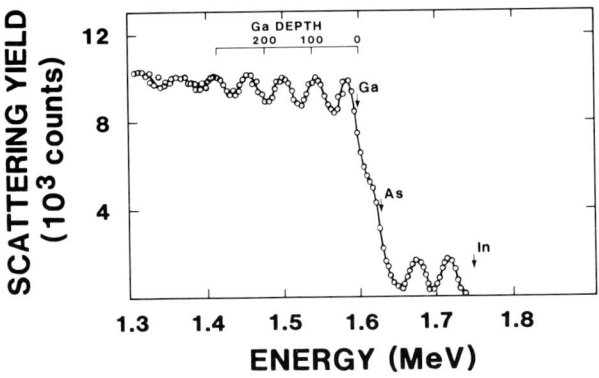

FIG. 7. Computer simulation (a) and experimental observation (b) for a 2-MeV He RBS spectrum for 160° scattering from an $In_{0.19}Ga_{0.81}As/GaAs$ superlattice (31 nm/layer).

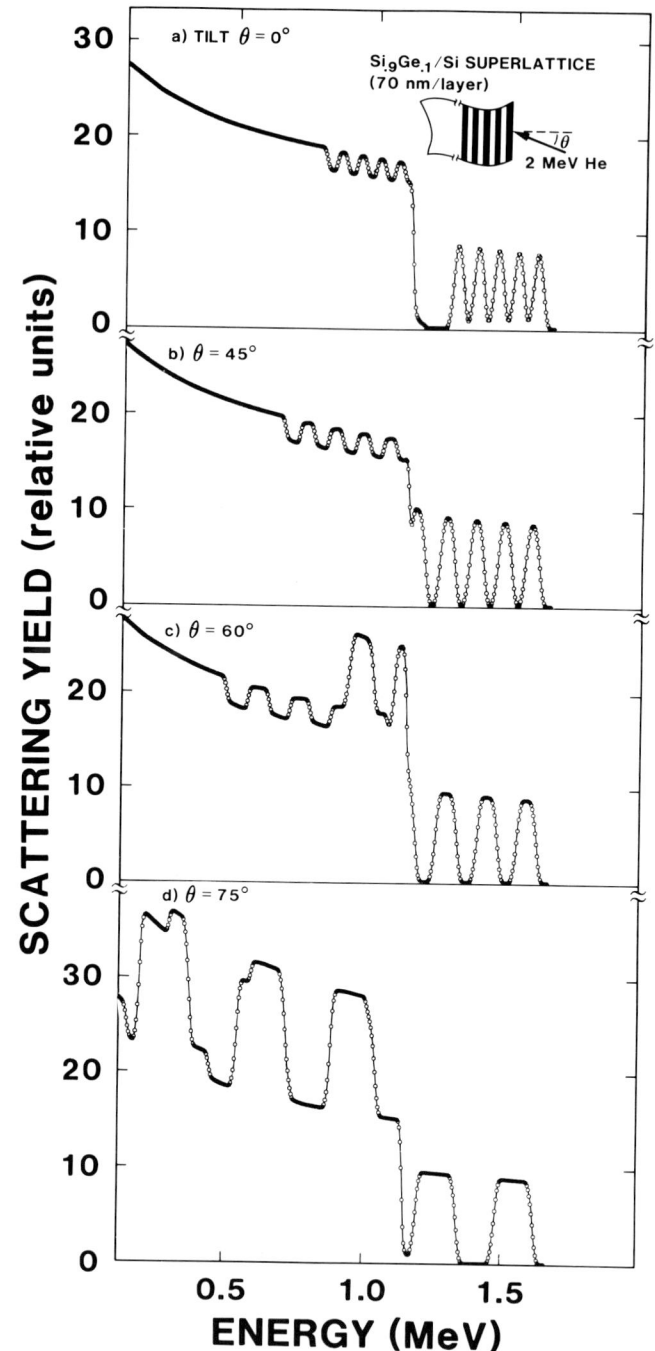

FIG. 8. Computer-simulated RBS spectra from a 5-layer $Si_{0.9}Ge_{0.1}/Si$ superlattice for 2-MeV He, 160° backscattering as a function of sample tilt angle from normal incidence: (a) $\theta = 0°$; (b) 45°; (c) 60°; and (d) 75°. The increased depth resolution as well as layer interferences are illustrated at the higher tilt angles.

spectra of Fig. 8 for a Si_xGe_{1-x}/Ge superlattice. From inspection of the spectra at $\theta = 60°$ and $75°$, one realizes that for superlattices with many layers, the selection of the scattering angle can minimize or maximize interferences between the different elements. For example, at $\theta = 75°$ in Fig. 8, the Ge signal from layer 5 (third superlattice period) exactly overlaps the Si signal of layer 2 (at 0.9-1.0 MeV). Quick estimates as to when major interferences are expected can be made based on the total energy shift corresponding to the thickness of a sequence of layers, $\Delta E_t = \Sigma_i \Delta x_i n \sec\theta_1 [S]_i$, relative to the kinematic shift in energy expected for the two masses present, $(k_a - k_b)E$. The relative values of these terms determine whether the scattering from one layer will fall within the valley of the next and tend to cancel the oscillations, or satisfy the opposite condition and maximize the observed oscillations.

In general, RBS analysis can determine the layer thickness of superlattice structures for layers $\gtrsim 5$ nm in thickness, and from the shape of the spectrum it can determine the composition of those layers, with both values determined to a relative accuracy of about $\pm 10\%$. The composition variation between layers required for such measurements depends on the particular elements involved. However, compositions down to about 1% for a high Z element in combination with lower Z elements in the layers can usually be resolved, whereas in the opposite condition of a small change in a lower Z element in alternating layers, they can result in sensitivity of an order of magnitude lower in unfavorable cases.

B. CHANNELING: LAYER STRAIN AND PERFECTION

Information about the crystal quality and the magnitude of the strain in strained-layer structures can be obtained by combining RBS analysis with ion-channeling measurements (Picraux *et al.*, 1983a, 1986, 1988a; Chu *et al.*, 1983; Fiory *et al.*, 1984; Hamdi *et al.*, 1985; Feldman *et al.*, 1987; Chami *et al.*, 1988). Ion-channeling analysis of materials (Feldman *et al.*, 1982) is based on the phenomenon of the steering of energetic particles by crystal rows or planes. These channeled particles do not approach the atomic nuclei closely enough to undergo Rutherford backscattering. This effect results in about a 30-fold reduction in scattering yield for channeling along major crystal axes and about a 5-fold reduction along major planes. Deviations from perfection, for example misfit dislocations in strain-relieved structures or changes in the direction of the crystal rows along inclined directions due to the strain-induced tetragonal distortions in the layers, will give rise to dechanneling and increased scattering of the ions. In this section, we first discuss methods by which the strain in layers can be determined, either quantitatively or qualitatively, using ion channeling. Then we discuss the influence of crystal imperfections on ion channeling in strained layers and superlattices.

Strain measurements by the RBS/ion-channeling technique in layered structures are based on the tetragonal distortions induced in the layers. These distortions imply a lattice constant, $a^\perp$, perpendicular to the layer, which differs from the in-plane lattice constant, $a^\|$, and which varies between layers. As a result, the crystal rows and planes that are inclined with respect to the surface normal will be slightly tilted in their direction from that expected for cubic structures. They will also undergo slight tilts in their direction at each interface where a change in strain occurs. This effect is illustrated schematically in Fig. 9a for a single commensurate layer under biaxial compressive stress. Here, the inclined [011] axis direction in the strained layer is slightly reduced from 45° relative to the [001] growth direction, since $a^\perp > a^\|$ in the layer. Likewise, in crossing the interface between the layer and the substrate, the [011] crystal rows undergo a slight increase in tilt, $\Delta\psi$. The ion-channeling method of measuring strain is therefore based on these small tilts for channeling along crystal rows or planes that are inclined to the layer normal.

There are three classes of channeling measurements that can be used to obtain information on the strain in strained-layer structures. First, angular scans through major channeling directions allow the crystal directions in the outermost layer, and thus the tilts, to be determined directly: Channeling dips (reduction in backscattering yields) occur for the incident ions aligned along the low-index crystal axes or plane. Second, for sufficiently large strains, the tilts between layers give rise to dechanneling along the inclined directions. Third, under special conditions, a superlattice with alternating tilts between layers can give rise to a resonance-channeling effect along crystal planes known as catastropic dechanneling.

Channeling angular scans provide the most convenient and easily quantified information on the strain in single layers and superlattices. The magnitude of the tilt in crystal directions due to the tetragonal distortion is directly given by the perpendicular and parallel lattice constants. For the usual case of the [001] direction corresponding to the layer normal (the crystal growth direction), any inclined direction $[i, j, k]$ in an epitaxial layer will be at an angle

$$\psi = \tan^{-1}\{ka_{\text{epi}}^\perp/(i^2 + j^2)^{1/2}a_{\text{epi}}^\|\}, \qquad (16)$$

where $a^\perp$ and $a^\|$ are the tetragonally distorted perpendicular and parallel lattice constants. Thus, the ratio of the perpendicular to parallel lattice constants in the layer can be determined directly from measuring the crystal directions, and the perpendicular to parallel strains are then given from Section II.A. by the relations $a^\perp/a^\| = (e^\perp + 1)/(e^\| + 1)$ and $e^\perp = -2c_{12}e^\|/c_{11}$. If the composition of the layer and substrate are independently determined by RBS measurements, then the equilibrium lattice constants (and lattice misfit) are known, and the observed strain in the layer can be compared to the expected strain for fully commensurate growth.

3. STRUCTURE AND CHARACTERIZATION OF SUPERLATTICES

STRAINED HETEROEPITAXIAL LAYER

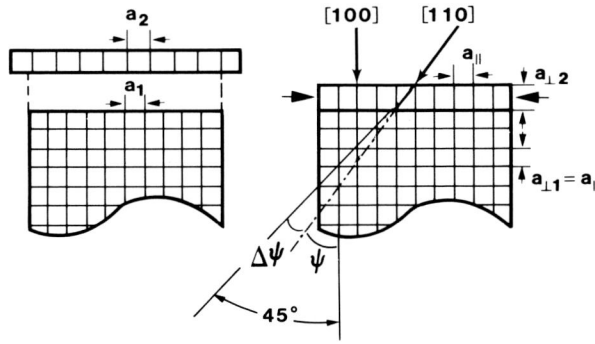

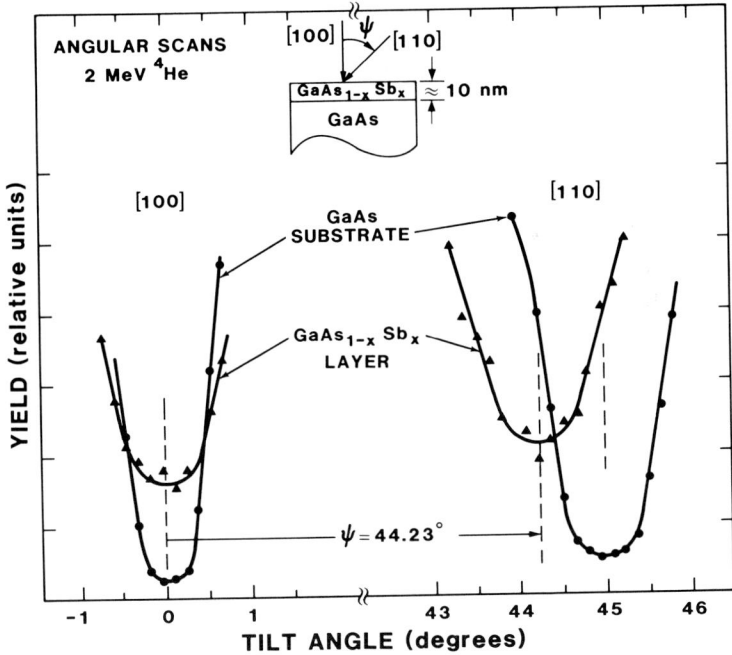

FIG. 9. (a) Schematic of a single strained layer grown commensurately on a substrate of slightly smaller lattice constant. (b) Channeling angular scans along [001] surface normal and inclined [011] axes for a 10-nm $GaAs_{1-x}Sb_x$ strained layer on a GaAs substrate. Reprinted with permission from Picraux (1988b).

The tilt angle between two commensurate epitaxial layers 1 and 2 therefore is given simply by the difference between their ψ values

$$\Delta\psi = \psi_2 - \psi_1, \tag{17}$$

which for the case of a single strained layer on a cubic substrate corresponds to

$$\Delta\psi = \tan^{-1}\{ka^{\perp}_{\text{epi}(2)}/(i^2 + j^2)^{1/2}a^{\parallel}_{\text{epi}(2)}\}$$
$$- \tan^{-1}\{ka^{\perp}_{\text{epi}(1)}/(i^2 + j^2)^{1/2}a^{\parallel}_{\text{epi}(1)}\}. \tag{18}$$

The [011] axis is a major channeling direction at 45° to the [001] direction and is most commonly used for inclined channeling studies. In this case, for commensurate layers there is a single in-plane lattice constant, $a^{\parallel}_{\text{epi}(1)} = a^{\parallel}_{\text{epi}(2)}$ and

$$\Delta\psi = \tan^{-1}\{a^{\perp}_{\text{epi}(2)}/a^{\parallel}\} - \tan^{-1}\{a^{\perp}_{\text{epi}(1)}/a^{\parallel}\}, \tag{19}$$

$$\approx \{a^{\perp}_{\text{epi}(2)} - a^{\perp}_{\text{epi}(1)}\}/2a^{\parallel} \tag{20}$$

to a very close approximation. For a superlattice where the parallel strains are equally shared between layers, Eq. (20) reduces to

$$\Delta\psi = \frac{e^{\perp}_{\text{SLS}(2)} - e^{\perp}_{\text{SLS}(1)}}{2(e^{\parallel} + 1)} \tag{21}$$

$$(e^{\perp}_{\text{SLS}(2)} - e^{\perp}_{\text{SLS}(1)})/2.$$

Thus, the tilt angle along the [011] direction between two layers is essentially a measure of the relative perpendicular strain between the layers. If the [011] direction of the top layer relative to the [001] layer normal is accurately determined, then the absolute strain in each layer is also obtained.

Typical lattice mismatches for coherent structures range from 0.1 to 2.0%, which leads to [011] tilt angles of $\Delta\psi \approx 0.1$ to $1.0°$. Examples of the tilt angles between fully commensurate layers in a superlattice based on published elastic constants and linear (Vegard's law) superpositions of these values for alloys are given in Table II.

An example of angular scans through channeling direction for a single strained epitaxial layer is shown in Fig. 9 (Picraux et al., 1988a). The $GaAs_{1-x}Sb_x$ layer grown by MBE on a thick GaAs substrate is under biaxial compressive strain, and thus the [011] axis in the layer is inclined slightly less than 45° relative to the [001] normal. The signal from the ≈ 10 nm $GaAs_{1-x}Sb_x$ layer is obtained from the Sb scattering signal in the RBS spectrum, and the GaAs substrate signal is taken from Ga and As scattering just below the GaAsSb–GaAs interface. From the centroid of the dips, the angle between the [011] and [001] crystal directions in the $GaAs_{1-x}Sb_x$ layer

3. STRUCTURE AND CHARACTERIZATION OF SUPERLATTICES

TABLE II

Lattice Misfit and Tilt Angles for Various Strained-Layer Superlattices; a_0, $a^{\|}$ and $a^{\perp}$ Are the Equilibrium, In-Plane and Transverse Lattice Constants of the Layers, f Is the Lattice Mismatch, $\Delta\psi$ the Tilt Angle for the Inclined [011] Axis (or {011} Plane for a (001) Growth Direction[a]

Superlattice	f (%)	a_0 (Å)	$a^{\|}$ (Å)	$a^{\perp}$ (Å)	$\Delta\psi$ (deg.)
$Al_{0.5}Ga_{0.5}As/GaAs$	0.08	5.6577/5.6532	5.656	5.660/5.651	0.04
$GaSb/InAs$	0.61	6.0954/6.0587	6.078	6.111/6.037	0.35
$GaAs_{0.2}P_{0.8}/GaP$	0.74	5.4910/5.4505	5.470	5.509/5.433	0.40
$Ge_{0.2}Si_{0.8}/Si$	0.83	5.4761/5.4307	5.453	5.494/5.414	0.42
$In_{0.2}Ga_{0.8}As/GaAs$	1.42	5.7432/5.6532	5.692	5.774/5.618	0.78
$Ge_{0.5}Si_{0.5}/Si$	2.07	5.5441/5.4307	5.484	5.590/5.390	1.05

[a]Calculated values correspond to commensurate superlattices of equal layer thickness grown on a substrate or buffer layer with lattice constant equal to the equilibrium ($a^{\|}$) lattice constant of the freestanding superlattice (zero average strain). Elastic constants (Simmons and Wang, 1971) used for C_{11} and C_{12} in units of 10^{11} dynes/cm^2 are, respectively: GaAs 11.877, 5.372; GaP 14.120, 6.253; GaSb 8.839, 4.033; InAs 8.329, 4.526; Si 16.578, 6.394; Ge 12.853, 4.826; and assuming for AlAs values 14.0, 6.0, similar to GaP.

is measured to be $\psi = 44.23 \pm 0.03°$. Thus, $\Delta\psi = 45° - \psi = 0.77°$, which corresponds to a perpendicular strain $e^{\perp} = 2.7\%$. This strain corresponds to a composition between $x = 0.18$ and 0.19 and a lattice mismatch of $f = 1.4\%$. The sensitivity of this measurement of strain is limited primarily by the angular resolution (0.01° in this case) of the goniometer used to orient the sample and the accuracy of the dip-centroid determinations. A sensitivity of about $\pm 0.03°$ in $\Delta\psi$ corresponding to perpendicular strains of $\pm 0.1\%$ is straightforward to achieve.

In the previous example, we determined the strain from the angle ($\psi = 45° - \Delta\psi$) between the [011] and [001] axes in the strained layer. If instead, we measure $\Delta\psi$ for the [011] rows between the $GaAs_{1-x}Sb_x$ layer and the GaAs substrate directly from the shift in the centroids of their dips, we obtain $\Delta\psi = 0.76°$ (versus 0.77° earlier). This value is slightly lower than the true $\Delta\psi$, due to the steering influence of the crystal rows in the overlying layer on the channeling-flux distribution as it enters the underlying GaAs substrate. The reduction due to channeling in the top layer is small here. However, deviations can be more significant for thicker layers, especially when the shift is near the critical half angle for channeling, or if layer thicknesses are near a multiple of the half wavelength of the most populated channeled-particle trajectories. By varying the energy of the incident beam, the angular width of the channeling dip and channeling wavelengths can be varied ($\psi_{1/2} \alpha [Z_1/E]^{1/2} \propto \lambda^{-1}$), so that interferences in the channeling measurements between the top and underlying layers can be reduced by appropriate selection of the beam energy and atomic number. Thus, with care, the angular shift between the layer and the substrate alone can provide a lower bound on the relative strain between layers, and the accuracy of this approach is fairly good for thin layers.

A strained-layer superlattice (SLS) is a simple extension of the single-layer case for the channeling angular scan technique. A schematic of a superlattice with alternating compressive and tensile strains is shown in Fig. 10a; a tilt, $\Delta\psi$, of equal and opposite sign occurs in the inclined rows and planes at each interface. The magnitude of the tilt is determined from Eq. (18) by the relative strains (tetragonal distortions) in each layer. As discussed earlier, the strain within the top layer of the superlattice is obtained directly by channeling angular scans of the [001] normal and [011] inclined crystal axes. To determine the relative strain between the top and second layers and to thus obtain the absolute strain in the second layer as well, it is necessary to measure the tilt, $\Delta\psi$, between layers. In Fig. 10b, we show an example of this measurement for an $In_xGa_{1-x}As/GaAs$ superlattice, where the relative strain between the top GaAs layer and the underlying InGaAs layer is determined from angular scans taken from the top two layers, as referenced to the signal from the fifth and sixth layers for the average direction. This technique is similar to measuring the angular shift between the layer and substrate and is therefore approximate, as discussed earlier. However, it is believed to give somewhat greater accuracy, since for equal layer thicknesses, the average direction can be determined fairly accurately by combining the signal from two adjacent layers at deeper depths. Here, one can observe that the angular shift is reduced for the second layer, and previous studies (Chu *et al.*, 1983) have followed this decay in shift with deeper layers in detail. In Fig. 10b, the value from the top layer shift of $\Delta\psi = 2 \times 0.32 = 0.64°$ for this case of equal layer thicknesses is consistent with the predicted value of $0.65°$ in this superlattice for commensurate growth.

The second method of characterizing the strain in SLS structures is to measure the dechanneling versus depth along inclined directions. Since the change in tilt angle, $\Delta\psi$, at each interface along the inclined [001] axis is a significant fraction of the critical angle for channeling, there is an increase in the dechanneling with depth that depends on the magnitude of the tilt. This dechanneling is shown in Fig. 11 for a single buried $Si_{0.17}Ge_{0.83}$ strained layer in Ge (Picraux *et al.*, 1988a). The signal for the [011] channeling direction increases rapidly at an energy corresponding to the depth at which the SiGe layer is reached. Then, after passing through the SiGe alloy layer with "tilted" inclined directions, the rate of dechanneling again returns to a value near that for the bulk Ge crystal. The increase in the dechanneling is a measure of the magnitude of the strain present and becomes greater with increasing tilt angle $\Delta\psi$. For layers that are thin compared to the half wavelength for channeling, the dechanneling will in addition increase with

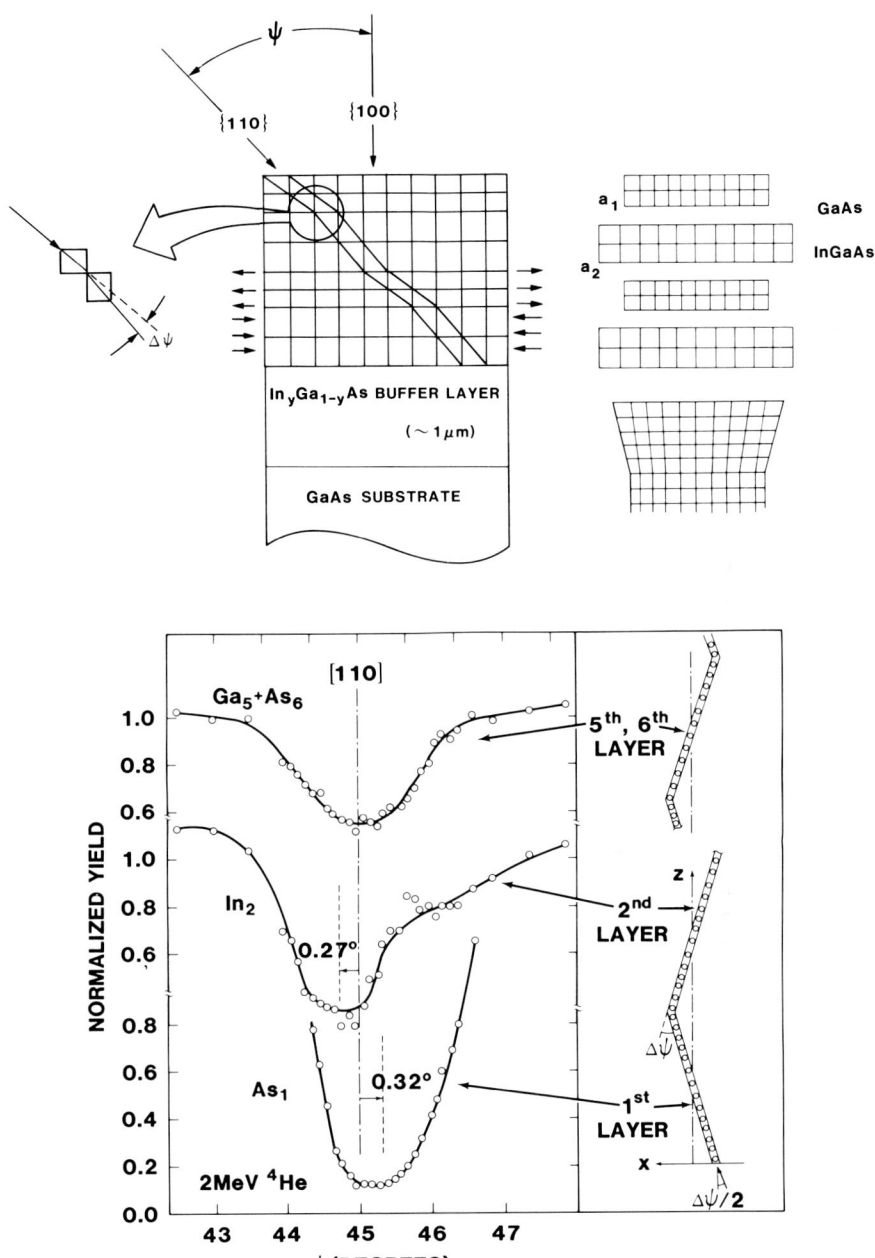

FIG. 10. (a) Schematic of a strained-layer superlattice grown commensurately on a buffer layer of lattice constant near the average value for the superlattice layers. (b) Channeling angular scans along the inclined [001] axis from the first, second, and fifth and sixth layers of a GaAs/In$_{0.15}$Ga$_{0.85}$As superlattice (38 nm/layer). After Picraux et al. (1983c).

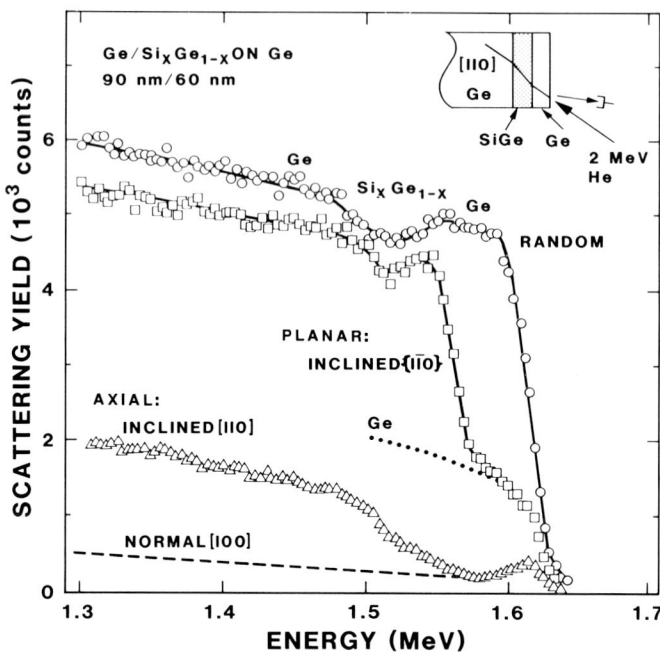

FIG. 11. Channeled and random spectra for a single strained $Si_{0.17}Ge_{0.83}$ 60 nm layer buried 90 nm below the surface in a Ge (001) substrate. The rapid increase in dechanneling at the depth of the buried SiGe layer is due to the strain-induced tilt in the inclined [011] axial and {011} planar directions, whereas there is no corresponding increase along the normal [001] direction for this commensurate structure. Reprinted with permission from *Nucl. Instr. and Meth.* **B33**, 891, Picraux, S. T., Dawson, L. R., Tsao, J. Y., Doyle, B. L. and Lee, S. R. (1988).

layer thickness. This occurs because for very thin layers, a substantial fraction of the channeled particle-flux distribution can pass through the layer before approaching and being dechanneled by the crystal rows. Thus, detailed analytical or computer-simulation calculations are required to quantify the strain measurements by dechanneling alone.

From combined RBS/channeling and x-ray DCD studies of the example in Fig. 11, no measurable strain relief has occurred, and this fully commensurate layer has the strains $\varepsilon_{SiGe}^{\perp} = -1.28\%$ and $\varepsilon_{SiGe}^{\parallel} = 0$ (Picraux and Lee, 1988). Figure 11 illustrates the additional fact that dechanneling along inclined planar directions, such as the {011} planes, is also a sensitive indicator of strain-induced interface tilts.

For this dechanneling approach, the same principles apply to a super-lattice structure. Figure 12 shows an example (Picraux *et al.*, 1983b) of dechanneling along the [011]-inclined direction in an $In_{0.13}Ga_{0.87}As$/GaAs superlattice. This superlattice had sufficiently thin layers (≈ 9 nm) so that the oscillations in RBS yield were not resolved in the spectra. The channeling for

3. STRUCTURE AND CHARACTERIZATION OF SUPERLATTICES

$In_x Ga_{1-x} As/GaAs$

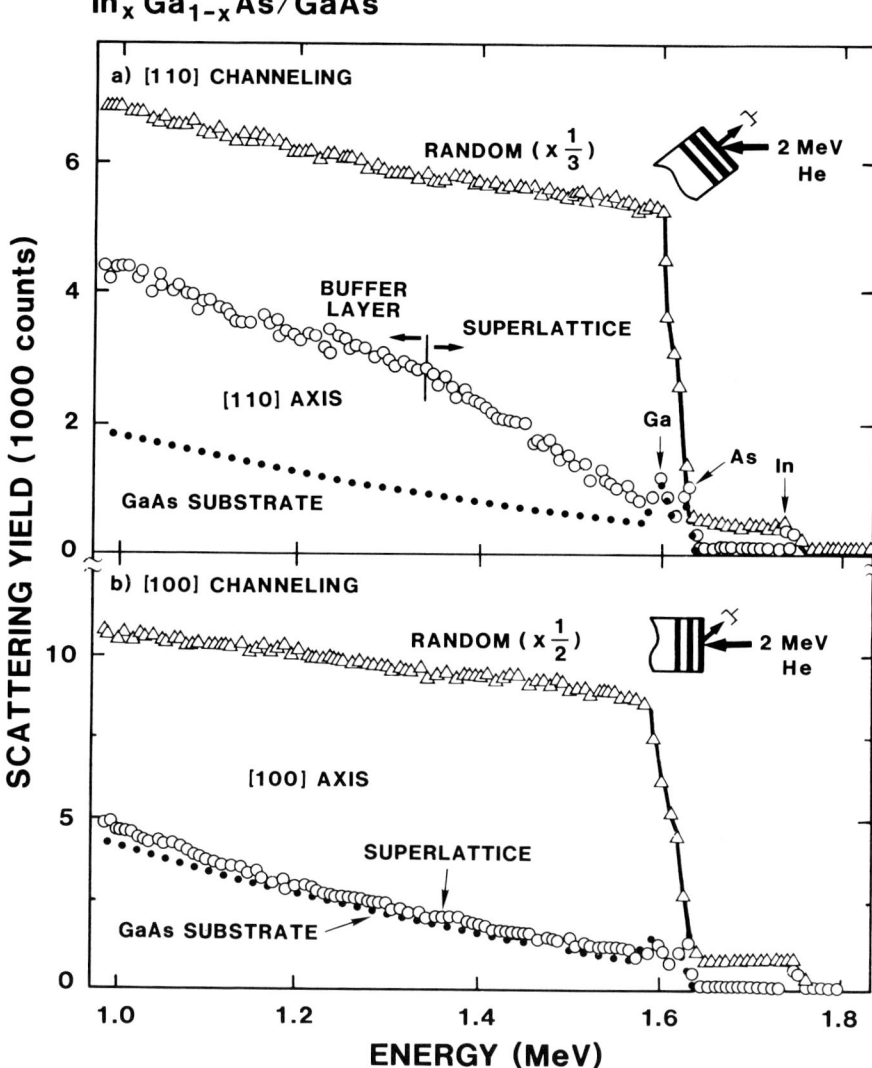

FIG. 12. Channeled and random RBS spectra for an $In_{0.13}Ga_{0.87}As/GaAs$ 9 nm/layer strained-layer superlattice for channeling along normal [001] and inclined [011] axes. Increased dechanneling along the [011] axis relative to a reference GaAs substrate is due to the alternating tilts at the interface for this commensurately grown structure. From Picraux et al. (1983b).

the [001] growth direction is similar to that for a bulk GaAs crystal, indicating no appreciable strain relief, whereas the dechanneling for the [011] increases rapidly as the beam passes through the 40-layer superlattice. In a bulk crystal, the channeling level for the [011] axis would be slightly lower than that for the [001] direction, and thus the significantly increased dechanneling along the [011] but not the [001] growth direction is quite a distinct signature for a strained-layer superlattice.

As discussed earlier, with increasing strain, the tilt angle increases and thus the rate of dechanneling also increases. However, over the range of typical superlattice layer thicknesses of 1–50 nm, the dechanneling rate also increases with increasing layer thickness. Again, this is because the total amount of dechanneling due to the change in direction of the crystal rows is not completed within a very thin layer before entering the next layer. The combined effect of both strain and layer thickness on the rate of dechanneling with depth can be accounted for by detailed Monte Carlo computer-simulation calculations for quantitative interpretation of the magnitude of the strain (Barrett, 1983).

Because of the above coupling of tilt angle and layer thickness in the dechanneling, quantitative measurement of the strain is less straightforward by this approach and is generally not done. However, the observed dechanneling is convenient for interpretation of the qualitative variation in strain between superlattices. In addition, it has the unique advantage of providing a direct depth-resolved measure of the strain within a single measurement, something that is not usually achieved by the other experimental techniques discussed here. Also, we note that the fundamental quantity being probed in this case is the relative strain between layers, and not the absolute strain within any given layer. Typically, the dechanneling measurements are sensitive in the region of lattice mismatches ranging from 0.3 to 3%. Examples of the lattice-strain mismatch and resulting tilt angle for various superlattice combinations are given in Table II.

The third method for measuring strain in strained-layer superlattices by ion channeling is the planar resonance effect (Picraux *et al.*, 1988b). In this case, the channeled beam is incident along inclined crystal *planes*, typically the {011} planes for (001)-grown structures. The channeled particles oscillate between the planes and have a classical trajectory wavelength that starts out with all the particles in phase. Since the wavelength varies only moderately for particles of different amplitude (a consequence of the nearly harmonic continuum potential), the trajectory phase coherence is retained to moderate depths. By adjusting the energy of the incident beam, the average channeled-particle wavelength may be matched to the superlattice period. Under these conditions, a new resonance phenomena called *catastropic dechanneling* is observed, which leads to a rapid increase in the dechanneling after a certain depth. An example of catastropic dechanneling for a $GaAs_{0.09}P_{0.91}/GaP$ superlattice is shown in Fig. 13. At shallow depths, the channeled yield is similar to that for a bulk crystal, whereas after penetrating a few layers, a very strong increase in the channeled signal occurs until almost all of the beam is dechanneled.

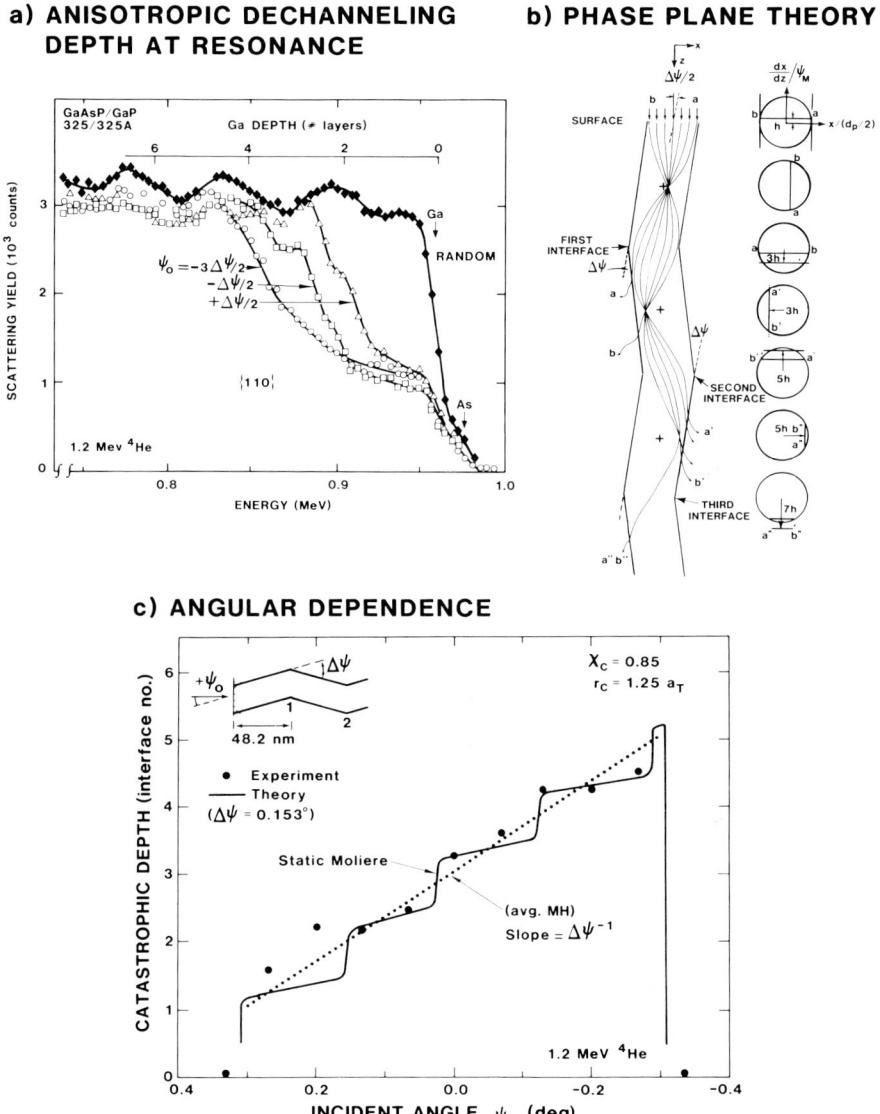

FIG. 13. Catastrophic-dechanneling measurements of strain in a $GaAs_{0.09}P_{0.91}/GaP$ (48.2 nm/layer) strained-layer superlattice based on the resonance matching between the planar channeled-particle trajectory wavelength and the superlattice period. (a) Random and {011} channeled spectra at several incident angles with respect to the first layer planes. (b) Phase-plane description (normalized channeled-particle angle versus position) of the channeled-particle evolution. (c) Catastrophic-dechanneling depth (for 85% dechanneled) versus incident angle for measured and calculated tilt angle between layers $\Delta\psi = 0.153°$. After Picraux et al. (1988b).

This behavior is readily understood in a phase-plane description, as shown in Fig. 13b. The incident beam is represented by a line on a diagram of normalized transverse momentum (i.e., angle) versus normalized lateral position in the channel. This line rotates as the channeled beam penetrates the crystal, and undergoes a vertical increase at each interface according to the sign and magnitude of the tilt $\Delta\psi$. As a result, in the simple modified harmonic model of Fig. 13b, the focus point of the beam moves closer to the channel wall in each layer until the particles have sufficient transverse energy to penetrate the planes. In penetrating each layer, the beam changes its direction, since this corresponds to traveling a half wavelength, but the crystal planes also change their direction by $\Delta\psi$. Thus, the angle of the channeled particles relative to the crystal planes increases at each interface until they become sufficiently large so that particles can no longer be steered by the crystal planes and are dechanneled. The layer in which the catastropic dechanneling occurs is directly related to the tilt angle $\Delta\psi$ at each interface and the incident angle of the beam with respect to the superlattice planes. By measuring the catastropic depth as a function of the incident angle, as shown in Fig. 13c, a well-defined asymmetric dependence on the incident angle is observed from which $\Delta\psi$ is determined. To a first approximation, the slope of the straight line through the staircased structure in Fig. 13c is just equal to the reciprocal of the tilt angle $\Delta\psi$.

The catastropic-dechanneling technique of strain measurement in superlattices is only suited for relatively thick layers. This constraint arises because the half wavelength for the channeled particles is typically of the order of 50 nm and must be made equal to the path length per superlattice layer along the inclined planar channeling direction. Of course, the wavelength can be tuned to a certain extent by varying the energy of the incident beam, since $\lambda \alpha E^{1/2}$, and also the path length can be varied by tilting within the inclined crystal planes. In spite of this constraint mentioned earlier, the catastrophic-dechanneling resonance technique is attractive for strain analysis at very low strains. Typically, the greatest sensitivity to $\Delta\psi$ will occur between 20% and 80% of the critical angle for channeling, which for planar channeling is $\lesssim 0.4°$.

The catastrophic-dechanneling technique is the most sensitive of the ion-channeling techniques for strain measurements. Typically, the values of the tilt angle, $\Delta\psi$, can be determined within an accuracy of as great as 0.01°, which corresponds to strains of 0.02%. In contrast, axial channeling angular scans provide the greatest versatility, since the layer strain can be determined over a wide range of strain magnitudes and layer thickness for both single-strained layers and superlattices. These measurements can typically be made with an accuracy of 0.03° corresponding to a sensitivity of about 0.1%.

In addition to strain measurements, the crystal quality of a strained layer can be assessed by ion-channeling measurements along the growth direction.

As shown in Figs. 9a and 10a for commensurate strained layers and superlattices, the rows along the growth direction (normal to the crystal planes) are straight, and channeling characteristics of perfect bulk crystals can be expected. However, under conditions of imperfect heteroepitaxial growth, defects such as stacking faults may form in the layers, and these will give rise to increased dechanneling in the layer. Of greater interest here, if the excess stress in the layers becomes sufficient for strain relaxation to have occurred, dechanneling along the growth direction will be observed due to dislocations or other strain-relieving defects. In this latter case, the strain around the dislocations gives rise to distortions of the crystal rows and planes that lead to the dechanneling. For example, a network of misfit dislocations near the interface will relieve the strain and give rise to an increase in the channeled level at that depth.

From a knowledge of the dechanneling cross section, σ_D, the density of misfit dislocations, and therefore the magnitude of strain relief, can be estimated. For a network of dislocations, the observed channeling yield, χ_D, due to the dislocations, will increase over that for a virgin crystal, χ_V, and the density of dislocations (N_D = projected length of dislocations per unit area) will be given by

$$N_D = \frac{1}{\sigma_D} \ln \frac{1 - \chi_V}{1 - \chi_D}, \tag{22}$$

where σ_D is the dechanneling cross section. For dislocations lying in the plane of the layers, the dechanneling cross section for the beam along directions normal to the layers (the growth direction) is approximately given by (Feldman et al., 1982)

$$\sigma_D = K(ab)^{1/2}/\psi_1, \tag{23}$$

where $K \approx 0.5$ and depends on dislocation orientation, a is the Thomas–Fermi screening distance, b is the magnitude of the Burgers vector, and $\psi_1 = (2Z_1, Z_2 e^2/Ed)^{1/2}$ for axial channeling and $(Z_1 Z_2 e^2 N d_p a/E)^{1/2}$ for planar channeling, with Z_1 and Z_2 being the average atomic numbers of the projectile and target, d the average spacing along crystal rows, E the beam energy, N the atomic density, and d_p the planar spacing. The magnitude of K depends on dislocation type and orientation and, for example, is reduced to zero if the beam were exactly lined up (end on) along the direction of a perfect edge-dislocation line. However, the variation is small for dislocations oriented normal to the beam direction, and, the dechanneling for 60°-inclined dislocations within the layers for example, is expected to be similar to that for perfect (90°) edge dislocations.

Putting values into Eq. (23) for a perfect edge dislocation in GaAs and 2-MeV He channeling gives $\sigma = 2.8$ nm for $\langle 001 \rangle$ axial dechanneling and

15.7 nm for (001) planar dechanneling. Then, a square array of 90° misfit dislocations sufficient to relieve a strain of 0.5% would correspond to a dislocation spacing of 80 nm (1.25×10^5 disl./cm) and would give rise to a step increase in the dechanneling, $\Delta\chi$, at the depth of the dislocations of about 7% and 23% for the $\langle 001 \rangle$ axial and (011) planar directions, respectively. Thus, appreciable strain relief by misfit-dislocation formation is reasily detected by ion dechanneling along the *growth* direction.

In addition, the cross section for dechanneling by dislocations is seen to increase with increasing incident beam energy ($\sigma_D \alpha E^{1/2}$). This behavior contrasts, for example, with stacking faults where the dechanneling cross section is essentially independent of energy and with isolated point defects or amorphous zones where the cross section decreases with increasing energy (Feldman *et al.*, 1983). Therefore, energy dependencies of the dechanneling can help confirm that the observed dechanneling is due to dislocations, and in favorable cases may even separate out that contribution in the presence of a small amount of additional disorder such as stacking faults.

An example of dechanneling due to strain-relieving dislocations in a single strained layer is shown in Fig. 14 for CdTe grown on $Cd_{0.97}Zn_{0.03}Te$ (Chami *et al.*, 1988), which corresponds to a lattice misfit of 0.18%. Here, for 750-nm-thick layers, partial strain relief (relaxation) has occurred, and the expected energy and incident projectile dependence of the dechanneling $\sigma \propto \sqrt{E/Z_1}$ for dislocations is found. However, the y intercept does not go through zero, which suggests that some additional defects of lower energy dependence, such as stacking faults, must also be present.

Another type of disorder that may be encountered in ternary and quarternary compound semiconductors is commonly referred to as alloy disorder. This type of disorder arises for a small but finite deviation (e.g., ~ 0.1 Å) in the atom positions due to the different bonding lengths of the binary compound components. In this case, a small continuous increase in dechanneling in the alloy due to these static displacements will be observed for displacements $\geqslant 0.1$ Å, analogous to the increase observed upon increasing temperature due to increased vibrational amplitude of the atoms. Alloy-disorder dechanneling is a bulk-alloy effect and so to first order is independent of layer thickness or number of interfaces. This type of disorder gives a relatively small additional increase over normal intrinsic contributions to the dechanneling along all crystal directions. It is usually not significant in the strain measurements discussed earlier.

In summary, one of the key benefits of the ion-channeling approach to the study of strained-layer structures is that it provides information on discrete layers, whereas the x-ray diffraction (XRD) technique, as discussed in the next section, gives the combined signal for all of the layers present. Moreover,

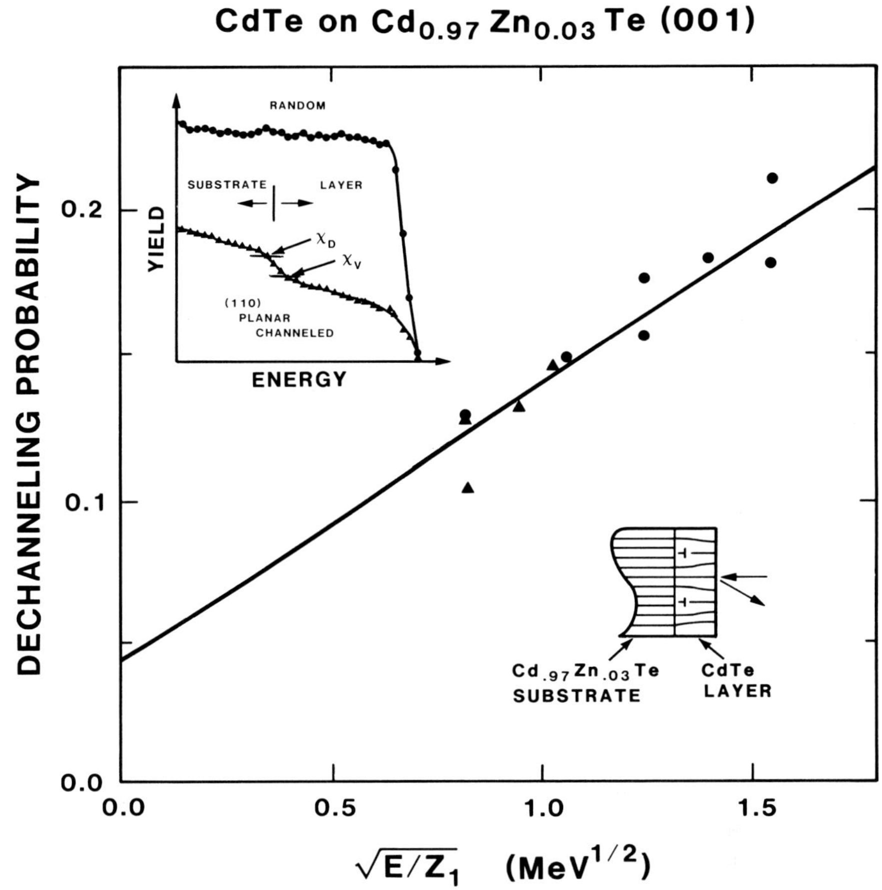

3. STRUCTURE AND CHARACTERIZATION OF SUPERLATTICES

CdTe on $Cd_{0.97}Zn_{0.03}$Te (001)

FIG. 14. Measured normal {011} planar dechanneling probability versus $\sqrt{E/Z_1}$ for incident protons (circles) and He (triangles), where E is the beam energy and Z_1 the beam-particle atomic number. The expected $\sqrt{E/Z_1}$ dependence due to dislocations is seen for this 750-nm-thick strain-relieved CdTe layer on $Cd_{0.97}Zn_{0.03}$Te substrate. Inset shows the increase in dechanneling at the layer–substrate interface due to misfit dislocations. After Chami et al. (1988).

good sensitivity can be retained for very thin single strained layers in ion-channeling measurements, for example for the strain in quantum-well structures. In contrast, x-ray rocking-curve measurements, for those superlattice cases where there is good layer uniformity and a sufficient number of layers, provide a much more accurate measurement of both the magnitude of the strains and the thickness of the layers.

IV. X-ray Diffraction Characterization

In the use of the RBS/channeling technique to analyze the structure of strained layers or superlattices, the perfection/imperfection of the crystal structure is examined in real space, i.e., $\{x, y, z\}$, where these coordinates specify the position of atoms within the crystal. As described in Section III, the effects of tilts in crystal rows due to strains in the layers are directly probed by ion channeling. In contrast, with x-ray diffraction (XRD) analysis, the crystal structure is investigated in reciprocal space, i.e., $\{G_{hkl}\}$, where these coordinates represent the reciprocal lattice points of the crystalline sample. The utility provided by such a reciprocal-space (or k-space) description of XRD will become clear in this section through our examination of the wave equations used to describe XRD from single crystals (e.g., the Laue equations).

XRD and RBS/channeling are among the most commonly used characterization tools in the structural study of epitaxial layers and superlattices. This utility stems from the fact that the advantages/disadvantages of one technique are usually compensated by the other, and therefore the combination of both analysis techniques is quite powerful. A third major complementary technique is transmission electron microscopy (see Table I). Whereas XRD and RBS/channeling give more accurate strain and composition measurements and do not require destructive layer-thinning sample preparation, TEM provides detailed information about the nature of strain-relieving dislocations. Thus, as discussed in Chapter 1 by Hull and Bean, TEM is particularly useful in investigations of the relaxation mechanisms in strained layers.

The objective of this section on the XRD characterization of strained layers is to present this technique at an intermediate level. Fundamental to this discussion is Bragg's law:

$$n\lambda = 2d \sin \theta_B, \qquad (24)$$

where n is the reflection order, λ is the wavelength of the incident x-rays, d is the spacing between diffracting planes, and θ_B is the angle between the incident x-rays and the diffracting plane. It is truly amazing how far the use of Bragg's famous equation can take a novice user in the understanding of XRD analysis of strained layers and superlattices.

Succinct and brief explanations of the XRD technique and theory will be given, and the reader is encouraged to make use of several excellent texts on this subject, including those by Kittel (1976), Warren (1969), Cullity (1978), and Zachariasen (theory) (1945), in addition to excellent scientific journal articles by Mathieson (1982), Speriosu (theory) (1981), and Speriosu and Vreeland (1984). We will start by describing the reciprocal spaces that result from substrates, simple epitaxial layers, and superlattices and then discuss how these reciprocal lattices are experimentally measured and interpreted with x-ray diffraction.

A. STRUCTURE OF THE RECIPROCAL LATTICE

As mentioned earlier, reciprocal lattices, i.e., Fourier representations of real-space crystalline layers or multilayers, are advantageous for XRD analysis and can provide considerable insight into the structure of such layers.

X-rays are scattered/diffracted by electrons in the sample material. Neglecting absorption and multiple scattering, the x-ray scattering amplitude, $\mathscr{A}$, is given by

$$\mathscr{A} = \int dV n(\mathbf{r}) \exp(-i\Delta\mathbf{k}\cdot\mathbf{r}), \qquad (25)$$

where dV is a differential volume element, $n(\mathbf{r})$ is the electron density, $\mathbf{r}$ is a position vector from the arbitrary origin to dV, and $\Delta\mathbf{k}$ is the scattering vector (i.e., the change of the x-ray's wave vector upon scattering). The scattering power is found by simply multiplying $\mathbf{A}$ by its complex conjugate. We use here the same notation and development found in Kittel's (1976) classic text.

The value of reciprocal lattice representation stems from the fact that if the electron number density, $n(\mathbf{r})$, is expressed as a Fourier series,

$$n(\mathbf{r}) = \sum_{\mathbf{G}} n_{\mathbf{G}} e^{i\mathbf{G}\cdot\mathbf{r}}, \qquad (26)$$

where $\mathbf{G}$ is a reciprocal lattice vector, the scattering amplitude can be rewritten as

$$\mathscr{A} = \sum \int dV n_{\mathbf{G}} \exp(i(\mathbf{G} - \Delta\mathbf{k})\cdot\mathbf{r}). \qquad (27)$$

It is straightforward to show that $\mathscr{A}$ is significant only when $\mathbf{G} = \Delta\mathbf{k}$ and the exponent in Eq. (27) vanishes. The $\mathbf{G}$ vectors, or reciprocal lattice points, for single-crystal substrates, strained layers, and SLS will be discussed in detail in the following section.

1. Single Crystals

The reciprocal lattice points of a single crystal are defined by

$$\mathbf{A} = 2\pi \frac{\mathbf{b}\times\mathbf{c}}{\mathbf{a}\cdot\mathbf{b}\times\mathbf{c}} \qquad \mathbf{B} = 2\pi \frac{\mathbf{c}\times\mathbf{a}}{\mathbf{a}\cdot\mathbf{b}\times\mathbf{c}} \qquad \mathbf{C} = 2\pi \frac{\mathbf{a}\times\mathbf{b}}{\mathbf{a}\cdot\mathbf{b}\times\mathbf{c}}, \qquad (28)$$

and

$$\mathbf{G} = h\mathbf{A} + k\mathbf{B} + l\mathbf{C}, \qquad (29)$$

where $\{h, k, l\}$ are the so-called Miller indices and are integers. Primitive real-space lattice vectors are normally used for **a**, **b**, and **c**, but in the case of the

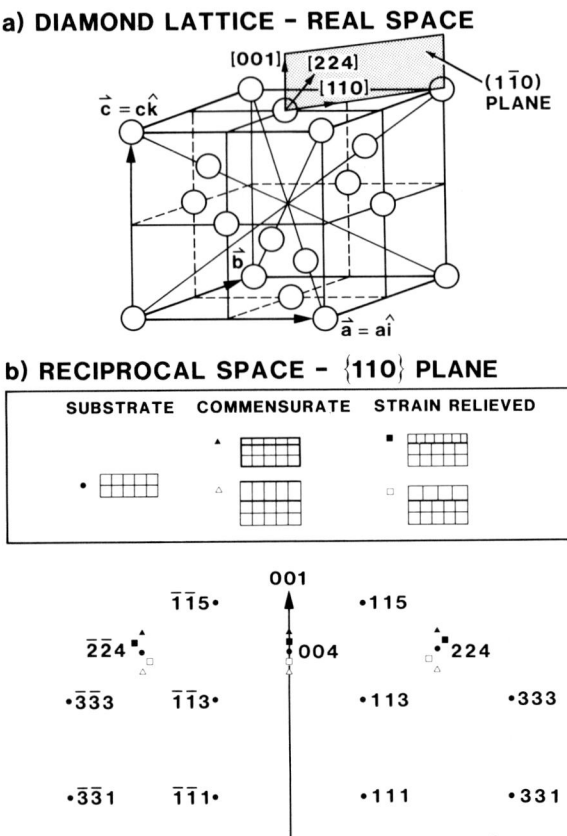

FIG. 15. Diamond crystal lattice and {110} plane illustrated in (a) real space, and (b) reciprocal space. The changes in reciprocal-space positions in the {110} plane are illustrated for unstrained, compressive, and tensile commensurately strained layers and strain-relieved layers for the 004 and 224 positions.

cubic-diamond lattice (all of our examples possess either this or the closely related zinc-blende structure), these vectors are not mutually orthogonal. A more convenient choice for **a**, **b**, and **c** are the sides of the conventional cell, which even for orthorhombic distortions, remain orthogonal. For a conventional cell oriented as shown in Fig. 15a, the reciprocal lattice points, $\mathbf{G}_{hkl}$, expressed in Eqs. (28) and (29) reduce to:

$$\mathbf{G} = h\frac{2\pi \mathbf{i}}{a} + k\frac{2\pi \mathbf{j}}{b} = l\frac{2\pi \mathbf{k}}{c}, \tag{30}$$

where **i**, **j**, and **k** are orthogonal unit vectors in both real and reciprocal spaces. In all of our examples, the crystals are oriented as shown in Fig. 15a with (001), i.e., $h = 0$, $k = 0$, $l = 1$, surface planes with normal vectors that point in the [001] direction in k-space. Furthermore, for perfect crystals/substrates, $a = b = c$, and our convention is that the substrate-surface normal is aligned with the [001] direction.

Continuing the development, the scattering amplitude can now be rewritten as

$$\mathscr{A} = N\mathscr{S}_G, \qquad (31)$$

where N is the number of conventional cells in the sample and $\mathscr{S}_G$ is the structure factor, which can be expressed as

$$\mathscr{S}_G = \sum_{j=1}^{n} f_j \exp[-i2\pi(x_j k + z_j l)], \qquad (32)$$

where $\{x_j, y_j, z_j\}$ represent the jth position of n (8 for diamond) atoms in the basis, and f_j is the form factor of the jth atom. For a diamond-cubic or zinc-blende lattice, the strongest diffraction conditions result when $\{h, k, l\}$ are all odd or when $h + k + l$ sums to an integer multiplied by 4. For these conditions, $\mathscr{S}_G = 8\langle f \rangle$, where $\langle f \rangle$ is the average form factor of the atoms that make up the conventional cell and can be determined through various compilations (Ibers and Hamilton, 1974).

A planar cut through the reciprocal lattice for a perfect diamond-cubic crystal is plotted as filled circles in Fig. 15b. The reciprocal lattice plane being plotted is the $\{h, h, l\}$ family of reflections, and only the most intense points, given by the recipe in the previous paragraph, are plotted. The inset at the top of Fig. 15b illustrates the real-space lattice schematically, which for this case is a perfect square lattice (labeled substrate).

It is important to recognize that these reciprocal lattice points are not δ-functions, but are smeared (slightly) in reciprocal space. This smearing is characterized by a peak density at a centroid position (this is the location of the reciprocal lattice "point"), which, using Eq. (31) for the amplitude, gives an intensity proportional to N^2, and widths in each of the three directions specified by $\{\mathbf{i}, \mathbf{j}, \mathbf{k}\}$ that equal $2\pi/(N_a a)$, $2\pi/(N_b b)$, and $2\pi/(N_c c)$ where N_a, N_b, and N_c are the number of conventional cells of the crystal or crystallite in the **i**, **j**, and **k**, or $\langle hkl \rangle$, directions. Note that because $N = N_a N_b N_c$, the "mass" (i.e., the integrated density) of a reciprocal-lattice point is proportional to N, the number of conventional cells in the sample, as expected.

2. Substrate Plus Epitaxial Layer

If a thin single crystalline layer is now grown on top of the substrate crystal described in Section IV.A.1, the resultant reciprocal lattice will reflect this change through the addition of extra points in k-space. Because epilayers are usually significantly thinner than the substrate, these points normally have large widths in the [001] direction, but we will nevertheless consider them as points. As discussed in Section II, two extreme cases can exist: (1) pseudomorphic growth where perfect coherence is maintained parallel to the growth direction (i.e., the epitaxial layer possesses the same in-plane lattice constant as the substrate) resulting in a tetragonally distorted epitaxial layer, and (2) a complete breakdown of pseudomorphic growth (i.e., perfect incoherence) where the epitaxial layer relaxes to its unstrained cubic-diamond lattice, which is completely independent, except in orientation, from that of the substrate. All lattice constants for the epitaxial layer between these limits are possible and depend on the "degree" of relaxation as discussed earlier.

The resultant reciprocal lattice for the substrate plus epitaxial layer becomes a simple superposition of the individual reciprocal lattices for the two systems that can be determined through the definitions given in Eq. (28) and the structure factors given in Eq. (32). Four cases are plotted in Fig. 15b for only the 004, 224, and $\overline{2}\overline{2}4$ reflections. The reciprocal lattice formed by open (filled) squares results from totally relaxed (i.e., unstrained) layers with lattice constants that are 10% greater (less) than the substrate lattice, as illustrated in the inset. From Eq. (28), the added reciprocal-space lattice for these two cases represent nothing more than simple $\pm 10\%$ scale changes (+ for the smaller lattice-constant case, − for the larger one) from that of the substrate lattice. The reciprocal lattice points in Fig. 15b, which are plotted as open (filled) triangles, result from the commensurate (coherent) growth of an epitaxial layer that, as with the previous two cases, possess unstrained lattice constants 10% greater (less) than that of the substrate (see inset). For these last two cases, we have assumed that the crystalline layer that is grown is incompressible (i.e., the Poisson ratio is 0.5 or $c_{11} = c_{12}$). For these tetragonally distorted layers, the reciprocal lattice maintains the same registry in the parallel ($hh0$) direction, but distorts in the perpendicular (001) direction by $\pm 20\%$ (+ for the smaller lattice constant, − for the larger one).

The common feature of the reciprocal lattices plotted in Fig. 15b is that of the substrate, and it is for this reason that almost all x-ray diffraction experiments, and particularly double-crystal diffraction, involve measurements that are relative to the substrate. It therefore becomes convenient to define "x-ray" strains that are relative to the substrate:

$$\varepsilon_{\text{epi}}^{\|} = \frac{a_{\text{epi}}^{\|} - a_{\text{sub}}^{0}}{a_{\text{sub}}^{0}} \qquad \varepsilon_{\text{epi}}^{\perp} = \frac{a_{\text{epi}}^{\perp} - a_{\text{sub}}^{0}}{a_{\text{sub}}^{0}}, \qquad (33)$$

where the superscripts indicate strains parallel (i.e., in the $\mathbf{i}$, $\mathbf{j}$ plane) or perpendicular (i.e., along $\mathbf{k}$, which is usually the surface normal) to the layers, expressed in terms of shifts in reciprocal space,

$$\varepsilon_{epi}^{\|} \approx -\frac{G_{epi}^{\|} - G_{sub}^{\|}}{G_{sub}^{\|}} \qquad \varepsilon_{epi}^{\perp} \approx -\frac{G_{epi}^{\perp} - G_{sub}^{\perp}}{G_{sub}^{\perp}}, \qquad (34)$$

for all *hkl* points where the substrate reciprocal lattice points have nonzero parallel and perpendicular components (e.g., the shifts from the 004 point cannot be used to determine parallel strains, and so on). These x-ray strains are easily related to the "true" parallel and perpendicular strains, e, discussed in Section II, by the equation

$$e_{epi}^{\perp,\|} = (\varepsilon_{epi}^{\perp,\|} + 1)(a_{sub}^0 / a_{epi}^0) - 1, \qquad (35)$$

where a_{epi} is the lattice constant of an unstrained epitaxial layer. The x-ray strains, both parallel and perpendicular, can be determined by measuring the shift of the layer's reciprocal lattice point(s) relative to that of the substrate, i.e., the solid points in Fig. 15b, whereas the true strains present in the layer are represented by the shifts measured relative to that of a hypothetical unstrained layer (i.e., shown by the filled and open squares in Fig. 15b).

Of course the real strains present in the epitaxial layer are not independent, but are related to each other through elastic constants as indicated in the Section II. The perpendicular and parallel x-ray strains are correspondingly related by the equation

$$\varepsilon_{epi}^{\perp} = (1 + 2c_{12}/c_{11})f - (2c_{12}/c_{11})\varepsilon_{epi}^{\|}, \qquad (36)$$

where

$$f = \left[\frac{a_{epi}^0 - a_{sub}^0}{a_{sub}^0}\right] \qquad (37)$$

is referred to as the lattice mismatch of the layer.

3. Strained-Layer Superlattices

Even more points are added to the reciprocal lattice for the case of a SLS. As indicated above, the SLS is characterized by the repetition of two layers, a and b. The layer widths are t_a and t_b, and therefore the period of the SLS is $P = t_a + t_b$. The layers are further specified by two structure factors, $\mathscr{S}_a$ and $\mathscr{S}_b$, and two sets of parallel and perpendicular x-ray strains, ε_a and ε_b. The buffer layer is characterized by a layer width, t_{buf}, and a structure factor $\mathscr{S}_{buf}$, in addition to parallel and perpendicular x-ray strains, ε_{buf}.

The reciprocal lattice points of the buffer layer will be exactly as prescribed in the previous section for an epitaxial layer where one additional point is added for each substrate point. The position of these points depends on the parallel and perpendicular x-ray strains present in the buffer layer.

For the SLS, the situation is more complicated due to the interferences that result because of the multiplicity of layers. Several excellent descriptions of x-ray diffraction from SLSs exist (Speriosu, 1981; Speriosu and Vreeland, 1984; Segmuller et al., 1977; Fewster, 1986), and the reader is referred to these articles for details; we will simply state results here. The added x-ray diffraction peaks that result from the SLS are called *satellites*, and we will refer to the added points in reciprocal space as satellites also. Figure 16 shows the reciprocal lattice points for a SLS system. As in Fig. 15b, this is an *hhl* cut in *k*-space where, for reasons of clarity, only points in the vicinity of the substrate 004, 224, and $\overline{2}\overline{2}4$ points (solid circles) are indicated. The buffer-layer points are plotted as open squares, and the SLS points are plotted as filled squares.

By convention, the SLS satellites are numbered $\{\ldots -2, -1, 0, 1, 2 \ldots\}$, where the ascending order corresponds to increasing $G_\perp$ and, as will be shown later, where an increasing angle of incidence of x-rays is to be diffracted. The point labeled 0, or the zeroth-order satellite, is located at a position characterized by the average perpendicular and parallel x-ray strains, $\langle \varepsilon_{SLS} \rangle^\perp$ and $\langle \varepsilon_{SLS} \rangle^\parallel$ given by

$$\langle \varepsilon_{SLS} \rangle^\perp = \frac{t_a \varepsilon_{sls(a)}^\perp + t_b \varepsilon_{sls(b)}^\perp}{P}, \qquad (38)$$

and likewise for $\langle \varepsilon_{SLS} \rangle^\parallel$. In other works, the zeroth-order satellite is found at the same position as that resulting from an epitaxial layer (not an SLS) that has perpendicular and parallel x-ray strains with these average values. The

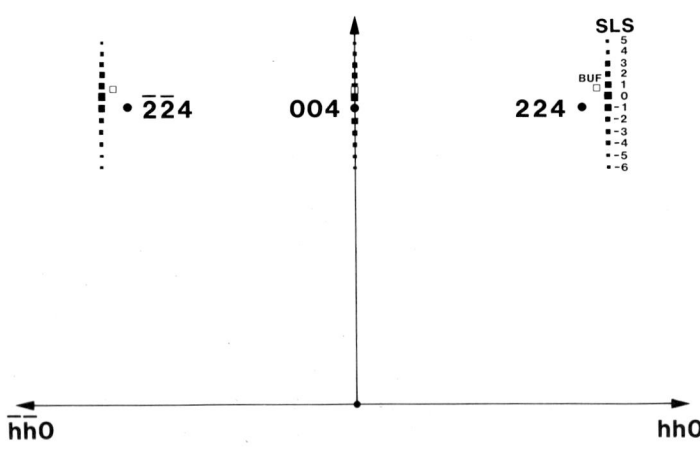

FIG. 16. Reciprocal-space lattice for a strained-layer superlattice on a nearly-matched buffer layer on substrate.

3. STRUCTURE AND CHARACTERIZATION OF SUPERLATTICES

remaining SLS satellites are separated by $2\pi/P$ in the $G_\perp$ or [001] direction. The intensity of these satellites will be discussed in Section IV.B.6.

4. Deviations from Perfection

In the previous sections, we assumed the substrate, epitaxial, buffer, and SLS layers to be perfect, i.e., each layer is a perfect crystal (no grain boundaries or variation in layer orientation or lattice constant, parallel or perpendicular). This situation is seldom encountered, and it is therefore important to discuss the perturbations that are applied to the reciprocal-space lattice when variations from perfection occur.

The effect of terracing is shown schematically in Fig. 17a. If a SLS is grown on a substrate that has a surface cut slightly off axis, which is usually the case, then the chemical modulation direction is not aligned with the [001] direction of the substrate, while the lattice vectors remain aligned. For the case in which this angle of miscut is α, Neumann et al. (1983, 1985) have shown that the SLS satellites are rotated about the zeroth-order satellite by the angle α, and that the zeroth-order satellite itself is rotated about the (000) point by an angle

$$\beta = 2\alpha f_b$$

with respect to the substrate lattice, where f_b is the mismatch strain (see Eq. (37)) of layer b with respect to the substrate. Layer a is identical to the substrate in Neumann's development.

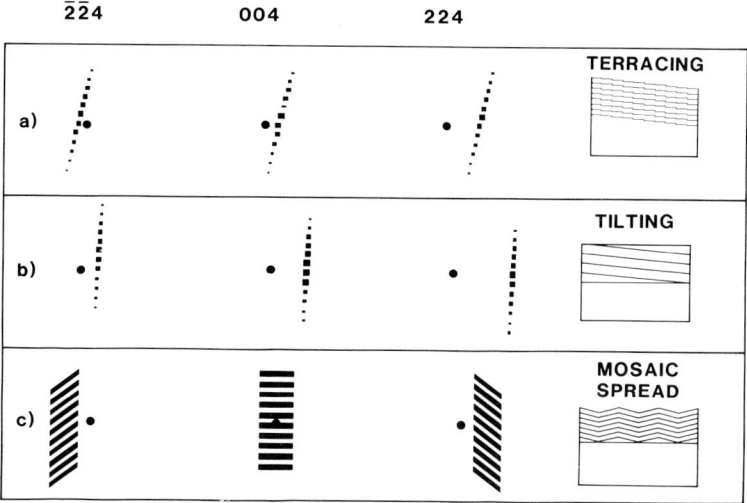

FIG. 17. Schematic of reciprocal-space diffraction intensities for a superlattice on a substrate with (a) terracing, (b) tilting, and (c) mosaic spread present for $\bar{2}\bar{2}4$, 004, and 224 directions.

The reciprocal lattice of a tilted SLS is shown in Fig. 17b. All of the SLS points can become tilted about the (000) position when the SLS completely loses registry from that of the substrate. This effect can occur for SLSs grown on substrates with large surface miscuts and represents an extreme case of terracing, where the lattice vectors of the SLS are actually rotated from that of the substrate by the miscut angle α.

The effect of mosaic broadening is depicted in Fig. 17c. A rotational smearing centered at (000) of the reciprocal lattice points results if the crystallites described in the previous paragraph are not perfectly aligned with respect to the perpendicular [001] direction.

If the individual layer thicknesses of the SLS fluctuate, two possible effects can modify the reciprocal lattice. If the fractional fluctuations are small, which is the case for relatively thick layers of ≥ 10 nm, for a system with a large number of periods, Speriosu and Vreeland (1984) have shown that the intensity of the satellites are affected in a fashion similar to a Debye–Waller factor. For the case where the percentage fluctuations become large, say, for layers of ≤ 10 nm, for SLS systems with large numbers of periods, Auvray et al. (1987) have shown that the satellites themselves can split into multiplets that are characteristic of the discrete distribution of SLS periods within the sample.

If the layer(s) is(are) not single crystalline, but rather composed of many crystallites, perfect in the perpendicular $\langle 001 \rangle$ direction but separated by low-angle grain boundaries laterally (i.e., subgrain boundaries), then the points in reciprocal space that correspond to this layer each broaden in the ($hk0$) plane to $2\pi/L$, where L is the average size of each crystallite.

B. Mapping of the Reciprocal Lattice

In this section, we will describe the experimental techniques by which XRD can be used to either directly or indirectly determine reciprocal-space-density maps in addition to outlining the methodology by which this data is analyzed to obtain the structure of an epitaxial layer or SLS.

Explained at a rudimentary level, XRD systems involved in SLS analysis are all quite simple. As shown in Fig. 18a, the procecure consists of positioning a sample at a variable angle ω with respect to an incident monoenergetic and highly collimated beam of x-rays and of detecting the diffracted x-ray intensity as a function of the scattering angle 2θ. The devices used both to provide the incident x-rays and to detect the diffracted x-rays are what distinguish the various techniques used in the XRD analysis of SLSs.

1. The Ewald Construction

A common feature of all XRD techniques is that they can be understood simply through the Ewald construction in reciprocal space. Such a construction is shown in Fig. 18b for the same reciprocal-space lattice shown in

a) SCATTERING GEOMETRY

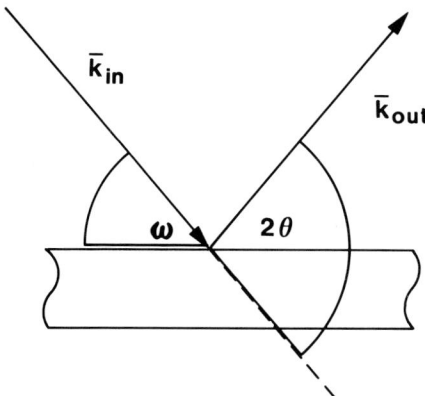

b) EWALD CONSTRUCTION

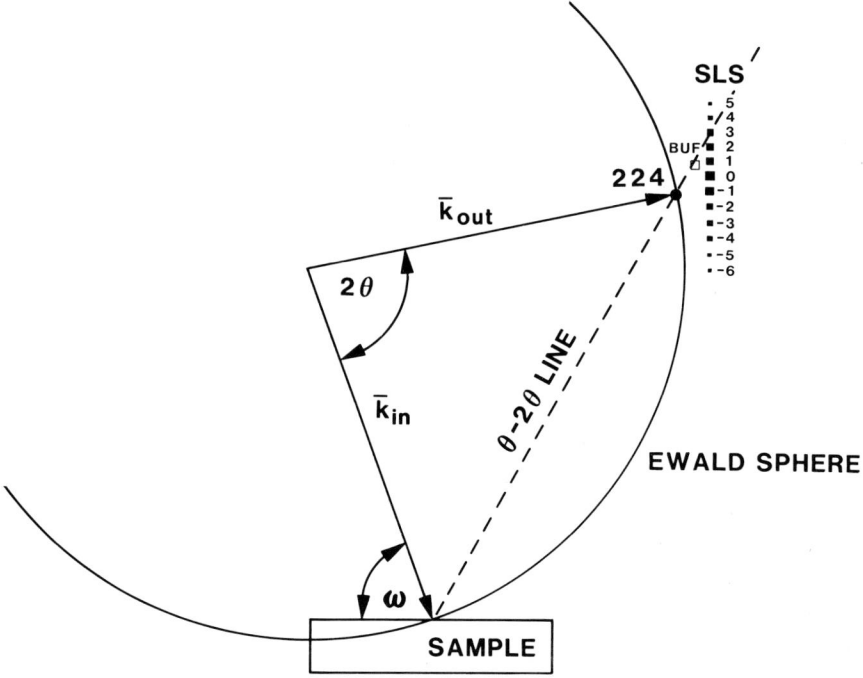

FIG. 18. Schematic illustration of (a) x-ray scattering geometry and (b) corresponding Ewald-sphere construction for a strained-layer superlattice.

Fig. 16—a substrate, buffer, and SLS system—and incident $CuK_{\alpha 1}$ x-rays. In an Ewald construction, the incident x-ray's **k**-vector, $\mathbf{k}_{in}$, is directed at the (000) reciprocal lattice point with an angle of incidence of ω with respect to the $hk0$ plane (this is the same ω angle of incidence in real space for crystals with no miscut). A sphere is then drawn that is centered at the origination point of the incident **k**-vector with a radius of $2\pi/\lambda$, the **k**-vector's length. Because x-ray diffraction involves elastic scattering and the fact that (usually) only monoenergetic x-rays are used in the analyses to be discussed, this "Ewald" sphere represents the locus of all possible out-going or diffracted x-ray **k**-vectors, $\mathbf{k}_{out}$, if they are considered to originate at the center of this sphere. By definition, the Ewald sphere passes through the 000 reciprocal lattice point, and diffraction will result along $\mathbf{k}_{out}$ only if it passes through any additional reciprocal lattice points. This diffraction occurs because, for this case, the scattering vector, $\Delta \mathbf{k} = \mathbf{k}_{in} - \mathbf{k}_{out}$, equals a reciprocal lattice vector $\mathbf{G}_{hkl}$, which is the condition for diffraction expressed in Eq. (27).

In Fig. 18b, the incident angle ω is selected so that the Ewald sphere passes through the 224 substrate point, and hence the strong 224 diffraction would be observed by a detector positioned along $\mathbf{k}_{out}$. All of the x-ray diffraction techniques to be discussed are made relative to the substrate, since this is usually the most intense and easily identifiable feature in the diffraction spectrum.

2. $\theta-2\theta$ Diffractometry

The first system to be discussed is a $\theta-2\theta$ diffractometer. With this method, the incident x-rays are usually not monochromatic but, through the use of filters or low-resolution monochromators, limited to the $K_{\alpha 1,2}$ x-ray doublet. The use of this doublet results in a splitting of the Ewald sphere drawn in Fig. 18b into two concentric spheres, where the inner sphere (not shown) has a radius corresponding to the weaker $K_{\alpha 2}$ line. The $\theta-2\theta$ mechanism maintains the orientation of the sample and the detector such that $\omega = \theta$, and therefore the locus of points examined in reciprocal space by this method lies along a straight line that passes through the (000) point. The dashed line in Fig. 18b shows the region of reciprocal space examined by a $\theta-2\theta$ system aligned to maximize the intensity of the substrate (224) reflection. A $\theta-2\theta$ diffractometer is not the best method to analyze a SLS because, even when the dual diffraction peaks resulting from the x-ray doublet can be resolved, optimal alignment of the sample is at times impossible. These diffractometers can be used to determine the period of SLSs using 004 reflections when terracing and/or tilts are negligible; the results of such an analysis are shown in Fig. 19 for a GaAsP/GaP superlattice. One advantage of this technique over some of the others is that sharp, well-defined diffraction peaks are obtained for (004) reflections even when crystallite-size or mosaic effects are appreciable. This is

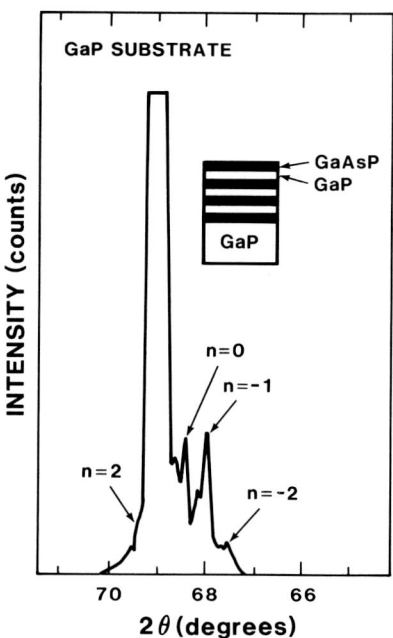

FIG. 19. X-ray diffraction scan for a θ–2θ diffractometer from a GaAs$_x$P$_{1-x}$/GaP superlattice.

because the line examined in reciprocal space lies perpendicular to that of these broadening distortions.

3. DCD Rocking Curves

Rocking-curve measurements with double crystal diffraction (DCD) is by far the most common XRD technique used to characterize the structure of both epitaxial layers and superlattices. The x-rays from a tube are passed through a single- or multiple-bounce crystal monochromator, which is set to diffract only the $K_{\alpha 1}$ x-rays. This highly monochromatized and parallel beam of x-rays then strikes the sample at an incident angle of ω, which can be set with an accuracy ranging from 0.1 to 1 arc sec. The range of ω is limited to a few degrees. There are two possible orientations (see Fig. 20) of the sample crystal: (1) In the so-called + − geometry, the x-ray diffraction is nondispersive (i.e., x-rays with different wavelengths will diffract at the same ω setting) if both the first (monochromator) and second (sample) crystals have the same lattice constant; (2) in the so-called + + geometry, the x-ray diffraction is dispersive. Therefore, the + − geometry is preferred. The x-rays that diffract from the sample are detected with a fixed detector, which typically has a 2θ

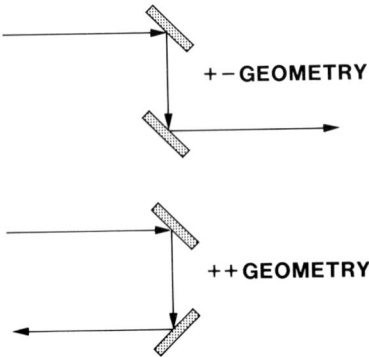

FIG. 20. Schematic of the two possible x-ray scattering geometries for the double-crystal diffraction measurements.

angular range of ~10°. Referring to Fig. 18b, with DCD, the reciprocal lattice points are rocked through the Ewald sphere, and the fixed detector integrates the overlap. The measurements are made relative to the substrate peak in the form of intensity versus $\Delta\omega$, with the substrate reflection being set to $\Delta\omega = 0$. An example of a DCD rocking-curve measurement and the analyzed strain profile for an InGaAs/GaAs superlattice is shown in Fig. 21. This SLS is similar to that depicted in the RBS analysis shown in Fig. 7. As many as 12 satellite superlattice peaks are resolved in Fig. 21, and good agreement between the fitted DCD rocking curve and the measured data points is achieved. However, one problem with DCD analysis is that, because only the integrated overlap of the Ewald sphere and k-space density is recorded, the broadening distortions caused by crystallite-size and-or mosaic spread can result in unresolvable SLS satellite peaks. It is common to perform several (at least two) DCD scans with different azimuthal rotation settings of the sample crystal (these are rotations about the sample normal) to ascertain any significant terracing or tilting effects on the rocking curves.

4. Triple-Axis System

Recriprocal-space grid maps are determined with a triple-axis diffractometer. The geometry is identical to a DCD, with the exception that a θ–2θ diffractometer (this is the third axis) is used for the final detection of the diffracted x-rays. This system provides a very-high-resolution (arc seconds) measurement of the 2θ angle of diffraction from the sample crystal. By mechanically varying both the ω and the 2θ setting of the diffractometer, one obtains a point-by-point density map of the points in reciprocal space. For measurements made relative to the substrate peak, i.e., diffraction intensity versus $\Delta\omega$ and $\Delta(2\theta)$ where the substrate peak occurs at $\Delta\omega = \Delta(2\theta) = 0$, it is

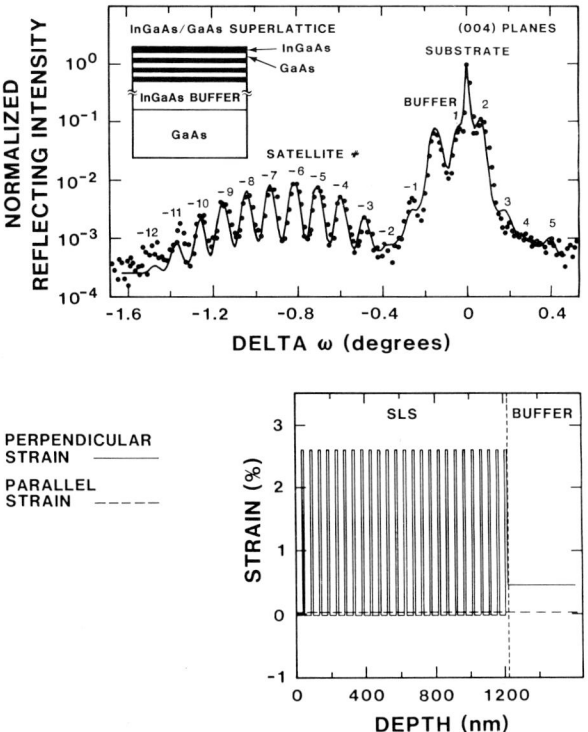

FIG. 21. Double-crystal diffraction x-ray rocking curve for (004) planes of an $In_{0.19}Ga_{0.81}As/GaAs$ (40 nm/8.7 nm) strained-layer superlattice on $In_yGa_{1-y}As$ buffer layer on GaAs substrate. Lower panel shows the x-ray strain profile (strain referenced to GaAs substrate) for perpendicular and parallel directions based on (004) and (224).

straightforward to show that

$$|\mathbf{G}| = (G_\perp^2 + G_\parallel^2)^{1/2} = \frac{4\pi \sin(\theta_B + \Delta\theta)}{\lambda}, \quad (39)$$

$$G_\parallel/G_\perp = \tan(\Phi - \Delta\theta + \Delta\omega), \quad (40)$$

where θ_B is the Bragg angle for hkl diffraction from the substrate calculated using Eq. (24) with $d = d_{hkl}$, the planar separation of the hkl planes, and Φ is the angle between the hkl planes and the surface. A k-space grid map is shown in Fig. 22, which depicts the effect of terracing in a GaAs/GaAsSb SLS structure. This figure shows x-ray satellites near the (004) and (002) reflections. "Forbidden" reflections such as the (002) reflection are at times useful and sensitive measures of the broken symmetry resulting from strains.

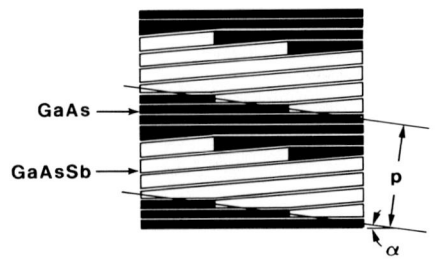

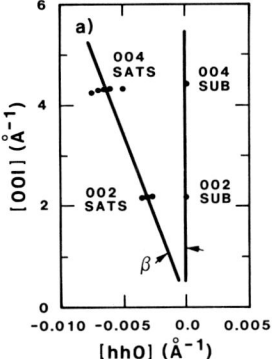 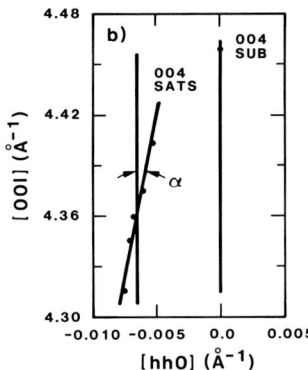

FIG. 22. Triple-axis x-ray diffraction measurements of a GaAsSb/GaAs strained-layer superlattice illustrating the effect of terracing on the superlattice diffraction spots. The map in (a) shows a large region of reciprocal space projected onto the [hhl] plane, while (b) is a closeup of the (004) substrate and satellites. A schematic diagram of the terraced SLS is also shown (top). Note that the terrace angle, α, can be directly determined in (b). The angle β determined in (a) depends on α and the amount of strain present in the SLS. After Neumann et al. (1985).

High-flux x-ray sources, such as a rotating anode or synchrotron are usually required to measure these weak reflections due to low signal strength.

5. The DCD + PSD System

A newly invented technique that can be used to determine reciprocal-space density maps is the use of a position-sensitive detector (PSD) as the x-ray detector in a conventional DCD system (Thompson and Doyle, 1989). Using a line-focus x-ray tube (40 μm beam width) and a curved position-sensitive proportional counter with a resolution of 60 μm, a $\Delta 2\theta$ resolution of 75 arc sec has been demonstrated. Although this is significantly poorer than the resolution in $\Delta\omega$, which is usually a few arc seconds, this DCD + PSD

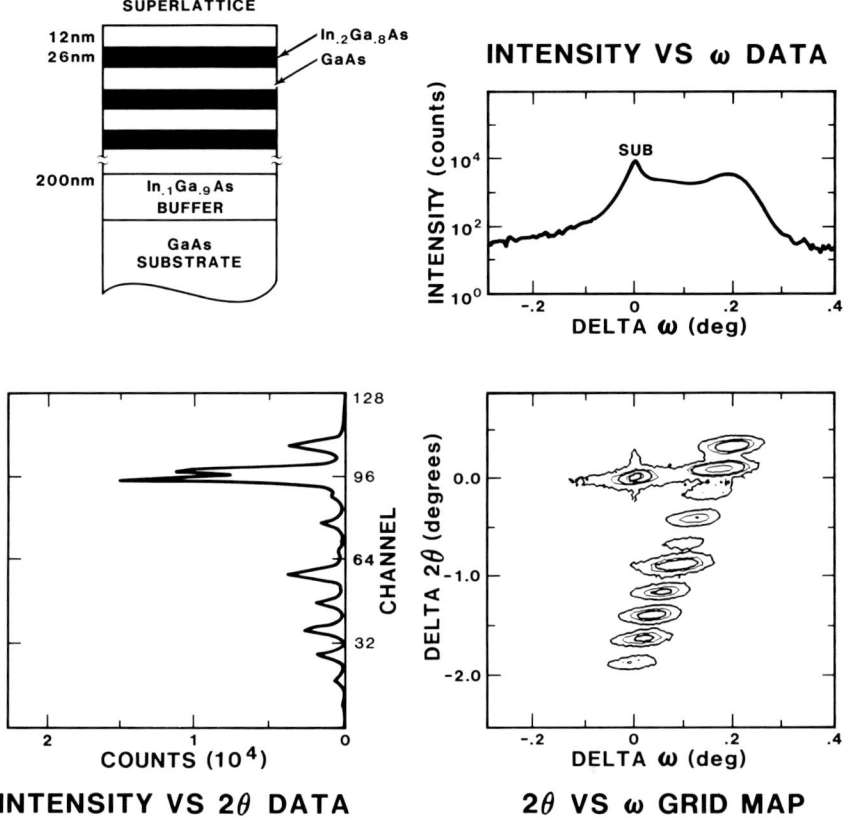

FIG. 23. Schematic of an $In_{0.2}Ga_{0.8}As/GaAs$ strained-layer superlattice (upper left) and the results of an x-ray double-crystal diffraction measurement with a position-sensitive detector to obtain a reciprocal-space density map. Upper and left-hand curves give the integrated intensity scans along the ω and 2θ directions, respectively. The individual superlattice satellite peaks lie in a straight line with the substrate peak to the left. Interpretation of the analysis of such structure is given in Section VIII.C.

method has proven to be quite valuable in analyzing the structure of SLSs and textured layers. K-space density maps are obtained through the $\{\Delta\omega, \Delta 2\theta\}$-to-$\{G_\parallel, G_\perp\}$ transform equations given in Eqs. (39) and (40). An example of an ω–2θ DCD + PSD analysis is given in Fig. 23, where a tilt effect is displayed for a GaAs/InGaAs SLS for (224) diffraction. In addition to providing a quick and easy determination of reciprocal-space density, the DCD + PSD technique can be used to resolve individual SLS satellite peaks for cases where normal DCD rocking curves cannot. An example of this situation is shown by the projections plotted in Fig. 23. The projection onto the $\Delta\omega$ axis is identical to a normal rocking curve, and the SLS satellites are not resolvable. The projection onto the $\Delta 2\theta$ axis gives a scan similar to that obtained with a θ–2θ diffractometer; it shows the individual satellites resolved but does not allow the tilt misorientation to be detected.

6. Analysis

The interpretation of XRD data would best proceed by first converting it into a reciprocal-space representation. This is rarely done, however, since most of the XRD analysis of SLSs is performed with either θ–2θ diffractometers or DCD rocking-curve systems. With either of these techniques, it is rather straightforward to convert the layer or superlattice-peak position/intensity data directly into real-space strain profiles using the kinematic theory of x-ray diffraction (Mathieson, 1982; Speriosu, 1981; Speriosu and Vreeland, 1984). In the following, we outline only the final results of such theoretical analysis. It is useful to define a parameter, C, which occurs several times in the development

$$C = \lambda|\sin(\omega_0)|/\sin(2\theta_B), \tag{41}$$

where ω_0 is the incidence angle for diffraction from the substrate.

The peak position $\Delta\omega_0$ relative to that of the substrate for an epitaxial or buffer layer is related to the change in hkl planar spacing, Δd, and the rotation of these planes, $\Delta\psi$, by the relation

$$-\Delta\omega_0 = \tan(\theta_B)\Delta d/d_{hkl} + \Delta\psi, \tag{42}$$

where the first term results from the differentiation of Bragg's law in Eq. (24), and $\Delta d/d_{hkl}$ and $\Delta\psi$ are expressed in terms of parallel and perpendicular x-ray strains by

$$\Delta d/d_{hkl} = \varepsilon_{epi}^{\perp} \cos^2 \psi + \varepsilon_{epi}^{\parallel} \sin^2 \psi, \tag{43}$$

and

$$\Delta\psi = \pm(\varepsilon_{epi}^{\perp} - \varepsilon_{epi}^{\parallel})\sin\psi \cos\psi, \tag{44}$$

where the $+$ or $-$ sign is selected in Eq. (44) according to whether the angle of incidence, ω_0, equals $\theta_B - \psi$ or $\theta_B + \psi$, respectively. Note that it is the planar rotation expressed in Eq. (44) with the $-$ sign that is measured using ion channeling.

For the analysis of an SLS, the shift $\Delta\omega_0$ of the zeroth-order satellite from that of the substrate is related to the thickness-averaged parallel and perpendicular x-ray strains, $\langle\varepsilon_{sls}^{\parallel}\rangle$ and $\langle\varepsilon_{sls}^{\perp}\rangle$, through relations analogous to those expressed in Eqs. (42)–(44) with the substitutions, $\varepsilon_{epi}^{\parallel} \to \langle\varepsilon_{sls}^{\parallel}\rangle$ and $\varepsilon_{epi}^{\perp} \to \langle\varepsilon_{sls}^{\perp}\rangle$.

Two different hkl diffraction measurements are therefore required to unambiguously determine either the average, in the case of a superlattice, or the absolute, in the case of a buffer or epitaxial layer, parallel and perpendicular x-ray strains. The most commonly used combinations are the (004) and (224) or the (004) and (115) when Cu or Fe x-rays are utilized to study semiconductor or compound semiconductor systems. In addition, it is possible to determine the planar rotation by combining measurements of (224) and ($\bar{2}\bar{2}4$) or other symmetric reflections.

The period of an SLS can be determined by measuring the angular separation of adjacent diffraction satellites, $\Delta\omega_P$, by

$$P = C/\Delta\omega_P. \tag{45}$$

For a perfect SLS where the x-ray satellites are easily resolved, the kinematic theory can be used to show that, for negligible x-ray attenuation in the sample, the integrated peak intensity of the nth SLS satellite, I_n, is given by

$$I_n \propto \sin(Ct_{sls(a)}\omega_{an})[\mathscr{S}_{sls(a)}/\omega_{an} - \mathscr{S}_{sls(b)}/\omega_{bn}], \tag{46}$$

where

$$\omega_{an} = \Delta\omega_n + K^\perp \varepsilon_a^\perp + K^\| \varepsilon_a^\|, \tag{47}$$

with

$$K^\perp = \cos^2\psi \tan\theta_B \pm \sin\psi \cos\psi, \tag{48}$$

$$K^\| = \sin^2\psi \tan\theta_B \pm \sin\psi \cos\psi, \tag{49}$$

and $\mathscr{S}_a$ and $\mathscr{S}_b$ are the structure factors for layers a and b. $\Delta\omega_n$ is the incident angle relative to the substrate-diffraction condition for the nth satellite. If the satellite intensities are reduced to fractional intensities relative to the zeroth satellite, then Eqs. (46)–(49) can be used to theoretically predict these intensity fractions as a function of six parameters t_a, t_b, $(\varepsilon_a, \varepsilon_b)^{\perp, \|}$. Nonlinear least-square fitting is used to obtain the optimal values for these parameters. Again, two different reflections must be analyzed to unambiguously separate the parallel and perpendicular x-ray strain components.

The kinematic theory must be applied in layer-by-layer fashion for the more general situation when the SLS is not perfect. In this case, each layer is described by a thickness, $\varepsilon_{sls(i)}^\|$, $\varepsilon_{sls(i)}^\perp$, and a structure factor $\mathscr{S}_{sls(i)}$. The resulting plane-wave solution for x-ray diffraction, which is usually broadened to account for experimental artifacts and distortions in the sample, is then compared with the experimentally determined rocking curve to obtain an optimal strain profile.

7. Pitfalls

While XRD has proven extremely valuable in determining the strains present in epitaxial layers and superlattices, both the techniques and the simplified analyses presented earlier do have restrictions. We shall discuss some of these limitations here.

It was indicated above that the diffraction intensity is proportional to the number of conventional cells examined in the sample. For a structure consisting of an epitaxial layer on a substrate, the diffraction intensity from the layer therefore becomes weaker as the layer thickness decreases. Furthermore, the diffraction peak of the layer sits on top of a background formed by the Lorentzian tail of the substrate peak, and therefore the detectability of the epitaxial layer peak depends not only on the power of the x-ray tube used in the experiment, but also on the amount of strain present in this layer. For example, for layers strained at the $\sim 1\%$ level, it is straightforward to detect the diffraction peak of layers only 50 nm thick using DCD with fine-focussed x-ray tubes with nonrotating anodes. On the other hand, if the layer strain is below 0.1%, the layer must be considerably thicker so that a diffraction peak separable from that of the substrate is formed. In addition, Fewster and Curling (1987) have shown that the simple previous analysis, which relates layer strains to diffraction peak positions, does not apply for systems that are nearly lattice matched when the layers are thinner than 1–2 μm because of dynamic diffraction effects.

Speriosu and Vreeland (1984) have pointed out that, because of superlattice imperfections such as strain and/or period variations, the satellite structure predicted by the analytical SLS diffraction theory rapidly breaks down if the number of superlattice periods is small. For cases like this, one must therefore use a more general x-ray diffraction theory, e.g., the kinematic theory of Speriosu (1981) or the dynamic theory of Wie *et al.* (1986), to determine the layer-strain profiles. Theories like these must also be used when the strain profiles are not periodic as in the case of implanted SLSs; and it is important to recognize that the profiles so determined are not always unique.

It is also very important to properly account for the rotations that result from terracing and tilting. This is automatically done when reciprocal-space density maps are measured, but the resultant peak shifts in the projected diffraction scans of the DCD measurements are easy to overlook. A commonly used technique (Myers *et al.*, 1986) is to make two DCD measurements for each reflection examined, where the sample is rotated 180° about its surface normal. These two azimuthal measurements are then used to determine the "average" position of the substrate peak with respect to the satellites and thereby account, at least to first order, for the shifts caused by terraces and tilts.

Finally, because RBS-channeling measurements are sometimes made in conjunction with XRD-based characterizations to determine the structure of strained layers, it is important to perform the XRD characterizations first. Although RBS does not damage the crystal to the extent that would alter a depth profile, it does reduce the crystalline quality of both the superlattice and the substrate, which results in diffraction peaks of reduced intensity and generally of increased width.

V. Other Characterization Techniques

In Table I, we give an overview and comparison of many of the characterization techniques that are useful for strained epitaxial layers and superlattices. As mentioned in the introduction, because of our own research interests, we have given primary emphasis in this chapter to ion-scattering and x-ray diffraction methods. However, many other powerful methods are in common use as well. For completeness, we shall discuss a number of additional techniques in this section, but shall not attempt a thorough treatment of these. A more detailed discussion for the specific technique of transmission electron microscopy can be found in Chapter 1 by Hull and Bean.

When applying these methods to the structural characterization of strained heteroepitaxial layers, a central issue is the quantitative measurement of the strain present in the layers. This requires techniques that are sensitive to lattice structure within thin layers. The first half of this section describes such techniques. At the same time, an epitaxial layer that is not fully commensurate with its substrate will have undergone strain relief by the formation of defects. Therefore, the amount of strain relief per defect and the defect density determine the strain remaining in the layer. Thus, techniques that detect and image strain-relieving defects in thin layers are equally important and provide a complementary approach to the strain measurements. In the second half of this section, we therefore discuss important methods for characterizing the defect structures in these materials. In some cases, additional defects, such as stacking faults, may be introduced during layer growth or subsequent processing. It is particularly important in such cases, to carry out both lattice-structure and defect-structure measurements.

A. LATTICE-STRUCTURE METHODS (RHEED, RAMAN SCATTERING)

We have discussed ion-channeling and x-ray diffraction methods for the lattice-structure characterization of strained epitaxial layers. Although these techniques are versatile and are frequently used, there are several other techniques that have proven useful for characterizing the state of strain in crystalline films. We discuss two of these here: RHEED and Raman spectroscopy. TEM is also applicable to structure measurements, but it is more powerful for characterizing laterally inhomogeneous systems. Thus, it is more typically used to image and characterize the defects present in heteroepitaxial films, as discussed in the second half of this section. An interesting feature of the RHEED and Raman techniques is that these methods have been found to be convenient for in situ characterization. In this way, it is possible to characterize the evolution of the strain in layers directly as their growth proceeds.

1. RHEED

RHEED is a powerful real-time diagnostic for characterizing surface structure and morphology and has been extensively used for monitoring of the qualitative evolution of surface structure during molecular-beam epitaxy (MBE) growth (Zhang et al., 1987; Cohen et al., 1986). In this technique, monoenergetic electrons (typically at 10 to 40 keV) are incident at low angles (1° to 5°) to the surface. The diffracted electron pattern is imaged and observed visually on a phosphor screen. The signal emanates primarily from the top few atomic layers, and so the intensity pattern of the diffracted electrons provides a measure of the surface and near-surface structure. Whereas qualitative measurements have primarily focused on the reconstructed state and "quality" of the surface, increasing use is being made of RHEED as a quantitative tool. For quantitative measurements, the position and/or intensity profiles of the diffracted beams are measured either by digitizing the phosphor image through video techniques or by directly scanning the electron intensities with a collimated detector.

The surface lattice spacing can be directly determined from the spacing of the diffracted beam through the usual diffraction condition for constructive interference,

$$n\lambda = a \sin \theta, \qquad (50)$$

where the electron wavelength at energy E is $\lambda = h/\sqrt{2mE}$, $n\lambda$ is an integral number of wavelengths, and ($a \sin \theta$) is the projection of the interatomic spacing along the diffracted direction of the beam. Under the approximation of diffraction by only the surface atoms, the diffracted beams correspond to Bragg rods in reciprocal space, which intersect the Ewald sphere at low angles and are detected on the screen as spots or streaky spots. The angular spacing between spots, $\Delta\theta$, then gives the atomic spacing along rows

$$a = \lambda/\sin \Delta\theta. \qquad (51)$$

This approach has been used to monitor the evolution of strained layers in situ during growth, where sensitivities of 0.1% or better were achieved. An example is shown in Fig. 24 for $In_{0.33}Ga_{0.67}As$ strained-layer growth on GaAs(001) by Whaley and Cohen (1988). The growth has been interrupted after each monolayer of deposition for the RHEED in-plane lattice parameter measurements. Here, we see that the in-plane lattice constant initially remains fixed at a strained value corresponding to the film-substrate lattice mismatch (2.3%) up to a critical growth thickness, after which a rapid increase occurs with thickness corresponding to increasing amounts of strain relief. Also, the onset of relaxation depends markedly on the growth temperature, demonstrating the metastable nature of these films. This behavior is consistent with the framework to be discussed in Section VI for thin strained films.

RHEED measurements are also sensitive to surface morphology through measurements of the diffraction-spot intensity profiles. In the approximation of surface atom scattering, the intensities of the Bragg rods perpendicular to the surface are determined by the surface-atom pair-correlation function. Although only particularly simple pair correlation functions have been determined exactly, the general trends are that a smooth single-layer surface gives narrow Bragg rods that are unmodulated along their length, resulting in the observation of sharp RHEED spots. In contrast, for a surface with many steps (in-plane roughness), the rods will broaden laterally, and for a surface distributed over many layers (out-of-plane roughness), scattering from the different layers will interfere, and the Bragg rods will become longitudinally modulated, resulting in broadened spots and streaks. Quantitative RHEED interpretation is difficult because the strong interactions of electrons with the atoms in a solid cause significant multiple scattering, giving rise to dynamical scattering effects, so that the usually applied kinematic scattering treatments

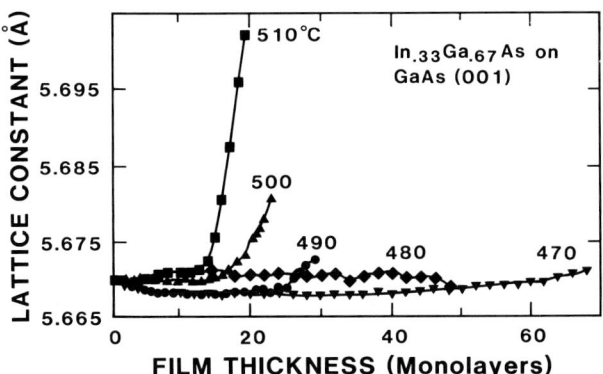

FIG. 24. In-plane lattice constant of $In_{0.33}Ga_{0.67}As$ epitaxial layers on GaAs (001) as a function of layer thickness and growth temperature as measured by RHEED. After Whaley and Cohen (1988).

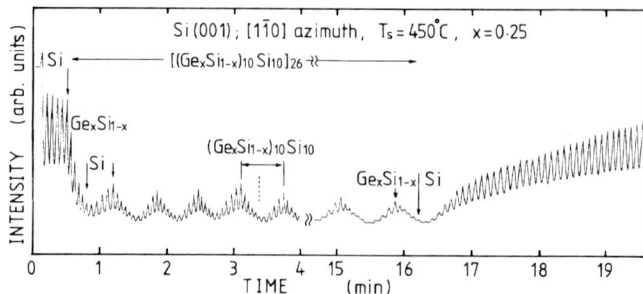

FIG. 25. RHEED intensity oscillations for $Ge_{0.25}Si_{0.75}$ and Si layer growth on Si(001) at 450°C measured along a $\langle 110 \rangle$ azimuth. As marked by the arrows, the magnitude of the oscillations increases during Si growth and decreases during Ge_xSi_{1-x} growth. From Sakamoto et al. (1987).

are approximate. However, kinematic treatments can give a first-order explanation of the observed intensity decrease and broadening of diffraction spots as surface roughness increases.

The in situ observation of surface roughening during strained-layer superlattice growth is illustrated in Fig. 25 by the work of Sakamoto et al. (1987). Here, the RHEED oscillations for Si and SiGe alloys are observed from the $\langle 110 \rangle$ azimuth and are due to the alternating growth of 2 × 1 and 1 × 2 reconstructed surfaces. Thus, each oscillation corresponds to biatomic layer growth. The superlattice region was grown for 26 periods with 10 atomic planes each in the Si and the alloy layers. The amplitude of the oscillations decay during the growth of the $Ge_{0.25}Si_{0.75}$ alloy layers, whereas the amplitude increases during the Si-layer growth. These results suggests a roughening during the alloy growth and recovery (smoothening) during pure Si growth. The above examples demonstrate how information on the morphology of layers, as well as their strain, can be obtained by RHEED measurements during heteroepitaxial growth.

2. Raman Scattering

Raman scattering is another nondestructive probe of the lattice structure of strained-layer superlattices. It has been used to measure layer strains and is also sensitive to lattice disorder in the near-surface region of crystalline semiconductors. In addition, well-defined behavior with confinement within the respective layers has even been observed for the longitudinal optic (LO) phonons in superlattices with individual layers as thin as one to six atomic planes (Ishibashi et al., 1986).

The Raman process is the inelastic scattering of a photon by a crystal, accompanied by the emission or adsorption of a lattice phonon. Thus, for first-order Raman scattering, as found in Si, GaAs, and others, $\omega = \omega' \pm \Omega$, where the frequencies ω and ω' are for the incident and scattered photons, and Ω is the frequency of the phonon that is created or destroyed.

In layers containing biaxial strain, the phonon frequency is shifted in accordance with the sign and magnitude of the layer strain. The Raman-scattering peaks also broaden with the presence of lattice disorder, although this measure of the disorder tends to be used primarily as a qualitative indicator in strained-layer studies. Since the optical phonon modes shift linearly in frequency with the strain in the diamond and zinc-blende semiconductors (Cerdeira et al., 1972), one can easily calibrate the frequency shifts. Thus, for uniform layers, the magnitude of the strain is determined directly from the observed frequency shifts.

An example of Raman-scattering spectra for strained and unstrained layers in the SiGe system is shown in Fig. 26. (Cerdeira et al., 1985; Bean, 1985a). The top panel shows the peaks for a thin $Ge_{0.65}Si_{0.35}$ strained layer grown commensurately on a Si(001) substrate (2.7% misfit strain), and the lower panel shows the peaks for an unstrained $Ge_{0.65}Si_{0.35}$ layer on Si(100),

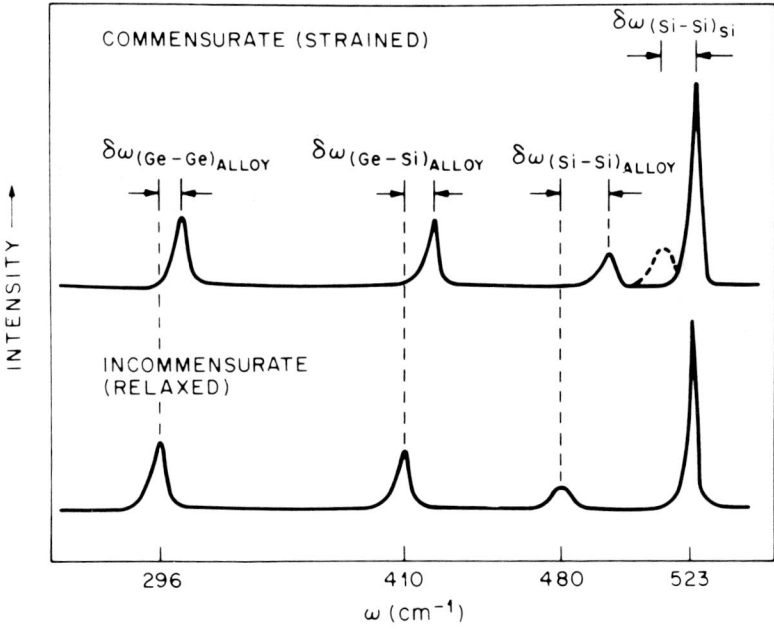

FIG. 26. Raman-scattering spectra of strained (top) and strain-relaxed (bottom) layers of $Ge_{0.65}Si_{0.35}$ grown on Si(001). Compressive displaced Si–Si substrate satellite peak at lower energy (dashed peak position) indicates the absence of significant tensile strain in Si. Reprinted with permission from the Materials Research Society, Bean, John C., "Molecular Beam Epitaxy of Ge_xSi_{1-x}/SiGe Strained-Layer Heterostructures and Superlattices," *Mat. Res. Soc. Symp. Proc.* **37**, 245 (1985).

which was grown thick enough to relieve the strain. Since the substrate Si has the smaller lattice constant, the $Ge_{0.65}Si_{0.35}$ strained layer is under biaxial compressive strain, and therefore the phonon lines are shifted to higher frequencies in this layer. Peaks corresponding to the Ge–Ge, Ge–Si, and Si–Si bonds in the alloy layer are all observed; the Si–Si peak due to the Si substrate at $523\,cm^{-1}$ is also observed. The Si layers of commensurate GeSi/Si superlattice structures on a Si substrate are unstrained, and so there is no satellite associated with the bulk Si–Si peak. However, for Si/GeSi superlattices grown on a SiGe buffer layer of different lattice constant than Si (or for a partly incommensurate superlattice), tensile strain in the Si layers gives rise to an additional Si–Si peak shifted to lower frequencies from the $523\,cm^{-1}$ Si substrate peak.

Similar Raman-scattering studies of single strained layers and superlattices have been carried out for III-V semiconductor systems. For example, Nakayama et al. (1985) have examined the strain-induced shifts for the longitudinal optical phonon modes in $In_xAl_{1-x}As/GaAs$ strained-layer superlattices with 1 to 20-nm-thick layers. In that case, $In_yAl_{1-y}As$ buffer layers were grown with a lattice constant close to the average value of that for the superlattice. Thus, the in-plane compressive and tensile strains in the $In_xAl_{1-x}As$ and GaAs layers were reflected in the corresponding frequency shifts to higher (InAs and AlAs bands) and lower (GaAs bands) values, respectively. Lattice strains of about 0.4% were readily characterized.

Raman scattering has several attractive features. The technique is clearly well suited for in situ measurements in a growth chamber. Also, the Raman scattering signals are sensitive to sample temperature. The phonon peaks both shift to lower frequency and broaden with increasing substrate temperature. Thus, the technique can in principle be used as a real-time, in situ monitor of layer strain or substrate temperature. However, this coupling is also a complication, particularly at very high growth temperatures, where signal-to-background difficulties may be encountered. Another particular feature of Raman scattering is that the measurement sums over the structure from the surface to depths of the order of 10 to 100 nanometers, depending on the light-absorption rate for the material system being studied. Thus, the probing depth and depth resolution coincide. For maximum depth-resolving power, it is desirable to use shorter-wavelength light so as to have higher adsorption coefficients. In contrast, RHEED measurements probe only the first several atomic layers, whereas the x-ray diffraction and RBS/channeling techniques probe depths of the order of 1 micrometer.

B. Defect-Structure Methods (TEM, XRT, EBIC, PLT)

In this section, we discuss other techniques that have proven useful for characterizing the nature and density of strain-relieving defects in superlattices and heterostructures. We consider first the use of these techniques for imaging dislocations present in strained layers that have undergone some relaxation. These techniques are extremely sensitive, because dislocations can be individually "counted." Then we briefly discuss nonimaging applications of these techniques to monitor the average density of the dislocations.

It should be emphasized that in most cases, it is possible to learn more from a combination of techniques rather than from a single technique. For example, because of the difference between depth sensitivities of x-ray topography (XRT) and EBIC, it has been shown that three-dimensional information can be deduced about dislocations both in the substrate and in strained epitaxial layers (see, e.g., Radzimski et al., 1988).

1. *Imaging Methods*

The techniques commonly used to image dislocations can be conveniently divided into two categories. In the first category are those techniques that are diffractive in nature and hence sensitive to the atomic structure. Image contrast is provided by differences in local diffraction due to the strain field around the dislocation. In this category, the two principal techniques are TEM (Thomas and Goringe, 1979) and XRT (Authier, 1967). TEM and XRT are complementary to each other in that TEM is useful for imaging high densities of dislocations ($> 10^3 \text{ cm}^{-1}$), whereas XRT is useful for imaging low densities of dislocations ($< 10^4 \text{ cm}^{-1}$).

TEM is usually only sensitive to local strain fields on the order of 10 nm away from the dislocation core and so can easily distinguish between closely spaced dislocations. It usually also has a small ($\approx 10 \, \mu\text{m}$) field of view, so that the dislocations must be spaced closely in order to be found with reasonable probability. An example of a series of plan-view TEM micrographs of an InGaAs strained layer grown on a GaAs substrate is shown in Fig. 27 (Kavanaugh et al., 1988). In this particular example, the InGaAs composition, and hence the strain, was laterally nonuniform. Therefore, the dislocation density varies markedly across the wafer. Note that a series of TEM micrographs were required in order to image dislocations over such a wide area.

In contrast, XRT has a large (wafer-scale) field of view, so the dislocations need not be spaced closely in order to be found. It also is sensitive to strain fields on the order of 1 μm away from the dislocation core and, therefore, cannot distinguish between dislocations spaced more closely than this. An example of an XRT image from a GaAsP/InGaAs SLS sample is shown in Fig. 28a (Radzimski, 1988). Note the much larger-size scale in Fig. 28 compared with Fig. 27.

We should emphasize that for extracting detailed atomic-scale information about defects, TEM is probably the most powerful technique available

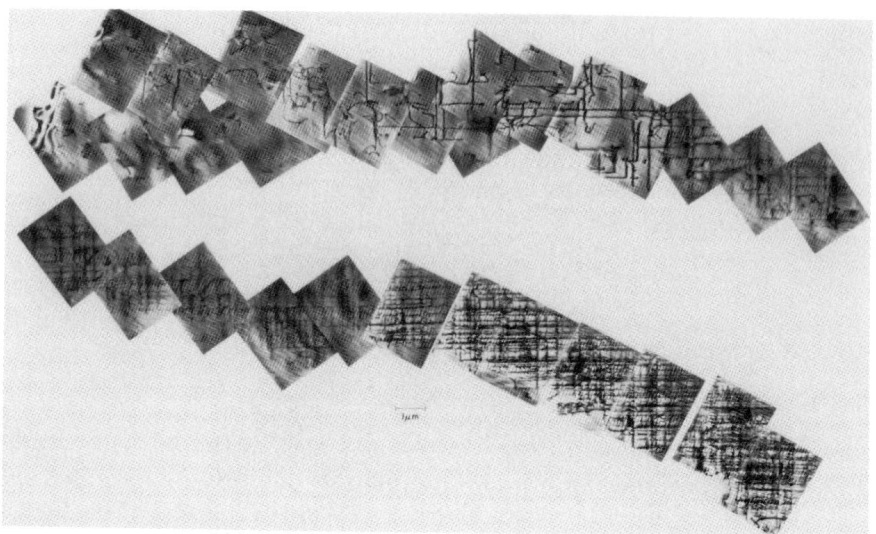

FIG. 27. Plan-view transmission electron microscopy images near the interface of a 1-μm-thick $In_xGa_{1-x}As$ layer grown on a GaAs(001) substrate with a lateral gradient in In composition ranging from $x = 0$ (upper left) to 0.15 (lower right). Micrographs illustrate the variation in dislocation density with lattice misfit. From Kavanaugh et al. (1988).

today. It is often possible, through diffraction-contrast analysis and appropriate sample tilting, to deduce the *type* of dislocation or defect that is being imaged. This type of information is crucial to an understanding of the mechanisms by which strain-relieving dislocations nucleate and subsequently multiply (see, e.g., Chapter 1 by Hull and Bean, or Eaglesham et al., 1989).

In the second category of imaging techniques are those for which the image contrast is provided by the electronic properties of the defect. In this category, the most common techniques are EBIC microscopy (Leamy, 1982), photoluminescence topography (PLT) (Nakashima and Shiraki, 1978) and etch-contrast microscopy (Stirland, 1988). EBIC is based on the synchronized measurement of currents induced by a rastered electron beam over a $p-n$ junction, as illustrated in Fig. 28b (Radzimski, 1988). By modifying the local recombination lifetime, dislocations decrease the local induced currents, and are imaged as dark stripes. Although EBIC is instrumentally compatible with scanning electron microscopy, its spatial resolution is much less ($\approx 0.5\,\mu$m), due primarily to lateral electron scattering within the substrate (the well-known "proximity effect" from e-beam lithography) (Greeneich, 1980).

PLT is based on the optical imaging of photoluminescence induced in a sample by photoexcitation. By modifying the local recombination lifetime, dislocations decrease the local emitted luminescence, and are imaged as dark

3. STRUCTURE AND CHARACTERIZATION OF SUPERLATTICES 197

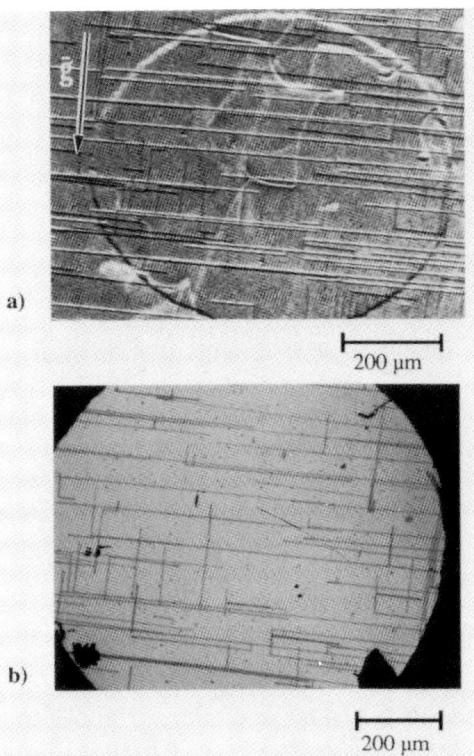

FIG. 28. (a) X-ray topography and (b) electron-beam-induced current images of a $GaAs_{0.84}P_{0.16}/In_{0.08}Ga_{0.92}As$ strained-layer superlattice on GaAs(001) illustrating the presence of misfit dislocations over a large area. From Radzimski et al. (1988).

stripes on a light background. Because the luminescence is optically imaged, the spatial resolution is limited to that for optical microscopy ($\approx 1\ \mu m$). PLT is especially valuable when information on the depth distribution of dislocations is desired, since the excitation wavelength can be chosen for various degrees of penetration into the structure (Gourley et al., 1985).

Finally, etch-contrast mocroscopy is based on differential reactivity of chemical etchants near and away from dislocations, with the imaging done by interference optical microscopy. This technique is among the oldest and most straightforward for defect delineation, but must be applied with great care. The etchant must be carefully calibrated, so that specific etch features can be correlated with the occurrence of specific defects. Furthermore, unless the differential reactivity of the chemical etchant is unusually great, surface layers thicker ($> 1\ \mu m$) than most common strained-layer structures must be etched away.

2. Nonimaging Methods

Another class of techniques is based on measurements of minority-carrier scattering and/or recombination to determine strain and average defect densities. For example, shifts in the peak position of photoluminescence lines can be used to deduce strain and/or composition, and broadening or weakening of photoluminescence lines is a sensitive indicator of crystal quality (Jones *et al.*, 1989). In practice, the peak position of photoluminescence lines also depends on composition and larger geometry (quantum-well) effects. Thus, the contribution due to strain is not always easily deconvolved.

Finally, low-temperature Hall mobilities have also been shown to be extremely sensitive to the average dislocation density (Fritz *et al.*, 1985). In situations where epitaxial quality is high except for occasional misfit dislocations, such electrical measurements are extremely valuable. However, electrical measurements are also sensitive to non-strain-relieving defects, such as impurities or clustering (in immiscible ternary III-Vs, for example).

VI. Stability, Metastability, and Relaxation

In the preceding sections, we described the two extreme cases of perfect coherence (fully strained layers) and perfect "incoherence" (fully strain-relieved layers) and then discussed various methods to characterize strained and strain-relieved structures. In many applications, perfectly coherent heterostructures will be the most useful, because they have the fewest defects. However, perfectly coherent heterostructures are not always attainable, and hence one may be interested in cases intermediate between the two extremes. In this section, we describe criteria by which fully strained heterostructures can be deduced to be stable or not. For the latter case, we will discuss our current understanding of the rates at which strain relaxation occurs through the introduction of misfit dislocations.

In simple terms, there are two independent parts to the problem. First, the driving force for relaxation must be determined. If the driving force is negative, then there is no tendency for relaxation, and the fully strained structure is thermodynamically stable. If the driving force is positive, then there is some tendency for relaxation, and the fully strained structure is not thermodynamically stable. Second, it has to be determined how the heterostructure responds (by plastic flow) to that driving force. If plastic flow is easy, then even a small driving force will drive relaxation. If plastic flow is difficult, then high driving forces (large deviations from equilibrium) will be required to drive relaxation.

A. Driving Force: Excess Stress

The stability of a strained heterostructure can be looked at as a balance between the homogeneous strain energy in the film and the strain energy around misfit dislocations that might accommodate the misfit. For very thin films, the homogeneous strain energy is low, but the strain energy of misfit dislocations at the interface is not, essentially because of the (nonlinear in film thickness) contribution of the dislocation core. Therefore, for thin enough films, the energy of the system is minimized by avoiding misfit dislocations, and pseudomorphic (commensurate) heterostructures are absolutely stable against strain relaxation.

Microscopically, the *mechanism* by which misfit dislocations are formed at the interface consists of motion of other dislocations (either threading dislocations from the substrate or dislocation half-loops nucleated at the surface). Then, it is nearly equivalent (as shown by Matthews and Blakeslee) to rewrite such an energy-balance criterion as a force-balance criterion. Furthermore, by doing so, it is possible to generalize the balance to a definition of the driving force for strained-layer relaxation.

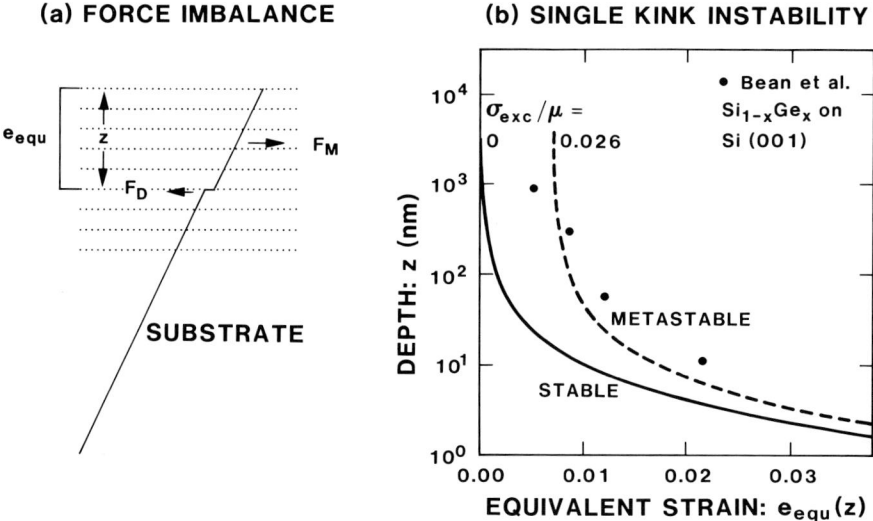

FIG. 29. (a) Schematic of single-kink instability for a strained layer on a substrate; the kinked diagonal line represents a threading dislocation. (b) Layer thickness (or depth) versus equivalent strain as defined by Eq. (53) showing stable and metastable regions of strained-layer films with the boundary demarked by the solid curve ($\sigma_{exc} = 0$). The data points are for Ge_xSi_{1-x} films grown on Si(001) by Bean (1985b), and the dashed line fit to this data is given by $\sigma_{exc}/\mu = 0.026$. After Tsao and Dodson (1988).

To establish a general force-balance criterion, it is important to distinguish the primary mode by which dislocation motion can lead to the formation of misfit dislocations at interfaces. This mode is shown schematically in Fig. 29. For definiteness, we consider a simple 60° dislocation threading through a (001) diamond-cubic biaxially strained layer. Along its length, the dislocation is subject to misfit-strain forces that are opposed by line-tension forces. In Fig. 29, the dislocation moves in response to these forces by expansion of a "single kink," thereby relieving strain in the entire structure above the kink.

The line tension associated with the misfit dislocations opposes dislocation motion (forming new lengths of misfit dislocation costs energy). The difference between those two forces, normalized by the appropriate thickness (the height of material above the kink for the single-kink mechanism) leads to the definition of the "excess" stress acting on the dislocations (Tsao and Dodson, 1988). For the single-kink mechanism this is

$$\sigma_{exc}^{SK}(z) = \frac{2F_{exc}^{SK}(z)}{zb} = \left| 2\mu \left(\frac{1+v}{1-v} \right) \int_0^z e(z') \frac{dz'}{h} \right|$$
$$- \frac{\mu}{2\pi} \left(\frac{1 - v\cos^2\beta}{1-v} \right) \frac{\ln(4z/b)}{z/b}. \tag{52}$$

This excess stress is equal to that of a uniform layer of thickness z having a single-kink *equivalent* strain

$$e_{equ}(z) = \int_0^z e(z')dz'/z \tag{53}$$

equal to the average strain of the structure between 0 and z.

Two contours of constant excess stress ("isobars") are shown in Fig. 29b, corresponding to thickness/equivalent strain combinations that result in the same excess stress, and hence the same deviation from equilibrium. The $\sigma_{exc}/\mu = 0$ isobar reproduces the original Matthews–Blakeslee criterion for absolute film stability. The $\sigma_{exc}/\mu = 0.026$ isobar represents the critical excess stress required for observable strain relief in SiGe layers grown at 550°C. The data points are from Bean (1985b).

Under certain special conditions, the dislocation may move by expansion of a "double kink," thereby relieving strain only in that part of the structure between the kinks. This condition may occur, for example, for an extremely deeply buried layer grown at a temperature such that there is not enough time for the structure to relax by the single-kink mechanism during growth, but such that it is subsequently subjected to a high temperature anneal once buried. Because the dislocation extends along the lower interface of the layer and then folds back along the upper layer in the double-kink mechanism, it requires a line tension on the dislocation that is twice as great as for the single-kink mechanism. Thus, this is less likely to occur, except for special cases such as the one just described. For the double-kink mechanism, the excess stress is given by

$$\sigma_{\text{exc}}^{\text{DK}}(z > 2h) = \frac{2F_{\text{exc}}^{\text{DK}}(z > 2h)}{hb} = \left| 2\mu'' \left(\frac{1+v}{1-v} \right) \int_{z-h}^{z} e(z') \frac{dz'}{h} \right|$$

$$- \frac{\mu}{\pi} \left(\frac{1 - v \cos^2 \beta}{1 - v} \right) \frac{\ln(4h/b)}{h/b}. \quad (54)$$

Similarly, this excess stress is equal to that of a uniform layer of thickness h having a double-kink *equivalent* strain $e_{\text{equ}}(z) = \int_0^z e(z')dz'/z$ equal to the average strain of the structure between z and $z - h$.

These excess stresses are measures of the driving forces for strain relief, and hence for the deviation from equilibrium. Note that the excess stresses are functions of position *within* a structure, rather than simply characteristic of the structure as a whole. If, for both mechanisms, the excess stresses are less than or equal to zero *everywhere* within the structure, then the structure is stable. If, for either mechanism, the excess stress is greater than zero *anywhere*

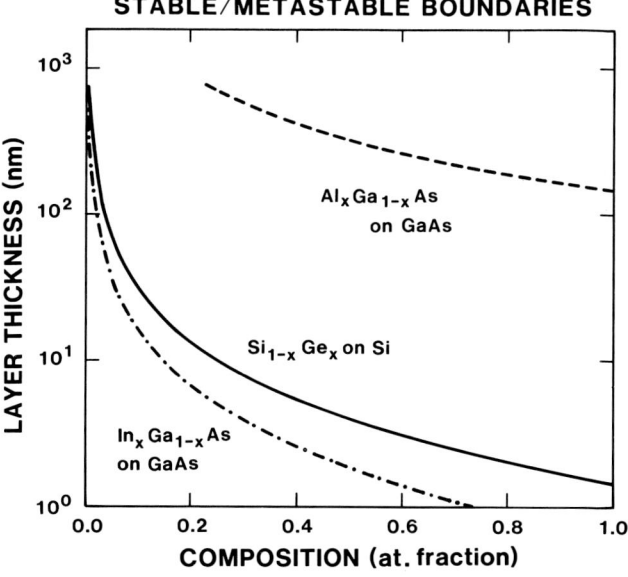

FIG. 30. Curves defining the stable–metastable boundaries (lower left/upper right) in terms of layer thickness versus composition based on zero excess stress for the single-kink mechanism (Eq. (52)) for $In_xGa_{1-x}As$ on GaAs(001), $Si_{1-x}Ge_x$ on Si(001), and $Al_xGa_{1-x}As$ on GaAs (001) systems.

in the structure, then the structure is unstable (or metastable) to dislocation motion by that mechanism.

We note that in both cases, the excess stress depends on both strain and layer thickness. Therefore, a given excess stress determines a family of strain/thickness combinations. For a particular materials system, this implies a family of composition/thickness combinations. In Fig. 30 are three such families for the $Al_xGa_{1-x}As/GaAs$, $Si_{1-x}Ge_x/Si$, and $In_xGa_{1-x}As/GaAs$ systems for a single-kink excess stress equal to zero. For each system, composition/thickness combinations to the left of the boundaries have excess stresses less than zero and are thermodynamically stable against strain relief. Composition/thickness combinations to the right have excess stresses greater than zero, implying a finite driving force for strain relief.

B. MATERIALS RESPONSE: STABILITY DIAGRAMS

Given a driving force (or excess stress) for strained-layer relaxation, it is still necessary to know what the response of the material is to that driving force. The reason is that relaxation of strained heterostructures is nearly always in competition with something else, e.g., film growth or even dopant activation during annealing in a furnace *after* growth. If relaxation is fast, then the film will have time to relax; if slow, the film will not relax.

These ideas are deceptively simple. In practice, the kinetics of plastic deformation is a notoriously complex subject. However, there is one greatly simplifying feature to plastic deformation; namely, no matter what the mechanism, there are really only two parameters that are important in determining plastic deformation. These are stress, which is basically the driving force for plastic flow, and temperature, because it determines how fast flow occurs in response to a driving force.

Indeed, a standard way of describing plastic deformation in the bulk is using stress/temperature diagrams, called deformation mechanism maps, illustrated in Fig. 31 for the case of silicon. The most important information on such a map is summarized by iso-strain-rate contours. On a given experimental time scale, observable strain relief will occur if the strain rate exceeds some minimum value, i.e., if it lies above a particular iso-strain-rate contour. Then, for a given growth temperature, in order to obtain this minimum strain rate, one requires a minimum "excess stress," as defined in the previous section. In a sense, deformation-mechanism maps offer the possibility of condensing in a convenient form the response of the heterostructure to a given excess stress.

Of course, it is not a priori clear that deformation-mechanism maps constructed for plastic flow in bulk materials will describe accurately plastic flow in thin epitaxial films, and such maps have not been constructed for many materials of interest. Therefore, a related tool has been developed,

3. STRUCTURE AND CHARACTERIZATION OF SUPERLATTICES   203

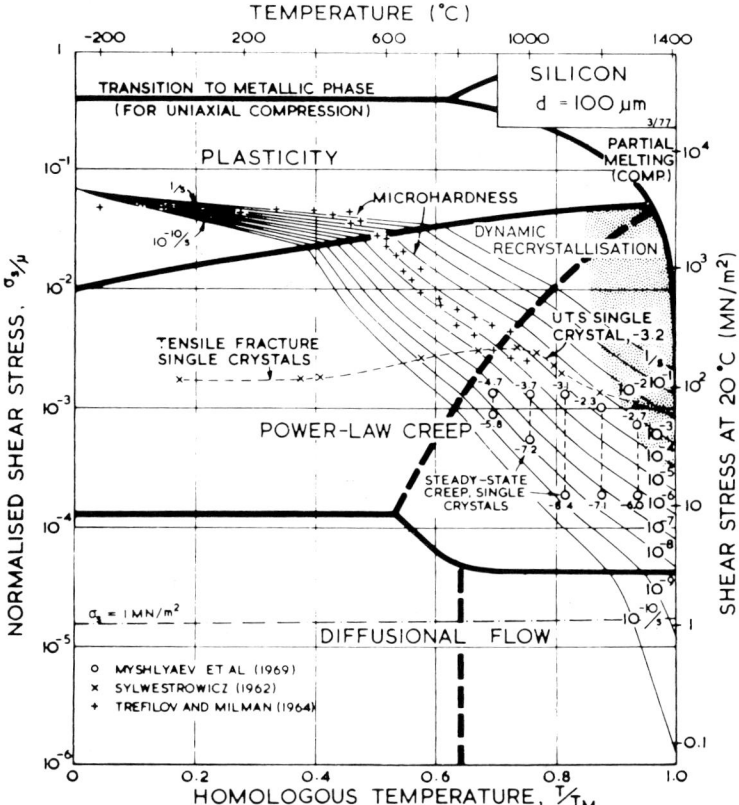

FIG. 31. Deformation-mechanism map for bulk Si of shear stress (normalized by the shear modulus) versus reduced temperature (normalized by the melting point) with contours of isostrain rates. Reprinted with permission from Frost and Ashby, © 1982, Pergamon Press plc.

called a "stability diagram," which essentially results from integration of the underlying deformation-mechanism map. On such a diagram, we draw contours of constant relaxation (rather than constant relaxation rate), which can be more conveniently compared to empirical data.

Such a stability diagram is shown in Fig. 32 for the case of SiGe single strained layers. Films with excess stress greater than zero, but less than that required for plastic deformation on the experimental time scale, are not observably strain relieved (although they may contain defects), and hence are metastable. Films with yet greater excess stress are observably strain relieved, although they may still be metastable.

Finally, we note that such stability diagrams can be estimated from surprisingly few measurements, using scaling relations developed recently by

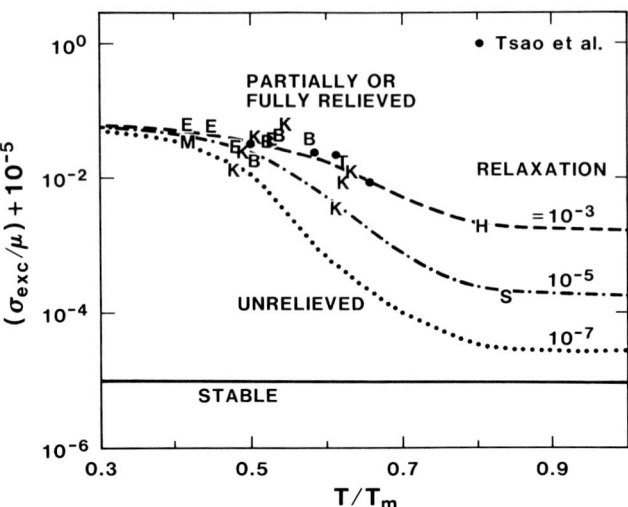

FIG. 32. Stability diagram for SiGe strained layers showing normalized excess stress versus reduced temperature with stable–metastable boundary (solid curve), iso-relaxation contours (dashed, dash–dot, dotted curves) and experimental data for various growth conditions. After Tsao et al. (1988).

Dodson and Tsao (1989b). In particular, if the amount of relaxation during growth, $\gamma(\sigma_0, T_0)$, is known for a particular intended excess stress, σ_0, and temperature, T_0, then the amount of relaxation during growth, $\gamma(\sigma_1, T_1)$, at another intended excess stress σ_1 and temperature T_1, is given approximately by

$$\frac{\gamma(s_1, T_1)}{\gamma(s_0, T_0)} = \left(\frac{\sigma_1}{\sigma_0}\right)^2 e^{[E(\sigma_0)/kT_0 - E(\sigma_1)/kT_1]}. \tag{55}$$

Here, the activation energy for dislocation glide is stress dependent and given by $E(\sigma) = E(0)[1 - \sigma/\tau]$, where $E(0)$ is the zero-stress glide activation energy, and τ is the zero-temperature flow stress (roughly 5–10% of the shear modulus in semiconductors). To derive this scaling relation, use has been made of the empirical observation that strain relaxation is proportional to the square of the initial (intended) excess stress. The parameters in the relation, $E(0)$ and τ, are expected to be similar to those found in studies of dislocation dynamics in the bulk, but should be regarded as fitting parameters to actual strained-layer relaxation experiments. For SiGe alloys, it has been found that $\tau = 0.1\,\mu$ and $E(0) = 16\,kT_m$ are consistent with observed relaxation behavior (Dodson and Tsao, 1988). Here, T_m is an effective alloy melting temperature, obtained by a linear weighting of the melting temperatures of the pure components.

VII. Application to Single Strained Layers

In strained-layer device technology, it is often true that the higher the strain, the better the device performance. The motivation is therefore strong to grow device structures as near to (and perhaps beyond) the stable–metastable boundary as possible. Thus, it becomes important to monitor structural integrity using characterization techniques that are sensitive to the onset of relaxation. Such characterization is also important at a more fundamental level as an aid to understanding the mechanisms by which strain relaxation takes place. In this section, we illustrate the use of various characterization techniques in determining the onset and mechanisms of strain relaxation in simple, single strained-layer structures.

A. The Coherent–Partially Relaxed Boundary

From a materials science point of view, perhaps the best-studied single strained-layer system thus far is the one with $Si_{1-x}Ge_x$ layers on Si. Much of the motivation for these studies stems from the early work of Kasper and coworkers (Kasper, 1986), and then later of Bean and co-workers (Bean, 1985b), in which it was demonstrated that coherent strained layers could be grown that were much thicker than expected from equilibrium considerations. Since then, much experimental work has been devoted to determining the dependence on growth temperature of the boundary dividing fully coherent from partially relaxed structures (Tsao et al., 1988; Hauenstein et al., 1989) and to determine the details of how dislocations nucleate (Kvam et al., 1988) and then propagate (Hull et al., 1988) during relaxation.

To illustrate the use of ion channeling to determine the boundary dividing fully coherent from partially relaxed structures, we show examples of our laboratory measurements on fully coherent and partially relaxed structures (Fig. 33). In both cases, random (nonchanneling) spectra are also shown, from which the indicated layer compositions have been deduced. Both structures consist of a Ge substrate, a thin Si_xGe_{1-x} strained-alloy layer, and a final Ge capping layer. In the fully coherent structure, the channeling minimum yield is 4%, and the dechanneling rate is $0.10 \mu m^{-1}$. In the partially relaxed structure, the channeling minimum is 6%, and the dechanneling rate is $0.33 \mu m^{-1}$.

Generally, the dechanneling rate is most sensitive to low densities of strain-relieving dislocations; this parameter is plotted in Fig. 34 versus the

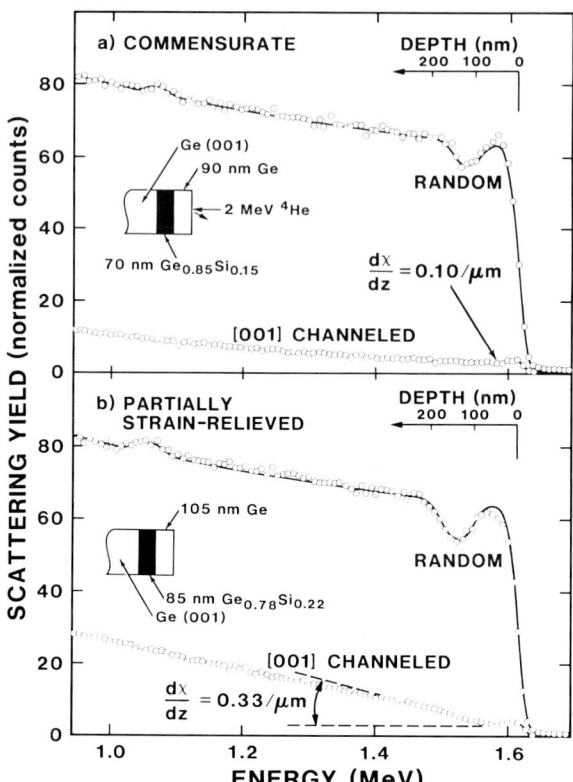

FIG. 33. Random and [001]-channeled RBS spectra for a buried single strained layer in Ge(001) for conditions where (a) the layer is commensurate (70 nm $Ge_{0.85}Si_{0.15}$) and (b) the layer is measurably strain relieved (85 nm $Ge_{0.78}Si_{0.22}$).

excess stress (as defined by Eq. (52) for various SiGe strained layers. For excess stresses less than zero, the structures are (thermodynamically) absolutely stable and have no driving force for plastic flow or dislocation formation. For excess stresses greater than zero, there is a driving force for plastic flow; however, if the excess stress is not too great and the growth temperature not too high, the structure may resist plastic flow on the time scale of the layer growth. Indeed, for the structures represented by the data in Fig. 34, the growth temperature (494°C) is low enough such that the dechanneling rate does not begin to increase perceptibly until a normalized excess stress of $\sigma_{exc}/\mu = 0.026$ has been reached.

At higher temperatures, SiGe structures are not as resistant to plastic flow during growth, which is also illustrated in Fig. 34 for a particular case. A

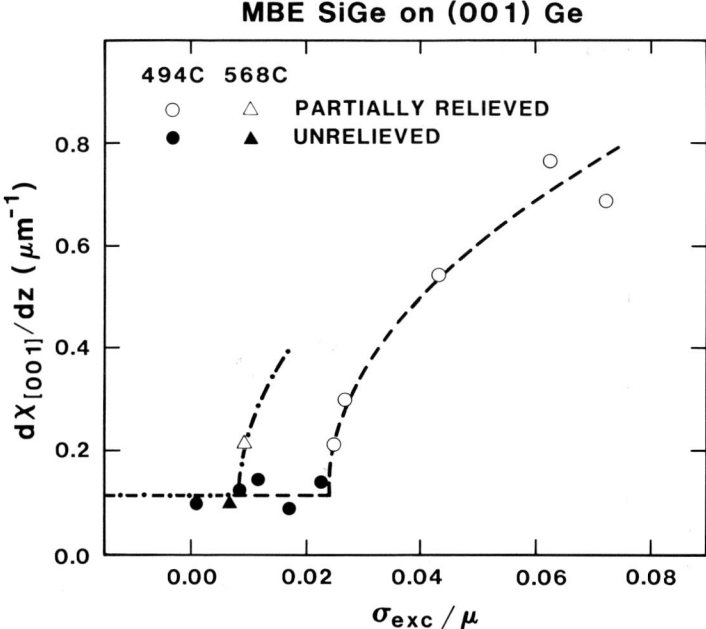

FIG. 34. Dechanneling rate versus normalized excess stress for buried $Si_{1-x}Ge_x$ layers in Ge(001) for MBE growth temperatures of 494 and 568°C. The intersection of the horizontal and vertical dashed curves demarcates the boundary between commensurate and partially strain-relieved layer conditions as measured by ion channeling, and defines the experimental data points shown in Fig. 32 (Tsao et al., 1988) on the $Si_{1-x}Ge_x$ stability diagram.

useful and compact way of depicting the temperature dependence of the boundary dividing fully coherent from partially relaxed structures is through the "stability diagrams" described earlier (illustrated in Fig. 32). On these stability diagrams, contours of constant relaxation are plotted on an excess-stress/temperature map. Data from measurements similar to those of Fig. 34 made over a range of temperatures were used to deduce the 10^{-3} iso-relaxation contour, since this is approximately the sensitivity of ion-channeling analysis.

The other iso-relaxation contours of Fig. 32 are extrapolations based on a recent model for plastic flow (Dodson and Tsao, 1988), using the same parameters that fit the measured 10^{-3} iso-relaxation contour. The 10^{-7} iso-relaxation contour represents essentially zero relaxation for practical device application, since it implies dislocations spaced nearly a centimeter apart.

With a more sensitive technique, it is possible to experimentally measure the intermediate iso-relaxation contours. For example, Fig. 35 plots dislocation densities as a function of excess stress, determined by direct counting

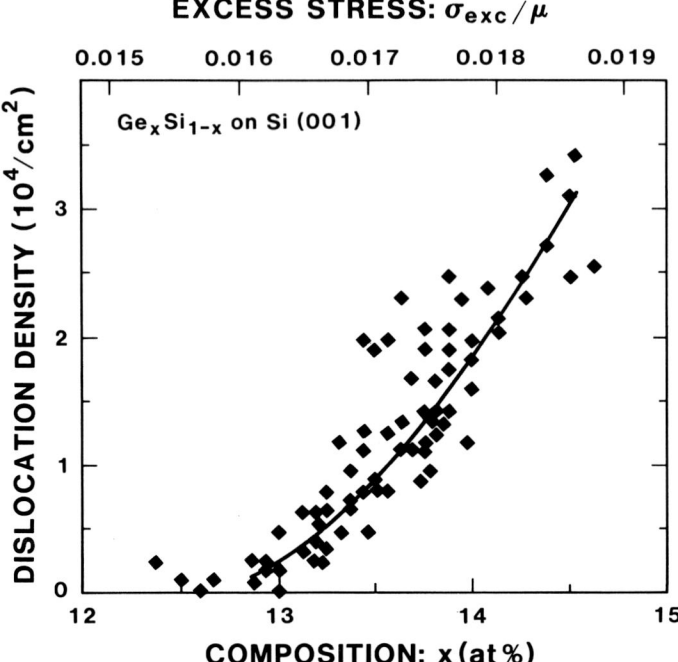

FIG. 35. Dislocation density versus composition as measured by x-ray topography for 180-nm-thick Ge_xSi_{1-x} strained layers grown on Si(001) at 550°C (after Eaglesham et al., 1989). The normalized excess stress is given at the top of the figure.

using XRT (Eaglesham et al., 1989). The sensitivity of this technique is limited only by the intrinsic (non-strain-relaxing) defect density in the structure (approximately $10^3/cm^2$ for currently grown SiGe layers), and accordingly, the excess stress at which partial strain relaxation just becomes observable is lower than that measured by ion channeling.

B. Dislocation Nucleation and Propagation

The ideas discussed earlier are, in a sense, extensions of our understanding of temperature-dependent plastic flow in bulk materials (Frost and Ashby, 1982). It is not clear that the initial stages of plastic flow are the same in extremely high-quality epitaxial layers as in bulk materials. For example, dislocations, normally present in bulk materials, may be absent. Therefore, there is a need to nucleate dislocations first, before they can propagate (and multiply). In fact, simple nucleation of dislocation half-loops at a perfectly terminated surface is thought to be extremely unlikely for energetic reasons (Dodson, 1988). Determining the nucleation sources has therefore been the aim of a number of recent experimental studies.

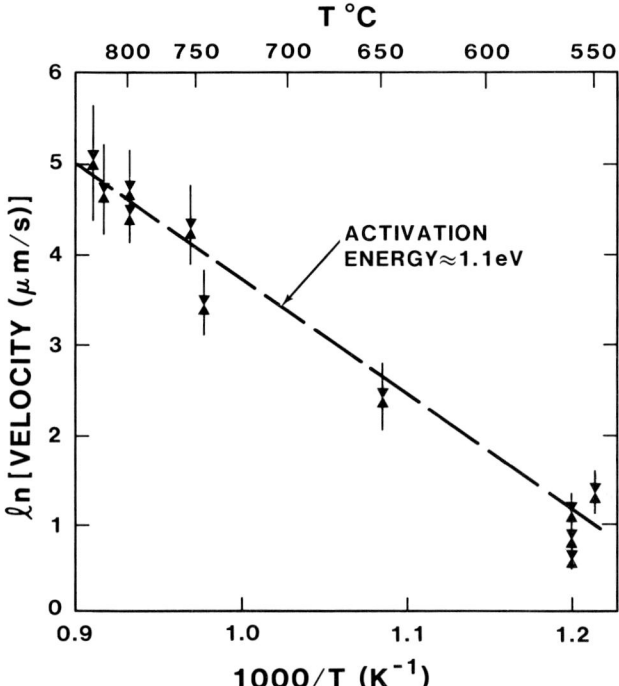

FIG. 36. The dislocation velocity versus anneal temperature as measured by real-time, in situ plan-view transmission electron microscopy in strained Ge_xSi_{1-x} layers on Si(001). After Hull et al. (1988).

For example, etch-contrast microscopy studies have shown that defects associated with the MBE process itself (spitting of the source materials, and so forth) can nucleate dislocations. TEM studies have also found strong evidence for an as-grown stacking-faultlike lattice defect that apparently participates in the nucleation process (Kvam et al., 1988).

Once nucleated, the propagation and multiplication of dislocations is also of great interest. Recently, it has become possible to observe such propagation and multiplication in real time, using in situ TEM (Hull et al), 1988). This is an especially difficult technique, since sample preparation must be compatible with a fragile strained-layer structure. However, the technique has made it possible to measure directly, for the first time, the kinetics of dislocation motion. As an example, Fig. 36 shows the temperature dependence of dislocation propagation in layers on SiGe/Si measured by this technique.

C. The Stable–Metastable Boundary

By now, it is clear that nonequilibrium structures may be grown that are resistant to relaxation, in spite of a finite driving force (finite excess stress) for plastic flow. These structures are metastable and will tend to relax if annealed at high enough temperatures. Even small amounts of relaxation can be detrimental to device performance; and hence metastable structures, which may have improved device characteristics, may also limit subsequent process options to low-temperatures techniques.

To illustrate this point, Fig. 37 shows (4 K) PL spectra from GaAs/InGaAs/GaAs single-quantum-well (SQW) structures (Peercy et al., 1988). Recent studies have shown direct correlation between PL line widths and transconductances in modulation-doped single-quantum-well field-effect transistors (FETs), and therefore PL spectra are a sensitive indicator of the electrical quality of the materials. Figure 37a shows spectra before and after rapid thermal annealing (to simulate dopant activation after ion implantation, for example) for a stable structure. Aside from a slight shift in the peak position, the PLM spectra are essentially identical. In Fig. 37b, spectra are shown before and after annealing a metastable structure. The PL signal has virtually disappeared after the anneal; plan-view TEM of similar structures confirms that dislocations have also been generated, and are likely responsible for the lack of PL signal. The above results are summarized in terms of their stability-diagram representation in Fig. 37c.

VIII. Application to Superlattices

In tailoring electronic properties of structures, the central focus is on the formation of fully commensurate superlattices. Further, electronic transport and optical properties are influenced by both the magnitude and the sign of the strain within the individual layers. These factors not only imply the need for precise control of the layer compositions and thicknesses during growth, but also necessitate establishing a surface with a certain in-plane lattice spacing on which the superlattice can be grown. Since the substrate is usually fixed by available high-quality crystals and other constraints, partially relaxed alloy buffer layers are often grown to establish a given lattice constant prior to strained-layer superlattice growth. Errors in the lattice constant achieved at the buffer surface will alter the strain in the subsequent layers. This can lead to a strain imbalance between layers, so that for thick superlattice structures, the strain will accumulate with each superlattice period and eventually give rise to relaxation.

3. STRUCTURE AND CHARACTERIZATION OF SUPERLATTICES

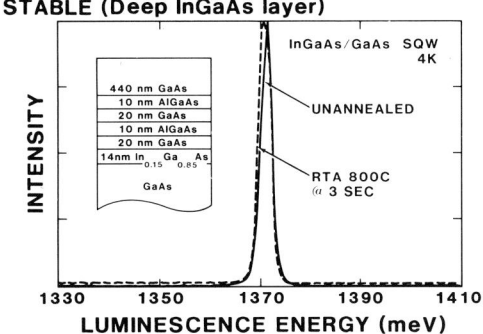

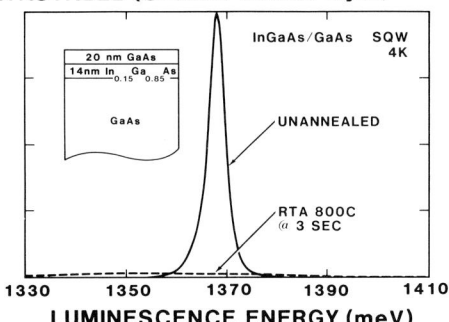

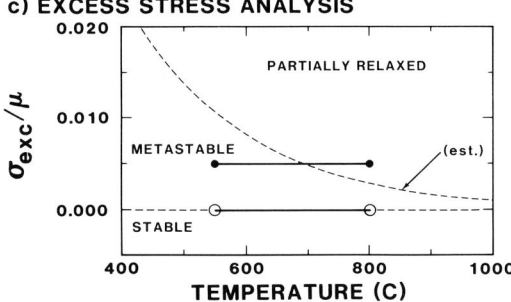

FIG. 37. Photoluminescence spectra for (a) a stable deeply buried $In_{0.15}Ga_{0.85}As$ strained layer in a AlGaAs/GaAs(001)-layered structure before and after 800°C rapid thermal annealing, and (b) a metastable shallow buried $In_{0.15}Ga_{0.85}As$ strained layer before and after a similar anneal sequence. Shown in (c) is the normalized excess stress versus temperature plot for these structures along with an estimated unrelaxed–partially relaxed metastable boundary. Adapted from Peercy, P. S., Dodson, B. W., Tsao, J. Y., Jones, E. D., Myers, D. R., Zipperian, T. E., Dawson, L. R., Biefeld, R. M., Klem, J. F., and Hills, C. R., *IEEE Electron. Device Letters* **EDL-9**, 621, © 1988 IEEE.

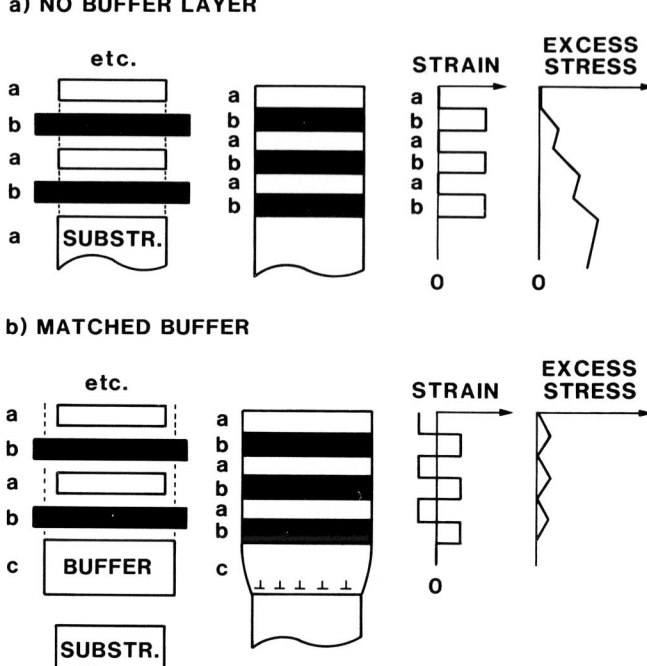

FIG. 38. Schematic diagram of strained-layer superlattice illustrating the layer stress and the excess stress profiles for the cases of (a) no buffer layer and the substrate corresponding to one of the superlattice layers, and (b) a buffer layer with lattice constant matched to the equilibrium in-plane lattice constant of the free-standing superlattice.

A. Coherent Structures and Buffer Layers

The role of the substrate or buffer layer in establishing the strain distribution in coherent superlattices is illustrated schematically in Fig. 38. In many cases, one of the layers of the superlattice will be the same material as the substrate. It is often convenient to grow the superlattice without a buffer layer (see Fig. 38a), so that for a coherent structure, the "a" layers will be unstrained, with the strain confined to the "b" layers. This configuration is usual for single-quantum wells or for a few strained layers. However, in this approach the relative strain between layers a and b cannot be adjusted, but is fixed by their relative lattice constants. Furthermore, the excess stress in the structure will increase with thickness, so that superlattices with many periods may exceed critical values and exhibit relaxation or "decoupling" from the substrate.

When a buffer layer is used (see Fig. 38b) prior to superlattice growth, the distribution of strain between layers a and b can be balanced. For example, by setting the lattice constant of buffer c to the in-plane lattice constant for

GaAs$_{0.9}$P$_{0.1}$/GaAs SUPERLATTICE

a) x-ray MEASUREMENTS

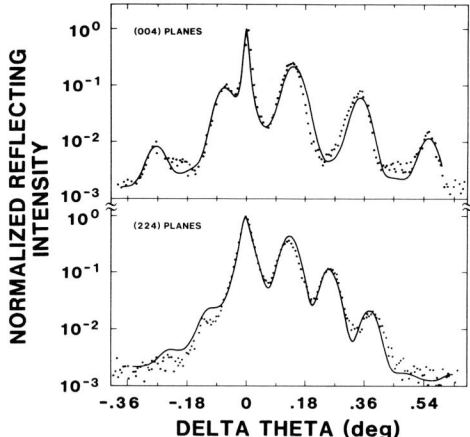

b) STRAIN PROFILE

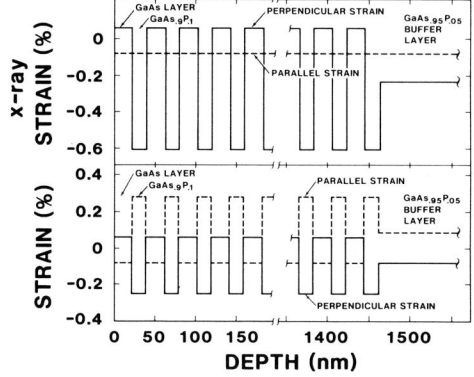

FIG. 39. Double-crystal x-ray diffraction rocking curves (a) for (004) and (224) planes of a GaAs$_{0.9}$P$_{0.1}$/GaAs superlattice on a matched GaAs$_{0.95}$P$_{0.05}$ buffer on GaAs(001) substrate, and (b) the determined x-ray strain profiles (all strain values referenced to the GaAs substrate) perpendicular and parallel to the layers. Also shown are the resulting "true" strain profiles referenced to the individual layers. The solid lines are the calculated diffraction intensity given by the strain profiles based on a best fit to the data.

the freely floating a/b superlattice, there will be no net accumulation of excess stress in the superlattice (Fig. 38b). However, buffer layers also have disadvantages. A given lattice constant is not easy to achieve, since the lattice constant of partially relaxed structures varies only slowly with thickness and can depend in detail on the growth parameters. Furthermore, misfit dislocations must be confined to the buffer layer or to the lower part of the superlattice structure for electronic applications. Thus, although desirable in many cases, the use of buffer layers increases the importance of accurate strain characterization.

In Fig. 39, measured strain distributions in a $GaAs_xP_{1-x}/GaP$ superlattice grown on a matched buffer layer are illustrated. X-ray double-crystal-diffraction rocking curves are shown for the normal (004) and inclined (224) crystal planes, with the most prominent peaks due to the GaP substrate and $GaAs_yP_{1-y}$ buffer layers. As discussed in Section IV, the layer thicknesses are determined by the spacing of the satellite peaks and the strain by the envelope of the peak intensities. We show both the "x-ray strain," which is measured relative to the GaP substrate, and the derived absolute strain in the layers. The buffer layer is seen to be fairly closely matched to the superlattice, so that the average excess stress accumulation per superlattice period is quite low. In contrast, the excess stress values that would be present if this superlattice were grown directly on GaP without a buffer would increase rapidly with superlattice thickness.

B. Nonperiodic Strain Profiles

By varying growth parameters or by subsequent processing, the periodic strain profile in a superlattice may not remain uniform in depth. Whereas in most cases uniform structures are encountered, it is sometimes necessary to be able to distinguish and characterize the more complex cases. Often these cases require some knowledge of the anticipated structure and can benefit from a combination of techniques.

An example of a periodic, but nonuniform, strain profile induced by Be ion implantation into a $GaAs_{0.15}P_{0.85}/GaP$ superlattice (Myers et al., 1986) is illustrated in Fig. 40. In the upper panel, the (004) x-ray double-crystal diffraction and $\langle 110 \rangle$ channeling angular scan data are shown for the as-grown superlattice. The strain profile that gives the best fit (solid line) to the x-ray data is shown in the upper center panel. Also, the relative strain of 0.91% between layers, as determined from the ion-channeling measurements, is in close agreement with the x-ray results.

Upon room temperature implantation with 75 keV Be to a fluence of $1 \times 10^{15}/cm^2$, the superlattice structure is still found to be commensurate. However, as seen from the x-ray rocking curve in the left panel of Fig. 40b, the strain distribution has been significantly altered. From ion-channeling measurements along the growth direction (right panel), the depth profile of the disorder is determined. Since ion-implantation disorder is known to induce lattice expansion in the covalent semiconductors, this measured disorder profile has been used here as the envelope of the strain profile. With this constraint on the superlattice strain profile (lower center panel of Fig. 40),

$GaAs_{.15}P_{.85}/GaP$ SUPERLATTICE on GaP (001)

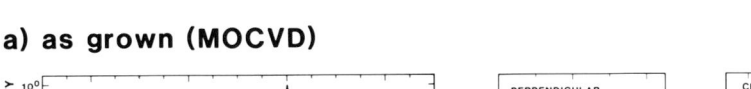

a) as grown (MOCVD)

b) Be implanted (75 keV Be 295 K)

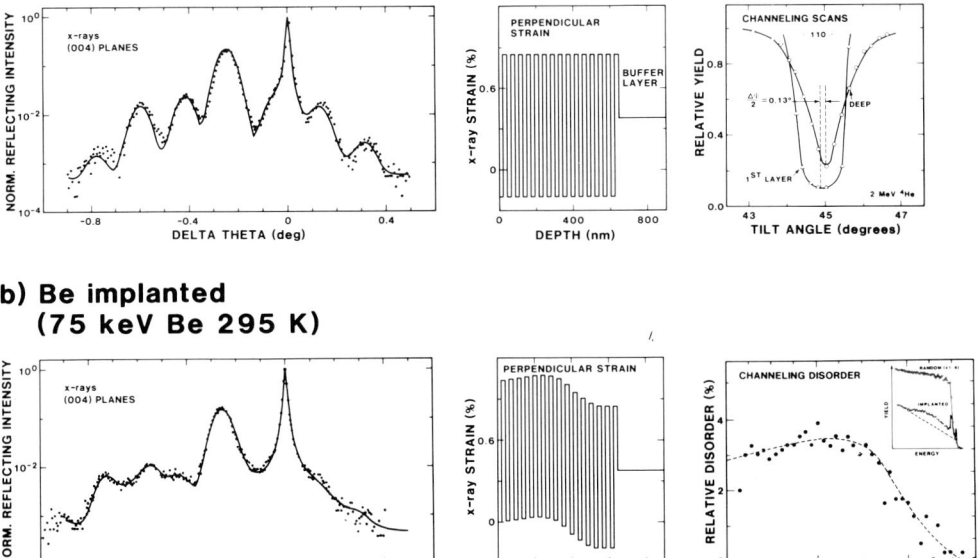

FIG. 40. X-ray diffraction (004) rocking curves and channeling measurements for a $GaAs_{0.15}P_{0.85}/GaP$ superlattice (a) as grown and (b) after 75-keV Be implantation (1×10^{15} Be/cm^2). The channeling angular scan measurements of the tilt angle between layers in (a) is in agreement with the x-ray-determined perpendicular strain shift between layers. The envelope of the x-ray strain profile in (b) was constrained to match the disorder depth profile determined by ion channeling (right side), and the solid lines in the x-ray rocking curves are the calculated values for the derived strain profiles. After Myers et al. (1986).

it is found that good agreement can be achieved between the calculated and measured x-ray rocking curves. We note that Monte Carlo calculations of the energy deposited into atomic displacements during Be implantation gave good agreement with the disorder profile observed by ion channeling, except that the calculated disorder was somewhat lower near the surface, and this relatively small difference in profiles could be equally well fit by the x-ray results. Further confidence in this interpretation of a commensurate but nonuniform strain profile was given by the fact that after annealing in a phosphine/arsine ambient, the original superlattice strain distribution was recovered and was indistinguishable by x-ray and channeling measurements from that of the as-grown superlattice.

Thus, by combining x-ray diffraction and ion-channeling measurements, this case of a nonuniform strain profile could be determined fairly accurately. In general, x-ray double-crystal diffraction results sum the contributions to the strain profile from the various depths, so that a unique correspondence between depth and strain is not necessarily inferred. Other techniques, such as ion channeling or Raman scattering, are often more sensitive to the strain distribution nearer the surface. However, by having information on the depth sequence of the layers grown and by correlating x-ray diffraction results with other measurements in the case of nonuniform arbitrary profiles, the strain distributions may be characterized with reasonable confidence. Of course, the most accurate results would be achieved by careful layer-removal techniques in combination with strain-sensitive measurements.

C. Imperfect Structures

Strain relaxation gives rise to imperfect superlattice structures and can significantly change the electrical and optical properties of these materials.

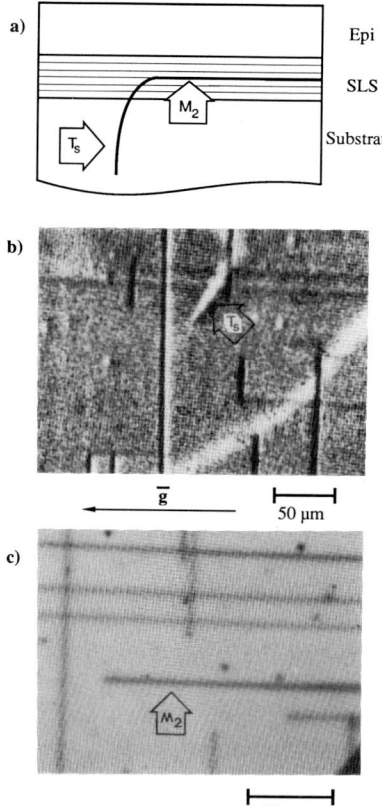

Fig. 41. Schematic (a), XRT (b) and EBIC (c) images illustrating a threading dislocation in the substrate (T_s) that turns over in a GaAsP/InGaAs superlattice to form a misfit dislocation (M_2). Also shown (d) are examples of other defect structures identified by combined XRT and EBIC imaging. From Radzimski et al. (1988).

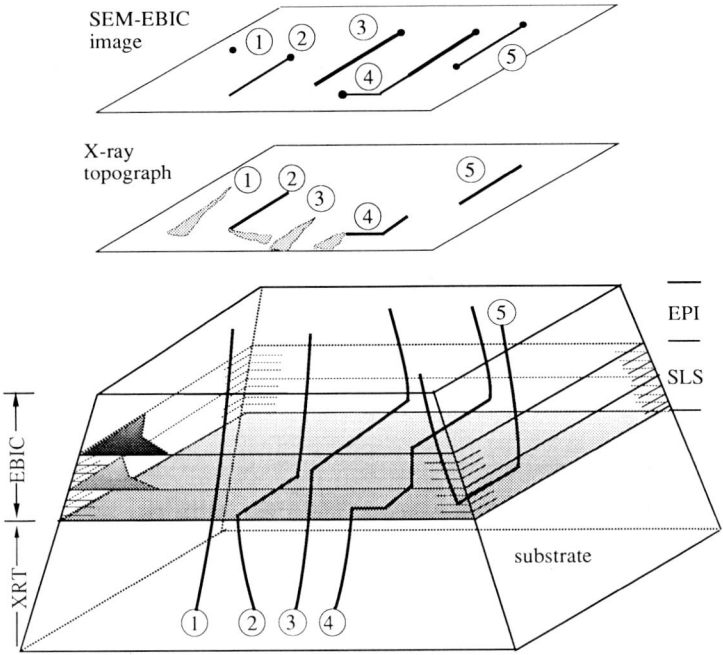

FIG. 41. Continued.

Techniques that can distinguish between fully commensurate and imperfect structures are therefore important in characterizing strained-layer superlattices. As discussed earlier in this chapter, defect-imaging techniques such as TEM, EBIC, PL microscopy, and XRT are particularly valuable in detecting the onset of relaxation. Furthermore, detailed TEM studies of the structure of defects present, and of their geometry and interactions with the superlattice, may give valuable insight into ways to avoid or minimize such defects in active regions of devices. In Fig. 41, an example is shown of schematic XRT and EBIC images of a threading dislocation that has been turned laterally to relieve strain at a superlattice layer interface. Various other types of characteristic dislocation geometries seen in these GaAsP/InGaAs strained-layer superlattices by EBIC and XRT are also illustrated in Fig. 41. In this case, the EBIC images provide an image of defects in the near-surface region, whereas the x-ray topography can provide depth-dependent information from the entire sample (based on different Bragg-angle diffraction for different layers of the sample). These techniques are well suited to imaging overall dislocation geometries and are applicable for dislocation densities below 10^4/cm, whereas for these lower densities, the TEM studies become difficult.

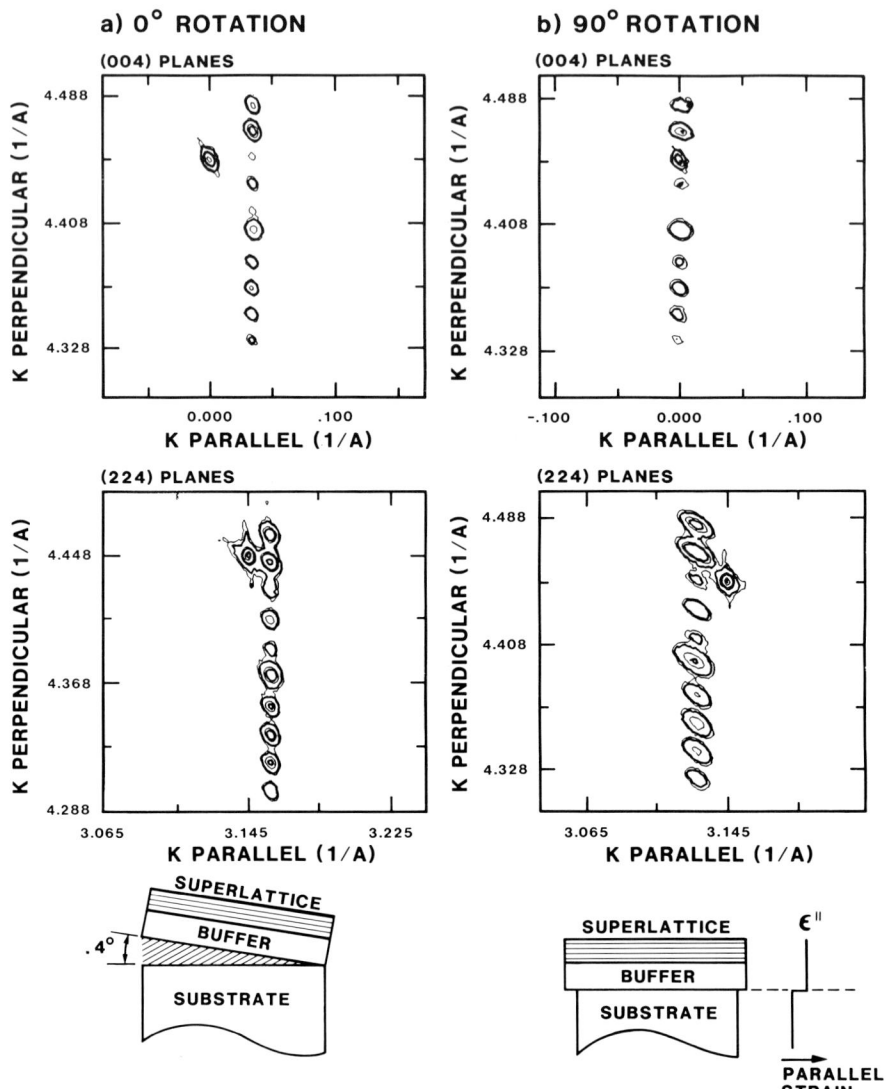

FIG. 42. X-ray reciprocal-space maps for an $In_xGa_{1-x}As/GaAs$ strained-layer superlattice and buffer layer on a GaAs(001) substrate showing intensity contours for (004) and (224) planes for (a) 0° azimuthal and (b) 90° azimuthal rotations of the sample. The separation of the substrate peak from the buffer and superlattice diffraction peaks arises from a 4° tilt of the superlattice–buffer layer structure, which is also relaxed relative to the substrate.

In contrast, TEM is very suitable at higher dislocation concentrations, giving more detailed information about the nature of the defects present.

Finally, we conclude with an example showing the use of the x-ray reciprocal-space maps developed in this chapter to detect tilt misorientations of superlattices. In this special case for an InGaAs/GaAs superlattice, we find that both the strained-layer superlattice and the buffer layer are appreciably tilted relative to the substrate. This effect is shown in the left panels of Fig. 42, where the row of superlattice-diffraction intensities (along with that of the substrate) are clearly shifted from that of the substrate-diffraction spot. Thus, a relaxation leading to a microscopic detachment of the buffer/superlattice from the substrate has occurred. Further evidence is given by the right-hand panels of Fig. 42, which show the same (004) and (224) reciprocal-space maps, but with the sample rotated 90° about the azimuth. Analysis of these scans indicates that the in-plane strain exhibits a discontinuous change at the substrate–buffer interface but then remains the same throughout the buffer/superlattice part of the structure. Thus, while the superlattice and buffer layer have remained a commensurate unit, the structure is nevertheless imperfect when the entire superlattice/buffer/substrate is considered. Although this case is unusual, it illustrates the power of layer orientation and strain measurements in fully understanding the nature of strained-layer structures.

IX. Conclusions

In this chapter, we have discussed a wide variety of techniques for characterizing the structure of strained-layer structures. Special emphasis was given to x-ray diffraction and ion-scattering analysis techniques for characterizing the composition, structure, and strain of these materials. When combined with TEM, EBIC, XRT, or other techniques that can image the strain-relieving dislocations and thus are particularly sensitive to relaxation effects, one has a powerful set of methods to characterize strained quantum wells and strained-layer superlattices. Also emphasized in this chapter is our current state of understanding of the stability, metastability, and relaxation of strained-layer structures. To provide a framework for understanding these phenomena, the driving force for relaxation is presented in terms of excess stress, and the materials response in terms of stability diagrams. Finally, we have given examples of the application of these concepts and methods to single strained layers and superlattices.

References

Anderson, N. G. (1985). *Appl. Phys. Lett.* **46**, 967.
Authier, A. (1967). In "Advances in X-Ray Analysis," Vol. 10, J. B. Newkirk and G. R. Mallet, eds. Plenum Press, New York, pp. 9–31.
Auvray, P., Baudet, M., Regreny, A., Poudoulec, A., and Guenais, B. (1987). *Proc. of 3rd Int. Conf. on Modulated Semiconductor Structures*, July 1987, Montpellier, France.
Barrett, J. H. (1983). *Phys. Rev.* **B28**, 2328.
Bean, J. C. (1985a). *Mat. Res. Soc. Symp. Proc.* **37**, 245.
Bean, J. C. (1985b). *Science* **230**, 127.
Cerdeira, F., Buchenauer, C. J., Pollak, F. H., and Cardona, M. (1972). *Phys. Rev.* **B5**, 580.
Cerdeira, F., Pinczuk, A., Bean, J. C., Batlogg, B., and Wilson, B. A. (1985). *J. Vac. Sci. Technol.* **B3**, 600.
Chami, A. C., Ligeon, E., Danielou, R., Fontenille, J., Lentz, G., Magnea, N., and Mariette, H. (1988). *Appl. Phys. Lett.* **52**, 1874.
Chu, W. K., Mayer, J. W., and Nicolet, M.-A. (1978). "Backscattering Spectrometry." Academic Press, New York.
Chu, W. K., Pan, C. K., and Chang, C.-A. (1983). *Phys. Rev.* **B28**, 4033.
Cohen, P. I., Pulcite, P. R., VanHove, J. M., and Lent, C. S. (1986). *J. Vac. Sci. Technol.* **A4**, 1251.
Cullity, B. D. (1978). "Elements of X-Ray Diffraction," 2d ed. Addison-Wesley, Reading, Massachusetts.
Dodson, B. W. (1988). *Appl. Phys. Lett.* **53**, 394.
Dodson, B. W., and Tsao, J. Y. (1988). *Appl. Phys. Lett.* **53**, 2498.
Dodson, B. W., and Tsao, J. Y. (1989a). *Annual Review in Materials Science* **19**, Annual Reviews Inc., Palo Alto, California.
Dodson, B. W., and Tsao, J. Y. (1989b). *Appl. Phys. Lett.*, **55**, 1345.
Doolittle, L. R. (1986). *Nucl. Instr. and Meth.* **B15**, 227.
Eaglesham, D. M., Maher, D. M., Kvam, E. P., Bean, J. C., and Humphreys, C. J. (1989). *Phys. Rev. Lett.* **62**, 187.
Feldman, L. C., Mayer, J. W., and Picraux, S. T. (1982). "Materials Analysis by Ion Channeling." Academic Press, New York.
Feldman, L. C., Berk, J., Davidson, B. A., Grossman, H. J., and Mannaerts, J. P. (1987). *Phys. Rev. Lett.* **59**, 664.
Fewster, P. F. (1986). *Philips J. Res.* **41**, 268.
Fewster, P. F., and Curling, C. J. (1987). *J. Appl. Phys.* **62**, 4154.
Fiory, A. T., Bean, J. C., Feldman, L. C., and Robinson, I. K. (1984). *J. Appl. Phys.* **56**, 1227.
Fritz, I. J., Picraux, S. T., Dawson, L. R., Drummond, T. J., Laidig, W. D., and Anderson, N. G. (1985). *Appl. Phys. Lett.* **46**, 967.
Fritz, I. J., Doyle, B. L., Schirber, J. E., Jones, E. D., Dawson, L. R., and Drummond, T. J. (1986). *Appl. Phys. Lett.* **49**, 581.
Frost, H. J., and Ashby, M. F. (1982). "Deformation Mechanism Maps." Pergamon Press, Oxford.
George, A., and Rabier, J. (1987). *Revue de Physique Appliquée* **22**, 941.
Gourley, P. L., Biefeld, R. M., and Dawson, L. R. (1985). *Appl. Phys. Lett.* **47**, 482.
Green, G. S., and Tanner, B. K. (1988). *Mat. Res. Soc. Symp. Proc.* **104**, 623.
Greeneich, J. S. (1980). "Electron-Beam Technology in Microelectronic Fabrication," G. R. Brewer, ed. Academic, New York, pp. 59–140.
Hamdi, A. H., Speriosu, V. S., Nicolet, M. A., Tandon, J. L., and Yeh, Y. C. M. (1985). *J. Appl. Phys.* **57**, 1400.
Hauenstein, R. J., Clemens, B. M., Miles, R. H., Marsh, O. J., Croke, E. T., and McGill, T. C.

(1989). *Proc. of the Conf. on Physics and Chemistry of Semiconductor Interfaces* (to be published in *J. Vac. Sci. Technol.* **B**).
Hull, R., Bean, J. C., Werder, D. J., and Leibenguth, R. E. (1988). *Appl. Phys. Lett.* **52**, 1605.
Ibers, J. A., and Hamilton, W. C. (1974). "International Tables for X-Ray Crystallography," Vol. IV. The Kynoch Press, Birmingham.
Ishibashi, A., Itabashi, M., Mori, Y., Kaneko, K., Kawado, S., and Watanabe, N. (1986). *Phys. Rev.* **B33**, 2887.
Jones, E. D., Zipperian, T. E., Lyo, S. K., Schirber, J. E., and Dawson, L. R. (1988). *Paper R2*, presented at the *1988 Electron. Mater. Conf.*, Boulder, Colorado, June 26–28.
Jones, E. D., Zipperian, T. E., Lyo, S. K., Schirber, J. E., and Dawson, L. R. (1989). To be published.
Kasper, E. (1986). *Surf. Sci.* **174**, 630.
Kavanagh, K. L., Capano, M. A., Hobbs, L. W., Barbour, J. C., Maree, P. M. J., Schaff, W., Mayer, J. W., Pettit, D., Woodall, J. M., Stroscio, J. A., and Feenstra, R. M., (1988). *J. Appl. Phys.* **64**, 4843.
Kittel, C., (1976). "Introduction to Solid State Physics," 5th ed. John Wiley, New York.
Kvam, E. P., Eaglesham, D. J., Maher, D. M., Humphreys, C. J., Bean, J. C., Green, G. S., and Tanner, B. K. (1988). *Mat. Res. Soc. Symp. Proc.* **104**, 623.
Leamy, H. J. (1982). *J. Appl. Phys.* **53**, R51.
Mathieson, A. McL., (1982). *Acta Cryst.* **A38**, 378.
Matthews, J. W. and Blakeslee, A. E. (1976). *Crystal Growth* **32**, 265.
Mayer, J. W., Ziegler, J. F., Chang, L. L., Tsu, R., and Esaki, L. (1973). *J. Appl. Phys.* **44**, 2322.
Myers, D. R., Picraux, S. T., Doyle, B. L., Arnold, G. W., and Biefeld, R. M. (1986). *J. Appl. Phys.* **60**, 3631.
Nakayama, M., Kubota, K., Kanata, T., Kato, H., Chika, S., and Sano, N. (1985). *J. Appl. Phys.* **58**, 4342.
Nakashima, H., and Shiraki, Y. (1978). *Appl. Phys. Lett.* **33**, 545.
Neumann, D. A., Zabel, H., and Morkoc, H. (1983). *Appl. Phys. Lett.* **43**, 59.
Neumann, D. A., Zabel, H., and Morkoc, H. (1985). *Mat. Res. Soc. Symp. Proc.* **37**, 47.
Peercy, P. S., Dodson, B. W., Tsao, J. Y., Jones, E. D., Myers, D. R., Zipperian, T. E., Dawson, L. R., Biefeld, R. M., Klem, J. F., and Hills, C. R. (1988). *IEEE Electron Device Letters* **EDL-9**, 621.
Picraux, S. T. (1974). *Japan. J. Appl. Phys.* Suppl. 2, Pt. 1, 657.
Picraux, S. T., and Lee, S. R. (1988). unpublished.
Picraux, S. T., Dawson, L. R., Osbourn, G. C., and Chu, W. K. (1983a). *Nucl. Instr. and Meth.* **218**, 57.
Picraux, S. T., Dawson, L. R., Osbourn, G. C., and Chu, W. K. (1983b). *Appl. Phys. Lett.* **43**, 930.
Picraux, S. T., Dawson, L. R., Osbourn, G. C., Biefeld, R. M., and Chu, W. K. (1983c). *Appl. Phys. Lett.* **42**, 1020.
Picraux, S. T., Chu, W. K., Allen, W. R., and Ellison, J. A. (1986). *Nucl. Instr. and Meth.* **B15**, 306.
Picraux, S. T., Tsao, J. Y., and Brice, D. K. (1987). *Nucl. Instr. and Meth.* **B19/20**, 21.
Picraux, S. T., Dawson, L. R., Tsao, J. Y., Doyle, B. L., and Lee, S. R. (1988a). *Nucl. Instr. and Meth.* **B33**, 891.
Picraux, S. T., Biefeld, R. M., Allen, W. R., Chu, W. K., and Ellison, J. A. (1988b). *Phys. Rev.* **B38**, 11086.
Radzimski, Z. J., Jiang, B. L., Rozgonyi, G. A., Humphreys, T. P., Hamaguchi, N., and Bedair, S. M. (1988). *J. Appl. Phys.* **64**, 2328.
Sakamoto, K., Sakamoto, T., Nagao, S., Hashiguchi, G., Kuniyoshi, K., and Bando, Y. (1987). *Japan J. Appl. Phys.* **26**, 666.

Saris, F. W., Chu, W. K., Chang, C. A., Ludeke, R., and Esaki, L. (1980). *Appl. Phys. Lett.* **37**, 931.
Segmuller, A., Krishna, P., and Esaki, L. (1977). *J. Appl. Cryst.* **10**(1); and new book.
Simmons, G., and Wang, H. (1971). "Single Crystal Elastic Constants," 2d ed. MIT Press, Cambridge, Massachusetts.
Speriosu, V. S. (1981). *J. Appl. Phys.* **52**, 6094.
Speriosu, V. S., and Vreeland, T., Jr. (1984). *J. Appl. Phys.* **56**, 1591.
Stirland, D. J. (1988). *Appl. Phys. Lett.* **53**, 2432.
Thomas, G., and Goringe, M. J. (1979). "Transmission Electron Microscopy of Materials." John Wiley, New York.
Thompson, L. R., and Doyle, B. L. (1988). *Mat. Res. Soc. Proc.* **EA-18**, 141.
Tsao, J. Y., and Dodson, B. W. (1988). *Appl. Phys. Lett.* **53**, 848.
Tsao, J. Y., Dodson, B. W., Picraux, S. T., and Cornelison, D. M. (1988). *Phys. Rev. Lett.* **59**, 2455.
Warren, B. E. (1969). "X-ray Diffraction." Addison-Wesley, Reading, Massachusetts.
Whaley, G. L., and Cohen, P. I. (1988). *J. Vac. Sci. Technol.* **B6**, 625.
Wie, C. R., Tombrello, T. A., and Vreeland, T., Jr. (1986). *J. Appl. Phys.* **59**, 3743.
Zachariasen, W. H. (1945). "Theory of X-ray Diffraction in Crystals." John Wiley, New York.
Zhang, J., Neave, J. H., Dobson, P. J., and Joyce, B. A. (1987). *Appl. Phys.* **A42**, 317.

CHAPTER 4

Group-IV Compounds

E. Kasper and F. Schäffler

DAIMLER-BENZ AKTIENGESELLSCHAFT
ULM, FEDERAL REPUBLIC OF GERMANY

I.	Introduction	223
II.	Material Properties	225
	A. Cubic Lattice	225
	B. Lattice Mismatch	228
III.	Basic Principles	230
	A. Mismatch Accommodation by Strain or Misfit Dislocations	230
	B. Tetragonal Distortion	232
	C. Dislocations	233
	D. Equilibrium Theory for Misfit Dislocations	235
	E. Metastability	238
IV.	Stability of Strained-Layer Superlattices	240
	A. Strain Symmetrization	241
	B. Buffer-Layer Design	242
V.	The $Si_{1-x}Ge_x/Si$ System	244
	A. Growth by MBE	246
	B. Band Offset and Two-Dimensional Carriers	257
	C. Zone-Folding Effects	274
	D. Device Structures	287
VI.	Outlook	303
	References	304

I. Introduction

Group-IV elements, compounds, and alloys cover a unique range of exciting properties. The list includes both the very precious diamond and silicon, the secondmost frequent element of the earth's crust. The hardest mineral, the best heat conductor, the highest hole mobility, and the large variation in semiconducting properties ranging from zero bandgap to near-isolating wide gap, are just a few of the highlights.

Germanium and especially silicon technology is very well developed. Also SiC is already being used for specific semiconductor applications. Research in diamond, SiC/Si heterojunctions, and SiGe alloys has reached interesting levels; research in GeSn compounds or alloys offers promising prospects.

The combination of different group-IV elements or compounds into superlattices is severely restricted by the wide variation in lattice spacings. Only the SiGe/Si superlattice system will be of acceptable quality in the near future. For the far-reaching future, techniques are required to overcome lattice mismatches of more than a few percent.

Today, microelectronics is strongly based on silicon as the semiconducting material. The annual production of microchips is now several thousand tons and will increase to several ten thousand tons in the 1990s. Heterojunction- and superlattice-based devices will furnish future electronic systems with higher speed, integrated optoelectronic functions, and multifunctionality, provided several problems can be overcome.

A schematic view of how such a future electronic system might look is shown in Fig. 1. For several physical, technical, economic, and ecological reasons, these superlattice devices have to be integrated with conventional integrated circuits on a silicon substrate. It is expected that the superlattices and heterostructures occupy only a minor part of the chip area, where they are required to define the ultimate performance of the device. The dominating percentage of the chip area will be made up of state-of-the-art silicon function units, such as memory cells, for example, which define the overall complexity of the device.

From today's point of view, the most promising heterosystems for a broad application are GaAs/Si and SiGe/Si. However, the relevance of silicon-based insulator and metal systems, e.g., silicide/Si and CaF_2/Si, shall be mentioned explicitly, although they are not covered by the scope of this chapter.

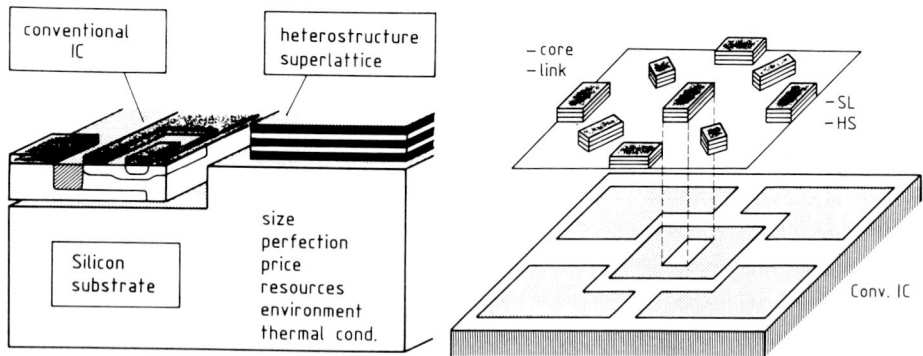

FIG. 1. Monolithic integration of a conventional integrated circuit (IC) with heterostructure and superlattice devices on a silicon substrate. Schematic view (left-hand side), and arrangement of heterostructure/superlattice devices in subchip cores and intrachip connections (right-hand side).

In the following sections, we will list some important material properties, treat basic principles of lattice-mismatch accommodation, and consider the stability of strained-layer superlattices. The main part will deal with results and applications concerning the SiGe/Si system.

II. Material Properties

In this section, the group-IV elements C, Si, Ge, Sn, and to some extent Pb are considered. Some key properties of these elements are listed in Table I.

A. Cubic Lattice

Silicon and germanium as well as tin at temperatures below 13°C (grey tin or α-Sn) crystallize in the cubic diamond lattice, which derives its name from the high-temperature, high-pressure modification of carbon. The only cubic structure (zinc-blende lattice) among various SiC polytypes is β-SiC. SiGe alloys are completely miscible between Si and Ge. Sn and Si alloys were found to be mixtures of the components (Hansen and Anderko, 1958). Epitaxial growth of diamond-type $Si_{1-x}Sn_x$ on Si substrates was observed for $x \leqslant 0.005$ (Aizaki, 1988). Lead and silicon do not form alloys. SnGe alloys with more than 0.6% Ge and more than 1% Sn were found to be two-phase.

The group-IV-element structures, stable at atmospheric pressure, show a nice trend from covalent to metallic bonding with increasing atomic number. Thus, we start with graphite, which has a coordination number CN = 3, as well as with the metastable diamond with CN = 4, both being predominantly covalent. We find the diamond structure to be the stable one for both Si and Ge. Sn marks the transition from the diamond structure, which is stable just below room temperature, to a metallic state with six-fold coordination, whereas Pb crystallizes only in a metallic FCC structure.

It had been anticipated that pressure would induce the same trend, which has been pretty much confirmed. Pressure is required to form diamond from graphite. Si and Ge convert at pressures above 100 kbar into a β-Sn-type modification that is highly conducting. Under pressure, Sn itself has been found to assume a body-centered tetragonal structure with $c/a = 0.91$, rather

TABLE I

Atomic Properties of Group-IV Elements

Element	C	Si	Ge	Sn	Pb
atomic number	6	14	32	50	82
molecular weight	12.01	28.09	72.59	118.69	207.19
main isotops	**12**,13	**28**,29,30	**74**,72,70	**120**,118	**208**,206,207

than the FCC structure of Pb. It should also be mentioned that Si and Ge can be found in modified structures that are some 12% denser than the ordinary forms. Si assumes a low-symmetry cubic structure that is still fourfold coordinated, but with a distortion of the tetrahedra that allows a higher packing density. For Ge, the CN also remains 4, but a still different grouping results in more tetrahedra per volume at the expense of irregularity in the tetrahedra. The precise location phase diagrams of these forms have not yet been determined, but they are readily obtained metastably after specimens are subjected to pressures in excess of 100 kbar (Kasper, 1969). A conversion of grey tin into white tin is inhibited, or at least delayed, by adding Ge and by applying a tensile force.

The growth regimes of cubic group-IV lattices and the associated tetragonal covalent radii ρ_T are shown in Fig. 2. Note that only Si and Ge are completely miscible, whereas SiSn and GeSn are only stable within a very small range. Little is as yet known about the epitaxial growth of strained GeSn alloys.

The diamond (zinc-blende) lattice consists of two FCC sublattices of the same (different) atomic species. The second sublattice is shifted in the [111] direction by $a_0\sqrt{3}/4$, where a_0 is the cubic lattice constant. The well-known ABC stacking of atomic {111} planes in the FCC lattice is changed into an AaBbCc stacking, where the upper case and lowercase letters refer to the atomic {111} planes of the two respective sublattices. The distances A–B, a–A, and A–b are given by $a_0/\sqrt{3}$, $a_0 \cdot \sqrt{3}/12$. and $a_0 \cdot \sqrt{3}/4$, respectively. Figure 3 shows the atomic arrangement and the bonding configuration for this lattice structure. The diamond cell contains eight atoms arranged in four monolayers (ML) along the [001] direction. Therefore, the height of one ML in this principal direction is given by $a_0/4$. Some relevant lattice data at room temperature are summarized in Table II. Generally, these data are only slightly temperature dependent, except for the thermal expansion coefficient α. The expansion coefficients for Si and Ge increase nearly parallel with temperature (800°C: $\alpha_{Si} = 5.8 \cdot 10^{-6} \, K^{-1}$; $\alpha_{Ge} = 8.7 \cdot 10^{-6} \, K^{-1}$), but dia-

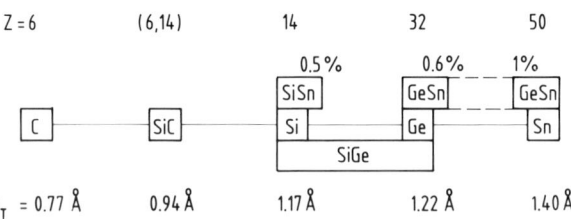

FIG. 2. Group-IV diamond lattices (SiC: zinc blende) ordered with increasing atomic number Z and increasing tetragonal covalent radius ρ_T.

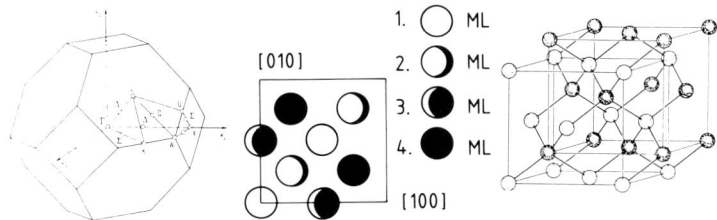

FIG. 3. Diamond lattice: First Brillouin zone of the diamond lattice (left), view in [001] direction onto the four monolayers within a cubic cell; the atoms of the first ML are drawn as open circles, those of the fourth ML as full circles (center). Atomic-bonding arrangement (right).

mond exhibits a very strong increase, reaching $\alpha_{dia} = 5.3 \cdot 10^{-6} \text{ K}^{-1}$ at 1000°C. Other sources (Maguire, 1987) assume even equal thermal expansion for Si and diamond for a growth experiment at 1100°C. SiC exhibits a sharp increase in α to a value of $4.5 \cdot 10^{-6} \text{ K}^{-1}$ at 300°C, but then levels off, reaching a mere $5 \cdot 10^{-6} \text{ K}^{-1}$ at 900°C.

The electronic band structure of the nonmetallic, cubic group-IV elements (except for α-Sn) is characterized by an indirect bandgap. The valence-band maximum is located at the center of the Brillouin zone (Γ-point). The conduction bands of diamond, SiC, and Si show six equivalent minima along the [100] axes (Δ). These minima are at $\Delta = 0.75X$ and $\Delta = 0.85X$ in diamond, and Si, respectively. The conduction band of Ge shows eight minima at the end points of the [111] axes (L-point). Unstrained $Si_{1-x}Ge_x$ alloys have a Si-like band structure for $x \leq 0.85$, and become Ge-like for higher Ge contents. The conversion into a Ge-like structure is strongly affected if the SiGe alloys become strained. It can be completely suppressed

TABLE II

Room-Temperature Cubic Lattice Data: Lattice Constant a_0. Thermal-Expansion Coefficient α, Direct Bandgap $E_g^{(d)}$, Indirect Bandgap $E_g^{(i)}$, Debye Temperature Θ_D, and Melting Point (Landolt-Börnstein, 1982).

Element	C	β-SiC	Si	Ge	α-Sn
lattice	diamond	zincblende	diamond	diamond	diamond
a_0 (Å)	3.5668	4.3596	5.431	5.657	6.489
α (10^{-6} K^{-1})	1.0	2.9	2.56	5.9	4.7
density (g/cm^3)	3.515	3.210	2.329	5.323	7.285
$E_g^{(d)}$ (eV)	7.3	6.0	3.4	0.80	0
$E_g^{(i)}$ (eV)	5.48	2.2	1.11	0.664	–
Θ_D (K)	1860	1270	645	360	240
melting point (K)	4100	(3103)[a]	1685	1210	(286)[b]

[a] decomposition
[b] transition temperature α-Sn/β-Sn.

even for pure Ge layers ($x = 1$) under sufficiently large compressive strain (see Section V. B.) The band structure of α-Sn is qualitatively different from those of the other group-IV elements, with the conduction-band minimum (CBM) and the valence-band maximum (VBM) being degenerate at Γ. Hence, α-Sn is a zero-gap semiconductor. Energies and symmetries of the bandgaps and the lowest direct transitions of the group-IV semiconductors are plotted in Fig. 4.

A superlattice (SL) built of two parent lattices reduces the symmetry of the constituents. This effect is listed in Table III for short-period SLs built of diamond sublattices. It is interesting to note that the SL symmetry depends strongly on whether the superlattice is composed of periods with two layers of even, odd, or mixed ML thickness (Alonso et al., 1989). This information is important, for example, for the selection rules of Raman scattering, which plays an important role in characterizing thin SL structures (see Section V.C.).

B. LATTICE MISMATCH

The large differences in covalent radii cause a strong mismatch between the group-IV diamond lattices. The mismatch η_0 between a thin-film material (lattice spacing a_2) and a thick rigid substrate (lattice spacing a_1) is commonly defined as the relative spacing difference $(a_2 - a_1)/a_1$. Many properties of group-IV elements are monotonically dependent on lattice constant or

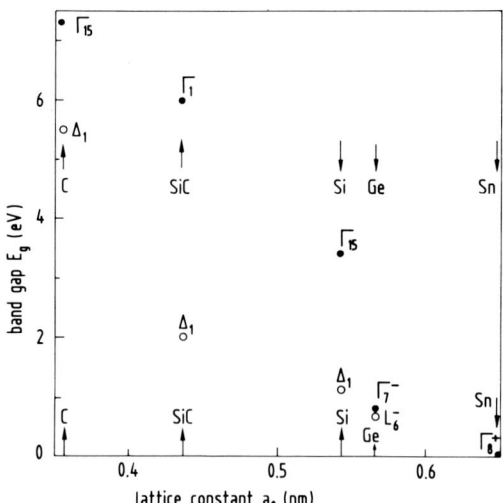

FIG. 4. Indirect bandgap $E_g^{(i)}$ (open circles) and lowest direct bandgap $E_g^{(d)}$ (full circles) versus lattice constant a_0. Data points are labeled by the symmetry of the respective conduction-band minimum.

TABLE III

Symmetry of (001) Superlattices. One Period Consists of m Monolayers of Material A and n Monolayers of Material B. Both Materials are Assumed to Crystallize in the Diamond Lattice. k is an Integer Number (Alonso et al., 1989).

Superlattice constituents	Lattice type	Space group
	diamond[a]	O_h^7 (Fd3m)
$m = 1$; $n = 1$	zinc blende[b]	T_α^2 (F$\bar{4}$3m)
m, n: even; $m \cdot n = 4k$	orthorombic[b]	$D_{2\alpha}^9$ (Pmma)
m, n: even; $m \cdot n = 4k + 2$	orthorombic[b]	D_{2h}^{28} (Imma)
m, n: odd; $m \cdot n = 4k + 2$	tetragonal[b]	$D_{2\alpha}^9$ (I$\bar{4}$2m)
m, n: odd; $m \cdot n = 4k$	tetragonal[b]	$D_{2\alpha}^5$ (P$\bar{4}$m2)
m: even, n: odd	tetragonal[b]	D_{4h}^{19} (I4$_1$/amd)

[a] $x \parallel [100]$; $y \parallel [010]$; $z \parallel [001]$
[b] $x \parallel [110]$; $y \parallel [1\bar{1}0]$; $z \parallel [001]$

mismatch. Elastic moduli, hardness, melting point, Debye temperature, bandgap, and effective mass decrease with lattice constant, whereas the dielectric constant increases. Table IV gives some of the relevant parameters in connection with the lattice mismatch η_0, which is referred to silicon.

TABLE IV

Important Material Data of Selected Group-IV Elements: Lattice Mismatch η_0, Elastic Moduli (c_{11}, c_{12}, c_{44}), Microhardness H_K (Knoop), Thermal Conductivity κ, Dielectric Constant ε, Electron Mobility μ_e, Hole Mobility μ_h, Effective Mass m^* of Electrons (Transvers $e^\perp$, Longitudinal $e^{\parallel}$) and Holes (Light h^l, Heavy h^h). All Values are at Room Temperature, Except m^*, Which Is at 1.3–4.2 K. (Landolt-Börnstein, 1982)

Element	C	β-SiC	Si	Ge	α-Sn
η_0 (%)	−34.32	−19.73	± 0	4.17	19.48
c_{11} (100 GPa)	10.764	5.00	1.658	1.240	0.690
c_{12} (100 GPa)	1.252	0.92	0.639	0.413	0.293
c_{44} (100 GPa)	5.774	1.68	0.796	0.683	0.362
H_K (kg/mm^2)	9000	3300	1150	800	
κ (Wm^{-1} K^{-1})	2000	(400)[a]	140	60	
ε	5.7	6.5	11.9	16.2	24[b]
μ_e (cm^2 V^{-1} s^{-1})	1800	900	1450	3900	1400
μ_h (cm^2 V^{-1} s^{-1})	1600	50[a]	450	1900	1200
m_e^* ($\perp$)		0.26	0.19	0.08	0.024
m_e^* ($\parallel$)		0.53–1.5	0.92	0.64	0.2–0.45
m_h^* (l)	0.7	1.0	0.15	0.043	
m_h^* (h)	2.18		0.54	0.28–0.38	

[a] 6H polytype
[b] doping dependent

The lattice mismatch of the $Si_{1-x}Ge_x$ alloys can be expressed to first approximation by Vegard's law. Dismukes et al. (1964) have shown that there are only minor deviations from Vegard's law, which can be expressed by a mixing correction factor C_m:

$$\eta_0 = (4.17 \cdot 10^{-2} \cdot x) C_m. \tag{1}$$

The mixing correction factor C_m is listed in Table V for a variety of x-values. It is smaller than unity over the full range of x.

The lattice mismatch is only slightly temperature dependent, because of the small thermal-expansion coefficients of diamond-type materials. The thermal-expansion coefficient α_{Si} of Si is increasing from $2.5 \cdot 10^{-6} \, K^{-1}$ at 300 K to $4.35 \cdot 10^{-6} \, K^{-1}$ at 800 K, with a mean value over this temperature range of $3.3 \cdot 10^{-6} \, K^{-1}$. α_{Ge} increases from $5.9 \cdot 10^{-6} \, K^{-1}$ at 300 K to $7.9 \cdot 10^{-6} \, K^{-1}$ at 800 K, with a mean value of $6.9 \cdot 10^{-6} \, K^{-1}$. The higher thermal-expansion coefficient of Ge leads to a slight increase of the lattice mismatch of Ge to 4.36% at 800 K as compared with 4.17% at room temperature. This effect has to be taken into account when designing strain-adjusted SiGe superlattices, which are generally grown at temperatures up to about 800 K.

III. Basic Principles

A. Mismatch Accommodation by Strain or Misfit Dislocations

Nature has two methods to accommodate lattice-mismatched films (Fig. 5), namely elastic accommodation by strain, and plastic accommodation by misfit dislocations lying in the interface plane. Elastic accommodation is the preferential mechanism for thin films up to a critical thickness t_c, whereas above the critical thickness, misfit dislocations are generated that relax the built-in strain. As a guideline for understanding nature's preference for strained-layer epitaxy below the critical thickness, one can consider thermodynamic equilibrium as is given by Van der Merwe (1972). Indeed, his theory predicts qualitatively correct the observed behavior by minimizing the total

TABLE V

Mismatch of $Si_{1-x}Ge_x$ Alloys. Mixing Correction Factor C_m Describing the Deviation from Vegard's Law (Eq. (1) with $C_m = 1$). C_m Values Are Calculated Using the Lattice Constants of Dismukes et al. (1964).

Ge content x	0.05	0.1	0.2	0.3	0.4	0.5	0.6	0.7	0.8	0.9
C_m	0.965	0.938	0.909	0.909	0.926	0.939	0.956	0.967	0.985	0.991

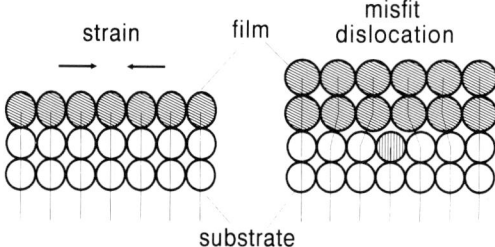

FIG. 5. Mismatch accommodation by strain (left) and by the creation of misfit dislocations (right).

energy of the system. However, the quantitative results of the equilibrium theory differ considerably from the results obtained by molecular-beam epitaxy (MBE). This can most clearly be seen by a plot of the critical thickness t_c versus lattice mismatch (Fig. 6). Unfortunately, the reader of the recent literature may be confused by theoretical curves covering a rather broad range (People and Bean, 1985; Kasper and Herzog, 1977). This is caused by various approximations given by Van der Merwe, but also simply by misprints. We strongly recommend the use of one set of data (Kasper et al., 1986), which is based on a careful analysis of the approximations valid for SiGe (Kasper and Herzog, 1977). Most probably, kinetic limitations (nucleation and glide) cause the deviations of the experiments from theory.

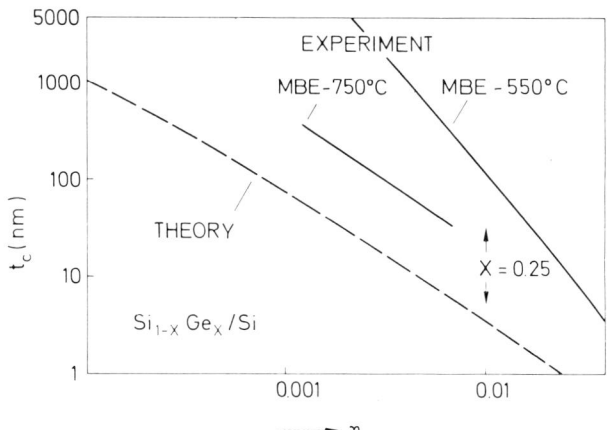

FIG. 6. Critical thickness t_c as a function of the lattice mismatch η_0. Results of the equilibrium theory are compared to experimental data derived at MBE growth temperatures of 550 and 750°C.

B. Tetragonal Distortion

Consider a cubic lattice cell that is in-plane strained (x, y) by the elastic film stress σ. The in-plane strain $\varepsilon \equiv \varepsilon_x \equiv \varepsilon_y$ causes a perpendicular strain ε_z, which is given by

$$\varepsilon_z/\varepsilon = -2v/(1-v), \qquad (2)$$

where v is Poisson's number.

The strains distort the cubic lattice cell to a tetragonal cell as can be measured, for example, by x-ray diffractometry (Fig. 7). The lattice spacing d as well as the inclination i of the lattice planes are changed by the tetragonal distortion. Figure 7 gives examples of the changes in lattice spacing and inclination; Δd and Δi, respectively. Table VI gives numerical values for the

TABLE VI

Elasticity Properties for $\{001\}$ Surfaces of Several Semiconductors: Poisson's Number v, Modulus of Elasticity M_E (Brantley, 1973).

Element	GaAs	GaP	Si	Ge
$M_E/(1-v)(10^{11}\,\mathrm{Pa})$	1.239	1.488	1.805	1.420
v	0.312	0.305	0.279	0.270

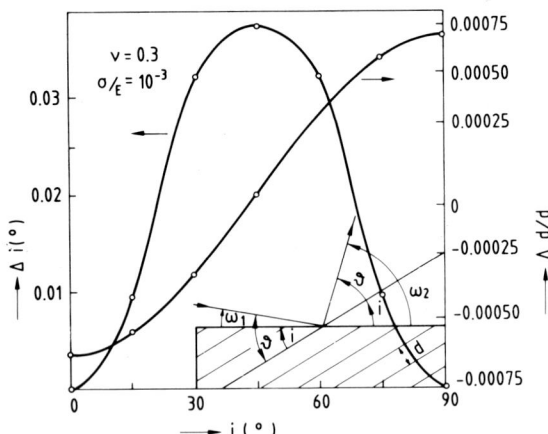

Fig. 7. Tetragonal distortion of the cubic lattice cell by a stress $\sigma = 10^{-3} \cdot M_E$ (M_E is the modulus of elasticity). Plotted are the relative change of the interplanar spacing $\Delta d/d$ and the change Δi of inclination versus the inclination of the lattice planes. The insert shows the principal arrangement of an x-ray diffractometric measurement.

modulus of elasticity M_E and for Poisson's number derived from anisotropic elasticity theory (Brantly, 1973).

The strained crystal stores elastic energy, whose density E is given for an isotropic crystal by

$$E = 2\mu \frac{1+v}{1-v} \cdot \varepsilon^2, \tag{3}$$

with shear modulus μ,

$$\mu = \frac{M_E}{2(1+v)}. \tag{4}$$

The areal energy E_h of an homogeneously strained film of thickness t is then given by

$$E_h = Et. \tag{5}$$

Note that a uniform, thin film on a thick substrate is always nearly homogeneously strained along the z-axis. Small deviations from homogeneous strain are caused by the strain-induced curvature k of the film/substrate couple. The curvature k (reciprocal of the curvature radius R) is given by

$$k = 6\frac{\varepsilon t}{t_s^2} \tag{6}$$

if $t \ll t_s$ (t_s is the thickness of substrate), and equal elastic constants of film and substrate are assumed. Due to the curvature, the substrate is also stressed, leading to a neutral plane at a distance $t_s/3$ from the backside of the substrate. The strain on the backside has the same sign as the film, whereas the substrate surface is strained in the opposite direction by the amount

$$\sigma_s(0) = -4\sigma t/t_s. \tag{7}$$

The film stress is lowered by the amount of σ_s at the interface but increases very slightly with distance z from the interface. The variation of strain along the z-axis of a uniform film of 1 μm thickness on top of a 250-μm-thick substrate is about 10^{-4} and can therefore be neglected for all practical purposes.

C. Dislocations

Before treating misfit dislocations themselves, let us summarize some general properties of dislocations. For more details, the reader is referred to monographs on this topic, for example, by Nabarro (1967), Hirth and Lothe (1968), or Kelly and Groves (1970).

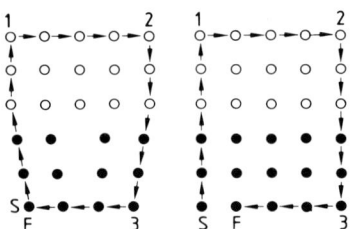

Fig. 8. Burgers circuits in a real crystal (left) and in a perfect reference crystal (right). The Burgers vector **b** defines the displacement field of a dislocation.

Dislocations are line defects whose displacement field is defined by their Burgers vector **b** (Fig. 8). The displacement discontinuity, which the Burgers vector measures, remains constant along a given dislocation line. The principle of conservation of the Burgers vector is a general one that implies that a dislocation cannot end inside a crystal. It must end at a surface, form a closed loop, or meet other dislocations. A point at which dislocation lines meet is called a node. The conservation of the Burgers vector in the latter case can be stated in the following way: if the directions of all the dislocation lines are taken to run out from the node, then the sum of the Burgers vectors of all the dislocations is zero. This is the dislocation analogue of Kirchhoff's law of the conservation of electric current in a network of conductors.

In general, a dislocation may lie at any angle to its Burgers vector or may be curved. A length of dislocation that lies normal to its Burgers vector is called edge dislocation, and one which lies parallel to its Burgers vector is called a screw dislocation. A dislocation is referred to as mixed if it has both edge and screw components.

Dislocation line element and Burgers vector define a slip plane. Dislocation motion in this slip plane is called glide. Dislocation motion outside the slip plane is called climb. When a dislocation glides, there is no problem of acquiring or disposing of extra atoms. Climb can be accomplished by adding or removing atoms. Therefore, a dislocation will be able to glide much more quickly than it can climb. Thus, a dislocation loop lying in its own glide plane can expand easily under the action of forces. Loops with Burgers vectors outside the loop plane are called prismatic loops because they could be produced by prismatic punching, in which case its structure corresponds to a penny-shaped disc of extra atoms in the plane of the loop.

Generally, one crystallographic plane is energetically favorable for gliding so that slip is effectively confined to this plane. For diamond-type lattices, the slip plane is of the $\{111\}$ type, which comprises four individual planes. The Burgers vector is of the $1/2 \cdot \langle 110 \rangle$ type, which contains 12 individual directions. Frequently the line direction of undisturbed dislocations is also of the ⟨110⟩ type, creating so-called 60°-dislocations with a 60° angle between line direction and Burgers vector. For the (111) slip plane, this could for example be the [110] and 1/2·[011] vectors as line direction and Burgers vector, respectively.

A free dislocation will move under the action of an applied stress field. Dislocations with opposite Burgers vector move in opposite directions. The dislocation will bend if it is pinned at a point, e.g., by a node or by an obstacle, like an inclusion, or by a line segment that is cross slipped. Cross slip means that a screw segment has glided off its original slip plane. Such pinning points are necessary for the operation of multiplication sources. A simple mechanism by which dislocations can multiply was suggested by Frank and Read. Its operation requires only that a dislocation in a slip plane be pinned at two points (Nabarro, 1967; Hirth and Lothe, 1968; Kelly and Groves, 1970). The Frank–Read sources and similar source types can generate any number of expanding loops under the action of sufficiently high stress fields.

The crystal lattice is distorted around a dislocation. The rather far-reaching strain field is proportional to the length of the Burgers vector **b**. It fades away with the reciprocal distance r from the dislocation. A considerable amount of strain energy is stored in the elastically distorted region around a dislocation line. The strain energy E_{ds} of a single dislocation is given for an isotropic medium by

$$E_{ds} = \frac{\mu \cdot \mathbf{b}^2}{4\pi} \ln(R/r_i) \tag{8}$$

for a screw dislocation, and by

$$E_{ds} = \frac{\mu \cdot \mathbf{b}^2}{4\pi(1-v)} \ln(R/r_i) \tag{9}$$

for an edge dislocation. (R is the range of the strain field, r_i is the inner cutoff radius for the dislocation core).

The difficulty in applying these equations is that the strain energy increases without limit as R increases or r_i decreases. Somewhat arbitrary cutoff radii for the strain-field range and the core have to be chosen to overcome this difficulty. In practice, the outer dimension of the crystal or half the distance to the next dislocation are chosen for the range R of the dislocation, and values close to the amount of the Burgers vector are chosen for the inner cutoff radius r_i.

D. Equilibrium Theory for Misfit Dislocations

By considering the energy of a film, or by taking the forces on dislocation segments, one can calculate the equilibrium status of a mismatched film. In this section, an example for the energy approach is given, following the route of Van der Merwe's calculation. For thin films below a critical thickness t_c, the lowest energy state is that of a strained film, and above t_c, a partly relaxed film with misfit dislocations at the interface is the stable one.

Kinetic reasons cause deviations of experiments from equilibrium theory. However, the knowledge of equilibrium is important for defining the range of metastability and for understanding the influence of postepitaxial thermal-processing steps.

A film can accommodate lattice mismatch to a substrate either by strain (tetragonal distortion) or by the formation of a misfit-dislocation network at the interface to the substrate. For the sake of simplicity, we consider a (001) surface with a network of misfit dislocations lying in the interface. From the fourfold symmetry of this surface, an orthogonal network of misfit dislocations is required, e.g., along the [110] and [1$\bar{1}$0] directions, respectively. For the threefold symmetric (111) surface, a network with three different dislocation directions, e.g., along [1$\bar{1}$0], [10$\bar{1}$], [01$\bar{1}$] would be required. The perfect misfit dislocation is of the edge type, although edge dislocations require more energy than screw dislocations for their generation (Eqs. (8), (9)). Screw dislocations will rotate the film against the substrate. Consider the effect of a parallel array of edge dislocations a distance p apart from another. Line direction and Burgers vector should lie in the interface as in Fig. 5. The parallel array accommodates a mismatch equal to b/p. The relation between mismatch η_0, film strain ε, and dislocation distance p is given by (Kasper and Herzog, 1977)

$$\eta_0 + \varepsilon = b/p, \qquad (10)$$

where b is the length of Burgers vector, the sign of b is the same as the sign of η_0.

A complete accommodation of mismatch ($\varepsilon = 0$) by perfect misfit dislocations would require a network with spacing p,

$$p = b/\eta_0. \qquad (11)$$

Van der Merwe (1972) calculated the energy E_d of a perfect misfit-dislocation network and the total energy E_{tot} of a film, whose mismatch is partly accommodated by strain and partly by misfit dislocations. The general equation, Eq. (9), is also valid for a misfit dislocation. For the range R of the strain field, the assumption was made that for very thin films, R approaches twice the thickness t of the film, whereas for very thick films, R approaches half the distance p. For medium values of t, a smooth interpolation formula was given

$$\begin{aligned} R &= p/2 & \text{for} \quad t \geqslant p/2, \\ R &= 2t/(1 + 4t^2/p) & \text{for} \quad t \leqslant p/2. \end{aligned} \qquad (12)$$

The inner cutoff radius r_i calculated for a Peierls-type dislocation core (Van der Merwe, 1972) is given for a (001) interface in the diamond lattice by (Kasper et al., 1986)

$$r_i = b(\pi/2\sqrt{2} \cdot e(1-v)), \tag{13}$$

where e is the base of the natural logarithm.

The energy E_d per unit area of an orthogonal network is given by

$$E_d = 2E_{ds}/p. \tag{14}$$

The total energy E_{tot} per unit area is given by

$$E_{tot} = E_h + E_d, \tag{15}$$

using Eqs. (3)–(5), (9), and (12)–(14) for calculating E_h and E_d.

Van der Merwe assumed that the equilibrium state is given by the minimum of the energy E_{tot}, thus neglecting entropy terms.

$$\frac{\partial E_{tot}}{\partial(1/p)} = 0 = \frac{\partial E_h}{\partial \varepsilon} b + 2E_{ds} + \frac{2}{p}\frac{\partial E_{ds}}{\partial(1/p)}. \tag{16}$$

By using Eqs. (5), (9), (12), and (13), we obtain

$$\frac{\partial E}{\partial \varepsilon} \cdot b = 4\mu \frac{1+v}{1-v} b \cdot t[(b/p) - \eta_0],$$

$$2E_{ds} = \frac{\mu \cdot b^2}{2\pi(1-v)} \ln(R/r_i),$$

$$\frac{2}{p}\frac{\partial E_{ds}}{\partial(1/p)} = -\frac{\mu b^2}{2\pi(1-v)} \quad \text{for} \quad t \geq p/2 \tag{17}$$

$$= -\frac{\mu b^2}{2\pi(1-v)} \frac{8(t/p)^2}{1+4(t/p)^2} \quad \text{for} \quad t \leq p/2.$$

Very thin layers up to a critical thickness t_c grow without misfit dislocations, a mode which is called strained-layer epitaxy or pseudomorphic or commensurate growth,

$$\varepsilon = -\eta_0; \quad p \to \infty \quad (\text{for} \quad t < t_c). \tag{18}$$

The critical thickness t_c is given by (Kasper et al., 1986)

$$t_c \cdot \eta_0 = \frac{b}{8\pi(1-v)} \ln\left(\frac{4 \cdot \sqrt{2} \cdot (1-v) \cdot e \cdot t_c}{\pi \cdot b}\right). \tag{19}$$

Assuming that $b = 0.384$ nm, $v = 0.3$, and by measuring t_c in nm, one obtains the following implicit relation for t_c,

$$\eta_0 \cdot t_c(\text{nm}) = 1.175 \cdot 10^{-2}(2.19 + \ln t_c). \tag{20}$$

The film strain ε in films thicker than t_c is relaxed by the formation of a misfit-dislocation network.

$$(t\varepsilon) = -\frac{b}{8\pi(1+v)}\left(\ln(R/r_i) - \frac{8(t/p)^2}{1+4(t/p)^2}\right) \quad \text{for} \quad t < p/2$$

$$= -\frac{b}{8\pi(1+v)}(\ln(R/r_i) - 1) \quad \text{for} \quad t \geqslant p/2. \tag{21}$$

Figure 6 compares the calculated equilibrium-theory values with experimental curves for MBE growth at 550°C and 750°C (Kasper, 1986). The equilibrium theory allows us to understand why thin films up to a certain thickness grow as strained layers, but quantitatively it predicts values that are too low for the critical thickness. The Van-der-Merwe formalism gives a rather raw estimate of the energy contribution of the dislocation core. It has been shown (Kasper and Herzog, 1977) that this is unimportant for critical thicknesses of many monolayers, but it should be considered for very thin thicknesses. For that regime, numerical calculations are available (Dodson and Taylor, 1986).

E. Metastability

The experimentally found critical thicknesses t_c of a mismatched layer are generally higher than predicted by an equilibrium theory. If one does not assume that the theoretical treatment is based on wrong assumptions, one has to consider kinetic reasons for this deviation from equilibrium. These deviations can come from the nucleation and motion of dislocations that are both hampered by the need for an activation energy. Matthews (1975) already argued that misfit dislocations that move to the interface by a glide process are not as ideal as perfect misfit dislocations. He considered a substrate dislocation that penetrates the surface (Fig. 9, top). The slip plane and the Burgers vector **b** are inclined to the interface. The dislocation is bent by the film strain and can create a straight misfit-dislocation segment if the thickness is raised above the critical thickness. But now, only the projection of the edge component of the Burgers vector onto the interface is able to accommodate mismatch instead of the full Burgers vector as in the case of an ideal edge-type misfit dislocation with the interface as its slip plane. Let us consider, for example, a (001) surface that is penetrated by a dislocation lying in a (111) slip plane with a $1/2 \cdot [011]$ Burgers vector **b**. The misfit-dislocation segment

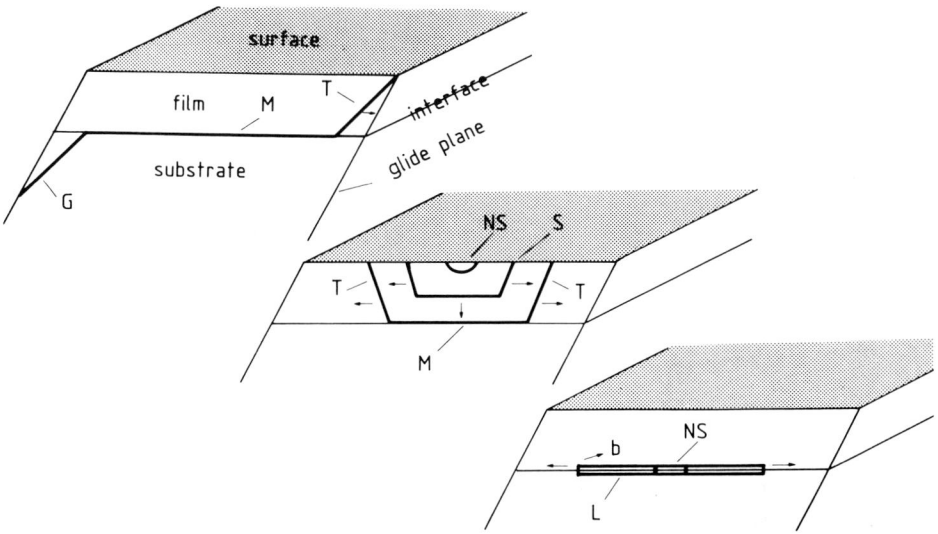

FIG. 9. Basic mechanism for misfit-dislocation generation. Top: Bending of a substrate dislocation. Center: Nucleation (NS) of a dislocation at the surface and subsequent movement to the interface. Bottom: Nucleation of a dislocation dipole at the interface.

extends along the [110] intersection of the slip plane with the interface. The projection of the edge component in the [110] direction amounts to half of the Burgers vector (**b**/2). For this nonideal misfit-dislocation arrangement, the relation between strain and misfit-dislocation array has to be reformulated,

$$\eta_0 + \varepsilon = \beta \cdot b/p$$
$$\beta \leq 1, \qquad (22)$$

with

$$\beta \cdot b = \mathbf{b}[\xi \times \mathbf{i}],$$

where **b** is the Burgers vector; ξ and **i** are the unit vectors along the dislocation line and perpendicular to the interface, respectively).

The weaker mismatch accommodation of this dislocation array shifts the predicted critical thickness to higher values. The formal treatment of Matthews uses a force balance instead of energy considerations as Van der Merwe has done, which has been proven to be equivalent by the definition of the force on a dislocation. Matthews's theory is closer to the experimental results than Van der Merwe's theory, especially for medium growth temperatures (e.g., 750°C MBE results). It should, however, be stated that it describes a metastable state, because principally ideal misfit dislocations can be created by the slower process, e.g., by climb or glide from the rim of the sample. The same result is obtained for glide of a dislocation half loop from a nucleation site at the surface (Fig. 9, center), which is a more probable process than the originally suggested bending of substrate dislocations, because of today's low dislocation-density substrates.

The kinetic reason for the increase in critical thickness at low growth temperatures (550°C MBE) is not yet clear. One can speculate that the barrier for nucleation controls the onset of accommodation by dislocations. Transmission eledcton microscopy (TEM) images (Herzog, 1984) of films grown at 550°C are compatible with the picture of nucleation of dislocation arrays at the interface itself (Fig. 9, bottom). A fitting curve for the low-temperature data (People and Bean, 1985) was given (see Fig. 6, 550°C MBE). In this interesting treatment, the authors compare the elastic energy E_h with what they call *areal energy density* of the dislocation. This may erroneously be taken as an indication for an equilibrium-theory treatment; however, it is probably more adequate to interprete E_h as an activation barrier for misfit-dislocation nucleation.

Recently, a Sandia Laboratory group (Tsao et al., 1987; Dodson and Tsao, 1987; Dodson, 1988) thoroughly investigated the metastable growth regime. They pointed out that most experimental techniques are only capable of detecting misfit dislocations above a certain density ($\approx 10^5 \, \text{cm}^{-1}$). A smaller density can already exist in the region claimed as metastable. This will critically depend on the density of pre-existing dislocations from the substrate and on the density of strain-inducing film defects. Although some understanding has been obtained, further investigations are clearly necessary.

IV. Stability of Strained-Layer Superlattices

In general, the elements of a strained-layer superlattice (SLS) comprise the thin, mismatched layers that are purely accommodated by strain and stacked to a superlattice, the buffer layer, and the substrate (Fig. 10). The mismatch between two layers with the lattice constants a_1, a_2 is given by

$$\eta = 2 \frac{a_2 - a_1}{a_1 + a_2}. \tag{23}$$

The elastic strains ε_1, ε_2 in the superlattice layers are connected to the mismatch η by

$$\eta + \varepsilon_2 - \varepsilon_1 = 0 \tag{24}$$

if no misfit dislocations are present in the superlattice interfaces as expected

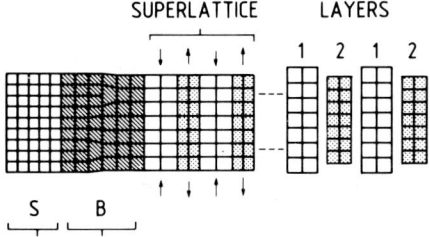

FIG. 10. Elements of a strained-layer superlattice (SLS). S: substrate; B: partly relaxed buffer layer.

for SLSs. That means that the layer thicknesses t_1, t_2 of the individual superlattice layers meet the requirements for pseudomorphic growth,

$$t_1, t_2 < t_c(\varepsilon_1, \varepsilon_2) \tag{25}$$

The driving force for the introduction of misfit dislocations is the strain energy stored in a layer. For the calculation of the critical thicknesses of individual superlattice layers, we should insert the strains ε_1, ε_2, instead of η_0, for the single-layer results (Fig. 6, Eq. (20)). The superlattice as a whole has the same in-plane lattice constant $a_\parallel$ (parallel to the interfaces).

$$a_\parallel = a_1(1 + \varepsilon_1) = a_2(1 + \varepsilon_2). \tag{26}$$

The strain changes abruptly by the amount $(\varepsilon_2 - \varepsilon_1)$ at each interface between adjacent superlattice layers. The strain distribution is determined by the choice of the in-plane lattice constant $a_\parallel$.

$$\varepsilon_1 = \frac{a_\parallel - a_1}{a_\parallel},$$

$$\varepsilon_2 = \frac{a_\parallel - a_2}{a_\parallel}. \tag{27}$$

A. Strain Symmetrization

The stability of SLSs was investigated for specific material systems as early as 1974 (Matthews and Blakeslee, 1974) and 1975 (Kasper et al., 1975). The general SLS concept was introduced by Osbourn in 1982. In a pioneering work (Bean, 1985), the concept of critical thickness t_{cs} of the superlattice as a whole was introduced that should be roughly equal to the critical thickness of a single layer of the same average composition. We will demonstrate that this definition applies only to specific structures.

One can give a very general definition of the stability of a superlattice as a whole against the introduction of misfit dislocations. For that purpose, we consider the strain energy E_h stored in a SLS. This is simply the sum over all the strain energies of the individual layers (with N being the number of periods with length L),

$$E_h = 2\mu \frac{1+\nu}{1-\nu} N \sum_{i=1}^{2} \varepsilon_i^2 t_i. \tag{28}$$

In Eq. (28), the same elastic constants for both materials are assumed, but it is straightforward to introduce different elastic properties and anisotropy of the material components. By using the relation between the ε_i (Eqs. (23), (24)), the minimum in strain energy is easily found by setting $\partial E_h/\partial \varepsilon_1 \equiv 0$. We get

$$\varepsilon_1 = \eta \frac{t_2}{L},$$
$$\varepsilon_2 = -\eta \frac{t_1}{L}. \tag{29}$$

For the most important case $t_1 = t_2 = L/2$, this condition reduces to

$$\varepsilon_1 = -\varepsilon_2 = \eta/2. \tag{30}$$

That means that the SLS is stable if the strain is symmetrized according to Eq. (30). We use the term strain symmetrization mainly for SLSs with equal layer thickness $t_1 = t_2$, but it means only replacing Eq. (30) by Eq. (29), without loss of generality if unequally thick layers are used. In the case of strain symmetrization, there is no limiting critical thickness of the superlattice as a whole ($t_{cs} \to \infty$). The situation changes if the strain is asymmetrically distributed. Then the superlattice can gain energy by the introduction of misfit dislocations at its base if the thickness of the superlattice ($N \times L$) exceeds the critical superlattice thickness t_{cs}. The critical superlattice thickness will decrease with increasing asymmetry of the strain distribution. The extreme case with the lowest critical thickness is achieved if only one material in the SLS is strained ($\varepsilon_1 = 0; \varepsilon_2 = -\eta$). This situation occurs, for example, if a SiGe/Si SLS is directly grown on a Si substrate without a strain-adjusting buffer layer (e.g., Bean, 1985; Bevk et al., 1987). If such an asymmetrical SLS exceeds its critical thickness t_{cs}, misfit dislocations are introduced, which change $a_\parallel$, hence shifting the strain distribution toward higher symmetry.

B. Buffer-Layer Design

A buffer layer between superlattice and substrate is grown for several reasons. The main functions of the buffer layer are:

(i) Transition from the substrate material to the superlattice material system, e.g., from an InP substrate to an InAs/GaAs superlattice via an InGaAs buffer layer, lattice-matched to the substrate.

(ii) Improvement of the crystal quality. Especially strained buffer layers or SLS buffers can effectively reduce the density of penetrating substrate dislocations by bending them into the interface plane.

(iii) Adjustment of the superlattice strain distribution by shifting the in-plane lattice spacing from the substrate value a_A to the superlattice value $a_\|$. This shift principally involves the generation of misfit dislocations in the buffer layer. The thick (several μm) buffer layers used usually with either linearly or step-graded composition suffer from being heavily dislocated, which might be an inherent problem for device applications. Therefore, growth of SLSs without such a strain-adjusting buffer layer was performed. However, high-quality growth is only obtained for specific material system where the lattice constant a_A of the substrate lies between the lattice constants a_1, a_2 of the superlattice materials (strain symmetrization by the substrate itself, Marzin et al., 1986); or for thin asymmetrically strained systems below the critical superlattice thickness (Bean et al., 1984, Yen et al., 1986).

We will discuss here the concept of a thin ($\leqslant 0.25\ \mu$m), homogeneous buffer layer, whose strain-adjusting misfit-dislocation network is mainly confined to the interface between substrate and buffer layer (Kasper et al., 1987b). The in-plane lattice spacing $a_\|$ is controlled (Fig. 11) by the lattice constant a_B of the buffer material and by the strain ε_B in the buffer. The latter depends on the

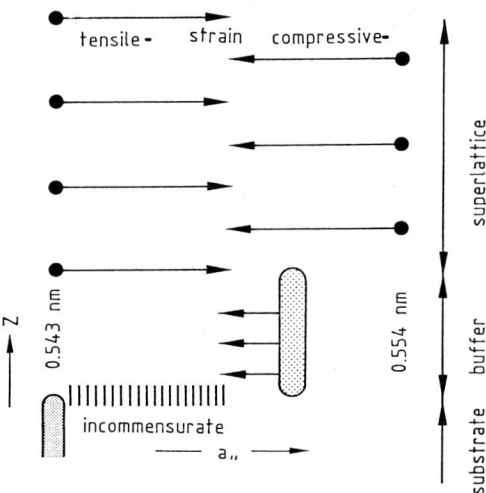

FIG. 11. Schematic view of strain symmetrization by a homogeneous, incommensurate buffer layer. Dots: natural lattice spacing. Arrows: distortion of the in-plane lattice spacing $a_\|$ by strain. The values refer to the situation of a $Si_{0.5}Ge_{0.5}/Si$ superlattice on a Si substrate.

mismatch η_0 between substrate and buffer layer, as well as on the buffer-layer thickness t_B.

$$a_\parallel = a_B(1 - \varepsilon_B) \tag{31}$$

The design of the buffer layer is explained here exemplarily for a $Si_{1-y}Ge_y$ buffer layer on a Si substrate. With the lattice mismatch between Ge and Si being 4.2%, and assuming Vegard's law, we get for the lattice constant of the buffer layer as a function of the chemical composition y,

$$a_B = a_A(1 + 0.042y) \tag{32}$$

The strain ε_B in thin SiGe layers grown at 550°C on Si substrates has been measured by He-backscattering (Bean et al., 1984). We use these data in the smoothed form given by Bean (1988) to get the relation between the in-plane spacing $a_\parallel$ and the buffer-layer lattice constant a_B. This relation is plotted in Fig. 12 for various buffer-layer thicknesses ranging from 10 to 250 nm. One can consider the couple substrate/buffer layer as a new, virtual substrate, which offers the in-plane lattice spacing $a_\parallel$. In this example, this would correspond to a SiGe substrate with an effective Ge content y^*, where y^* is defined by

$$\begin{aligned} a_\parallel &= a_A(1 + 0.042\,y^*), \\ y^* &= y + 23.8\varepsilon_B \end{aligned} \tag{33}$$

Note that the effective Ge content y^* is smaller than the actual Ge content y of the buffer layer because of the remaining compressive strain ε_B (negative sign) in the buffer layer. For complete strain relaxation (dashed curve in Fig. 12), which is approached with increasing buffer-layer thickness, y and y^* become identical.

V. The $Si_{1-x}Ge_x$/Si System

Concerning group-IV elements, undoubtedly the most important SLS systems consist of the two "classical" semiconductor elements Si and Ge, and the $Si_{1-x}Ge_x$ alloys, where x ranges from zero (pure Si) to one (pure Ge). Successful growth of $Si_{1-x}Ge_x$/SiSLSs was already reported back in 1975 (Kasper et al., 1975), and increasing effort has been dedicated to the investigation of growth conditions and physical properties of such SLSs ever since. Meanwhile, the quality of the epitaxial layers and interfaces, as well as improvements of doping techniques, have reached a state that allows for device applications. Promising results of several device structures based on the electronic and optical properties of such heterostructures have been published in the past few years. It is expected that SiGe SLSs will play a major role in future electronic and optoelectronic device concepts, since they

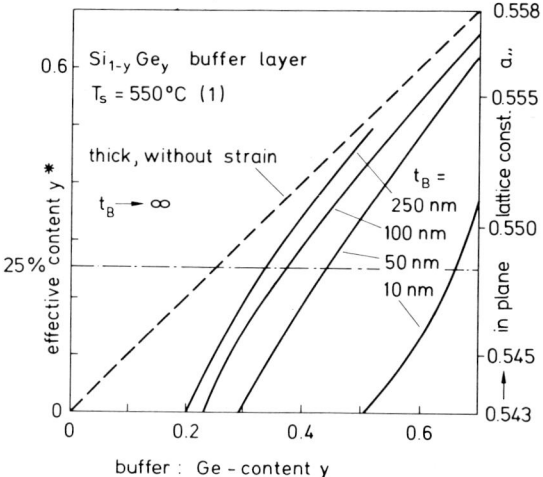

FIG. 12. Design of $Si_{1-y}Ge_y$ buffer layer on a Si substrate. Plotted is the effective Ge content y^* of the virtual substrate (Si substrate + buffer) versus the actual Ge content y of the buffer layer for various buffer-layer thicknesses t_B.

offer the advantage of integration with state-of-the-art Si–very large scale integrated circuits (VLSIC).

The demonstration of device-relevant features has become possible by extensive studies of the influence of strain on the electronic-band structure of SiGe SLSs. Furthermore, zone-folding effects of thin, periodic SiGe superlattices are under intensive theoretical and experimental investigation. These latter studies gain a lot of impact from the perspective of a zone-folding-induced conversion of SiGe SLSs into a direct gap material, a possibility that has already been considered in Gnutzmann and Clausecker (1974).

Because of its importance, this section is entirely dedicated to the $Si_{1-x}Ge_x/Si$ system. We will concentrate on epitaxial SLSs grown by MBE, which make up the dominant part of experimental data published so far. It should, however, be mentioned that there are several groups working on alternative growth techniques, which are based on chemical vapor deposition (CVD). CVD, the standard technique for industrial Si homoepitaxy, has been used directly to grow SiGe layers in the earlier work of Manasevit et al. (1982), but modifications were found necessary for the growth of Si/SiGe heterostructures and superlattices with abrupt interfaces. Two such methods have been developed. Myerson (1986) describes a low-temperature CVD process that takes place at low pressures in an ultrahigh-vacuum (UHV) chamber. The other modification called limited reaction processing (LRP) has been introduced by Gibbons et al. (1985). It differs from the standard CVD process mainly in providing the facilities for rapidly heating and cooling the substrates. Since the growth rate in the CVD process is controlled by the substrate temperature, growth can rapidly be switched on or interrupted. This allows changing the gas flow and/or the gas composition without sacrificing the sharpness of a hetero interface, for example. The principal feasibility of LRP was demonstrated by Gronet et al. (1987) and Hoyt et al. (1990), who grew strained Si/SiGe superlattices, and more recently by Gibbons et al. (1988), who reported on fully operational Si/SiGe heterobipolar transistors grown by LRP (see Section V.D.4). Hence, at least LRP has already reached a state that allows, to some extent, the growth of device-quality Si/SiGe layers. Nonetheless, many fundamental questions concerning the advantages and limitations of LRP and other CVD-derived techniques have yet to be addressed.

In the following sections, we will treat some of the basic aspects of growing SiGe SLSs by MBE, give a review of the actual knowledge of band-ordering, two-dimensional carriers and zone-folding effects, and close with a description of device structures based on these special physical properties.

A. GROWTH BY MBE

MBE is the technique of choice for the epitaxial growth of SLSs, because it is a low-temperature growth technique that offers excellent control over the relatively low deposition rates. Both of these features are essential for the growth of SiGe/SiSLSs with sharp (on an atomic level) interfaces between adjacent layers. The low deposition rates allow abrupt changes in the composition of $Si_{1-x}Ge_x$ layers by simple shutter operation, whereas low substrate temperatures conserve the abruptness of these interfaces. A third traditional strong point of MBE is its unique capability of selectively doping individual layers of almost arbitrary thickness. In contrast to III–V MBE, however, special techniques are necessary to achieve this goal. These, and other basic aspects of the Si-MBE technique will be briefly discussed in the following section. For a comprehensive treatment of this subject, the reader is referred to a recent monograph on Si MBE (Kasper and Bean, 1988).

1. UHV System and Evaporation Sources

Figure 13 shows the cross-sectional view of a modern Si-MBE machine, which is based on an AEG design that incorporates the special demands of SiGe epitaxy (Bean and Kasper, 1988). The system was built by ATOMIKA and became operational in 1987 at the AEG research center (Kasper et al., 1988a). The basic unit of the system consists of the growth chamber and a storage chamber, which are interconnected by a large-diameter gate valve. Additional units can be attached to the growth chamber opposite to the storage vessel. The two chambers of the basic unit are individually pumped, using a combination of a turbo molecular pump and a titanium sublimation

4. GROUP-IV COMPOUNDS

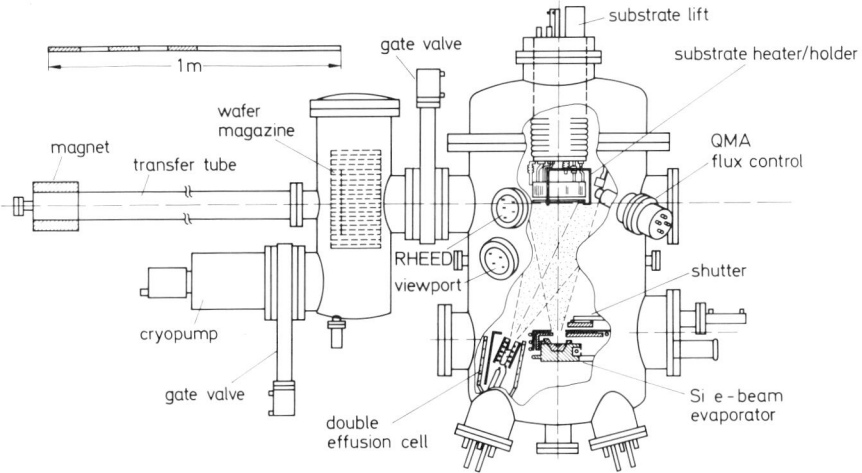

FIG. 13. Cross-sectional view of a modern Si-MBE machine consisting of a growth chamber (right) and a storage chamber for a cassette-type substrate magazine. For reasons of clarity, the pumping system of the growth chamber is not shown.

pump for the growth chamber, and a cryopump for the storage chamber. Base pressures of $\leqslant 3 \times 10^{-11}$ and $\leqslant 1 \times 10^{-9}$ torr are reached in the growth and storage vessels, respectively, the latter after a pump-down time of less than 12 hours. Up to 25 substrates of 150 mm diameter (or less) can be stored in a cassette-type magazine at one time. They are transferred to the growth position by means of an automated, magnetically coupled transfer mechanism. In this position, the substrate is suspended underneath a resistively heated graphite meander, which has been optimized for temperature uniformity.

The substrate is facing down to where the evaporation sources are located. The vertical arrangement of substrate and sources (Fig. 14), which differs from the horizontal setup widely used in III-V MBE machines, is due to the necessity of using an electron-beam evaporator for the low-vapor-pressure material Si. Pyrolytic-boron-nitride (PBN) effusion cells for the dopants are grouped around the central Si evaporator. In principle, Ge, the second matrix material, can be evaporated from such an effusion cell, but the temperatures necessary for reasonable growth rates are close to the temperature limit where PBN begins to decompose. This results in an enrichment of boron in the Ge content, which causes an ever increasing level of p-type doping in the Ge-containing epilayers (Kibbel, 1987). To avoid this problem, a second e-beam evaporator is frequently employed for Ge.

High-purity Si ingots are used as the base material for the inside coating of the cold-walls encompassing the e-beam evaporators. As a matter of fact, all

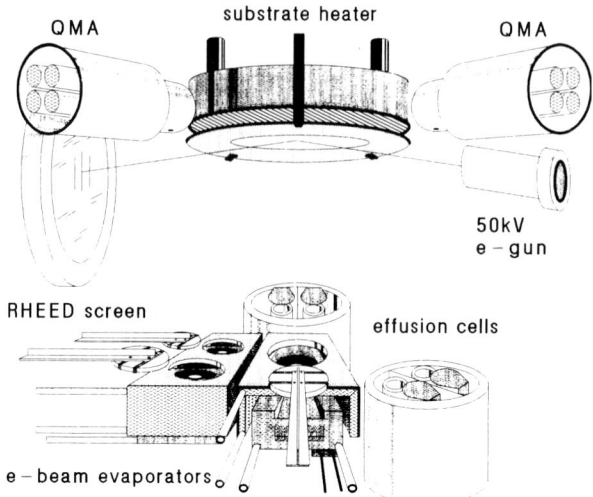

FIG. 14. Typical arrangement of evaporation sources (e-beam evaporators, effusion cells), substrate heater, flux control sensors (here quadrupole mass analyzers (QMA)) and analyzing equipment (RHEED system)).

surfaces that are directly exposed to charged particles during e-beam evaporation (i.e., primary and high-energy secondary electrons as well as ionized source atoms) are Si shielded to avoid beam and source contamination by sputtering. The cold-walls and the walls of the growth chamber are water cooled, rather than LN cooled, to reduce flaking off of matrix-material deposits. Another reason is the residual gas component CH_4, which would be physisorbed, and hence accumulated, at LN-cooled panels, from where it could be desorbed by electron impact during e-beam evaporation. Thus, the only LN-cooled part in the growth chamber is the cold-panel of the titanium sublimation pump.

2. Flux Control

For reproducible fluxes of the matrix materials Si and Ge, it is usually not sufficient to run the e-beam evaporators at constant electron emission current. Instead, closed-loop regulation is employed, which necessitates an adequate sensor for real-time measurement of the actual flux. Electron impact emission spectrometers (EIES) (Gogol and Cipro, 1985), or quadrupole mass spectrometers (QMS), which are adapted to particle-beam measurements by using so-called cross-beam ion sources, are widely used for that purpose. Generally, the QMS has a higher sensitivity, which is advantageous for monitoring the relatively low deposition rates usually employed for the growth of ultrathin SLSs. On the other hand, the principle of EIES makes it relatively insensitive to space charge effects within the effective area of the sensor head. This is not the case for QMSs, which need more or less complicated apertures to prevent deposition of the high-resistivity matrix materials on the walls of the cross-beam ion source, where they could easily charge up. Moreover, efficient electrical shielding of ion source and quadrupole setup have to be provided to minimize the influence of charged particles created in the e-beam evaporators. Despite these additional precautions necessary for proper operation, the gain in sensitivity is rated higher for SiGe application. Consequently, the ATOMIKA unit shown in Fig. 13 is equipped with a QMS and an electronic multiplexer to allow the simultaneous control of the Si and Ge fluxes. The QMS signals are calibrated against the absolute fluxes at the position of the substrate by ex situ film-thickness measurements of test epilayers deposited at a constant rate. A calibrated quartz-crystal microbalance is periodically utilized for in situ checking of the QMA calibration.

3. Layer-Thickness Control by RHEED Oscillations

The simplest way of controlling the thickness of an epilayer is by evaporation at a constant (actively stabilized) rate and timed shutter actuation. This straightforward technique is fully acceptable for most purposes. There are, however, applications where growth termination at a completed atomic layer is desirable, e.g., in the case of ultrathin Si/Ge SLSs with individual layer thicknesses of just a few MLs.

It has been demonstrated recently that such an accuracy is indeed possible in some cases. T. Sakamoto *et al.* (1985; 1986) showed that intensity oscillations of the specularly reflected beam in the reflection high-energy electron diffraction (RHEED) pattern, which are well known from GaAs MBE, can also be observed during Si MBE under certain conditions.

Depending on the azimuth of the incident electron beam, oscillations with a periodicity corresponding to one ([100] azimuth) or two ([110], [1$\bar{1}$0] azimuths) atomic layers can be observed on (001)-oriented substrates. The dominating part of these RHEED oscillations is related to the surface reconstruction of the atomically clean Si (001) surface. Such a surface, which is routinely achieved after thermal desorption of the natural (or intentionally created) oxide layer at temperatures of $\geq 850°C$, usually shows a mixed 2×1 and 1×2 reconstruction. It directly reflects the formation of dimer rows along the $\langle 110 \rangle$ directions (Hamers *et al.*, 1986), two equivalent ones of which exist within the (001) surface plane. Extended heat treatment above 900°C, however, produces single domains large enough to dominate the RHEED signal. Since deposition of one ML reverses the reconstruction of such a domain (i.e., 2×1 becomes 1×2 and vice versa), RHEED oscillations with a two-ML period are observed along the [110] and [1̄10] azimuths. The one-ML periodicity along the [100] azimuth, with intensity maxima occurring whenever an atomic layer is completed, results directly from the symmetry of the (001) surface (see Fig. 15). The RHEED oscillations on properly heat-treated Si (001) surfaces are quite long-lived. The observation of several thousand periods has been reported recently (Sakamoto, T. et al., 1988).

The data suggest that in principle the RHEED oscillations allow layer thickness control with an inherent accuracy of a fraction of an atomic layer (Sakamoto, et al., 1986). By simply counting periods and by actuating the source shutters in the respective intensity maximum of the specular beam, it should be possible to grow SiGe SLS with an unprecedented period definition. The successful application of such a technique for the epitaxial growth of a $(Si_{0.75}Ge_{0.25})_{10}/Si_{10}$ SLS with 26 periods (the suffix denotes the number of MLs) has recently been reported (Sakamoto, et al., 1988). Part of such a RHEED pattern is depicted in Fig. 16 as a function of growth time.

Despite the elegance of this technique and the convincing data demonstrating its capability, a word of caution is necessary. As mentioned, a prolonged heat treatment is a prerequisite for the observation of intense,

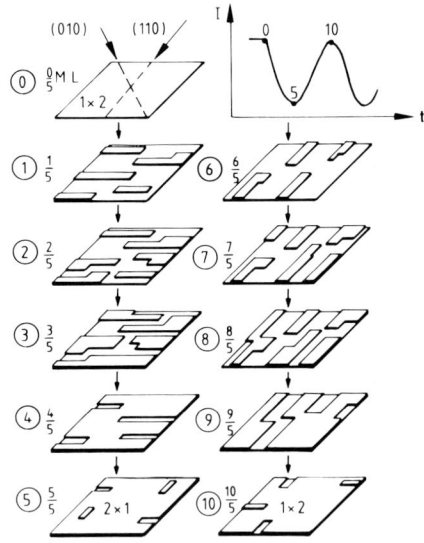

FIG. 15. Layer-by-layer growth of a single-domain Si(001) surface leading to intensity oscillations of the specular RHEED beam. 1×2 and 2×1 reconstructed surfaces appear alternately upon deposition of one atomic layer. (Adapted from T. Sakamoto, D. Sakamoto, G. Hashiguchi, N. Takahashi, S. Nagao, K. Kuniyoshi and K. Miki, *Proc. 2nd Int. Symp. on Silicon MBE* **88-8**, 285 (1988). This paper was originally presented at the Fall 1987 Meeting of The Electrochemical Society, Inc. held in Honolulu, Hawaii.)

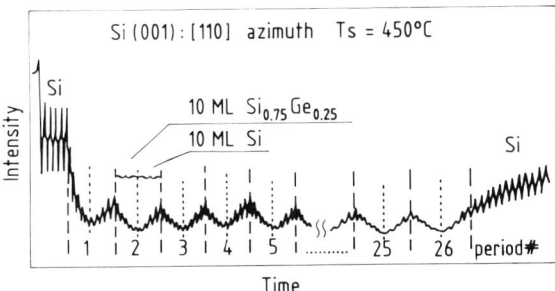

FIG. 16. RHEED intensity oscillations during the growth of a $Si_{0.75}Ge_{0.25}/Si$ strained-layer superlattice on Si(001). 26 periods of 20 monolayers were grown without growth interruption. Substrate temperature during deposition was 450°C. (Adapted from T. Sakamoto, D. Sakamoto, G. Hashiguchi, N. Takahashi, S. Nagao, K. Kuniyoshi and K. Miki, *Proc. 2nd Int. Symp. on Silicon MBE* **88-8**, 285 (1988). This paper was originally presented at the Fall 1987 Meeting of The Electrochemical Society, Inc. held in Honolulu, Hawaii.)

long-lived oscillations. However, such an annealing process is not always desirable or possible. For example, it is generally not acceptable to apply such high temperatures to $Si_{1-y}Ge_y$-buffer layers, which are frequently used as effective substrates for the subsequent growth of symmetrically strained superlattices. Such a procedure would lead to Ge depletion concomitant with a roughening of the surface rather than the formation of a single-domain surface.

But even in the case of an appropriately treated single-domain Si substrate, some limitations apply when using RHEED oscillations for the thickness control of SiGe SLSs, as can be seen in Fig. 16. Growth of a lattice-mismatched SiGe layer on a single-domain Si substrate leads to a rapid decrease of the amplitude of the RHEED oscillations with SiGe coverage. It is remarkable enough that the oscillations gain intensity after growth is switched back to Si, but the example shows that shutter operation based on RHEED oscillations is limited to SiGe/Si superlattices with relatively short period lengths.

There is still another conclusion that can be drawn from Fig. 16. The aforementioned precision of interface definition made possible by RHEED oscillations seems only to apply when going from Si to SiGe. The rapid deterioration of the oscillations after SiGe growth commences indicates defective two-dimensional growth, if not the beginning of a three-dimensional growth mode. Consequently, the subsequent interface between the SiGe layer and the following Si layer is expected to be somewhat smeared out. Since Si deposition seems to flatten out the surface morphology left behind by the SiGe layer (hence the recovery of the RHEED oscillations), a natural asymmetry between the abruptness of the two interfaces appears likely.

This example shows that RHEED is a very powerful in situ instrument that is capable of giving information about the growth kinetics of SLSs. Its application as a layer-thickness monitor, however, appears to be somewhat restricted.

4. Doping of Si and SiGe Layers

Doping of Si and SiGe layers grown by MBE has been—and still is—the subject of intense study. In contrast to GaAs/GaAlAs MBE, where dopants for simple coevaporational doping are readily available, most dopants have only a low incorporation probability in the case of Si MBE. For these reasons, a consequent use of ion-implantation doping has been proposed and was implemented in a few systems (Bean and Sadowski, 1982; Bean and Leibenguth, 1988). These attempts were quite successful. Doping by ion implantation at typical ion energies between about 500 and 3000 eV has been shown to produce sharp doping structures (on the order of a few tens of atomic layers) with a high degree of electrical activation (demonstrated for As and B) at growth temperatures down to 550°C (Bean, 1984). An early concern were the growth temperatures required for a complete annealing of the damage induced by ion implantation. Ota (1980) showed that growth temperatures $\geqslant 750°C$ may be required to completely anneal out damage induced by ion implantation. However, device applications have meanwhile been demonstrated with substrate temperatures reduced to 550°C (Pearsall et al., 1986a).

Despite the unquestionable merits of ion implantation, the proliferation of implanters connected to Si-MBE machines has as yet been quite limited. This has several reasons, the most relevant probably being the bulky and very expensive installations necessary. Moreover, there are still problems with the available ion implanters, which, besides the problems of maintaining adequate UHV conditions, are mainly the quite stringent restrictions in available ion-current densities. Hence, the dominating percentage of Si-MBE machines still relies on thermal dopant sources. This will most likely not change in the near future, especially since new techniques based on conventional dopant sources have been developed in the last few years, which allow very sharp doping profiles. Because of their inherent differences, n- and p-type doping will be treated separately in the following.

a. *n-Doping.* Almost all group-V elements act as shallow donors if substituted for Si (or Ge) atoms at regular lattice sites. Except for a few attempts to use P (Kubiak *et al.*, 1987) and As (Itho *et al.*, 1985), Sb is most commonly employed in Si MBE. The main reason is the convenient vapor pressure of Sb, which is low enough to keep unintentional doping from contaminated parts of the growth chamber (e.g., shutters) within an acceptable limit, but high enough for easy evaporation from conventional effusion cells. The main disadvantage of Sb (which also applies to As and Bi) is its pronounced segregation behavior: At the commonly used epitaxy temperatures between 550°C and 750°C, coevaporated Sb segregates almost completely at the surface of the growing epilayer. Under these conditions, doping levels of only about 10^{15} cm^{-3} (at 550°C) are achievable, which may be acceptable as background doping but are hardly adequate for device structures.

To overcome this problem, Jorke and Kibbel (1985), and independently Kubiak et al. (1985a, 1985b), developed a technique they called doping by secondary implantation (DSI), and potential enhanced doping (PED), respectively, which allows Sb doping levels in excess of 10^{19} cm^{-3}. Conceptually, DSI utilizes the strong surface-segregation behavior to create a steady-state Sb adlayer of typically 5×10^{13} to 1×10^{14} cm^{-2} (i.e., less than about two tenth of a monolayer). This adlayer is typically pre-built-up while growth is temporarily interrupted, and kept constant during the subsequent growth phase by Sb coevaporation, which is adjusted to compensate the losses of thermal desorption and incorporation (Metzger and Allen, 1984). If doping above the mentioned background level is desired, Si$^+$ ions are accelerated toward the substrate by simply applying a voltage of a few hundred volts. Absorbed Sb atoms that suffer a direct hit of a Si$^+$ ion become recoil implanted into the Si or SiGe epilayer (Fig. 17). Because of the low energy of

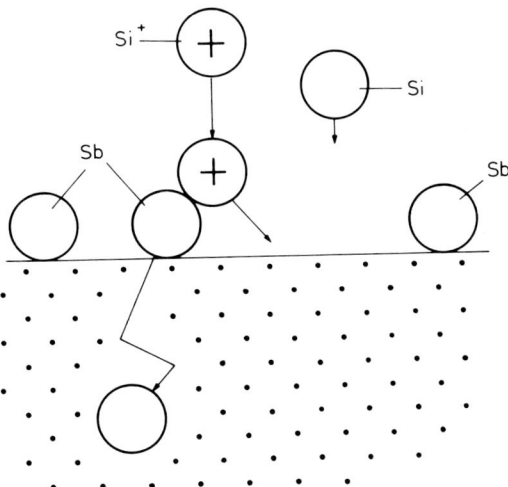

FIG. 17. Simplified model for doping by secondary implantation: Sb atoms segregating at the growing episurface become recoil implanted by Si$^+$ ions, which are accelerated to between 100 and 1000 eV. Very shallow doping profiles are achieved by this technique.

the Si$^+$ ions, a very shallow implantation profile results, which has been estimated to be just some 10 to 20 Å deep (Jorke and Herzog, 1985). Also, implantation damage is negligible down to substrate temperatures below 550°C.

In the simplest case, the natural amount of Si$^+$ ions present in the e-beam-evaporated Si flux is directly utilized for the DSI technique. However, an independent ionizer, arranged between Si source and substrate, is recommended (Jorke and Kibbel, 1985), since the ratio between Si$^+$ ions and Si$^+$ neutrals increases with the total Si flux. (To first approximation, the total Si flux Φ_{Si} is proportional to the electron emission current I_e, whereas the flux of Si$^+$ ions Φ_{Si}^+ is proportional to $\Phi_{Si} \cdot I_e$, i.e., proportional to I_e^2.) This means that without an additional ionizer, the doping efficiency depends on growth rate, which can cause problems especially at the lower rates usually employed for the growth of SLSs. As further refinements, electrostatic extractor screens and focusing ion lenses (Jorke et al., 1989a) have been implemented to improve the density and radial homogeneity of the Si$^+$ ion flux at the position of the substrate. Such means are also expected to reduce doping inhomogeneities that can be induced by charging-up of structural components in the growth chamber, which usually become more and more coated by a highly resistive layer of amorphous Si.

Although at present, DSI is one of the most important n-doping techniques available to Si MBE, there seems to be an alternative that might become relevant in the near future. It is a low-temperature doping (LTD) technique, which exploits the kinetical limitation of Sb segregation at low temperatures. Based on his own experimental results and on preceding investigations of Metzger and Allen (1984) and of Barnett and Greene (1985), Jorke (1988) proposed a two-state exchange model, which assumes distinct energy levels for Sb atoms at a surface and at a subsurface site, respectively. An Sb atom in a subsurface position has to surmount an activation barrier in order to arrive at an energetically favorable surface site. This results in a critical temperature T^* below which surface segregation becomes kinetically limited. The model calculation and experimental data are depicted in Fig. 18, which shows the segregation ratio n_s/n and the related profile broadening Δ for several growth rates as a function of temperature, where n_s is the area density of segregated, n is the volume density of incorporated Sb atoms. Despite the simplicity of the model, excellent agreement with experimental data is found. The most remarkable aspect of kinetical limitation is the prediction of negligible segregation at growth temperatures $T_s \leqslant 400°C$. In that temperature range, doping by simple coevaporation should be possible. First investigations show indeed a high incorporation probability at low temperatures, although with a somewhat reduced electrical activation (Jorke et al., 1989b). Further studies concerning the activation and the crystal

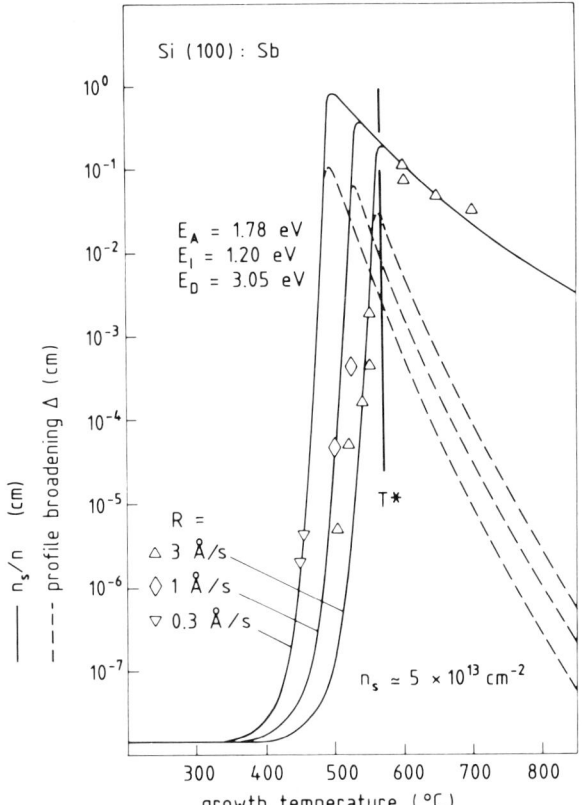

FIG. 18. Calculated Sb segregation ratio n_s/n (solid lines) and profile broadening Δ (dashed lines) for several growth rates as a function of substrate temperature T_s. n_s is the area density of segregated, n the volume density of incorporated Sb atoms. For $T_s < T^*$, segregation is kinetically limited, which results in strongly enhanced spontaneous incorporation. In this temperature range, n_s/n becomes identical with Δ because of the negligibly small thermal desorption of segregated Sb. The theoretical curves are in excellent agreement with experimental data (from Jorke, 1988).

quality at such low growth temperatures are clearly required. Nonetheless, it is worth mentioning that fully operational, LTD-doped SiGe modulation-doped field-effect transistors (MODFETs) have already been fabricated (Daembkes, 1988; see also Section V.D.2).

Figure 18 reveals also that in principle doping is possible at higher substrate temperatures, where the energetically higher subsurface state becomes more and more thermally populated. This regime has been studied in detail by Metzger and Allen (1984), who reported sharp doping profiles by using the aforementioned pre-build-up, and also flash-off (thermal desorption) techniques. A severe shortcoming of Sb doping at high temperatures, however, is the dominance of thermal desorption, which requires high Sb-doping fluxes to be compensated. This is not desirable, because of the Sb contamination of the growth chamber. Moreover, SiGe applications frequently do not allow for high-temperature processes due to possible interdiffusion mechanisms.

b. *p* Doping. For many years, Ga was the most frequently used element for *p*-doping of Si and SiGe layers. Again, the vapor pressure, which allows evaporation from a conventional PBN cell at temperatures below about 850°C, was a main reason for selecting Ga. However, even though Ga is still widely used, there is a clear-cut tendency toward finding a superior substitute. These attempts are triggered by the strong segregation behavior of Ga concomitant with the propensity to form clusters in the segregated adlayer. Moreover, secondary ion impact, which works so well for recoil implantation of Sb, is hindered by a much larger cross section for sputter desorption in the case of Ga. Hence, Ga-DSI is restricted to Si^+-ion energies below about 100 eV, which makes it a more tedious task to achieve adequate ion densities over the substrate area than in the case of Sb, where ion energies in excess of 1000 eV are acceptable (Schäffler and Jorke, 1990). Nevertheless, the higher spontaneous incorporation probability of Ga (as compared to Sb) allows doping levels up to about 10^{18} cm^{-3} in MBE-grown layers. Within this range, Iyer *et al.* (1981) were able to demonstrate remarkably sharp doping profiles, which were achieved by a clever combination of Ga adlayers pre-built-up during growth interrution and thermal desorption of excess Ga. Higher doping levels can be achieved by solid-phase (SP) recrystallization of amorphous films doped by Ga coevaporation (Streit *et al.*, 1984; Ahlers and Allen, 1987). So far, SP MBE has only been demonstrated for Si, not for SiGe layers.

As a substitute for Ga, boron appears to be the most favored candidate, despite some attempts to use Al (Becker and Bean, 1977) and In (Knall *et al.*, 1988). The reason for the popularity of B is its high incorporation probability, which sets it apart from the other group-III elements (Kubiak *et al.*, 1984). However, the low vapor pressure of elemental boron, which demands evaporation temperatures in excess of 1600°C, causes appreciable problems in finding adequate thermal sources. Since such temperatures are beyond the decomposition temperature of PBN crucibles, substitutes for the effusion-cell material had to be found, too. Attempts were made to evaporate elemental B from resistively heated boats made of refractory metals or of graphite (Kubiak *et al.*, 1985c). Graphite has turned out to be superior because of its limited reactivity with the boron slug. Consequently, a pyrolytic graphite crucible is used in a new effusion-cell design introduced by Andrieu *et al.* (1988), which combines radiative heating with electron bombardment.

Although first results with elemental B doping are promising, more detailed investigations of potential contaminants released at the high operation temperatures of the cell are necessary. It is also an open question whether the relatively high vapor pressure of graphite, which is only about three orders of magnitude lower than that of boron, has any effect on crystal growth.

To circumvent the high temperatures required for B evaporation, boron sources utilizing boron compounds rather than elemental boron have been implemented. The use of sintered BN, which contains a certain amount of B_2O_3 (Aizaki and Tatsumi, 1985), the direct use of B_2O_3 (Ostrom and Allen, 1986), and the use of metaboric acid (HBO_2) (Tatsumi et al., 1988a) have been reported. All of these compounds have a relatively high vapor pressure in common, which permits evaporation from PBN effusion cells. The experiments of Tatsumi et al. (1988a) suggest that B_2O_3 as well as HBO_2 evaporate as complete molecules, which means they have to decompose at the episurface. Moreover, the growth temperature has to be high enough to desorb the oxygen released. This condition appears to be fulfilled at $T_s \geq 750°C$, while at $650°C$ about the same amounts of B and oxygen were detected by secondary-ion mass spectroscopy (SIMS) analysis of B-doped layers (Tatsumi et al., 1988a). Even though relatively sharp doping profiles with an onset gradient of at least one order of magnitude per 200 Å were reported, the demand for substrate temperatures in excess of 750°C limits the value of boron-compound sources for SiGe applications. These limitations are somewhat relieved if a certain amount of oxygen is tolerable in the p-doped layers. There are, for example, reports on heterobipolar transistors with a SiGe base doped by an HBO_2 source (Tatsumi et al., 1988b).

B. Band Offset and Two-Dimensional Carriers

One is not always aware of the fact that it is actually a system of two-dimensional (2D) carriers that passes electronic information in our modern digital integrated circuits. In fact, the carriers in the conducting channel of a MISFET (metal-insulator-semiconductor field-effect transistor, which in connection with silicon-based technology is usually called MOSFET, where the O refers to the oxide insulator used) are truly 2D in a strict quantum-mechanical sense. The extension of the conducting inversion or accumulation layer perpendicular to the plane of the device (z-direction) is comparable to the de-Broglie wavelength of the carriers. This leads to a momentum quantization in the z-direction, which leaves only the two in-plane directions for the carriers to behave like free electrons with effective mass $m_\parallel^*$. The discrete values of the momentum in the z-direction, k_z, which result from a self-consistent solution of the Poisson and Schrödinger equations (e.g., Ando et al., 1982), define the energy eigenvalues of the so-called subbands,

$$E_i(k_\parallel) = E_i(k_z) + \hbar^2 k_\parallel^2 / 2m_\parallel^*, \tag{34}$$

where i is the integer subband index, and $E_i(k_z) = \hbar^2 k_i^2/2m_z^*$. That means, there is a 2D carrier system associated with every eigenvector k_i. The 2D character of inversion or accumulation layers at the insulator–semiconductor interface of a MOSFET was predicted by Schrieffer back in 1957 (Schrieffer, 1957) and confirmed experimentally by Fowler et al. (1966).

In the case of a MOSFET, the 2D carriers are trapped in a potential well, which results from the large band offset at the insulator/semiconductor interface on one side and the space charge region of ionized dopant atoms on the other side (Fig. 19a). To first approximation, this potential well is triangular. (For the exact shape, the carriers trapped in the well have to be taken into account in a self-consistent calculation (e.g., Stern, 1972).)

Another realization of a 2D carrier system has been made possible by MBE-based heteroepitaxy, i.e., by growing crystals consisting of more than one semiconducting material. In such systems, the potential well results from the different bandgaps of the (usually two) semiconductors involved, which have to be accommodated by distinct band offsets at the respective interfaces. Figure 19b shows a section of a GaAs/GaAlAs SL, the by now "classical" hetero system: GaAlAs has the larger bandgap, and its absolute value depends on the Al content. The bandgap difference is accommodated by a valence—as well as by a conduction—band offset, with the conduction-band offset accounting for about 60% of the total bandgap difference. The signs of the offsets are such that the smaller energy gap of GaAs is entirely contained within the larger gap of GaAlAs, i.e., the GaAs layers are energetically favorable for electrons as well as for holes. Such relations are referred to as Type-I band offset. The rectangular shape of the potential well is—similar to the triangular well discussed earlier—modified by the 2D carrier gas, which induces a curvature into the bottom of the well (Fig. 19b). For wide wells, this curvature leads to a spatial splitting of the lowest subbands into two separate two-dimensional electron gases (2 DEGs), which are located next to the two

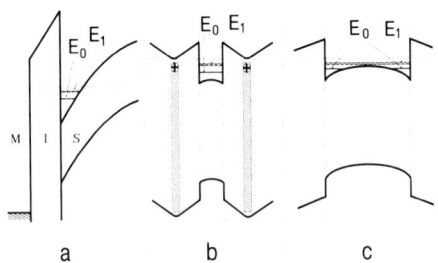

FIG. 19. Various potential wells associated with two-dimensional carriers: (a) triangular potential of a MOSFET, (b) narrow rectangular potential wells of a Type-I superlattice, (c) wide rectangular potential well leading to a splitting into two 2D carrier gases.

barriers. Under these conditions, each of the two potential wells is similar to the triangular-shaped well of a MOSFET (Fig. 19c).

One of the most remarkable features of carrier transfer in SLs is the inherent potential for enhanced carrier mobilities. If doping is restricted to the layers of the wider gap material, carrier transfer into the smaller gap layers is concomitant with a spatial separation of mobile carriers and ionized dopant atoms. This reduces ionized impurity scattering, a mobility-limiting mechanism that becomes dominant at low temperatures. Indeed, extremely high low-temperature electron mobilities are reported for modulation-doped GaAs/GaAlAs heterostructures, which allowed, for example, the detection of the fractional quantum Hall effect (Tsui et al., 1982).

In the following, we will discuss 2D carrier properties of the SiGe/Si heterosystem, which differ quite substantially from the by now well-understood GaAs/GaAlAs system. Because of the relatively large lattice mismatch of about 4.2% between Si and Ge (which is almost absent in the case of GaAs/AlAs), the individual layers of a SL are strained as long as the layer thicknesses remain below a critical value t_c (see Section III). It is the strain distribution between the layers that affects the effective band offsets and hence provides an additional degree of freedom in tailoring the electronic properties of SiGe/Si SLSs.

1. *Asymmetrically Strained Layers*

The first observation of 2D carriers in modulation-doped SiGe/Si SLSs were reported by People et al. (1984). Their heterostructures consisted of a pseudomorphic $Si_{0.8}Ge_{0.2}$ layer sandwiched between two mirror-symmetrical Si layers. The structure, which is schematically depicted further down in Fig. 21, was grown on a n^--Si substrate to suppress a conducting parallel channel. An asymmetric strain distribution results from the use of a Si substrate, i.e., the pseudomorphically grown SiGe layer accommodates all of the strain induced by the lattice mismatch, while the Si cladding layers remain unstrained.

Three different doping configurations were tested: (i) uniform p-type (boron) doping throughout the complete heterostructure at a level of $\approx 10^{18}\,cm^{-3}$; (ii) only the Si cladding layers were intentionally doped; (iii) same as (ii), except for two Si spacer layers of about 100 Å thickness that were left undoped on either side of the SiGe layer. Measurements of the Hall mobility showed strong carrier freeze-out at about 50 K for the homogeneously doped sample, which is typical for a $10^{18}\,cm^{-3}$ doped bulk sample. The two modulation-doped samples, however, had higher mobilities at all temperatures and no freeze-out behavior at low temperatures. Especially the sample with the undoped spacer layers showed a steady increase in mobility with decreasing temperature, with a maximum value of $3300\,cm^2/Vs$ at 4.2 K being the lowest temperature reached in the experiment (Fig. 20). The mobility behavior of the two modulation-doped samples is indicative of a 2D hole gas (2DHG) in the SiGe layer, which is the result of charge transfer from the doped Si cladding layers. Because of the spatial separation between ionized dopants and the 2DHG, which is largest for the sample containing the spacer layers, ionized impurity scattering is strongly suppressed, leading to the observed low temperature increase in mobility.

To unambiguously substantiate the existence of a 2DHG, People et al. (1984) performed magnetoresistance measurements. A 2D carrier system is characterized by Shubnikov–deHaas (SdH) oscillations, whose amplitude shows a $\cos\Theta$ behavior, where Θ is the angle between the direction of the magnetic field B and the normal to the 2D carrier sheet (von Klitzing et al., 1974). That means a 2D carrier system requires the SdH oscillations to vanish if B is parallel to the epilayers, which was indeed observed in the experiment. From the temperature dependence of the SdH amplitude (Adams and Holstein, 1959), an effective carrier mass of $m^* = 0.30 \pm 0.02 m_0$ was derived at a magnetic field of $B = 3.5$ T. This value is smaller than expected from a linear interpolation of the heavy hole mass at the Ge content $x = 0.2$, possibly due to strain effects.

After proving the existence of a 2DHG in the SiGe layer of their test structure, People et al. (1985) systematically studied the influence of the doping level in the cladding layers and of the width of the undoped Si spacer layers on sheet carrier density and mobility of the 2DHG. The results of these low-temperature experiments were in qualitative agreement with expectations derived from earlier studies on the GaAlAs/GaAs system (Delescluse et al., 1981; Drummond et al., 1982). The 2D density of carriers in the SiGe well (n_s) decreases monotonically with decreasing doping level in the cladding layers as well as with increasing spacer thickness. Both effects have a

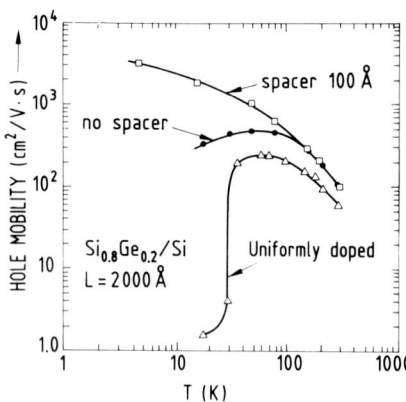

FIG. 20. Temperature-dependent Hall mobilities for holes in SiGe/Si heterostructures for different doping conditions: The lowest curve is from a homogeneously doped structure that shows the typical freeze-out behavior of a bulk sample. The upper curves are from modulation-doped structures with and without a spacer layer between doping layer and channel (People et al., 1984).

straightforward explanation when taking into account that the charge transfer from the cladding layers into the well leaves behind a space charge layer in the cladding layers. Charge transfer ceases as soon as the valence band, which varies linearly in the undoped spacer layer and parabolically in the doped region, reaches the Fermi level E_F. E_F, on the other hand, is determined by the filling of the subbands in the potential well, which is self-consistently derived from the charge transfer from the cladding layers. This means that there is—for a given spacer thickness L_s—a maximum doping level N_a in the cladding layers, which can be completely transferred into the well. At higher doping concentrations, a parallel channel of low-mobility carriers within the cladding layers develops. Of course, the maximum charge that can be transferred decreases with increasing L_s.

Somewhat more complicated appears the behavior of the hole mobility μ_h of the 2DHG, which shows a maximum as a function of L_s. Drummond *et al.* (1982) gave an interpretation that is based on the combination of two different mechanisms: With increasing spacer thickness the ionized dopants are spatially more and more separated from the 2DHG, which reduces ionized impurity scattering, hence leading to an increase in μ_h. Simultaneously, the carrier density in the channel decreases, which means that predominantly states with small k vectors parallel to the layers remain populated within the lowest subband(s). This leads to enhanced Coulomb scattering, which is most effective at small k vectors. Thus, at large L_s, the mobility enhancement becomes reversed, resulting in the observed maximum of the μ_h versus L_s curve.

In addition to varying doping concentration and spacer thickness, People *et al.* (1985) also studied the influence of the SiGe well width L_w on the 2DHG mobility. As long as the SiGe layers were relatively thick (500 to 2000 Å), they found no significant influence on μ_h. At such large values of L_w, the 2DHG is split into two parallel channels, which are located on either interface to the adjacent Si layers (Fig. 19c). Further reducing the well width, however, leads to a situation similar to that shown in Fig. 19b, i.e., the 2DHG is trapped in a only slightly distorted rectangular potential well. Under these conditions (at $L_w = 100$ Å), a noticeable reduction in mobility was observed at low temperatures, even though doping level and spacer thickness remained the same. This finding was interpreted in terms of enhanced surface roughness scattering at one of the two interfaces. The suspected asymmetry of the interface quality is consistent with the amplitude behavior of the RHEED oscillations discussed in Section V.A.3).

In order to maintain pseudomorphic growth over the relatively large thicknesses of their structures, People and coworkers chose a $Si_{1-x}Ge_x$ well region with a rather small x-value of 0.2. Although they observed the mobility enhancement expected from modulation doping, the maximum achievable channel mobility is quite limited owing to the notoriously low hole mobility in Si, which is only slightly increased in the still Si-like $Si_{0.8}Ge_{0.2}$ alloy. Moreover, an additional mobility-restricting scattering mechanism, namely alloy scattering, becomes relevant at low temperatures. Hence, it was only consequent when Ostrom et al. (1988) proposed and actually grew modulation-doped SLs with hole channels consisting of pure Ge layers, whereas the doped layers were $Si_{1-x}Ge_x$ alloys with $x \approx 0.6$. The principal advantages of such SiGe/Ge SLSs are clear cut, since Ge has the highest hole mobility of the commonly used semiconductors ($1900\,cm^2/Vs$). However, the extremely small critical thickness of Ge on Si, which is only on the order of three or four monolayers, does not allow the growth of such SLSs directly on Si substrates. Although in principle it is possible to grow on Ge substrates (or possibly on GaAs substrates, which have almost the same lattice constant), technological reasons imply the use of Si substrates for any device-relevant SiGe structure. Consequently, Ostrom and coworkers based their SLSs on a thick, partly relaxed $Si_{1-y}Ge_y$ buffer layer on a Si(001) substrate with an effective lattice constant of a $Si_{0.25}Ge_{0.75}$ alloy. The subsequent $Si_{0.4}Ge_{0.6}/Ge$ SL was kept below the individual and overall critical thicknesses, leading to a strain distribution where both constituents are strained (see also Section V.B.2). The alloy layers of the SLSs were modulation doped with boron that was evaporated from a supersaturated piece of silicon (Ostrom and Allen, 1986).

The SiGe/Ge SLSs of Ostrom and coworkers showed partly 2DHG character and room-temperature mobilities of $520\,cm^2/Vs$, which corresponds to bulk Ge doped to a level of $5 \times 10^{17}\,cm^{-3}$. This value exceeds the hole mobility of bulk Si at that doping level by more than a factor of two and is also superior to the room-temperature hole mobilities reported by People et al. (1984; 1985) for their asymmetrically strained SiGe/Si SLS. On the other hand, a background doping of $5 \times 10^{17}\,cm^{-3}$ appears quite high for a modulation-doped SL. For comparison, People et al. (1985) determined in their SiGe/Si SLs a background doping of a few times $10^{15}\,cm^{-3}$. The high background is most likely the dominating factor responsible for the observed premature saturation of the hole mobility with decreasing temperature. This is not the case for the SiGe/Si structures of People and coworkers, which actually reveal a higher hole mobility at liquid-helium temperature as compared to the samples of Ostrom and coworkers. Nevertheless, the room-temperature performance of the SiGe/Ge SLSs already demonstrates its principal superiority. A full exploitation, however, requires substantial improvements of the growth and doping conditions.

People et al. (1984; 1985) were only able to observe a 2DHG, even though they also grew similar samples where the cladding layers were phosphorus (n-type) doped. No experimental indication for a 2D EG could be found. (However, no attempts were made to dope the SiGe layer, which would be necessary for a modulation-doped 2DEG system in a superlattice with a staggered (Type II) band offset, see Section V.B.2.) From these findings, they concluded that the band offset at their $Si_{0.8}Ge_{0.2}/Si$ interfaces has to occur mainly in the valence band, whereas the conduction-band offset δE_c has to be much smaller, in fact too small to observe a confined 2DEG. Consequently, People and coworkers suggested the qualitative band-edge variation depicted in Fig. 21 for their structures. Despite some disagreement about the sign of δE_c (see Section V.B.3), it is meanwhile widely accepted that $\delta E_v \gg \delta E_c$ under the asymmetric strain conditions that characterize the heterostructures of People and coworkers.

2. *Symmetrically Strained Layers*

In 1985, Jorke and Herzog (Jorke and Herzog, 1985) reported convincing evidence for enhanced electron mobilities in $Si_{0.55}Ge_{0.45}/Si$ SLSs grown on a 2000-Å-thick $Si_{0.75}Ge_{0.25}$ buffer rather than on the Si substrate. The thickness of the buffer is beyond t_c, i.e., the buffer is relaxed, except for a small residual strain component. Hence, the SLS is grown on an effective substrate whose lattice constant lies in between those of the two SL constituents. This leads to a symmetric strain distribution over the SLS, with the Si layers tensilely and the equally thick Ge layers compressively strained (see Section IV). The SLSs consisted of 10 periods of 120 Å width, each of them being doped at the same relative position by a narrow ($\approx$ 50-Å-wide) Sb doping spike. The absolute position of the doping spike within a SLS period was shifted in a series of subsequent samples from the center of the SiGe layer to the center of the Si layer.

Hall mobility measurements of the samples revealed very high carrier densities of up to 3×10^{12} cm^{-2} per period. Even more interesting was the observed mobility behavior: The electron mobility μ_e becomes maximum if the respective centers of the SiGe layers are doped, and minimum for the doping spikes centered in the Si layers. This result was quite surprising at that time, since it was strong evidence for a staggered band lineup (Type II) with

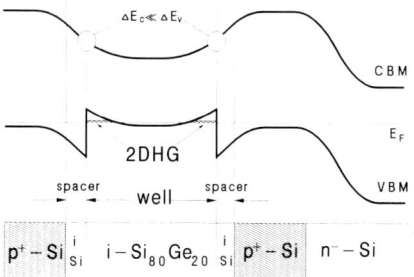

FIG. 21. Band-edge variation associated with the SiGe/Si heterostructures investigated by People et al. (1984; 1985). The 2000-Å-thick SiGe channel leads to a splitting into two 2DHG layers confined to the respective heterointerfaces.

the Si conduction-band edge being energetically below the band edge in SiGe. Under these conditions, electrons will be transferred into the Si layers, where they suffer the maximum ionized impurity scattering, resulting in the lowest mobility when the centers of the Si layers are doped. Vice versa, the maximum mobility is reached when the ionized impurities are as far as possible removed from the mobile electrons, which is the case if the doping spike is centered within the SiGe layers.

Since the Sb doping spikes were realized by applying the DSI technique (see Section V.A.4), there was a slight chance that the mobility results could have been affected by the—then unknown—implantation depth. For these reasons, the doping experiment was repeated with a period width of 200 Å. Apart from an increase in the maximum mobility observed, the second series confirmed the first and, in addition, allowed a deduction of the aforementioned implantation depth of 10 to 20 Å. Room-temperature mobility measurements of both series are depicted in Fig. 22.

The clear evidence for a 2DEG in the Si layers of a symmetrically strained SiGe/Si superlattice was further confirmed by the observation of enhanced low-temperature mobilities (Fig. 23) and by SdH experiments (Abstreiter *et al.*, 1985), which showed the characteristic $\cos\Theta$ behavior of a 2D carrier system (Fig. 24); see also Section V.B.1). In addition, Abstreiter and coworkers derived an effective mass of $0.2m_0$ for the 2DEG from cyclotron resonance measurements at 890.7 GHz. This value is also observed for the twofold degenerate electrons in the inversion layer of a Si MOSFET (Abstreiter *et al.*, 1976), again indicating a charge transfer into the Si layers.

From optical-phonon Raman spectra of the above SLSs, Abstreiter *et al.*

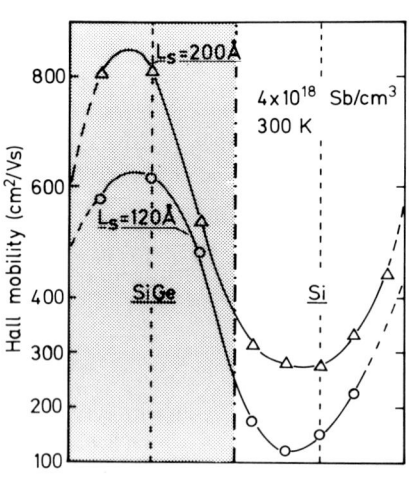

FIG. 22. Room-temperature Hall mobility of a symmetrically strained SiGe/Si superlattice as a function of the position of the *n*-type doping spike within one period. Data for two different period lengths are depicted. Note that the mobility becomes maximum if the centers of the SiGe layers are doped. Adapted from H. Jorke, H.-J. Herzog, *Proc. 1st Int. Symp. on Silicon MBE* **85-7**, 352 (1985). This paper was originally presented at the Spring 1985 Meeting of The Electrochemical Society, Inc. held in Toronto, Ontario, Canada.

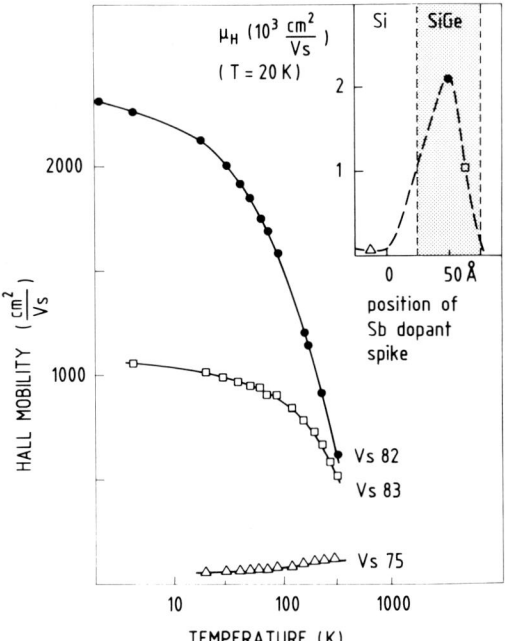

FIG. 23. Temperature-dependent electron Hall mobility for modulation-doped, symmetrically strained SiGe/Si SLs. The different curves correspond to variations of the doping spike within the SL period, as indicated in the insert.

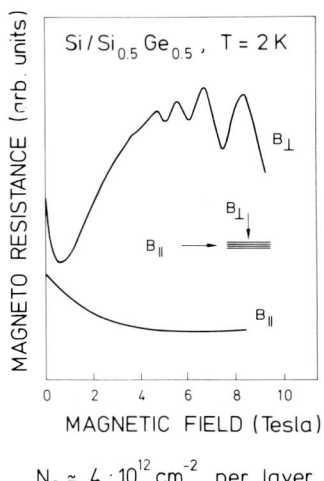

FIG. 24. Magnetoresistance curves of the high-mobility sample in Fig. 23. Shubnikov–deHaas oscillations are only observed if the magnetic field is applied perpendicular to the layers, as is expected for a 2DEG (Abstreiter et al., 1985).

(1985) were able to deduce absolute values of the built-in biaxial (in-plane) strain within the SiGe and Si layers, which turned out to be quite substantial (on the order of 1.7 GPa). Based on these strain measurements, Abstreiter and coworkers proposed a semiquantitative model for the band lineup in SiGe/Si SLSs, which could reconcile the seeming discrepancy between the experimental results of People *et al.* (1984) and of Jorke and Herzog (1985). In People's case, i.e., for unstrained Si layers and compressively strained SiGe layers, $\delta E_v \gg \delta E_c$, hence only a 2DHG can be observed. For the symmetric strain situation of Jorke's SLSs, the strain splitting of the sixfold degenerate (Si-like) conduction band becomes important. The measured biaxial strains can easily be converted into the sum of a hydrostatic and a uniaxial component, where the latter is perpendicular to the layers (i.e., in the [001] direction) and has the inverse sign of the original biaxial component. This allows utilization of studies by Baselev (1966), who investigated the effect of a uniaxial strain on the conduction band (CB). In the present case of a uniaxial strain in the [001] direction, the six-fold degenerate CD splits into a four-fold and into a twofold degenerate component. The doublet is associated with the two CB minima perpendicular to the layers ([001] direction), whereas the four-fold degenerate component contains the four in-plane valleys. For the symmetrically strained SLS, the built-in strains lead to downward movement of the two-fold and an upward movement of the four-fold degenerate conduction bands in the silicon layers, whereas the direction of the energy shifts is reversed in the SiGe layers. By using the strain values derived from the Raman experiments, the absolute splitting within either of the two CBs has been estimated to be on the order of 150 mV. Since the strain-induced energy shift is twice as large for the twofold degenerate level as compared with the four-fold degenerate level, it becomes plausible that for a range of Ge contents, the lowest CB level in Si (two-fold) can be energetically lower than the lowest CB state in the SiGe layers (four-fold). This is obviously the case for the $Si_{.55}Ge_{.45}$/Si SLSs of Jorke and Herzog. The situation is schematically depicted in Fig. 25 (see also Fig. 27).

Even though Abstreiter and coworkers were able to give experimental values for the CB splitting within each layer, their investigations did not give an absolute value for the conduction-band offset, which also contains the "natural" offset of the weighted average CB levels. A similar situation applies to the valence-band offset, which is connected to the CB lineup via the strain-affected energy gaps. Theoretical and experimental attempts to derive the absolute band lineup as a function of strain distribution over the SLS will be discussed in the following section.

3. *Band Alignment in SiGe/Si Strained-Layer Superlattices*

Calculating the valence- and conduction-band offsets at a strained heterointerface is fairly complicated. Besides the valence-band offset, the strain

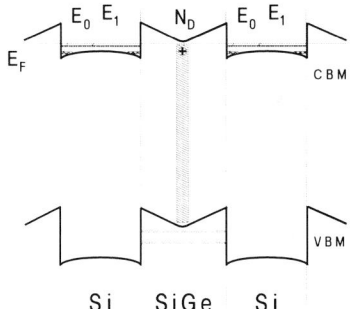

FIG. 25. Proposed Type-II band lineup for the symmetrically strained SiGe/Si superlattice: The strain-induced conduction-band splitting results in the CB minima being located in the Si layers, while the VB maxima occur in the SiGe layers (Abstreiter et al., 1985).

splitting of valence and conduction band and the overall strain-induced change of the bandgap energy have to be computed. In addition, these calculations would necessarily have to be done for a variety of different strain conditions, since the strain component in both layers can be determined by choosing the proper buffer composition and by adjusting the compositions of the two layers separately. Hence, the most general case is that of a $Si_{1-x}Ge_x/Si_{1-y}Ge_y$ SLS on a buffer layer with the lattice constant of an unstrained $Si_{1-z}Ge_z$ alloy.

Although a complete set of calculated values is presently not available, interpolation between some key situations calculated by Van de Walle and Martin (1985) allow an estimate of the trends with relatively high accuracy. Van de Walle and Martin performed self-consistent calculations based on local density functional and ab initio pseudopotentials in order to determine the minimum-energy configurations and the relative position of the Si and Ge bands. They treated the extreme conditions of Si/Ge interfaces either on a Si substrate (i.e., only the Ge layer is (compressively) strained) or on a Ge substrate (the Si layer is tensilely strained), and additionally on a $Si_{0.6}Ge_{0.4}$ substrate (both layers are strained, the Si layer tensilely, the Ge layer compressively). These calculations were performed for the (001) interface plane, the two asymmetrically strained cases also for the (111) plane, and subsequently (Van de Walle and Martin, 1986), for the (110) plane.

The calculations of Van de Walle and Martin (1985, 1986) basically consist of two parts: First, the average potentials V_{Si} and V_{Ge} are determined from a self-consistent supercell calculation using ab initio pseudopotentials. These potentials are used as reference levels for the band lineups, which finally result from band-structure calculations of the two bulk materials strained according to their respective interface situation. The overall accuracy of the calculations is not easily estimated, since it contains the computational uncertainty of the pseudopotential calculations in addition to those of the density functional band-structure calculations. Especially the latter are infamous for severely underestimating bandgaps while describing the band dispersions quite accurately. Hence, the absolute bandgaps are frequently adjusted to experimental values by shifting the bands against each other by a constant amount. This was also done by Van de Walle and Martin (1986). They were able to show that the same shift for both materials is adequate, since the corrections necessary to reproduce the experimental bandgaps are nearly the same for Si and Ge. This, and other test calculations, led Van de Walle and Martin to estimate an overall accuracy of their band lineups of ≈ 100 meV.

The main results of Van de Walle and Martin are depicted in Fig. 26, which shows the interpolated variations of the conduction- and valence-band levels for strained $Si_{1-x}Ge_x$ layers on a Si(001), and a Ge(001) substrate, respectively. Besides the strain-split CB levels, and strain- plus spin-orbit-split valence-band (VB) levels, the respective weighted average energies are drawn as dashed lines.

Figure 26 reveals that asymmetrically strained $Si_{1-x}Ge_x$ layers on unstrained Si have a conduction-band offset that is smaller than the accuracy of the calculation, i.e., less than 100 meV. For the limiting case of pure Ge on Si, a Type-II offset is predicted (see also Section V.C.2), whereas for $x < 0.6$, a transition to Type-I behavior might occur. For the aforementioned asymmetrically strained $Si_{0.8}Ge_{0.2}/Si$ double heterostructures of People et al. (1985), Fig. 25 gives a flat CB situation, i.e., vanishing offset, whereas the VB offset amounts to ≈ 150 meV. Hence, the calculation is in agreement with the condition $\delta E_v \gg \delta E_c$ that People and coworkers derived from their experiments. Van de Walle and Martin (1986) did not publish theoretical curves for symmetrically strained layers but gave explicitly values for the symmetrically strained $Si_{0.55}Ge_{0.45}/Si$ SLSs of Jorke and Herzog (1985) mentioned earlier.

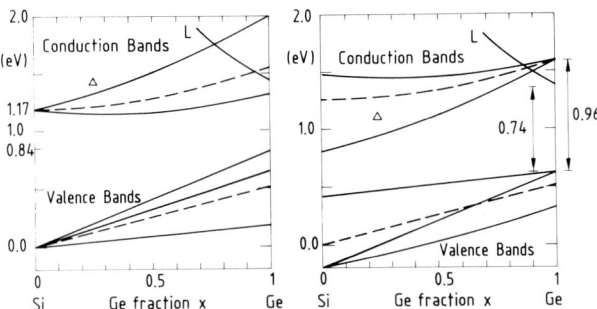

FIG. 26. Valence and conduction bands in strained $Si_{1-x}Ge_x$ alloys matched to a Si(001) substrate (a), and to a Ge(001) substrate (b). All energies are referred to the top of the valence band in Si. Dashed lines indicate the weighted averages of the valence bands and the Δ conduction bands (Van de Walle and Martin, 1985).

They find a Type-II CB offset of about 130 meV, which is consistent with the 2DEG observed experimentally in the Si layers of the SLS. Thus, it can be concluded that despite the relatively large uncertainty of some 100 meV, the theory of Van de Walle and Martin is capable of describing the strain-induced band lineup correctly.

People and Bean (1986) used the valence-band offsets of Van de Walle and Martin (1985), and linear inter- and extrapolations of these, to independently construct the CB offsets. For this purpose, the bandgaps of the strained constituent layers were estimated (People, 1985) by using the phenomenological deformation potential theory of Kleiner and Roth (1959). These calculated bandgap variations were found to be in excellent agreement with photocurrent spectroscopy data of strained $Si_{1-x}Ge_x$ layers on Si substrates (Lang et al., 1985). With the combination of the two theories, People and Bean arrived at a VB offset of ≈ 150 meV and a Type-I CB offset of ≈ 20 meV for unsymmetrically strained $Si_{0.8}Ge_{0.2}$ layers on Si substrate, whereas a VB offset of ≈ 300 meV and a type-II CB offset of ≈ 150 meV was deduced for the symmetrically strained SL of Jorke and Herzog (1985). These values are in good agreement with the respective data of Van de Walle and Martin (1986), even though the theoretical formalisms for deriving the bandgaps under strain were entirely different. There is, however, a small systematic difference between the two calculations, which results from the valence-band offset values in Van de Walle and Martin's 1985 paper that were used by People and Bean (1986). These values, which were computed for pure Ge on SiGe, have been corrected in the 1986 paper of Van de Walle and Martin by 100 meV to properly account for the spin-orbit splitting. Of course, this value has to be scaled down to the actual x of a $Si_{1-x}Ge_x/Si$ SLS, but it might well account for the minor difference in the results of People et al. (1986) and Van de Walle and Martin (1986).

The theoretical treatments of Van de Walle and Martin (1985, 1986), and of People and Bean (1986) fully confirm the semiquantitative model proposed by Abstreiter et al. (1985), which predicted a strain-induced Type-II-band ordering for symmetrically strained $Si_{1-x}Ge_x/Si$ SLs. The band offset in the case of unstrained Si layers and compressively strained $Si_{1-x}Ge_x$ layers, on the other hand, is restricted mainly to the valence band, with only a small percentage occurring in the conduction band. While the latter component leads to a Type-II band lineup at higher x-values, it is not entirely clear whether a transition to Type-I ordering takes place at small Ge contents.

So far, the discussion was restricted to calculations that directly give the band lineups at $Si_{1-x}Ge_x/Si$ interfaces. There is, however, an interesting theoretical approach to determine the CB offset at such interfaces indirectly from a comparison between experimental data and theoretical predictions concerning the properties of the 2DEG in $Si_{1-x}Ge_x/Si$ SLS. Zeller and

Abstreiter (1986) performed detailed electronic subband calculations, with special attention being paid to the strain situation within the respective constituent layers of the SL. Figure 27 schematically shows the strain-induced splitting of the conduction-band edges at a SiGe/Si interface. Both layers are assumed to be strained, which leads to the downward shifts of the twofold degenerate component in the tensilely strained Si layers and the fourfold degenerate CB levels in the compressively strained Ge layers, while the fourfold and twofold components of Si and Ge, respectively, are shifted to higher energies. The absolute value of the shifts is about twice as large for the doublets (Baselev, 1966), as was mentioned earlier. Now, Zeller and Abstreiter introduced the offset of the weighted average CB energies δE_c^*, which is treated as a parameter, since it is not easily accessible experimentally. (δE_c^* comprises the intrinsic CB alignment of the separated constituents, the dipole terms induced by forming the heterojunction and the strain effects on the weighted average of the respective CBs.) Hence, Zeller and Abstreiter self-consistently calculated the subband levels and the charge distribution in a SLS as a function of δE_c^*. For the computation, they chose a $Si_{0.5}Ge_{0.5}/Si$ superlattice with equally thick layers and a period length of 120 Å. 20 Å-wide doping spikes providing 4×10^{12} cm^{-2} carriers per period and located in the center of the SiGe layers were assumed. These parameters correspond actually to the aforementioned samples used by Jorke and Herzog (1985). In addition, the values for the CB strain splitting that were experimentally determined on these samples (Abstreiter et al., 1985) were used. Results of the computations are depicted in Fig. 28, which shows the subband levels and the charge distribution as a function of δE_c^*. Negative values of δE_c^* correspond to a Type-I lineup of the averaged CB levels (see Fig. 27.). The subband calculations reveal that for values of $\delta E_c^* < -40$ meV, the strain-induced splitting of the CBs is not sufficient to pull the Si doublet below the fourfold degenerate SiGe level. Consequently, under these conditions, all the carriers are in the SiGe layers. For $\delta E_c^* > -40$ meV, there is a sudden onset of carrier spilling into the lowest subband of the Si doublet, but there still remain

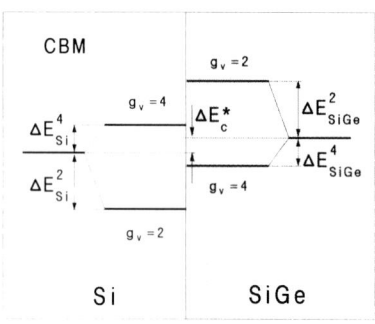

FIG. 27. Conduction-band offset at a Si/SiGe heterointerface with a tensilely strained Si and a compressively strained SiGe layer. The strain-induced CB splitting in fourfold (δE^4) and twofold (δE^2) degenerate valleys, and the virtual CB offset of the weighted-average CB energy (δE^*) are depicted (Zeller and Abstreiter, 1986).

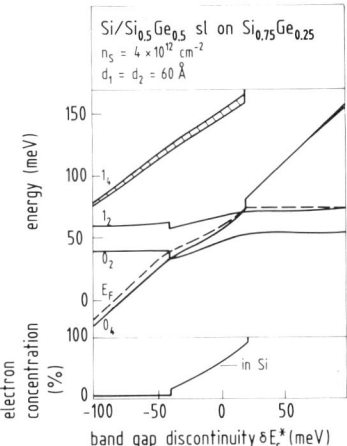

FIG. 28. Subband energies and carrier distribution for a symmetrically strained SiGe/Si SL as a function of the weighted average CB offset δE^*. In addition to the CB splittings, it takes a Type-II CB offset of $\delta E^* \geq 20$ meV to transfer all carriers into the Si layers (Zeller and Abstreiter, 1986).

carriers in the SiGe layers. It takes $\delta E_c^* > +20$ meV until all carriers are transferred into the Si layers where, by then, the population of the second subband has already begun. (Note the dashed line that shows the variation of the Fermi level E_F at $T = 4.2$ K.) It is interesting that the δE_c^* range where the electrons are spread over both layers is terminated at either end by discontinuities in subband energies, and hence in carrier distribution. This steplike behavior is attributed to many-body effects, which were included in the self-consistent calculations using the density functional approximation of Kohn and Sham (1965). That formalism had been successfully applied to Si inversion layers before (Ando, 1976).

From their calculation, Zeller and Abstreiter (1986) concluded that the experimental results of Abstreiter *et al.* (1985) and Jorke and Herzog (1985) can be consistently explained with a weighted average conduction-band offset δE_c^* of at least $+20$ meV (Type II). Their additional statement that δE_c^* might be as large as $+200$ meV, however, was based on an asymmetrically strained sample (Abstreiter *et al.*, 1986), which later turned out to be partly relaxed due to a layer thickness slightly beyond the critical value t_c (Kasper, 1987). Therefore, this upper limit of $\delta E_c^* = +200$ meV seems to be overestimated from today's point of view.

The limiting values of δE_c^* determined by Zeller and Abstreiter allow an estimate for the total CB offset δE_{CB} in the symmetrically strained situation. According to Fig. 27, the minimum CB offset occurs between the down-shifted twofold degenerate Si CB states and the fourfold degenerate SiGe levels. Hence, we get

$$E_{CB} = \delta E_c^* + \delta E_{Si}^2 - \delta E_{Ge}^4. \tag{35}$$

With the experimental value $\delta E_{Si}^2 - \delta E_{Ge}^4 \approx 50\,\text{meV}$ (Abstreiter et al., 1985) and the above values of δE_c^*, the resulting total CB offset becomes

$$70\,\text{meV} \leqslant \delta E_{CB} < 250\,\text{meV}.$$

For quite some time, these briefly outlined theoretical treatments of the SiGe/Si band lineup provided the only consistent data available. There were some experimental values for the valence-band offset of Si/Ge heterojunctions determined by photoemission spectroscopy (PES), but they disagreed by quite a large margin (Margaritondo et al., 1982; Mahowald et al., 1985). The conflicting values are mainly attributed to the undefined strain situation in the investigated layers, the relevance of which had not been recognized at the time the experiments were performed.

Meanwhile, a new approach has been undertaken by Ni et al. (1987; 1988), who used in situ x-ray PES to determine VB and CB offsets under carefully defined strain conditions. The essential aspect of their new method was to grow three $Si_{1-x}Ge_x/Si$ junctions for every studied x-value, which were p^+-doped, n^+-doped and unintentionally doped, respectively. These triplets of samples have the Fermi level at the VB, at the CB, and near midgap, respectively. By measuring the relative positions of the directly accessible Ge $3d$ and Si $2p$ core level signals with respect to the Fermi level E_F, Ni and coworkers were able to derive the CB and VB offsets according to

$$\delta E_c = (E_c - E_F)_{SiGe} - (E_c - E_F)_{Si} \equiv (E_{Ge3d}(c) - E_{Ge3d}(i))$$
$$- (E_{Si2p}(c) - E_{Si2p}(i)),$$
$$\delta E_v = (E_F - E_v)_{Si} - (E_F - E_v)_{SiGe} \equiv (E_{Si2p}(i) - E_{Si2p}(v)) \quad (36)$$
$$- (E_{Ge3d}(i) - E_{Ge3d}(v)),$$

where the indices c, v, and i refer to the kinetic energy values of the denoted core levels measured at the p^+, n^+, and intrinsic samples, respectively.

Since a surface-band bending within the electron escape depth ($\approx 25\,\text{Å}$ in the experiments) would introduce a systematic error, the surfaces were covered by about one ML of Sb or In. Such an adlayer has been shown to reduce surface-band bending to less than about 100 meV (Rich et al., 1987). The remaining experimental uncertainty together with the slight bandgap narrowing due to the heavy doping of the samples have been estimated by Ni et al. (1987) to be within $\pm 60\,\text{meV}$. This is comparable to the error margin given by Van de Walle and Martin (1986) for their calculated band offsets. In fact, a comparison between Ni's experimental and Van de Walle's theoretical values for various strain conditions, which are listed in Table VII, shows remarkably good agreement. This might, however, be fortuitous to some extent, given the respective error margins, which are quite substantial with

TABLE VII

THEORETICAL AND EXPERIMENTAL CONDUCTION-BAND OFFSETS FOR VARIOUS $Si/Si_{1-x}Ge_x$ HETEROJUNCTIONS ON $Si_{1-y}Ge_y$ BUFFER LAYERS. ENERGIES ARE GIVEN IN UNITS OF meV; ENERGIES ARE POSITIVE FOR TYPE-II BAND OFFSET, i.e., THE Si CONDUCTION-BAND MINIMUM IS LOWER IN ENERGY THAN THAT OF $Si_{1-x}Ge_x$.

x	y	δE_c^a	δE_c^b	δE_c^c	δE_c^d	δE_c^e	δE_v^a	δE_v^b	δE_v^d
0.25	0	±0	−20	−	±0	−	+170	+150	+180
0.5	0	−20	−20	−	+30	−	+380	+370	+360
0.5	0.25	+130	+150	<+240	+130	+165	+280	+300	+240

[a] theoretical values of Van de Walle and Martin (1986); estimated error ±50 meV.
[b] theoretical value of People et al. (1986).
[c] upper limit given by Zeller and Abstreiter (1986).
[d] experimental values of Ni et al. (1987); estimated error ±60 meV.
[e] experimental value of Jorke and Sawodny (1989).

respect to the absolute offset values. Especially the principal question of whether or not the CB offset converts from Type II to Type I, when going from the symmetrically strained to the asymmetrically strained situation, remains unsettled.

An entirely different experimental access to determining the CB offset has been reported by Jorke et al. (1987). They used vertical current-transport measurements on multiple-quantum-well structures to derive the lowest CB offset δE_c. Several $Si_{0.5}Ge_{0.5}/Si$ SLSs on a symmetrizing buffer layer and with varying period length were grown for this purpose. The number of periods was fixed to five. The SLs (including the symmetrizing buffer layer) were clad between heavily n-doped Si-layers, which were used to furnish ohmic contacts.

Transport across the layers is thermally activated as long as tunneling processes can be neglected, which is the case for the relatively thick SiGe barrier layers investigated. With this in mind, Jorke coworkers solved the one-dimensional Fokker–Planck equation (Smoluchowski equation) to model the activated transport in the presence of the periodic SL potential. They found an analytical expression connecting the derivative of the I-V characteristics in the limiting case of vanishing applied voltage with the CB offset δE_c,

$$(dI/dV|_{V=0})^{-1} \sim \exp(\delta E_c/kT). \tag{37}$$

Using this expression, Jorke and coworkers deduced the CB offsets of their SL samples from room-temperature I-V curves. For strain symmetrized $Si_{0.5}Ge_{0.5}/Si$ SLs, they found a value of $\delta E_c = 185 \pm 15$ meV, whereas a drastic reduction to $\delta E_c \approx 60$ meV occurred for layer thicknesses exceeding the critical thickness t_c. Subsequently, Eq. (37) was evaluated as a function of temperature, which led to a slightly lower value of $\delta E_c \approx 165\,\text{meV}$ for the pseudomorphic $\text{Si}_{0.5}\text{Ge}_{0.5}/\text{Si}$ SLs (Jorke and Sawodny, 1989). This value is higher than the theoretical and photoemission-derived CB offsets, but still lies within the respective upper error margins (see Table VII). It is also compatible with the CB-offset range estimated by Zeller and Abstreiter, but is definitely lower than their upper limit of 240 meV.

In conclusion, the Type-II band offset of symmetrically strained $\text{Si}_{1-x}\text{Ge}_x/\text{Si}$ is well established. So far, however, the uncertainty in the absolute CB offset is still high and definitely requires additional precision experiments. Under asymmetric strain conditions resulting from pseudomorphic growth on a Si substrate, the CB offset is generally small. Although δE_c is also of Type II for higher Ge contents (and for pure Ge on Si; see also Section V.C.2), it is not entirely clear whether it will slightly convert into Type I for low x-values.

C. Zone-Folding Effects

The band structure and the phonon dispersion of a single crystalline solid depend on the size and the symmetry of the unit cell, which define the Brillouin zone of the reciprocal lattice. In the case of a superlattice, the original crystal symmetry is disturbed by the presence of longer-range periodicity superimposed on the regular lattice in the direction perpendicular to the SL layers (z-direction). The effect of the SL periodicity can be accounted for by extending the unit cell in real space to accommodate a complete SL period. In the reciprocal lattice, this leads to a corresponding reduction of the Brillouin zone length in the k_z-direction. Since crystal momentum is only defined modulo G (where $G \equiv 2\pi/L$ denotes the reciprocal lattice vector, and L is the real-space unit-cell length), the dispersion curves in the k_z-direction can be constructed to first approximation by folding back the original Brillouin zone. Hence, electronic, optical, and phonon effects resulting from the superimposed periodicity of a SL are frequently referred to as "zone-folding effects." In the following, we will discuss such effects for the case of SiGe/Si SLSs.

1. Zone-Folded Acoustic Phonons

The observation of folded longitudinal acoustic (LA) phonons by Brugger *et al.* (1986a) was probably the first experimental evidence for superlattice effects in the SiGe/Si system. In this study, which was extended in subsequent publications (Brugger *et al.*, 1986b; Dharama-wardana *et al.*, 1986; Lockwood *et al.*, 1987), Raman-backscattering experiments on a variety of strained SiGe/Si SLs were performed in the regime of the acoustic phonons (Fig. 29). Due to the reduced Brillouin zone length $2\pi/L_{\text{SL}}$ (L_{SL} is the SL

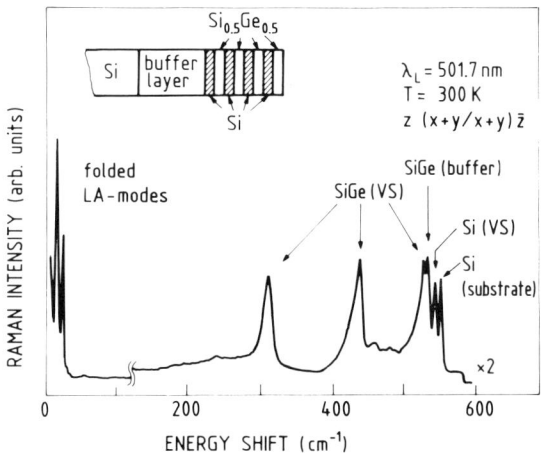

FIG. 29. Typical Raman spectrum of optical and folded acoustic modes in a $Si_{0.5}Ge_{0.5}/Si$ strained-layer superlattice. The arrows indicate the energy positions of optical phonon modes originating from the various layers. A schematic cross-sectional view of the sample is depicted in the insert (Brugger et al., 1986b).

period length) in k_z-direction, phonons propagating perpendicular to the layers with wave vectors,

$$q = q_s + m(2\pi/L_{SL}); \quad m = 0, \pm 1, \pm 2, \ldots, \quad (38)$$

can become Raman active. Here q_s denotes the scattering wave vector, which is conserved in a one-phonon Raman process modulo a reduced reciprocal lattice vector $G_{LS} = 2\pi/L_{SL}$. In backscattering geometry, q_s is given simply by two times the photon momentum of the incoming laser light within the sample, i.e., $q_s = 2(2\pi n/\lambda_L)$; λ_L is the laser wavelength, and n is the refractive index of the sample at the laser wavelength. Hence, for a SLS with a relatively long period length, almost the entire reduced Brillouin zone may become accessible by varying the laser wavelength over the visible spectral range. Even more, it becomes possible to excite phonons with scattering wave vectors beyond the reduced Brillouin zone via an umklapp process, which is normally not possible because $q_s \ll G$ in regular crystals (Lockwood et al., 1987; Brugger et al., 1986b).

As an example, the folded LA phonon dispersion of an asymmetrically strained $Si_{0.5}Ge_{0.5}/Si$ SLs with a period length of 163 Å is shown in Fig. 30. Folded LA phonon signals up to the $m = -4$ branch of the dispersion have been observed. Open symbols represent relatively weak signals, which occur for even values of m in SL with almost identical layer thicknesses (Colvard et al., 1985). Full symbols stand for very intensive signals, which can be comparable to, or even larger than, the optical phonon modes (Brugger et al.,

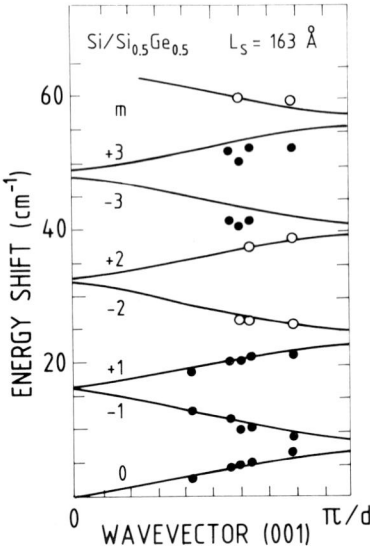

FIG. 30. Dispersion of folded acoustic LA phonons for a $Si_{0.5}Ge_{0.5}/Si$ SLS with a period length of 163 Å. Solid lines are calculated by using a simple elastic continuum model. Data points are derived from Raman experiments at various excitation wavelength (Brugger et al., 1986b).

1986a). The lines in Fig. 30 represent the branches of theoretical dispersion curves based on the elastic continuum model of Rytov (1956). This approximation is, despite its simplicity, obviously fully capable of describing the experimental data within the energy range of Fig. 30. In the continuum limit, where the acoustic phonon dispersions are approximately linear and hence entirely characterized by the sound velocities, the energy dispersion of a SL is given by (Rytov, 1956)

$$\cos(qd) = \cos(\omega \cdot L_{SL}/2v_1) - ((1 + r^2)/2r) \cdot \sin(\omega \cdot L_{SL}/2v_1) \cdot \sin(\omega \cdot L_{SL}/2v_2), \quad (39)$$

where v_1, v_2 are the sound velocities of the two constituent layers of the SL and $r \equiv v_1 d_1/v_2 d_2$; d_1, d_2 are the densities of the two respective layers.

It can be shown that in the case of equally thick layers, the average sound velocity v of the SL is directly related to the observed doublet splitting at a given scattering momentum q_s: $\delta\omega = 2q_s \cdot v$. For q_s not too close to the Brillouin zone boundary, the doublet splitting is independent of the SL period length L_{SL}, thus allowing a straightforward determination of the sound velocities of arbitrary $Si_{1-x}Ge_x$ alloys. On the other hand, by fitting the data of a SL to the continuum dispersion relation, a precise evaluation of the SL period becomes possible. This appears to be an attractive and relatively simple technique, which is an alternative to the usual period determination by x-ray rocking analysis (Baribeau et al., 1988).

The calculated dispersion curve in Fig. 30 reveals that the superimposed

SL periodicity does not only fold back the original dispersion into the reduced Brillouin zone, but that energy gaps open between adjacent branches at the zone boundaries. Brugger et al. (1986b) showed that these gaps can actually be seen in a Raman experiment. For this purpose, a $Si_{0.5}Ge_{0.5}/Si$ superlattice with a relatively large period length of $L_{SL} = 280$ Å was grown, which shifts the zone boundary π/L_{SL} into a q_s-range that is easily accessible with standard laser lines. By tuning q_s through the zone boundary, a finite energy gap of $\approx 1\,cm^{-1}$ between the Brillouin mode ($m = 0$) and the first folded mode ($m = -1$) could unambiguously be observed (Fig. 31). The experimentally found gap is in very good agreement with Rytov's theory, again demonstrating the adequateness of this simple model.

Recently, the Raman studies of folded-phonon properties were extended to Si/Ge SLS with extremely narrow period widths of only 5 to 20 ML (Kasper et al., 1988b). Such SLSs became popular mainly because of their predicted electronic and optical properties, which will be discussed in some detail in the next section. Although the short period length L_{SL} limits the range of the reduced Brillouin zone accessible by Raman excitation with visible laser light, the observation of the first LA, and also of the first transversal acoustic (TA) doublet ($m = \pm 1$), suffices to determine the period length precisely. Moreover, the optical phonons measured simultaneously give valuable information about the strain components within the respective layers. It is interesting to note that, in contrast to the acoustic phonons, the optical phonons develop modes confined to the individual layers. This is due to the missing overlap of the (relatively flat) optical-phonon dispersions of Si and Ge, which does not allow propagation within the respective other SL constituent (Friess et al., 1989).

2. Optical Properties of Ultrathin Si/Ge Strained-Layer Superlattices

An almost forgotten paper by Gnutzmann and Clausecker (1974) was entitled

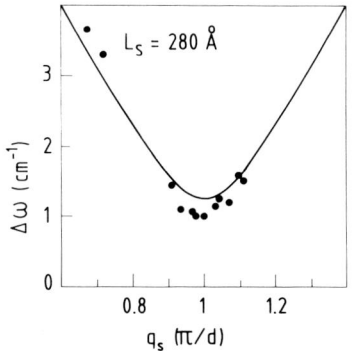

FIG. 31. Energy gap between the first folded LA phonon mode ($m = -1$) and the Brillouin mode ($m = 0$) near the reduced Brillouin-zone boundary. The solid line is based on an elastic-continuum calculation. Data points are derived from Raman experiments (Brugger et al., 1986b).

"Theory of direct optical transitions in an optical indirect semiconductor with a superlattice structure." In this contribution from the early days of the superlattice concept, the authors discuss the superlattice-mediated conversion of an indirect-gap semiconductor into a quasidirect-gap material by appropriate zone folding of the electronic band structure. Although this paper was well ahead of its time, it is interesting that Gnutzmann and Clausecker explicitly mentioned $Si_{1-x}Ge_x$ SL structures as potential candidates for the implementation of their proposal. Indeed, such a conversion, if it will turn out to be possible, could have tremendous impact in terms of device applications. It might finally provide optoelectronic capabilities to the celebrated Si technology, which could readily be integrated with standard VLSICs.

Jackson and People readdressed this subject in 1986 (Jackson and People, 1986), i.e., at a time when the growth of high-quality SiGe superlattices had already been demonstrated by several groups. The basic idea of the SL-induced direct-gap transition is as simple as the related zone folding of LA phonons. The superimposed SL reduces the widths of the Brillouin zone in the k_z-direction from $\pm 2\pi/a_0$ (a_0 is the cubic lattice constant) to $\pm \pi/L_{SL}$. Like the phonon dispersion, the electronic band structure is folded back into the reduced Brillouin zone. Within this simple model, L_{SL} has to be chosen appropriately to fold the CB minimum back into the Γ-point. This is shown for the case of Si (and Si-like $Si_{1-x}Ge_x$ alloys with $x < 0.8$) in Fig. 32. It takes $L_{SL} = 5a_0/2$ ($= 10$ ML) to fold the Δ_1 minimum, which is located at about 80% of the distance between the Γ- and X- point in the original CB structure, into the Γ-point.

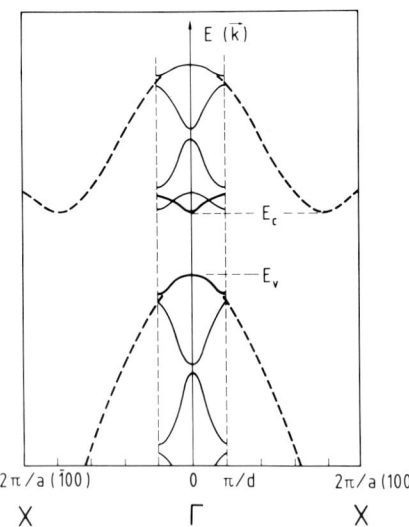

FIG. 32. Zone folding of the Si conduction-band dispersion introduced by a superimposed SL periodicity. In this simple model, the Δ_1 CB minimum is folded back into the Γ-point for SL period length of five times the half-lattice constant a_0 of Si.

It is obvious from the earlier discussion of band offsets in SiGe/Si strained-layer superlattices that the electronic band structure is drastically affected not only by the presence of a heterointerface, but also by the strain within the constituents of the SLS. Hence, it can not be expected a priori that the concept of simply folding back the unperturbed Si CB will adequately describe the CB properties of such a SL. Meanwhile, a variety of theoretical approaches have been published, with the aim of giving a more complete picture of the electronic band structure in ultrathin Si/Ge SLSs. The underlying models and computational techniques differ quite substantially, reaching from an effective-mass envelope-function model incorporating a phenomenological treatment of strain effects (People and Jackson, 1987), over tight-binding calculations (Brey and Tejedor, 1987), to self-consistent local-density calculations (Froyen et al., 1987, 1988; Hybertsen and Schlüter, 1987). Despite the differences in the model assumptions and also in the absolute energy values resulting from these calculations, there are basic trends that can be seen in most of these theoretical treatments. Exemplarily, we shall outline those aspects following the results and notations of Froyen et al. (1987, 1988).

Consider a Si/Ge SLS consisting of n atomic layers of Si and m layers of Ge which will be referred to as $(n \times m)$ SL in the following. The SL is assumed to be pseudomorphic, i.e., the respective layers are strained to an extent determined by the lattice constant of the effective substrate. As we know from the simple zone-folding picture, the modifications of the electronic band structure relevant for the optical properties of the SL occur in the CB. Since the SL periodicity implies a reduction of the Brillouin zone only in the k_z-direction, the conduction-band states may be divided into two categories, reflecting their origin in terms of the bulk properties of the two constituents. The first group consists of states from around the X-point in direction perpendicular to the SL-layers ($X_\perp$-states), which are zone-folded into the Γ-point of the SLS. These states are strongly confined to the Si wells, both due to the Type-II band lineup and the heavy longitudinal electron effective mass ($m_l(\text{Si}) = 0.98 m_0$). This group of states provides new direct transitions (in k-space, not necessarily in real space), which cannot be observed in a corresponding alloy. The other group consists of states from those parts of the Brillouin zone that are not folded by the superlattice periodicity (transversal states with respect to the growth direction). They are located at the X- and L-points of the SLS Brillouin zone, thus allowing only indirect transitions. The transversal states are extended over both layers due to their low effective masses ($m_t(\text{Si}) = 0.19 m_0$; $m(L, \text{Ge}) = 0.12 m_0$). Although these states are not zone-folded, they appear as pairs that are split by the different potentials in the Si and Ge regions. This splitting shows some kind of an oscillatory behavior as a function of the period length of the $(n \times m)$ SL.

The two groups of states resulting from the calculations by Froyen *et al.* (1987; 1988) are schematically depicted in Fig. 33 for the asymmetrical strain situation induced by a Si substrate (i.e., only the Ge layers are strained). For reasons of clarity, only the energy states at the critical points Γ, X, and L are shown for several $(n \times m)$ SLs with equally thick layers $(n = m)$. The energy dispersion of the respective bands is indicated by hatched areas. Note that some of the bands show downward dispersion, i.e., the depicted critical point level is not the lowest energy level in those bands. It should also be mentioned that the absolute energy values given in the figure are not corrected for the systematic underestimation of the bandgap, which is typical for the local-density approximation. This shortcoming has already been discussed in Section V.B.3 in connection with the self-consistent band-offset calculations of Van de Walle and Martin (1985, 1986). The same argument applies here;

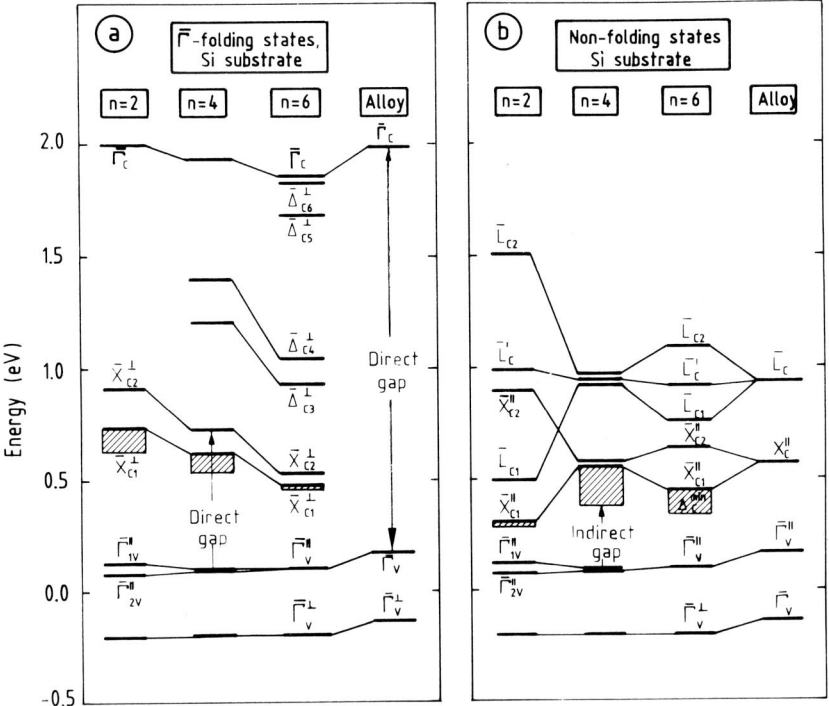

FIG. 33. Calculated variation of conduction-band states for various asymmetrically strained $n \times n$ Si/Ge SLs on a Si substrate. (a): Zone-folding states at the Γ-point. (b) Nonfolding states. Only folding states introduce new direct transitions. Hatched areas indicate a downward dispersion of the respective level. The energy values of the local-density calculation are not corrected to give the experimental bandgaps (Froyen *et al.*, 1988).

i.e., a quite accurate description of the absolute energies can be expected by simply adding a value of approximately 0.65 eV to the respective bandgap given in the figure.

The overall behavior of the SLS depends on the energetic arrangement of these two groups of levels, which is determined by two competing mechanisms, namely (i) zone folding, which makes the SLS more direct in k-space with increasing n, and (ii) compressive in-plane strain in the Ge layers, which pushes down the transverse states. For the case of $(n \times n)$ Si/Ge SLSs on a Si substrate (i.e., only the Ge layers are tetragonally distorted), Froyen et al. (1987; 1988) found the lowest CB state to be a non-zone-folded state, namely the lower of the two $X_c^{\|}$ states (Fig. 33). Hence, the lowest optical transition is expected to be indirect under these strain conditions. This remarkable result was confirmed by similar calculations of Hybertsen and Schlüter (1987) and is also in agreement with the effective-mass approach of People and Jackson (1987).

Since the transversal states are strongly affected by the strain distribution within the $(n \times m)$ SL, one might hope to achieve a direct-gap semiconductor when reducing the compressive strain in the Ge layers. Froyen et al. (1988) as well as Hybertsen and Schlüter (1987) have studied that possibility by calculating the $(n \times m)$ SL band structures for various strain conditions. The respective lowest zone-folded and transversal states resulting from the computations of Froyen and coworkers are depicted in Fig. 34. Three different strain situations are shown: (i) for a Si substrate (only the Ge layers are strained), (ii) for a $Si_{0.5}Ge_{0.5}$ substrate (symmetrical strain distribution), and (iii) for a Ge substrate (only the Si layers are strained). The mentioned trend is clear-cut, i.e., the strain-induced lowering of the $X_c^{\|}$ state is coupled to the tetragonal distortion of the Ge layers. This leads to a crossing of the two lowest CB states at around the symmetric strain distribution. If the Ge layers remain completely unstrained, the zone-folding (direct) state lies clearly below the lowest transversal state.

This is a very interesting result, which might turn out be a key element in the search for the indirect/direct gap transition. There remains, however, a principal problem: As was mentioned earlier, the lowest zone-folding state shows a downward dispersion (hatched areas in Figs. 33 and 34). It becomes smaller with increasing period width $n + m$ and has recently been shown to virtually vanish at $n + m = 10$ ML and 14 ML (Hybertsen et al., 1988a). This effect can already be seen in the simple zone-folding picture (Fig. 32), reflecting just the fact that the bulk CB minimum is not exactly at the X-point. The necessary increase of the period length, however, is expected to reduce the oscillator strength for a "direct" (in k-space) optical transition, because the transition becomes more and more indirect in real space with increasing period length. This is a consequence of the strong confinement of

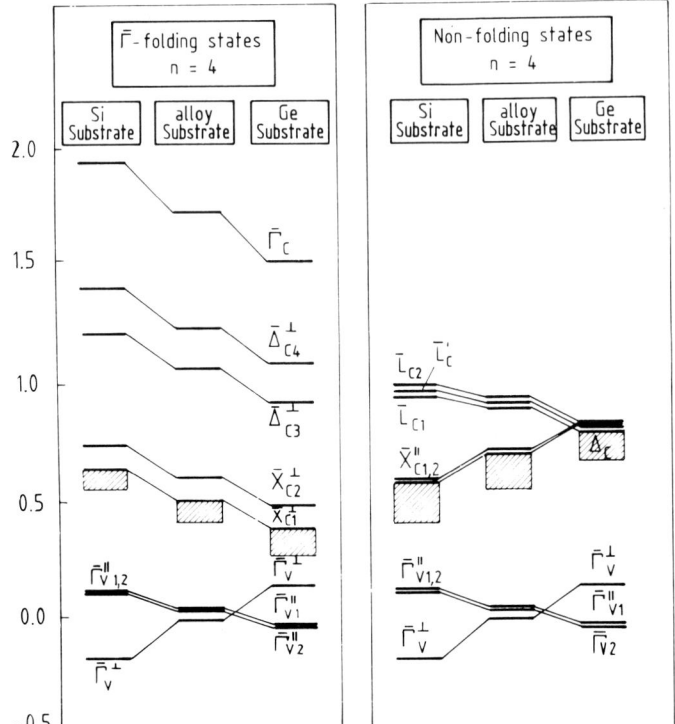

FIG. 34. Variation of the conduction-band states of the 4 × 4 Si/Ge SLS in Fig. 33 as a function of the substrate lattice constant, which affects the strain distribution. An increasing substrate lattice constant (i.e., increasing tensile strain in the Si layers) pulls the lowest zone-folding state below the first nonfolding state. Energy values are not corrected (Froyen et al., 1988).

the zone-folding states to the Si layers, whereas the upper VB states are mainly localized in the Ge layers (Type-II band lineup).

The question of whether a $(n \times m)$ Si/Ge SLS with quasidirect bandgap and reasonable oscillator strength can be realized will need experimental investigations in order to finally become settled. Such experiments are available and shall briefly be discussed next, even though their outcome is yet far from giving a complete picture.

Structural and optical characterizations of $(n \times m)$ Si/Ge SLSs were first reported by Pearsall *et al.* (1987) and by Bevk *et al.* (1987). On Si substrates, they grew a series of pseudomorphic Si/Ge SLSs with equal layer thicknesses $(n = m)$ and period lengths $2n$ of 2, 4, 8, and 12 ML. For the asymmetrically strained Si/Ge SL of Pearsall and coworkers, the maximum period length is naturally given by the critical thickness of about 6 ML for pure Ge on Si. Moreover, the total thickness is limited to less than about 100 Å, i.e., the critical thickness of an equivalent $Si_{0.5}Ge_{0.5}$ alloy on an unstrained Si buffer. The (4 × 4) SL reported by Pearsall et al. (1987), for example, consisted of just five periods, leading to a total thickness of only 51 Å.

High-resolution transmission electron microscopy (HRTEM) micrographs and in situ RHEED studies were indicative of abrupt interfaces with no evidence for islanding or interdiffusion. To characterize the optical properties of these very thin SLS wells, electroreflectance spectroscopy was employed, which can be up to a factor of 10^4 more sensitive than absorption or luminescence measurements (e.g., Cardona, 1969; Seraphin, 1972). In addition, this technique is only sensitive to first-order optical transitions, i.e., to transitions that do not require a secondary excitation such as a phonon. First-order optical transitions are commonly (but not generally, as we will see) associated with direct energy gaps, which is especially advantageous considering the intention of the experiments. For adapting this kind of spectroscopy, the SLs were capped with a 1000-Å-thick Si epilayer and a semitransparent Schottky contact for applying the required modulation voltage.

Figure 35 shows electroreflectance spectra of the Si/Ge SLS samples. Optical-transition energies are closely related to the minima in the spectra

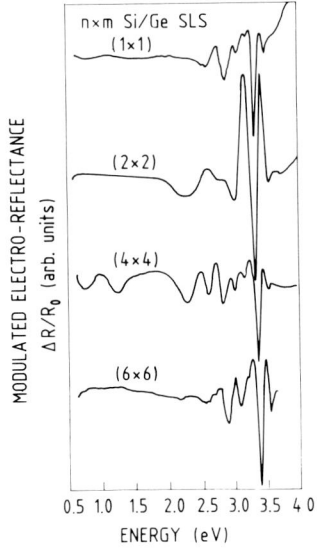

FIG. 35. Experimental electroreflectance data for several $n \times n$ Si/Ge superlattices. Note the structurally induced low-energy transition in the spectrum of the 4 × 4 SL (Bevk et al., 1987).

and can be determined precisely by means of a numerical routine that adjusts for amplitude, phase, and line width (Pearsall et al., 1987, and references therein). The spectra reveal a rich variety of features resulting from first-order optical transitions. Due to the Si cap layer, all spectra show the direct Si gaps at 3.37 eV (E_1) and 3.10 eV (E'_0), but in most cases, the other transitions differ significantly from sample to sample and, moreover, are entirely different from a reference spectrum taken on a $Si_{0.45}Ge_{0.55}$-alloy well (Pearsall et al., 1987). A very special behavior is found for the 4 × 4 SL, which is the only one that shows transitions in the infrared range at 0.76 eV and at 1.25 eV. These two, and a third transition at 2.31 eV, are all lower in energy than the lowest transition observed in the random alloy. This striking behavior of the (4 × 4) SL was subsequently also investigated by photocurrent measurements performed on similar structures, which revealed absorbtion onsets at 0.78 and 0.90 eV (Hybertsen et al., 1988a). Although these experiments confirmed the low transition energies, the shape of the observed absorption edges was not indicative of a direct optical transition.

The interpretation of the experiments of Pearsall et al. (1987) were the subject of extensive theoretical efforts. Actually, most of the theoretical work mentioned was triggered by the electroreflectance results of the (4 × 4) SL. Since almost all of the model calculations found the lowest optical transition to be indirect (in k-space) for the strain situation of the grown structures, the discussion concentrated on how to explain the observed first-order transition.

Hybertsen and Schlüter (1987; Hybertsen et al., 1988a, 1988b) attribute the energetically lowest feature in the electroreflectance spectrum of the (4 × 4) SL to the narrow overall width of the SLS of only 51 Å. Together with the two adjacent Si cladding layers, the complete SLS can be seen as a quantum well. Hybertsen and Schlüter calculated the band lineup between the SLS well and the Si cladding layers by using an average potential technique similar to that of Van de Walle and Martin (1986). The result is a strong confinement of holes in the SLS well, whereas the SL-CB states are higher in energy than the CB edge of the Si cladding layers. According to this picture, the lowest transition would be between confined-hole states in the SLS and the Δ minima in the Si layers. This model is corroborated by the good agreement between experimentally observed photocurrent edges at 0.78 eV and 0.90 eV (Hybertsen et al., 1988a) and the calculated values of 0.77 eV and 0.87 eV (Hybertsen et al., 1988b). Within this picture, the lowest transition would be indirect in k-space as well as in real space. Nevertheless, it could be observed as a first-order transition owing to the relaxation of k_z-conservation, which results from the break of the (Si) lattice periodicity in the SLS region. An additional mechanism, however, is necessary to explain the relatively high oscillator strength observed experimentally. Hybertsen et al., (1988b) suggest an electro-optical enhancement induced by the electrical modulation field present in the electroreflectance technique.

The interpretation of Hybertsen and coworkers is not generally accepted and has been questioned especially by Wong et al. (1988). They presented their own calculation based on an empirical pseudopotential formalism. Although their lowest transition is also indirectly due to a small downward slope in the dispersion of the lowest CB state, they find a significant increase in the dipole transition matrix element, if the Ge–Ge bond length in the longitudinal direction is slightly stretched in order to make them more bulklike. (The in-plane bond length remains fixed by the Si substrate). Such an increase in bond length is suggested by the results of the first-principle calculations of Van de Walle and Martin (1986), which gave a roughly 1% larger bond length as compared with the usual value derived from minimizing the macroscopic elastic energy of the SLS. Hence, without invoking any relaxation of momentum conservation, Wong and coworkers were able to explain most of the strength of the electroreflectance signal by minor adjustments of the bond lengths.

Because of the inherent thickness restrictions for $(n \times n)$ Si/Ge SLSs grown on a Si substrate, it is hard to verify experimentally whether momentum nonconservation or bond-length adjustment is the dominating factor responsible for the relatively large low-energy electroreflectance signals. The thickness restriction also strongly reduces potential applications of such a structure, almost independently of the principal character of the transition. Unfortunately, this also applies to the other extremely asymmetric situation, where an effective Ge substrate is used. As we have seen (Fig. 34), a direct optical transition (in k-space) is predicted for that configuration. A device-relevant Si/Ge superlattice, however, has to overcome the thickness limitation, which, as we have seen, is possible by SL growth on a symmetrizing buffer layer. Within certain restrictions, such a buffer can well be Ge-rich, provided the average composition $(n + m)/2$ of the SL is adequately adjusted. This can be done by increasing the thickness of the Ge layers with respect to the Si layers (i.e., $m > n$). With this concept of a symmetrizing buffer, $(n \times m)$ Si/Ge SLSs can be grown to an arbitrary overall thickness, as long as the individual layers remain below their respective critical thickness, which, of course, depends on the effective lattice constant of the substrate.

The first experiments utilizing $(n \times m)$ Si/Ge SLSs grown on a symmetrizing buffer became available recently (Kasper et al., 1988b). These initial attempts were still restricted to Si-rich $Si_{1-y}Ge_y$ buffer layers with an effective y value of 0.4. Consequently, the $(n \times m)$ Si/Ge SLSs were grown with different period length $m \times n$, but with the constant ratio $m/n = 3/2$. (3×2), (6×4), and (12×8) SLs were grown to an unprecedented thickness of 2000 Å each and characterized by Raman scattering. The observation of folded LA and TA phonon modes were a strong indication for the excellent period definition of these SLSs (see Section V.C.1). The overall thickness of the SL layers allows a variety of experimental techniques for the characterization of the optical properties to be applied. So far, only luminescence measurements have been reported, which have been performed by Zachai et al. (1988, 1990). Their preliminary results show a strong, asymmetric luminescence signal at around 0.85 eV for the case of the (6 × 4) Si/Ge SLS (Fig. 36). A weaker signal at lower energies is found in the (12 × 8) SL. The position of this signal agrees roughly with that of a very weak signal found in a reference alloy sample of the same average composition as the SLs. This signal has tentatively been attributed to deep levels associated with dislocations and/or impurities. The unique signal of the (6 × 4) SL shifts to higher energies with increasing excitation power density, a behavior that is expected for a band-edge-related transition.

It is quite remarkable that the distinct behavior of the (6 × 4) Si/Ge SLS coincides with the aforementioned theoretical prediction of Hybertsen et al. (1988b), namely the vanishing downward dispersion of the lowest zone-folding state for $n + m = 10$ ML. According to their theoretical treatment, this condition depends mainly on the period length and not much on the individual layer thickness.

The implementation of thick, strain-symmetrized Si/Ge SLSs and the

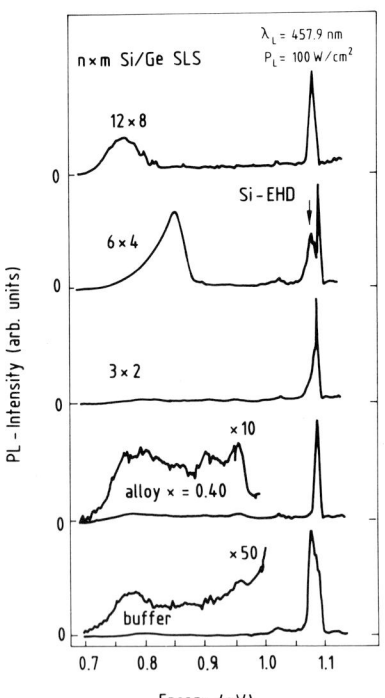

FIG. 36. Photoluminescence spectra of a series of $(n \times m)$ Si/Ge SLSs grown on a symmetrizing buffer layer. The (6 × 4) SLS shows a strong, asymmetric luminescence signal in the infrared region around 0.85 eV, which is attributed to a band-to-band transition (Zachai et al., 1988).

promising luminescence results are definitely important steps toward a comprehensive experimental characterization of the optical properties of such structures. This can, however, be just a beginning. Various techniques, e.g., electroreflectance spectroscopy, electroluminescence, or absorption spectroscopies, have to be applied to get a complete picture of the character of the optical transition observed in the luminescence data. Moreover, SLs grown on Ge-rich buffers are necessary to test the theoretical predictions. The great technological and physical interest in a SL-mediated indirect/direct gap transition will surely remain a strong motivation for such experiments in the near future.

D. Device Structures

As we have seen in the two preceding sections, the basic understanding of many of the key features of the strained SiGe/Si system, e.g., the strain-dependent band lineup, has been gathered in the past three or four years. Most of these achievements became only possible by the improvements and breakthroughs in Si-MBE techniques, like the development of suitable doping methods. On the other hand, there has always been a large impact from a device aspect of SiGe SLSs, which is based on the prospects of integrating novel electronic and optoelectronic features with the mature Si technology. Despite these strong driving forces, it is still amazing how many device structures and demonstrators for device functions have been developed in the few years since about 1984. In any case, the aims were to drive Si-based devices beyond their contemporary physical restrictions concerning switching speed and optoelectronic response in the infrared regime. Admittedly, such applications benefited from the pioneering achievements in the field of MBE-grown GaAs/GaAlAs heterodevice structures, but an even higher degree of sophistication can be expected from the additional freedom in band-structure engineering that is provided by proper strain adjustment in the SiGe/Si system.

In the following, we shall discuss some of the most promising devices based on SiGe/Si SLSs and heterostructures. Emphasis will be on infrared detectors in the technologically important wavelength range between 1.3 and 1.6 μm, and on high-speed devices ranging from MODFETs over resonant-tunneling (RT) structures to heterobipolar transistors (HBT).

1. *Infrared Detectors*

It is well known that Si technology is unable to produce an on-chip long-wavelength IR detector for modern fiber-optical communication. This shortcoming has its origin in the relatively large bandgap of Si, which lies above the photon energy range transmitted in silica fibers (1.3 to 1.6 μm). Hence, employing Si integrated-circuit (IC) technology for fiber-optical communication necessitates hybrid solutions with detector units made from Ge or InGaAsP on a separate chip. A first step to overcome this rather inefficient practice was made by Luryi et al. (1984), who fabricated a Ge p–i–n photodiode on a silicon substrate. The p–i–n diode was grown by MBE on a thick, incommensurate Ge buffer layer, which incorporated thin, coherently strained $Si_{1-x}Ge_x$ layers (Fig. 37). The purpose of these layers (so-called glitches), which are not part of the active detector volume, is to reduce the number of threading dislocations penetrating through the Ge p–i–n region (Kastalsky et al., 1985), thus reducing the leakage current of the photodiode. The principal feasibility of this concept was underlined by the high internal quantum efficiency achieved ($\approx 41\%$ at 1.45 μm, 300 K), which is as high as that of commercial Ge detectors. Nevertheless, this design is not entirely satisfactory, since the epitaxy is Ge terminated, which limits further vertical Si integration. It would also result in nonideal avalanche photodetection (APD) due to the low ratio of the ionization coefficients for holes and electrons in Ge. The latter would lead to an excess noise factor. Another disadvantage is a required thickness of several μm for the active Ge layer in a standard vertical absorption layout.

All of these restrictions of the conventional Ge p–i–n design have been overcome by the subsequent development of narrow waveguide diode structures, with the active layer consisting of a pseudomorphic $Si_{1-x}Ge_x$/Si SLS (Temkin et al., 1986a; 1986b). The required sensitivity in the infrared range transmitted by optical fibers is achieved by utilizing the strain-induced bandgap narrowing in the $Si_{1-x}Ge_x$/Si SLS (People, 1985; Van de Walle and Martin, 1986), which was discussed in Section V.B.3. Thanks to this strain shift, efficient photodetection at 1.3 μm becomes already possible at a Ge

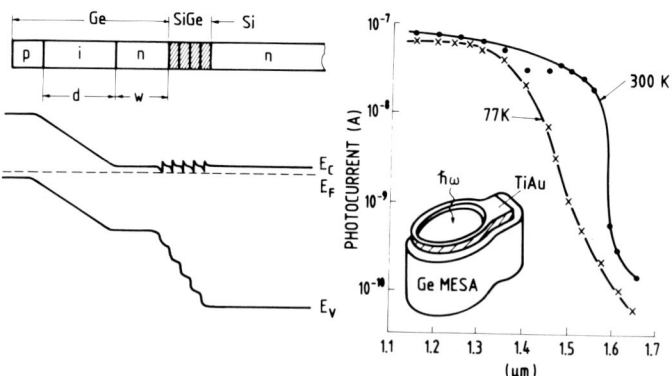

FIG. 37. Ge p–i–n photodetector grown on a Si substrate via a thick SiGe buffer layer. Left: cross-sectional view and band-edge variation; right: photoresponse at 300 K and 77 K (Luryi, S., Kastalsky, A., and Bean, J. C. IEEE Trans. Electron. Devices **ED-31**, 1135, © 1984 IEEE).

content of $x = 0.6$ in the SLS layer (Fig. 38). The critical thickness for this Ge fraction, however, is only about 100 Å. Hence, in order to maintain the pseudomorphic growth conditions on the Si substrate and to simultaneously achieve a reasonable absorption length, Temkin et al. (1986a) adapted a rib waveguide geometry, where the light is absorbed laterally (Luryi et al., 1986). A typical $p-i-n$ photodiode structure consists of about 20 undoped periods of 60 Å $Si_{0.4}Ge_{0.6}$ and 290 Å Si, for example, which are sandwiched between about 1 μm-thick Si n- and p-layers. The larger index of refraction of the SLS acts as an efficient, buried waveguide, which is laterally defined in one dimension by etching a rib structure. The single detectors (or parallel arrays) are cleaved perpendicular to the rib to expose a front surface suitable for launching the incident light from an optical fiber into the device. This design features efficient absorption despite the relatively thin active SLS layer, full compatibility with Si technology because of the pseudomorphic growth condition, and straightforward implementation of low-noise APD operation. The latter operating condition was demonstrated by Temkin et al. (1986b) by adding an undoped Si avalanche layer to the SLS waveguide layout (Fig. 39).

Very promising results have been achieved with the waveguide SLS SiGe photodetectors, especially with the version featuring avalanche gain. For a typical device length (= absorption length) of ≈ 0.5 mm, a maximum internal quantum efficiency of about 47% was estimated at a wavelength of 1.23 μm for a device with Ge content $x = 0.6$ (Temkin et al., 1986a; see also Fig. 38). By adding avalanche gain, external quantum efficiencies of 400% were reported at 1.3 μm, which correspond to a dc responsivity of 4 A/W (Temkin et al., 1986b). There was some initial concern about possible speed limitations due to hole-trapping effects in the active SLS layer. Such an effect might result from the large valence-band offset that amounts to almost the complete

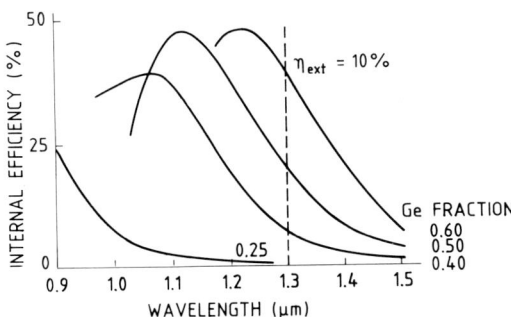

FIG. 38. Room-temperature spectral response of a SLS $p-i-n$ photodiode as a function of the Ge content x of the active layer. A Ge content of at least 60% is necessary for efficient operation at 1.3 μm (Temkin et al., 1986a).

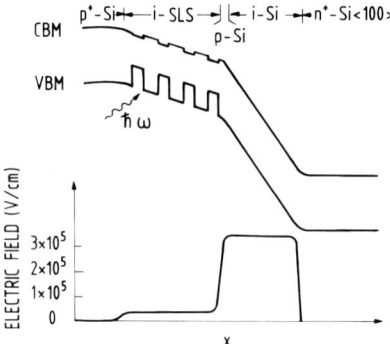

FIG. 39. Band diagram and electric-field variation of a SiGe/Si SLS avalanche photodiode. Light is absorbed in the low-field SL region while avalanche multiplication occurs in the adjacent high-field Si layer (Temkin et al., 1986b).

bandgap difference in the present unsymmetrical strain situation (Fig. 39; see also Section V.B.3). Therefore, Temkin et al. (1986b) performed high-frequency-response measurements at different APD gains by illuminating the device with a train of 50-ps-long mode-locked pulses from an InGaAsP laser operated at 430 MHz. At a gain of 6, a response pulse width at half maximum of 136 ps was found, with the leading edge rising in 50 ps from 10 to 99%. The high speed of the rib-waveguide APD is corroborated by a measured -3 dB bandwidth in excess of 8 GHz at a gain of 6, i.e., the gain-bandwidth product amounts to $\geqslant 48$ GHz. Given these excellent data, Temkin et al. (1986b) tested the performance of the SiGe waveguide APDs in transmission experiments at a laser wavelength of 1.275 μm. A single-mode optical fiber of 45 km length was coupled to the devices utilizing a lens and a polarization controller to assure TE polarization. A pseudorandom bit pattern at a rate of 800 Mb/s was transmitted, resulting in a bit error rate $< 10^{-9}$ at an incident light power of 13.4 μW. Under similar conditions, but with the length of the optical fiber reduced to 7.5 km, the transmission rate could be increased to 1.7 Gb/s, without affecting the bit error rate.

The reported performance of the SiGe SLS APDs compare quite well with state-of-the-art InGaAs $p-i-n$ detectors. Further improvements are expected upon increasing the efficiency of the fiber/waveguide coupling, which was limited in the experiments to about 18%, and by reducing the leakage currents in the devices, which restrict the achievable APD gain. But even in the present state, these devices are very promising high-speed, high-sensitivity photodetectors for applications near 1.3 μm, which offer the exciting potential for monolithic integration with Si readout electronics.

2. Modulation-Doped Field-Effect Transistors

The most important three-terminal device in the ever increasing field of digital electronics is undoubtedly the MOSFET, which was briefly introduced at the beginning of Section V.B. Its dominance is based on the excellent qualities of the SiO_2/Si interfaces, which are characterised by very low trap densities and abruptness on the scale of a few atomic layers. On the other hand, semiconductor materials with higher mobilities than Si would be desirable for high-speed applications. III-V semiconductors, like GaAs, are very well suited for that purpose; however, at this time, there is no insulator available that can provide the required interface quality. To overcome this problem, heterojunction FETs based on the GaAlAs/GaAs interface were developed, which are controlled by Schottky gates. Like in the MOSFET, the electrical transport in a heterointerface field-effect transistor (FET) (see Fig. 19c) is via a 2D carrier gas, which led to the expression TEGFET (2DEG FET). In contrast to the MOSFET, however, the heterojunction between two semiconductors allows for selective doping of the larger gap material, which results in an additional mobility enhancement in the channel due to the associated reduction of ionized impurity scattering. This concept is often referred to as high-electron-mobility transistor (HEMT), or MODFET.

MODFETs based on the GaAlAs/GaAs junction have very successfully driven microelectronics to higher speeds. A severe drawback, however, is the incompatibility with Si-based technology, which triggered worldwide feasibility studies of the growth properties of GaAs on Si. A more straightforward way toward Si-based high-speed devices might result from the exploitation of the strained SiGe/Si system, which appears as a "natural" extension of the Si homosystem. Consequently, the demonstration of modulation-doping effects for holes and electrons in SiGe/Si, which were discussed in some detail in Section V.B, led almost instantaneously to the development of p- and n-channel MODFETs.

a. *p*-Channel MODFET. Based on the aforementioned work of People *et al.* (1984; 1985), p-channel MODFETs were fabricated and characterized by Pearsall *et al.* (1985) and by Pearsall and Bean (1986). Figure 40 shows the valence- and conduction V-band edge variation along the growth direction of a typical *p*-MODFET. The 2DHG is confined to a $Si_{0.8}Ge_{0.2}$ layer of about 250 Å thickness, which is directly grown on an unstrained p^- Si buffer layer. p^+-doping is located in a narrow layer between the hole channel and the Schottky gate, which is separated from the gate as well as from the channel by p- spacer layers on either side of the doped region. Standard VLSI processing techniques were employed for device definition, which included BF_2 implantation for the source and drain contacts. Annealing of the implanted structures was limited to temperatures below $\approx 700°C$ in order to keep defect formation and interdiffusion in the SiGe layer low. Both depletion- and enhancement-mode devices were fabricated, the latter by thinning the region underneath the gate by reactive ion etching prior to gate deposition.

The first *p*-MODFETs reported (Pearsall and Bean, 1986) showed well-

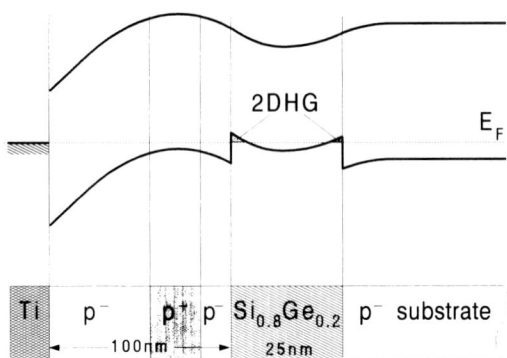

FIG. 40. Band diagram of the first p-channel MODFETs. The 2DHG is confined to a 250-Å-thick $Si_{0.8}Ge_{0.2}$ layer, which is separated from the doped Si layer by an undoped spacer layer. The MODFET is controlled by a Schottky gate (Pearsall and Bean *IEEE Electron. Device Lett.* **EDL-7**, 308, © 1986 IEEE).

behaved characteristics. Except for the incomplete pinch-off behavior, which was attributed to faulty surface passivation, the characteristics are quite comparable to those of state-of-the-art p-MOSFETs. Maximum transconductance values in excess of 3 mS/mm for depletion- as well as enhancement-type devices were found for gate lengths around 2 μm. These values are limited by the relatively high parasitic lead resistances that originate from the low hole mobilities at room temperature and the relatively low 2DHG concentrations of only about 2.5×10^{11} cm^{-2}. It is expected that further improvements in device fabrication and optimized 2DHG densities will lead to superior room-temperature transconductance values. Such devices might at some time replace the p-MOSFETs in complementary MOSFET (CMOS) ICs, which, due to the low hole mobility in Si, occupy two to three times the area of their n-channel counterparts on the chip.

b. *n-Channel MODFET.* As was mentioned in Section V.B, molulation-doping effects for electrons can not be observed in the above situation of a commensurate, asymmetrically strained SiGe layer on a Si substrate, because of the negligible CB offset associated with that strain condition. Hence, a prerequisite for the n-channel MODFET is an adequate exploitation of the strain dependence of the CB lineup, which requires a strain-symmetrizing SiGe buffer layer. Following this route, an AEG group designed and fabricated the first n-channel MODFETs (Daembkes et al., 1985, 1986). A schematic cross-sectional view of this device is shown in Fig. 41. The active layers are grown on a partly relaxed $Si_{0.7}Ge_{0.3}$ buffer layer whose thickness and Ge content are chosen to result in the effective lattice constant of a free-standing $Si_{0.75}Ge_{0.25}$ alloy. The channel layer consisting of a nominally undoped Si layer of ≈ 200 Å thickness is grown on the SiGe buffer layer,

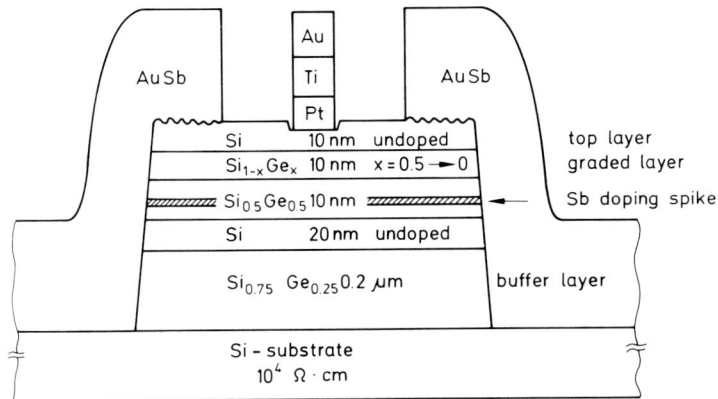

FIG. 41. Schematic cross-sectional view of an *n*-channel MODFET. Carrier transfer from the doped $Si_{0.5}Ge_{0.5}$ layer into the Si channel is made possible by a strain-symmetrizing $Si_{0.7}Ge_{0.3}$ buffer layer that induces Type-II band lineup (Daembkes, H., Herzog, H. J., Jorke, H., Kibbel, H., and Kasper, E., *IEEE Trans.* **ED-33**, 633, © 1985 IEEE).

followed by an *n*-doped $Si_{0.5}Ge_{0.5}$ layer. To further reduce ionized impurity scattering, the Sb doping is restricted to a narrow ($\approx$ 20-Å-thick) layer centered in the $\approx$ 100-Å-thick SiGe layer, hence leaving an undoped spacer of about 40 Å thickness next to the Si channel. A graded $Si_{1-x}Ge_x$ layer, with *x* continuously varying from $x = 0.5$ to $x = 0$, and a Si cap layer separate the doped layer from the Schottky gate. The purpose of the graded layer is to suppress a second electron channel between the Schottky gate and doping layer. The energetic variation of valence- and conduction-band edges are depicted in Fig. 42. Strain symmetrization leads to a CB offset at the SiGe/Si interface (see Section V.B.3), which separates the 2DEG from the ionized dopants in the adjacent SiGe doping layer.

Device fabrication utilized standard lithography and dry-etching techniques for mesa isolation. Special care was taken to minimize heat treatment during processing. For these reasons, the ohmic source and drain contacts were alloyed at 330°C for 30 seconds rather than ion implanted and annealed, as were the aforementioned *p*-MODFETs.

Several batches of *n*-channel MODFETs were fabricated, with 2D carrier concentrations in the range of 1 to $6 \cdot 10^{11} cm^{-2}$. All of the investigated samples led to devices with good characteristics that showed complete pinch-off behavior and distinct ohmic and saturation regions. Indications for traps were found in some of the devices, which showed a small amount of bias-dependent shift of the output characteristics toward higher currents. Excellent values of the extrinsic transconductance g_m of up to 50 mS/mm for 1.6 μm gate length were reported (Daembkes, 1988), despite the relatively high parasitic resistances induced by the simple alloying step used for the

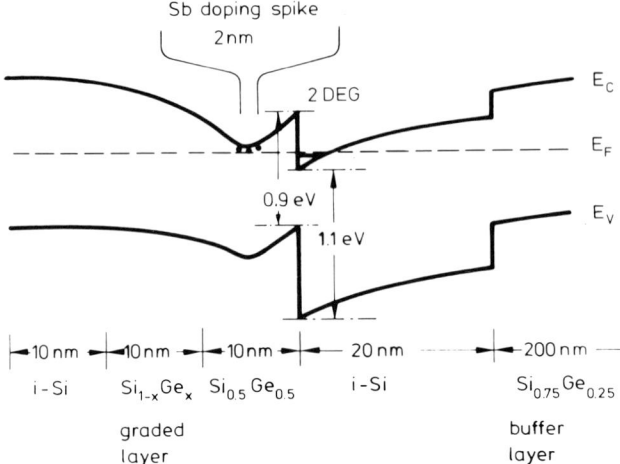

FIG. 42. Proposed band-edge variation of the n-MODFET depicted in Fig. 41. The purpose of the graded SiGe layer is to suppress a second conducting channel (Daembkes, H., Herzog, H.-J., Jorke, H., Kibbel, H., and Kasper, E., *IEEE Trans.* **ED-33**, 633, © 1985 IEEE).

source and drain contacts. It has been estimated from the measured parameters of the existing devices that g_m-values of about 180 mS/mm should be achievable upon further optimizing the structures with respect to 2D carrier density and gate length.

c. *Further Developments*. The successful, albeit not yet optimized implementation of p- as well as n-channel MODFET structures allows further developments of these concepts in several directions. Two promising examples shall be briefly discussed, even though only fragmentary or preliminary results are available at present.

The first example concerns a proposal about how to reduce the relatively high parasitic series resistances between the ohmic source and drain contacts and the gate-controlled channel region. This resistance contributes the dominating part to the limitation of the transconductance values observed in the p- and n-MODFETs. It is determined by the product of the 2D carrier density and the associated mobility and by the lateral distance of the ohmic source and drain contacts from the gate area. For a given geometry, this series resistance can be reduced by increasing the channel mobility and by increasing the 2D carrier density. As we have seen in Section V.B.1, these two parameters are not independent of each other. In addition, the maximum density of carriers transferable from the doped layer into the channel is limited by the respective band offset together with the width of the spacer layer (People *et al.*, 1985). For these reasons, the highest mobility values are achievable at relatively low 2D carrier densities, which do not allow exploitation of the density range controllable by the Schottky gate. However, the modulation-doped quantum-well concept allows an increase of the 2D carrier density without sacrificing the maximum mobility, by simply adding quantum wells with their associated doping layers. This means that in such a multiple-quantum-well MODFET structure, mobility and carrier density can independently be optimized, which is expected to result in increased saturation currents, higher transconductances, and reduced parasitic resistances. First steps in that direction were reported by Daembkes (1988).

The second example concerns the proposal for a CMODFET structure (Daembkes, 1988). In Si-based VLSIC technology, CMOS are widely used for digital applications. An optimized implementation of n- and p-MODFETs is expected to result in superior high-frequency characteristics. However, the designs for these two structures are presently based on different SiGe/Si layer sequences, which means a severe disadvantage for the integration of both types on the same chip. This problem could be overcome by the concept shown in Fig. 43. Both n- and p-channel MODFETs use the same undoped Si/Si$_{0.5}$Ge$_{0.5}$/Si layer sequence, which has to be grown on a symmetrizing SiGe buffer with the lattice constant of a relaxed Si$_{0.75}$Ge$_{0.25}$ alloy. With the symmetric strain distribution, the active layers arrange in a Type-II band lineup, which provides an electron quantum well in the Si layer and a hole quantum well in the SiGe layer (see Section V.B.3). Since the layers are undoped, both channels are not occupied by carriers. Nevertheless, MODFET operation is possible, if electrons or holes are injected from n^+- and p^+-source contacts, respectively. Hence, the device type is entirely defined by the doping type of the source/drain contact layers, which allows for a quite simple VSLI layout. Although the prospects of this concept are

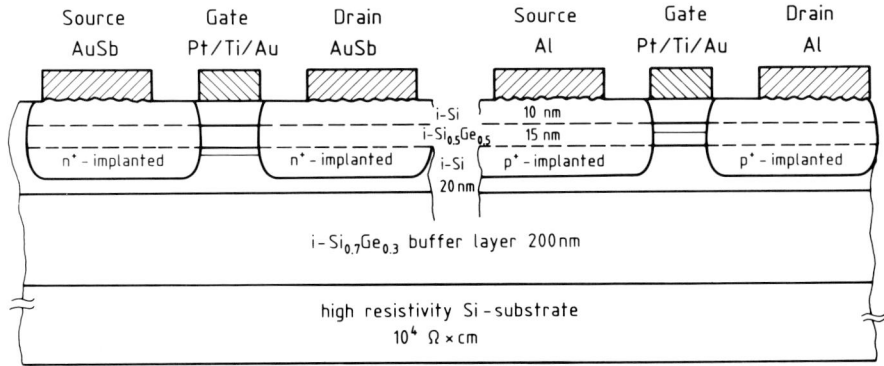

FIG. 43. Proposed structure of a complementary SiGe/Si MODFET. The active layers are undoped, i.e., transistor action relies on carrier injection from the implanted source contacts (Reprinted with permission from the publisher, The Electrochemical Society, from Daembkes, 1988).

quite exciting, its implementation will depend on the successful application of ion-implantation and annealing techniques. Such techniques have already been employed for the fabrication of the first p-MODFETs (Pearsall and Bean, 1986), but it remains to be clarified whether the annealing step is compatible with the presence of a metastable, symmetrizing SiGe buffer layer.

3. Two-Terminal Devices

Two-terminal devices are advantageous for high-frequency oscillator applications in the mm-wavelength range ($f > 30\,\text{GHz}$). Whereas at around 30 GHz GaAs impact-avalanche transit time (IMPATT) diodes gain importance, the higher frequency range of up to more than 100 GHz is dominated by Si homojunction IMPATT oscillators, which are by far the most powerful solid-state devices available at those frequencies (Luy, 1987). There are at present two trends for the further development of such two-terminal oscillators. The first research activity aims toward the monolithic integration of two-terminal devices with receiving and transmitting antenna networks based on the microstrip technique. Very promising results with 100 GHz Si mm-wave integrated circuits (SIMMWIC) were recently published by Büchler et al. (1988). The second trend, on which we will concentrate in the following, concerns the introduction of strained Si/SiGe layers to either improve certain properties of available IMPATT designs or to implement new concepts based on quantum effects. Preliminary results for both of these concepts are available, which are again an indication for the widespread activities in the Si/SiGe field.

a. *Heterojunction MITATT.* Transit time devices for high frequency oscillators require a mechanism for carrier injection into the drift zone. In the case of an IMPATT diode, injection is based on carrier multiplication in the avalanche zone of the device. The TUNETT concept uses a tunneling process for this purpose, which can be combined with avalanche injection to give the so-called MITATT (mixed tunneling avalanche transit time) device. The latter device type was recently implemented by Luy et al. (1988), who added a narrow (≈ 100-Å-thick) SiGe layer between the n- and p-type drift zones of a conventional double-drift IMPATT diode (Fig. 44). Under reverse bias, the SiGe well operates as a mixed tunneling and avalanche region of very-well-defined dimensions. Moderate output-power levels of about 20 mW at 75 GHz, and 1 mW at 100 GHz were obtained with the first, not yet optimized, device structures. The outstanding feature of these devices, however, were the noise properties, which are by far superior to those of comparable IMPATT diodes. Hence, the heterojunction MITATT appears as a promising concept for low-noise mm-wave applications.

b. *Resonant Tunneling Devices.* The principle of resonant carrier tunneling accompanies molecular-beam epitaxy since its early days (Tsu and Esaki,

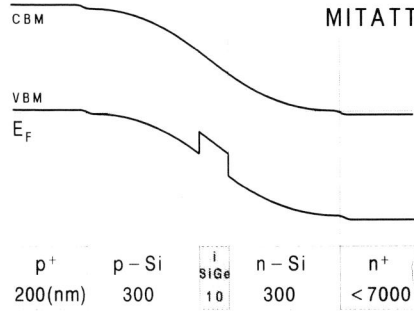

FIG. 44. Mixed-tunneling and avalanche-transit-time (MITATT) diode utilizing a thin SiGe layer. The layer sequence and the associated band-edge variation are depicted schematically. Layer thicknesses are not to scale (Reprinted with permission from IEE, from Luy et al., 1988).

1973; Chang et al., 1974). The basic idea is illustrated in Fig. 45 for the case of electrons. A narrow potential is well defined by two thin barriers, which are introduced by the band offset of appropriate heterolayers. Applying a bias voltage will induce a small current that becomes possible by quantum-mechanical tunneling of the carriers through the thin barriers. The voltage drop across the barriers and the well will cause the energy of the injected carriers to increase (relative to the averaged bottom of the quantum well) with increasing bias voltage. Thus, the overall current through the device reaches a maximum whenever the injection energy is in resonance with an energy eigenstate of the quantum well. Because the tunneling process conserves lateral momentum, only a very narrow range of the in-plane subband dispersion around $k_{\parallel} = 0$ is accessible, leading to sharp resonance peaks in the I-V characteristics of the device. Each of these peaks, whose number and spacing depend on the barrier height and well width, is followed by a region of negative differential resistance. It is clear that such a behavior can be exploited for high-frequency oscillators and detectors (Sollner et al., 1983).

The advantage of utilizing the quantum-mechanical tunneling mechanism is its extremely high intrinsic speed, which is basically given by the widths of

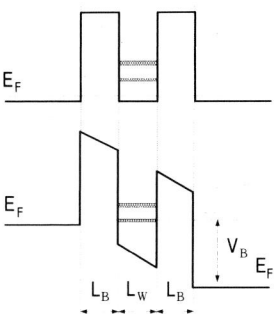

FIG. 45. Principle of resonant tunneling through a double-barrier heterostructure. A maximum in the I–V characteristics occurs whenever the energy of the injected carriers is in resonance with an eigenstate of the quantum well.

the energy states in the well region via Heisenberg's uncertainty principle. Switching times on the order of picoseconds are expected and were recently confirmed experimentally with AlAs/GaAs double-barrier structures (Whitaker et al., 1988).

Resonant tunneling structures based on double- and triple-barrier designs have reached a high standard in III-V systems, for example, with room-temperature operation reported at peak-to-valley current ratios of up to 30 (Broekaert et al., 1988). Liu et al. (1988) were the first to report resonant hole tunneling in the Si/SiGe system. The structures consisted of an asymmetrically strained SiGe well between two narrow Si barriers. Since the structure was grown on a Si substrate, SiGe cladding layers are required, which have to be thick enough to allow the thermalization of the holes injected from the substrate layer. For a well width of 34 Å, Liu and coworkers observe two regions of negative differential resistance (Fig. 46). Their best peak-to-valley ratio was reported to be larger than 3 at a temperature of 4.2 K. By varying the well width, Liu and coworkers hoped to identify the observed resonances in terms of bound states in the well. However, tunneling experiments in a magnetic field perpendicular to the layers revealed that most likely both heavy and light holes participate in the tunneling current (Liu et al., 1989). Because of the complicated valence-band structures of Si and SiGe, which in addition are affected by the built-in strain, no satisfying identification of the resonance peaks is presently available. This applies also to the hole tunneling structures of Rhee et al. (1988), which were grown either on a symmetrizing buffer layer or on bulk SiGe substrates. By these means, the strain distribution between well and barrier can be adjusted.

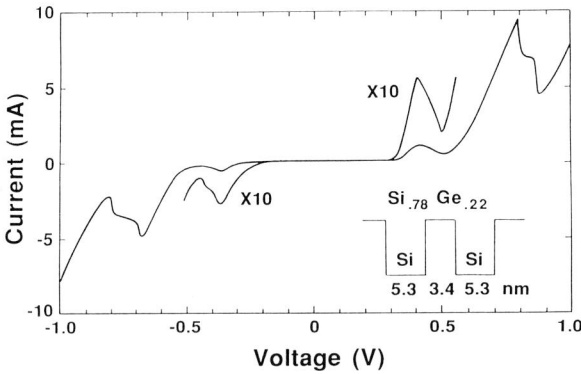

FIG. 46. I–V characteristics of a SiGe/Si double-barrier resonant tunneling structure measured at 4.2 K. Two areas of negative differential resistance, which appear almost symmetrically in forward and backward direction, are clearly resolved in the I–V curve. The layer sequence is depicted in the insert (Liu et al., 1988).

Although the peak-to-valley ratio is not yet competitive with the best III–V data reported, and room-temperature operation has still to be demonstrated, the results on hole tunneling in Si/SiGe structures appear quite promising when taking into account that they were derived from the very first devices grown.

With the demonstration of resonant tunneling, a novel concept for a transit-time device based on quantum-well injection has chances to become implemented in the SiGe/Si system. This so-called QWITT (quantum-well injection and transit-time) diode was proposed by Kesan *et al.* (1987) and consists basically of a resonant tunneling injector with an adjacent undoped drift zone. Such a device is expected to combine low-noise tunneling injection with superior high-frequency behavior. There are also speculations about higher efficiencies as compared with conventional transit-time devices that might become possible with the QWITT design. Figure 47 shows a preliminary low-temperature dc I-V characteristics of a SiGe/Si QWITT diode, which reveals, besides the expected asymmetry, several regions of negative differential resistance (Luy and Büchler, 1989).

4. *Heterobipolar Transistors*

The concept of the heterobipolar transistor (HBT) is almost as old as the bipolar transistor itself (Shockley, 1951; Kroemer, 1957). The basic idea is to use two semiconductors with different bandgaps for the emitter and base regions of a bipolar transistor, i.e., to introduce a heterojunction at the emitter/base interface. If the emitter is made from the material with the larger energy gap, injection of base majority carriers into the emitter is greatly reduced due to the potential barrier at the interface that is equal to the

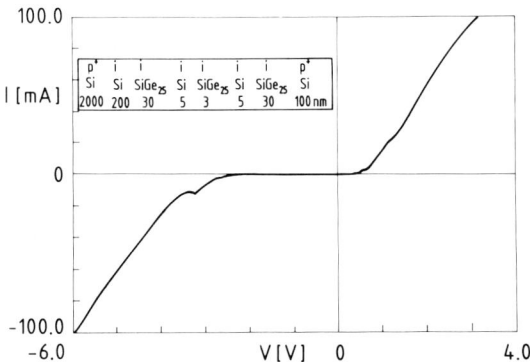

FIG. 47. Layer sequence and 77 K I–V characteristic of a *p*-type QWITT structure. Regions of negative differential resistance are clearly observed (Luy and Büchler, 1989).

respective band offset. As an example, a $n-p-n$ HBT with a graded Type-I heterotransition is depicted in Fig. 48. It can easily be shown (Kroemer, 1982) that under these conditions, the maximum current gain β_{max} depends exponentially on the bandgap difference δE_g;

$$\beta_{max} = I_n/I_p = N_e/P_b(w_e D_b)/(w_b D_e)\exp(\delta E_g/kT). \tag{40}$$

I_n and I_p denote the electron and hole currents in the base region, N_e is the (uniform) n-type emitter doping, P_b the p-type base doping level, w_e, w_b are the (effective) widths, D_e and D_b the diffusion constants in the emitter and base regions, respectively.

Hence, the wide-gap emitter results in a significant enhancement of β_{max}, provided the bandgap difference exceeds energies of a few kT. This additional degree of freedom can either be used to fabricate high-β transistors or, more important, to reduce the emitter-to-base doping ratio N_e/P_b, which has to be much larger than 1 in the conventional design. A high enough δE_g allows even a reversal of the usual doping ratio, i.e., a low-doped emitter combined with a highly doped base region becomes feasible. This combination of doping levels is especially favorable for high-frequency applications, since the low doping level of the emitter reduces the emitter-base capacitance, while the high doping level of the base region reduces the base-series resistance. Both effects contribute to an enhanced frequency response.

When the HBT concept was introduced, it was far ahead of the technological capabilities available at that time. MBE and other modern epitaxy techniques (e.g., metalorganic chemical vapor deposition (MOCVD)) allowed the implementation of this promising concept, and very impressive results have been achieved in the GaAlAs/GaAs system (Wang et al., 1986). The technological relevance of Si-based devices and the mature integration techniques available in this system have led to several attempts to fabricate Si-based HBTs. Initially, amorphous Si:H (Ghannam et al., 1984), and amorphous SiC:H (Sasaki et al., 1985) were used for the wide-gap emitter regions, but were not found optimum with respect to resistivity, interface state densities, and hydrogen desorption upon thermal treatment. Much more promising is the SiGe/Si system. As was discussed in Section V.B, the

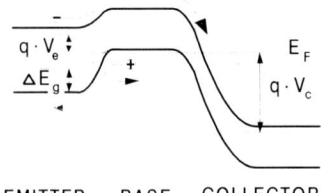

FIG. 48. Band diagram of a $n-p-n$ heterobipolar transistor with wide-gap emitter. Note the repelling effect for holes due to the additional potential step at the heterointerface. Recombination currents are not shown.

bandgap difference is almost completely adapted by the valence-band offset in the case of a compressively strained SiGe layer on an (unstrained) Si substrate. Hence, an n–p–n HBT can be implemented by sandwiching a narrow-gap SiGe base layer between two wide-gap Si layers that provide the emitter and collector functions. For fully exploiting the bandgap difference δE_g, the $Si_{1-x}Ge_x$ base has to be grown pseudomorphically, i.e., its thickness has to be kept below the critical thickness t_c corresponding to the chosen Ge content x and the maximum expected processing temperature.

Several groups are currently working on the fabrication of MBE-grown, strained SiGe/Si HBTs. First results were recently published, demonstrating the enormous potential of this concept. Iyer et al. (1988) report $Si_{1-x}Ge_x$/Si HBT characteristics for devices grown completely and without interruption by MBE. The Ge content of the base layer was varied between 6% and 12%, and a similar homotransistor ($x = 0$) was grown for comparison. The 1000-Å-thick base layers were Ga doped to a level of $\approx 10^{18}$ cm^{-3}. Care was exercised to keep the processing temperatures for the definition of the devices below 550°C. The HBTs were fully functional, showing improved current gains as compared with the reference homodevice. As expected from Eq. (40), the current gains were increasing with the Ge content of the base layer and also with decreasing temperature.

An advanced SiGe/Si HBT concept has been reported by Tatsumi et al. (1988), who adapted a collector-top design, which can have several advantages as compared with the conventional emitter-top design (Kroemer, 1982). Layer sequence and device definition were realized by two subsequent MBE processes utilizing so-called differential epitaxy (Kasper et al., 1987b). This technique is based on MBE growth on a Si substrate that is partly covered by a patterned oxide layer. Epitaxial growth is only achieved in the areas where the oxide is removed, whereas a polycrystalline Si layer is deposited on the oxidized area. By these means, lateral device definition becomes possible during the epitaxy itself. Several devices with pure Si and with $Si_{0.7}Ge_{0.3}$ base layers, and base-doping levels of 8×10^{17} and 5×10^{19} cm^{-3}, were fabricated. An HBO_2 source for boron doping was used (see Section V.A.3). Although the thickness of the base layers (3000 Å) was beyond the critical thickness, which is expected to degrade the performance due to formation of misfit dislocations at the interface, a direct comparison of the HBT and homodevices revealed quite significant improvements in current gain: A current gain of 15 (as compared with 1 for the homodevice) has been reported for the HBT with a base doping of 5×10^{19} cm^{-3}. The gain ratio for the 8×10^{17} cm^{-3} doped devices was 250 to 100.

The HBTs of Tatsumi and coworkers demonstrated transistor operation with base-doping levels of up to 5×10^{19} cm^{-3}. However, the doping ratio between emitter and base layer N_e/P_b (see Eq. (40) was still on the order of one, due to utilizing the highly n^+-doped substrate for the emitter function in their collector-top design. A significant improvement has recently been achieved by Schreiber et al. (1989). They report a dc-current gain of up to 17 for a base-doping level 20 times higher than the emitter doping. On similar structures with equal base and emitter doping, β_{max}-values in excess of 400 could be achieved (Narozny et al., 1989). The structures were grown by MBE, using an elemental boron doping source similar to the one described by Andrieu et al., (1988). This type of source does not imply restrictions on the growth temperature T_s (see Section V.A.4), hence allowing a reduction of T_s during growth of the 750 Å-thick $Si_{0.8}Ge_{0.2}$ base layer. Doping levels of up to $1 \times 10^{19}\,cm^{-3}$ were realized in the base layer while emitter doping was kept at $5 \times 10^{17}\,cm^{-3}$. The complete structure was grown in one MBE run and processed to a double-mesa structure. Selective etching, which uses the SiGe base layer as an etch-stop, and reactive ion etching were employed for the definition of the two mesas. A schematic view of the complete device is depicted in Fig. 49 together with typical output characteristics.

Even more striking results have been reported by Gibbons et al. (1988), who find β_{max}-values of up to 400 for HBTs with similar doping profiles as those of Schreiber et al. (1989). Their improved dc gains were partly due to an increased Ge content of 31% in the base layer concomitant with a reduced base width of 200 Å. It is interesting to note that the HBTs of Gibbons et al. (1988) were not grown by MBE but by limited reaction processing (LRP see Section V.4), which demonstrates the potential of this novel technique.

From the temperature dependence of the collector current, Gibbons et al. (1988) determined the effective band offset for several Ge contents in the base layer. They found good agreement with the values given by People (1985), which indicates that their HBTs completely exploit the valence-band offset.

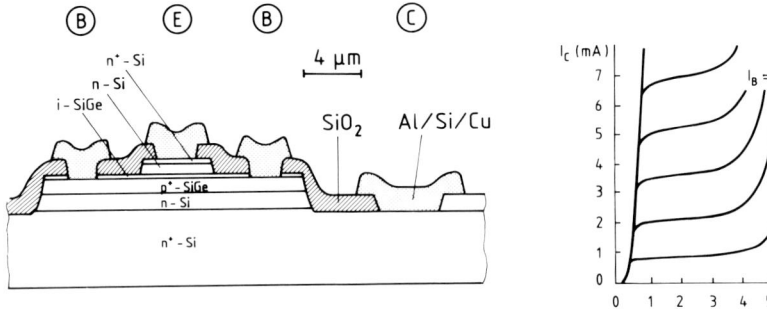

FIG. 49. Cross-sectional view and typical dc output characteristics of a SiGe/Si heterobipolar transistor with inverted emitter/base doping ratio. Current gains of about 17 are achieved at room temperature for collector currents $\geqslant 3\,mA$. (Reprinted with permission from IEE, from Schreiber et al., 1989).

Schreiber et al. (1989) estimated the band offset by using Eq. (40). By assuming a diffusivity ratio D_B/D_E of 2/3 and by neglecting recombination losses at the interface, they derived $\delta E_v = 124\,\text{meV}$ for their Ge content of 20%. This value is somewhat lower than theoretically expected (see Fig. 26 and Table VII in Section V.B.3), which is partly attributed to neglecting recombination losses.

These few examples can only give a faint impression of the extremely rapid development in the SiGe HBT field. Within about one year, the devices advanced from the mere demonstration of a transistor function to HBTs that fully exploit the available band offset. Especially the recent improvements leading to substantial current gains even with inverted emitter/base doping ratios promise a significant extension of the present high-frequency limits of Si-based bipolar technology. According to an estimate of Smith and Welbourne (1987), that limit is expected to shifted to values exceeding 40 GHz, i.e., into a frequency range that so far has been the domain of GaAs devices.

VI. Outlook

SiGe/Si superlattices are now available in a quality that is at least high enough for the realization of discrete device types. The tailoring of basic material properties in these artificial, man-made semiconductors has impressively been demonstrated. Increase of the room-temperature electron mobilities by more than a factor of five for doped SiGe/Si superlattices with a device-relevant average doping of $10^{18}\,\text{cm}^{-3}$ as compared to bulk Si of the same doping, new folded acoustic and confined optical phonon modes, and strong photoluminescence from ultrathin Si/Ge superlattices are just a few examples that mark the route to new silicon-based systems. Such systems can define the ultimate performance of future electronics if fabrication and device quality meet the demands of an industrial application. A rapid development of Si MBE equipment, and a rigorous understanding of misfit-dislocation nucleation and motion are necessary prerequisites. An especially spectacular result would be the realization of optoelectronic functions in a silicon-based system, which might become possible by the quasidirect character of a zone-folded indirect-energy gap. In the near future, however, the heterobipolar transistor will most likely play the dominating role concerning device applications. The recent progress that has been made in this field within a very short period of time demonstrates that the SiGe/Si system has the potential for fulfilling the industrial demands for such a Si-based high-speed device.

An exciting field for long-term research would be the investigation of metal/group-IV-semiconductor superlattices (e.g., $NiSi_2/Si$) or of insulator/group-IV-semiconductor superlattices (e.g., CaF_2/Si). Surprising properties and completely new applications can be expected. Techniques for the growth of group-IV superlattices with extremely large lattice mismatch ($>4\%$), and for metastable alloy superlattices are urgently needed to realize fascinating group-IV superlattice combinations outside the familiar Si/Ge system.

References

Abstreiter, G., Kotthaus, J. P., Koch, J. F., and Dorda, G. (1976). *Phys. Rev.* **B14**, 2480.
Abstreiter, G., Brugger, H., Wolf, T., Jorke, H., and Herzog, H. J. (1985). *Phys. Rev. Lett.* **54**, 2441.
Abstreiter, G., Brugger, H., Wolf, T., Jorke, H., and Herzog, H. J. (1986). *Surf. Sci.* **174**, 640.
Adam, E. N., and Holstein, T. D. (1959). *J. Phys. Chem. Solids* **10**, 254.
Ahlers, E. D., and Allen, F. G. (1987). *J. Appl. Phys.* **62**, 3190.
Aizaki, N. (1988). private communication.
Aizaki, N., and Tatsumi, T. (1985). *Proc. 17th. Int. Conf. Solid State Devices and Materials, J. Soc. Appl. Phys.*, Tokyo, p. 301.
Allen, F. G., Iyer, S. S., and Metzger, R. A. (1982). *Appl. Surf. Sci.* **11/12**, 517.
Alonso, M. I., Cardona, M., and Kanellis, G. (1989). *Solid State Comm.* **69**, 479.
Ando, T. (1976). *Phys. Rev.* **B13**, 3468.
Ando, T., Fowler, A. B., and Stern, F. (1982). *Rev. Mod. Phys.* **54**, 437.
Andrieu, S., Chroboczek, J. A., Campidelli, Y., André E., and Arnaud, d'Avitaya, F. (1988). *J. Vac. Sci. Technol.* **B6**, 835.
Baribeau, J.-M. (1988). *Appl. Phys. Lett.* **52**, 105.
Barnett, S. A., and Greene, J. E. (1985). *Surf. Sci.* **151**, 67.
Baselev, J. (1966). *Phys. Rev.* **143**, 695.
Bean, J. C. (1984). *J. Cryst. Growth* **70**, 444.
Bean, J. C. (1985). *Proc. 1st. Int. Symp. Silicon MBE* (J. C. Bean, ed.), Proc. Vol. **85-7**. Electrochemical Society, Pennington New Jersey, p. 337.
Bean, J. C. (1988). In "Silicon Molecular Beam Epitaxy" (E. Kasper and J. C. Bean, eds.). CRC Press, Boca Raton, Florida; chapter 11.
Bean, J. C., and Kasper, E. (1988). In "Silicon Molecular Beam Epitaxy" (E. Kasper and J. C. Bean, eds.). CRC Press, Boca Raton Florida; chapter 14.
Bean, J. C., and Leibenguth, R. E. (1988). *Proc. 2nd. Int. Symp. Silicon MBE* (J. C. Bean and L. J. Schowalter, eds.), Proc. Vol. **88-8**. Electrochemical Society Pennington, New Jersey.
Bean, J. C., and Sadowski, E. A. (1982). *J. Vac. Sci. Technol.* **20**, 137.
Bean, J. C., Feldman, L. C., Fiory, A. T., Nakakaro, S., and Robinson, I. K. (1984). *J. Vac. Sci. Technol.* **A2**, 436.
Becker, G. E., and Bean, J. C. (1977). *J. Appl. Phys.* **48**, 3395.
Bevk, J., Ourmazd, A., Feldman, L. C., Pearsall, T. P., Bonar, J. M., Davidson, B. A., and Mannaerts, J. P. (1987). *Appl. Phys. Lett.* **50**, 760.
Brantley, W. A. (1973). *J. Appl. Phys.* **44**, 534.
Braunstein, R., Moore, A. R., and Hermann, F. (1958). *Phys. Rev.* **109**, 695.
Brey, L., and Tejedor, C. (1987). *Phys. Rev. Lett.* **59**, 1022.
Broekaert, T. P. E., Lee, W., Fonstad, C. G. (1988). *Appl. Phys. Lett.* **53**, 1545.
Brugger, H., Abstreiter, G., Jorke, H., Herzog, H. J., and Kasper, E. (1986a). *Phys. Rev.* **B33**, 5928.

Brugger, H., Reiner, H., Abstreiter, G., Jorke, H., Herzog, H. J., and Kasper, E. (1986b). *Superlattices and Microstructures* **2**, 451.
Büchler, J., Kasper, E., Luy, J.-F., Russer, P., and Strohm, K. (1988). *Digest of Papers, IEEE Microwave and Millimeterwave Monolithic Circuits Symposium*, New York, p. 67.
Cardona, M. (1969). In "Modulation Spectroscopy," Solid State Physics Supplement **11**. Academic Press, New York.
Chang, L. L., Esaki, L., and Tsu, R. (1974). *Appl. Phys. Lett.* **24**, 593.
Chang, S. J., Huang, C. F., Kallel, M. A., Wang, K. L., Bowman, R. C., Jr., and Adams, P. M. (1988). *Appl. Phys. Lett.* **53**, 1835.
Colvard, C., Gant, T. A., Klein, M. V., Merlin, R., Fisher, R., Markoc, H., and Gossard, A. C. (1985). *Phys. Rev.* **B31**, 2080.
Daembkes, H. (1988). *Proc. 2nd. Int Symp. Silicon MBE* (J. C. Bean and L. J. Schowalter, eds.), Proc. Vol. **88-8**. Electrochemical Society Pennington, New Jersey, p. 15.
Daembkes, H., Herzog, H.-J., Jorke, H., Kibbel, H., and Kasper, E. (1985). *IEDM Technical Digest IEEE*, New York, p. 768.
Daembkes, H., Herzog, H.-J., Jorke, H., Kibbel, H., and Kasper, E. (1986). *IEEE Trans.* **ED-33**, 633.
Delescluse, P., Laviron, M., Chaplart, J., Delagebeaudeuf, D., and Linh, N. T. (1981). *Electron. Lett.* **17**, 342.
Dharma-wardana, M. W. C., Lockwood, D. J., Baribeau, J.-M., and Houghton, D. C. (1986). *Phys. Rev.* **B34**, 3034.
Dismukes, J. P., Ekstrom, L., and Paff, R. J. (1964). *J. Phys. Chem.* **68**, 3021.
Dodson, B. W. (1988). *Appl. Phys. Lett.* **53**, 394.
Dodson, B. W., and Taylor, P. A. (1986). *Appl. Phys. Lett.* **49**, 642.
Dodson, B. W., and Tsao, J. Y. (1987). *Appl. Phys. Lett.* **51**, 1325.
Drummond, J., Kopp, W., Keever, M., Morko, H., and Cho, A. Y. (1982). *J. Appl. Phys.* **53**, 1023.
Fowler, A. B., Fang, F. F., Howard, W. E., and Stiles, P. J. (1966). *Phys. Rev. Lett.* **16**, 901.
Friess, E., Brugger, H., Eberl, K., Krötz, G., and Abstreiter, G. (1989). *Solid State Comm.* **69**, 899.
Froyen, S., Wood, D. M., and Zunger, A. (1987). *Phys. Rev.* **B36**, 4547.
Froyen, S., Wood, D. M., and Zunger, A. (1988). *Phys. Rev.* **B37**, 6893.
Ghannam, M., Nijis, J., Mertens, R., and Dekeersmaecker, R. (1984). *IEDM Techn. Digest, IEEE*, New York, p. 746.
Gibbons, J. F., Gronet, C. M., and Williams, K. E. (1985). *Appl. Phys. Lett.* **47**, 721.
Gibbons, J. F., King, C. A., Hoyt, J. L., Noble, D. B., Gronet, C. M., Scott, M. P., Rosner, S. J., Reid, G., Laderman, S., Nauka, K., Turner, J., and Kamins, T. I. (1988). *IEDM Tech. Digest, IEEE*, New York, p. 566.
Gronet, C. M., King, C. A., Opyd, W., Gibbons, J. F., Wilson, S. D., and Hull, R. (1987). *J. Appl. Phys.* **61**, 2407.
Gnutzmann, U., and Clausecker, K. (1974). *Appl. Phys.* **3**, 9.
Gogol, C. A., and Cipro, C. (1985). *Proc. 1st. Int. Symp. Silicon MBE*, (J. C. Bean, ed.), Proc. Vol. **85-7**. Electrochemical Society Pennington, New Jersey, p. 415.
Hamers, R. J., Tromp, R. M., and Demuth, J. E. (1988). *Phys. Rev.* **B34**, 5343.
Hansen, M., and Anderko, K. (1958). "Constitution of Binary Alloys," 2d ed. McGraw-Hill, New York.
Herzog, H.-J. (1984). Unpublished.
Hirth, J. P., and Lothe, J. (1968). "Theory of Dislocations." McGraw-Hill, New York.
Hoyt, J. L., King, C. A., Noble, D. B., Gronet, C. M., Gibbons, J. F., Scott, M. P., Laderman, S. S., Rosner, S. J., Nauka, K., Turner, J., and Kamins, T. I. (1990). *Thin Solid Films* **184**, 93.
Hybertsen, M. S., and Schlüter, M. (1987). *Phys. Rev.* **B36**, 9683.
Hybertsen, M. S., Schlüter, M., People, R., Jackson, S. A., Lang, D. V., Pearsall, T. P., Bean, J. C., Vanderburg, J. M., and Bevk, J. (1988a). *Phys. Rev.* **B37**, 10195.

Hybertsen, M. S., Friedel, P., and Schlüter, M. (1988b). *Proc. 19th. Int. Conf. Phys. Semicond.*, (W. Zawadzki, ed.), Warsaw, Poland.
Itho, T., Shinomura, K., and Kojima, H. (1985). *Proc. 1st. Int. Symp. Silicon MBE* (J. C. Bean, ed.), Proc. Vol. **85-7**, Electrochemical Society Pennington, New Jersey, p. 158.
Iyer, S. S., Metzger, R. A., and Allen, F. G. (1981). *J. Appl. Phys.* **52**, 5608.
Iyer, S. S., Patton, G. L., Delage, S. L., Tiwari, S., and Stork, J. M. C. (1888). *Proc. 2nd. Int. Symp. Silicon MBE* (J. C. Bean and L. J. Schowalter, ed.), Proc. Vol. **88-8**, Electrochemical Society Pennington, New Jersey, p. 114.
Jackson, S. A., and People, R. (1986). In *Materials Research Society Symposium Proceedings* (J. M. Gibson, G. C. Osbourn, and R. M. Tromp, eds.). Materials Research Society, Pittsburgh, Pennsylvania, Vol. **56**, p. 365.
Jorke, H. (1988). *Surf. Sci.* **193**, 569.
Jorke, H., and Herzog, H.-J. (1985). *Proc. 1st. Int. Symp. Silicon MBE* (J. C. Bean, ed.), Proc. Vol. **85-7**, Electrochemical Society Pennington, New Jersey, p. 352.
Jorke, H., and Kibbel, H. (1985). *Proc. 1st. Int. Symp. Silicon MBE* (J. C. Bean, ed.), Proc. Vol. **85-7**, Electrochemical Society Pennington, New Jersey, p. 194.
Jorke, H., and Sawodny, M. (1989). Unpublished.
Jorke, H., Herzog, H.-J., Kasper, E., and Kibbel, H. (1987). *J. Crystal Growth* **81**, 440.
Jorke, H., Casel, A., Kibbel, H., and Herzog, H.-J. (1989a). *J. Electrochem. Soc.* **135**, 254.
Jorke, H., Kibbel, H., Schäffler, F., Casel, A., Herzog, H.-J., and Kasper, E. (1989b). *J. Cryst. Growth* **95**, 484.
Kasper, J. S. (1969). *Trans. American Cryst. Assoc.* **5**, 1.
Kasper, E. (1986). *Surf. Sci.* **174**, 630.
Kasper, E. (1987). Unpublished.
Kasper, E., and Bean, J. C. eds. (1988). "Silicon Molecular Beam Epitaxy." CRC Press, Boca Raton, Florida, (two volumes).
Kasper, E., and Herzog, H. J. (1977). *Thin Solid Films* **44**, 357.
Kasper, E., Herzog, H. J., and Kibbel, H. (1975). *Appl. Phys.* **8**, 190.
Kasper, E., Herzog, H.-J., Dämbkes, H., and Abstreiter, G. (1986). *Mat. Res. Soc. Symp. Proc.* **56**, 347.
Kasper, E., Herzog, H.-J., Jorke, H., and Abstreiter, G. (1987a). *Superlattices and Microstructures* **3**, 141.
Kasper, E., Herzog, H. J., and Wörner, K. (1987b). *J. Cryst. Growth* **81**, 458.
Kasper, E., Kibbel, H., Schäffler, F. (1988a). *Proc. 2nd. Int. Symp. Silicon MBE* (J. C. Bean and L. J. Schowalter, eds.), Proc. Vol. **88-8**. Electrochemical Society Pennington, New Jersey, p. 590.
Kasper, E., Kibbel, H., Jorke, H., Brugger, H., Friess, E., and Abstreiter, G. (1988b). *Phys. Rev.* **B38**, 3599.
Kastalsky, A., Luryi, S., Bean, J. C., and Sheng, T. T. (1985). *Proc. 1st. Int. Symp. Silicon MBE* (J. C. Bean, ed.), Proc. Vol. **85-7**. Electrochemical Society Pennington, New Jersey, p. 406.
Kelly, A. and Groves, G. W. (1970). "Crystallography and Crystal Defects." Longman, London.
Kesan, V. P., Neikirk, D. P., Streetman, B. G., and Blakey, P. A. (1987). *IEEE, Electron. Device Lett.* **EDL-8**, 129.
Kibbel, H. (1987). Unpublished.
Kleiner, W. H., and Roth, L. M. (1959). *Phys. Rev. Lett.* **2**, 334.
Klitzing, K. von, Landwehr, G., and Dorda, G. (1974). *Solid State Comm.* **14**, 387.
Knall, J., Hasan, M.-A., Sundgren, J.-E., Rockett, A., Markert, L., and Greene, J. E. (1888). *Proc. 2nd. Int. Symp. Silicon MBE* (J. C. Bean and L. J. Schowalter, eds.), Proc. Vol. **88-8**, Electrochemical Society Pennington, New Jersey, p. 470.
Kohn, W., and Sham, L. J. (1965). *Phys. Rev.* **140**, A1133.
Kroemer, H. (1957). *Proc. IRE* **45**, 1535.

Kroemer, H. (1982). *Proc. IEEE* **70**, 13.
Kubiak, R. A. A., Leong, W. Y., and Parker, E. H. C. (1984). *Appl. Phys. Lett.* **44**, 878.
Kubiak, R. A. A., Leong, W. Y., and Parker, E. H. C. (1985a). *J. Vac. Sci. Technol.* **B3**, 592.
Kubiak, R. A. A., Leong, W. Y., and Parker, E. H. C. (1985b). *Proc. 1st. Int. Symp. Silicon MBE* (J. C. Bean, ed.), Proc. Vol. **85-7**. Electrochemical Society Pennington, New Jersey, p. 230.
Kubiak, R. A. A., Leong, W. Y., and Parker, E. H. C. (1985c). *Proc. 1st. Int. Symp. Silicon MBE* (J. C. Bean, ed.), Proc. Vol. **85-7**. Electrochemical Society Pennington, New Jersey, p. 169.
Kubiak, R. A. A., Newstead, S. M., Leong, W. Y., Houghton, R., Parker, E. H. C., and Whall, T. E. (1987). *Appl. Phys.* **A42**, 197.
Landolt-Börnstein, (1982). "Numerical Data and Functional Relationships in Science and Technology." N. S., Group III, Vol. **17a**, Springer-Verlag, Berlin.
Lang, D. V., People, R., Bean, J. C., and Sergent, A. M. (1985). *Appl. Phys. Lett.* **47**, 1333.
Liu, H. C., Landheer, D., Buchanan, M., and Houghton, D. C. (1988). *Appl. Phys. Lett.* **52**, 1809.
Liu, H. C., Landheer, D., Buchanan, M., and Houghton, D. C., D'Iorio, M., and Kechang, S. (1989). *Superlattices and Microstructures* **5**, 213.
Lockwood, D. J., Dharma-wardana, M. W. C., Baribeau, J.-M., and Houghton, D. C. (1987). *Phys. Rev.* **B35**, 2243.
Luryi, S., Kastalsky, A., and Bean, J. C. (1984). *IEEE Trans. Electron Devices* **ED-31**, 1135.
Luryi, S., Pearsall, T. P., Temkin, H., and Bean, J. C. (1986). *IEEE Trans. Electron. Device Lett.* **EDL-7**, 104.
Luy, J.-F. (1987). *IEEE Trans.* **ED-34**, 1084.
Luy, J.-F., and Büchler, J. (1989). *MIOP: Proc. 4th Conf. Ultra High Frequency Technology*, Network, Hagenburg, FRG, paper 3A.2.
Luy, J.-F., Jorke, H., Kibbel, H., Casel, A., and Kasper, E. (1988). *Electron. Lett.* **24**, 1386.
Maguire, H. (1987. Private communication.
Mahowald, P. H., List, R. S., Spicer, W. E., Woicik, J., and Pianetta, P. (1985). *J. Vac. Sci. Technol.* **B3**, 1252.
Manasevit, H. M., Gergis, I. S., and Jones, A. B. (1982). *Appl. Phys. Lett.* **41**, 464.
Margaritondo, G., Katnani, A. D., Stoffel, N. G., Daniels, R. R., and Zhao, Te-Xiu (1982). *Solid State Comm.* **43**, 163.
Martin, J. Y., Goldstein, L., Glas, F., and Quillec, M. (1986). *Surf. Sci.* **174**, 586.
Matthews, J. W., ed. (1975). "Epitaxial Growth." Academic Press, New York.
Matthews, J. W., and Blakeslee, A. E. (1974). *J. Cryst. Growth* **27**, 118.
Metzger, R. A., and Allen, F. G. (1984). *J. Appl. Phys.* **55**, 931.
Myerson, B. S. (1986). *Appl. Phys. Lett.* **48**, 797.
Nabarro, F. R. N. (1967). "Theory of Crystal Dislocations." Clarendon Press, Oxford.
Narozny, P., Dämbkes, H., Kibbel, H., and Kasper, E. (1989). *IEEE Trans. Electron. Devices* **ED-36**, 2363.
Ni, W.-X., Knall, J., and Hansson, G. V. (1987). *Phys. Rev.* **B36**, 7744.
Ni, W.-X., Knall, J., and Hansson, G. V. (1888). *Proc. 2nd. Int. Symp. Silicon MBE* (J. C. Bean and L. J. Schowalter, eds.), Proc. Vol. **88-8**. Electrochemical Society Pennington, New Jersey, p. 68.
Osbourn, G. C. (1982). *J. Appl. Phys.* **53**, 1985.
Osbourn, G. C. (1983). *Phys. Rev.* **B27**, 5126.
Ostrom, R. M., and Allen, F. G. (1986). *Appl. Phys. Lett.* **48**, 221.
Ostrom, R. M., Allen, F. G., and Vasudev, R. K. (1988). *Proc. 2nd. Int. Symp. Silicon MBE* (J. C. Bean and L. J. Schowalter, eds.), Proc. Vol. **88-8**, Electrochemical Society Pennington, New Jersey, p. 85.
Ota, Y. (1980). *J. Appl. Phys.* **51**, 1102.
Pearsall, T. P., and Bean, J. C. (1986). *IEEE Electron. Device Lett.* **EDL-7**, 308.
Pearsall, T. P., Bean, J. C., People, R., and Fiory, A. T. (1985). *Proc. 1st. Int. Symp. Silicon MBE* (J. C. Bean, ed.), Proc. Vol. **85-7**. Electrochemical Society Pennington, New Jersey, p. 400.

Pearsall, T. P., Temkin, H., Bean, J. C., and Luryi, S. (1986a). *IEEE Electron. Device Letters* **EDL-7**, 330.
Pearsall, T. P., Pollak, F. H., Bean, J. C., and Hull, R. (1986b). *Phys. Rev.* **B33**, 6821.
Pearsall, T. P., Bevk, J., Feldman, L. C., Bonar, J. M., Mannaerts, J. P., and Ourmazd, A. (1987). *Phys. Rev. Lett.* **58**, 729.
People, R. (1985). *Phys. Rev.* **B32**, 1405.
People, R., and Bean, J. C. (1985). *Appl. Phys. Lett.* **47**, 332; Erratum: *Appl. Phys. Lett.* **49**, 229 (1986).
People, R., and Bean, J. C. (1986). *Appl. Phys. Lett.* **48**, 538.
People, R., Jackson, S. A. (1987). *Phys. Rev.* **B36**, 1310.
People, R., Bean, J. C., Lang, D. V., Sergent, A. M., Störmer, H. L., Wecht, K. W., Lynch, R. T., and Baldwin, K. (1984). *Appl. Phys. Lett.* **45**, 1231.
People, R., Bean, J. C., and Lang, D. V. (1985). *J. Vac. Sci. Technol.* **A3**, 846.
Rhee, S. S., Park, J. S., Karunasiri, R. P. G., Ye, Q., and Wang, K. L. (1988). *Appl. Phys. Lett.* **53**, 204.
Rich, D. H., Samsavar, A., Miller, T., Lin, H. F., Chiang, T.-C., Sundgren, J.-E., and Greene, J. E. (1987). *Phys. Rev. Lett.* **58**, 579.
Rytov, S. M. (1956). *Akust. Zh.* **2**, 71 (Sov. Phys. Acoust. **2**, 68).
Sakomoto, K., Sakamoto, T., Nagao, S., Hashiguchi, G., Kuniyoshi, K., and Bando, Y. (1986). *Japan J. Appl. Phys.* **26**, L666.
Sakamoto, T., Kawai, N. J., Nakagawa, T., Ohta, K., and Kojima, T. (1985). *Appl. Phys. Lett.* **47**, 617.
Sakamoto, T., Kawai, N. J., Nakagawa, T., Ohta, K., and Kojima, T. (1986). *Surf. Sci.* **174**, 651.
Sakamoto, T., Sakamoto, K., Hashiguchi, G., Takahashi, N., Nagao, S., Kuniyoshi, K., and Miki, K. (1988). *Proc. 2nd. Int. Symp. Silicon MBE* (J. C. Bean and L. J. Schowalter, eds.), Proc. Vol. **88-8**. Electrochemical Society Pennington, New Jersey, p. 285.
Sasaki, K., Furukawa, S., and Rahman, M. M. (1985). *IEDM Techn. Digest, IEEE*, New York, p. 294.
Schäffler, F., and Jorke, H. (1989). *Thin Solid Films* **184**, 75.
Schreiber, H.-U., Bosch, B. G., Kasper, E., and Kibbel, H. (1989). *Electron. Lett.* **25**, 185.
Schrieffer, J. R. (1975). In "Semiconductor Surface Physics." University of Pennsylvania Press, Philadelphia, Pennsylvania.
Seraphin, B. O. (1972). In "Semiconductors and Semimetals" (R. K. Willardson and A. E. Beer, eds.). Academic Press, New Jersey, Vol. 9, p. 1.
Shockley, W. (1951). U.S. Patent No. 2,569,347.
Smith, C., and Welbourne, A. D. (1987). *Proc. IEEE Bipolar Circuits and Techn.* Minneapolis, Minnesota.
Sollner, T. C. L. G., Goodhue, W. G., Tannenwald, P. E., Parker, C. D., and Peck, D. D. (1983). *Appl. Phys. Lett.* **43**, 588.
Solomon, P. M. (1982). *Proc. IEEE* **70**, 489.
Stern, F. (1972). *Phys. Rev.* **B5**, 4891.
Streit, D., Metzger, R. A., and Allen, F. G. (1984). In "VLSI Science and Technology 1984" (K. E. Bean and G. A. Rozgonyi, eds.). Electrochemical Society, Pennington, New Jersey.
Tatsumi, T., Hirayama, H., and Aizaki, N. (1988a). *Proc. 2nd. Int. Symp. Silicon MBE* (J. C. Bean and L. J. Schowalter eds.), Proc. Vol. **88-8**. Electrochemical Society Pennington, New Jersey, p. 430.
Tatsumi, T., Hirayama, H., and Aizaki, N. (1988b). *Appl. Phys. Lett.* **52**, 895.
Temkin, H., Pearsall, T. P., Bean, J. C., Logan, R. A., and Luryi, S. (1986a). *Appl. Phys. Lett.* **48**, 963.

Temkin, H., Antreasyan, A., Olsson, N. A., Pearsall, T. P., and Bean, J. C. (1986b). *Appl. Phys. Lett.* **49**, 809.
Tsao, J. Y., Dodson, B. W., Picraux, S. T., and Cornelison, D. M. (1987). *Phys. Rev. Lett.* **59**, 2455.
Tsu, R., and Esaki, L. (1973). *Appl. Phys. Lett.* **22**, 562.
Tsui, D. C., Störmer, H. L., and Gossard, A. C. (1982). *Phys. Rev. Lett.* **48**, 1559.
Van de Walle, C. G., and Martin R. M. (1985). *J. Vac. Sci. Technol.* **B3**, 1256.
Van de Walle, C. G., and Martin, R. M. (1986). *Phys. Rev.* **B34**, 5621.
Van der Merwe, J. H. (1972). *Surf. Sci.* **31**, 198.
Wang, K. C., Asbeck, P. M., Chang, M. F., Miller, D. L., and Sullivan, G. J. (1986). *GaAs IC Symp. Tech. Dig., IEEE*, New York, p. 159.
Whitaker, J. F., Mourou, G. A., Sollner, T. C. L. G., and Goodhue, W. D. (1988). *Appl. Phys. Lett.* **53**, 385.
Wong, K. B., Jaros, M., Morrison, I., and Hagon, J. P. (1988). *Phys. Rev. Lett.* **60**, 2221.
Yen, M. Y., Madhukar, A., Lewis, B. F., Fernandez, R., Eng, L., and Grunthaner, F. J. (1986). *Surf. Sci.* **174**, 606.
Zachai, R., Friess, E., Abstreiter, G., Kasper, E., and Kibbel, H. (1988). *Proc. 19th. Int. Conf. Phys. Semicond.,* (W. Zawadzky, ed.), Warsaw, Poland.
Zachai, R., Eberl, K., Abstreiter, G., Kasper, E., and Kibbel, H. (1990). *Phys. Rev. Lett.* **64**, 1055.
Zeller, Ch., and Abstreiter, G. (1986). *Z. Physik* **B64**, 137.

CHAPTER 5

Molecular-Beam Epitaxy of IV-VI Compound Heterojunctions and Superlattices

Dale L. Partin

PHYSICS DEPARTMENT
GENERAL MOTORS RESEARCH LABORATORIES
WARREN, MICHIGAN

I.	Introduction	311
II.	Growth by Molecular-Beam Epitaxy	313
	A. Fundamental Material Properties	313
	B. MBE Effusion-Cell Sources.	314
	C. Alloying the IV-VI Compounds	315
	D. In Situ Evaluation Utilizing RHEED	322
	E. Conductivity Control via Incorporation of Impurity Species	323
III.	Use of Lattice-Mismatched Substrates	325
	A. GaAs and Other III-V Compound Substrates	325
	B. Alkaline-Earth Fluoride-Compound Substrates.	325
	C. Group-IV Substrates	326
	D. IV-VI Substrates	327
	E. Other Substrates	328
IV.	Strained-Layer Superlattices and Heterojunctions	329
	A. Energy Band Offsets	329
	B. Magnetic-Semiconductor Superlattices	332
V.	Conclusions	333
	Acknowledgments	333
	References	333

I. Introduction

Major new advances in epitaxial growth techniques have contributed to an improved understanding of the properties IV-VI materials and their alloys. The creation of sophisticated heterojunctions, quantum wells, and strained-layer superlattices can be attributed to their successful epitaxial growth by molecular-beam epitaxy (MBE). The objective of the chapter is to provide an overview of the research activity in the area of IV-VI-based structures and to highlight the significant achievements.

The discussion of IV-VI compounds is mainly concerned with the properties of PbS, PbSe, PbTe, alloys of these with Sn and Ge chalcogenides, and with other elements (of group IIa, IIb, Mn, and of certain rare-earth metals). Since oxides are excluded in the discussion, the interested reader is referred to Dalven (1969) for the properties of PbO. Chalcogenides of the other group-IV elements containing C and Si are also excluded.

The IV-VI materials of interest center on the lead chalcogenide, or lead-salt compounds, which are named after their rock-salt (face-centered cubic) crystal structure. These materials have been in use for over a century (Braun, 1874), since single crystals were found native in sufficiently high purity for fabrication of simple diodes. These binary compounds have energy bandgaps of 0.2–0.4 eV and thus are useful for infrared detectors and sources. So far, the lead-salt materials have excelled in the device area of diode lasers, where they are used for spectroscopic studies and other applications in the 2.5–30 μm wavelength range (Fig. 1). There has been a long period of development since the first lead-salt diode laser was made (Butler et al., 1964), and they appear to be inherently able to lase at higher temperatures than III-V or II-VI compound devices for wavelengths longer than about 3 μm (Partin, 1988). For other reviews of lead-salt diode lasers, see Preier (1979) or Horikoshi (1985).

Polycrystalline thin films of lead-salt materials have long been of interest for infrared detector applications (Dalven, 1969). These films were prepared by vacuum deposition or by precipitation from aqueous solution. Recently, higher-performance devices have been fabricated from epitaxial films consisting of a PbTe-doping superlattice grown on a lattice-mismatched BaF_2 substrate (Jantsch et al., 1987); and photovoltaic PbTe detectors have been

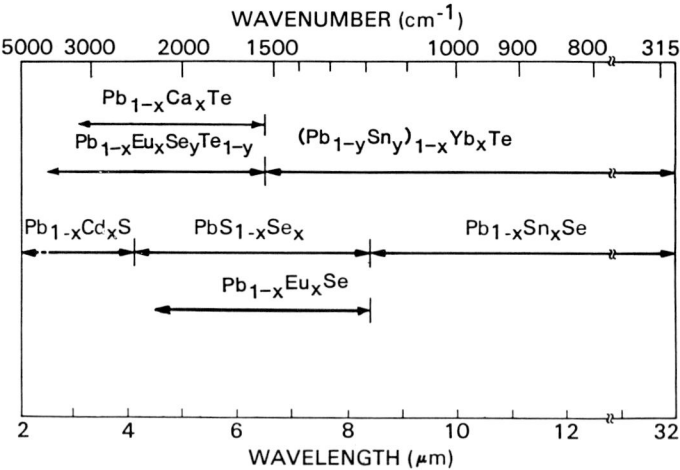

FIG. 1. Wavelength coverage of lead-salt materials systems.

formed on similarly prepared substrates (Maissen et al., 1988). Longer-wavelength photovoltaic detectors have been grown on PbSnTe substrates with a lattice-matched PbSnTe/PbSeTe structure (Rotter et al., 1982). Whereas most very-high-quality mid-infrared detectors are made from II-VI compounds such as HgCdTe, difficulties in controlling the Hg concentration could lead to additional consideration of PbSnTe or PbSnSe because of the relative ease of controlling the Sn concentration, which in turn controls the energy bandgap, and hence a detector's cutoff wavelength.

From a materials-oriented point of view, much of the recent research activities have centered on epitaxial growth techniques, especially MBE, metalorganic chemical vapor deposition (MOCVD), and hot-wall epitaxy (HWE), development of new alloys with higher energy bandgaps, exploration of magnetically active alloys, growth on lattice-mismatched substrates, and the growth of superlattices that are lattice mismatched or have components with different crystal structures. The goal of this chapter is to highlight significant advances in these exciting new areas of materials science, physics, and engineering.

II. Growth by Molecular-Beam Epitaxy

A. Fundamental Material Properties

The properties of the lead-salt compounds and their classicalloys have been extensively reviewed (Dalven, 1969; Harman and Melngailis, 1974; Kressel and Butler, 1977; Holloway and Walpole, 1979; Preier, 1979; Horikoshi, 1985). Some of their properties that distinguish them from II-VI compounds will be briefly discussed. As previously mentioned, the lead-salt compounds have the rock-salt crystal structure, as do some of the other IV-VI compounds. Thus, they have (100) cleavage planes and tend to grow in the (100) growth orientation, although they can also be grown in the (111) orientation (as will be described later). The thermal-expansion coefficients of the lead salts are $2 \times 10^{-5} °C^{-1}$, which is much larger than that of most commonly available single-crystal substrates, such as Si ($2.6 \times 10^{-6} °C^{-1}$) or GaAs ($6.86 \times 10^{-6} °C^{-1}$), and which tends to present some difficulties.

The lead salts are direct-energy-gap semiconductors with the band extrema at the four equivalent L points of the Brillouin zone. Addition of Sn to PbTe or PbSe causes the bandgap to decrease and go through zero at certain concentrations, with L-band inversion at higher Sn concentrations. In contrast with most semiconductors, the energy bandgap increases with increasing temperature below this inversion point (i.e., for PbTe- or PbSe-rich alloys). The static dielectric constant and index of refraction are relatively large, with the index of refraction being a fairly strong function of alloy composition, temperature, and carrier concentration (a typical value for PbTe is 6). If a PbTe diode laser is emitting photons at a free-space wavelength of 5μm, the optical cavity need only have a thickness of the order of 1 μm, which is a convenient value for MBE growth.

Because the conduction and valence bands at the L points are near mirror images of each other, the electron and hole effective masses and mobilities are nearly equal. These carrier mobilities are typically around $1000\,\text{cm}^2/\text{Vs}$ at room temperature and are relatively unchanged with increasing carrier concentrations because of the shielding effects of the large dielectric constant ($\varepsilon_0 \sim 400$ for PbTe at 300 K). Other aspects of carrier mobility and doping will be discussed later. Finally, the lead salts become mechanically softer as the chalcogen becomes heavier, making pure bulk PbTe rather difficult to work with. This problem, however, is greatly reduced by alloying with another major constituent, or by doping.

B. MBE Effusion-Cell Sources

The IV–VI compounds and certain alloys sublime predominantly as molecules, $MX(s) \rightarrow MX(g)$. The strong intramolecular IV–VI bonding (and the lack of it in II–VI compounds) arises from the "inert-electron-pair" effect, in which outer s-shell electrons in the group-IV metal are precluded from participating in the intramolecular bond. Greater overlap of the outer p-shell electrons of both group-IV and VI atoms (Farrow, 1985) is induced. Small concentrations of dissociated vapor species (M, X, X_2, M_2X, M_2X_2) were also observed in this study. Thus, the use of binary lead salts for source compounds is convenient, since they sublime at moderate temperatures (~ 500–$600°C$) and allow near-stoichiometric growth. In one case, if the PbTe, which is used as MBE source material, is initially somewhat nonstoichiometric, its composition will shift during sublimation to give congruently evaporating material (Smorodina and Tsuranov, 1980). These materials are not extremely reactive with vacuum-system components, althouth elemental sulfur vapor appears to react rapidly with copper at ~ 100–$200°C$.

The rare-earth (e.g., Eu, Yb) and alkaline-earth (Ca, Sr) chalcogenides sublime at a temperature of $\sim 2000°C$, such that elemental sources are required for evaporation of these metals. They typically sublime at $\sim 500°C$, and either graphite or pyrolytic boron nitride crucibles may be used. The formation of a visible surface oxide/hydroxide film on these elements, which occurs upon exposure to air during MBE system reloading, is not a significant problem. After outgassing H_2O in an ultrahigh-vacuum (UHV) environment, only elemental metallic species are observed with a mass spectrometer. Metal atoms readily diffuse through any remaining oxide crust; no oxide sublimes at normal operating temperatures. Caution is needed when removing partially sublimed Yb source material from an UHV system. If at least 15 minutes are not allowed for an adequately thick oxide to form, touching the Yb with metallic tweezers will cause the Yb to explosively oxidize if a static electrical discharge occurs.

Other alloys may be grown by using binary or ternary sources. GeTe source material may be used to grow PbGeTe (Partin, 1982), and either PbSnTe or SnTe may be used to grow PbSnTe alloys. In one study, a single source oven with separate compartments was used, whereby controlling the aperture over each binary provided excellent control of the vapor composition (Holloway and Walpole, 1979).

Elemental bismuth or bismuth chalcogenides (e.g., Bi_2Te_3) are nearly universally used as a MBE source for n-type dopant species. Tl_2Te is commonly used as a p-dopant source for PbTe, since elemental Tl oxidizes very rapidly in air. Elemental silver appears to be the preferred dopant for MBE-grown PbSe. Finally, elemental Te or Se source ovens are used for stoichiometry control. By using these sources, it is generally possible to maintain vacuums in the low to mid 10^{-9} torr range during the MBE growth of PbTe- and PbSe-based alloys.

C. ALLOYING THE IV–VI COMPOUNDS

The most important classical alloys of the lead-salt semiconductors are PbSnTe and PbSnSe, where using Sn to reduce the energy bandgap makes it possible to make long-wavelength devices. As discussed in the section on MBE effusion-cell sources (Section II.B), it is possible to grow these alloys with well-controlled composition by using either a single PbSnTe source oven or separate PbTe and SnTe sources, either incorporated into the same oven or into separate ovens. Similar techniques may be used for PbSnSe, and in all cases, the sources sublime mainly as molecules. An additional source oven is often used containing Te or Se to adjust the stoichiometry. Although Hg has commonly been used to reduce the energy bandgap of CdTe, it has received relatively little use in reducing the energy bandgap of lead-salt semiconductors. Evaluation of the properties of PbHgTe bulk crystals has so far shown that they are inferior to the properties of PbSnTe crystals (Vujatovic et al., 1982; Jain et al., 1984).

Another important classic lead-salt alloy is $PbS_{1-x}Se_x$ (Dalven, 1969; Harman and Melngailis, 1974). The lattice constant, carrier effective masses, and energy gap are nearly linear functions of composition. This material may be grown by MBE by using binary source ovens (PbS and PbSe) with a Se oven to adjust stoichiometry (Preier, 1979), where the use of elemental sulfur is avoided because of its high vapor pressure. However, elemental sulfur was used recently by one group in the growth of PbCdSSe (Koguchi et al., 1987).

Whereas its MBE background vacuum was originally near 10^{-10} torr, it was in the 10^{-8} torr range after sulfur was introduced into the system.

Many elements have been used to increase the energy bandgap of lead-salt semiconductors, either for shorter-wavelength devices or for heterojunctions with longer-wavelength devices. In early work, Cd and Ge were used to increase the energy gap of bulk crystals of PbTe and PbS (Kressel and Butler, 1977). In the case of PbTe, at a MBE growth temperature of 400°C (near the high end of the range), the estimated solid solubility of Cd is small (2%) (Rosenberg et al., 1964). Cd is also a fast-diffusing donor in PbTe (Silberg et al., 1981). In PbS, the solubility of Cd is much larger, and dopant action of Cd is not a problem. Recently, PbCdSSe has been successfully grown by MBE (Koguchi et al., 1987). In PbTe, it may be possible to overcome the donor action of fast-diffusing, low-concentration Cd with a p-type dopant such as Tl. However, the solid solubility of Cd is so low that the best approach to achieving higher energy gaps in the PbCdTe material system may be the use of CdTe/PbTe heterojunctions or superlattices (Yoshino et al., 1987; Takagi et al., 1985; Yamada et al., 1986); this possibility will be further discussed later.

In the case of Ge, the solid solubility of GeTe in PbTe is about 7% at 400°C (Hohnke et al., 1972). An additional difficulty is that the vapor pressure of GeTe is so high that GeTe tends to re-evaporate from the sample surface during MBE growth (Partin, 1982). An additional issue is that PbGeTe undergoes a ferroelectric phase transition from a cubic to a rhombohedral crystal structure as the temperature is lowered and the GeTe concentration is increased (Katayama and Murase, 1980).

Mn chalcogenides are also potentially useful for increasing the bandgap of lead-salt semiconductors. The solid solubility of Mn (2%) at 400°C is quite small (Vanyarkho et al., 1970). However, it was recently shown that $Pb_{1-x}Mn_xTe$, with x up to 0.10, can be grown by MBE (Yoshino et al., 1987). More Mn-rich compositions were phase separated. The previous solubility studies suggest that MBE-grown compositions in the range $0.02 < x < 0.10$ are metastable. We note that PbMnS has also recently been studied (Karczewski et al., 1981).

In the aforementioned high-energy-gap alloys, a lead-chalcogenide binary compound is alloyed with a Cd, Ge, or Mn chalcogenide, which has a different crystal structure, resulting in a limited solid solubility. This has led to the consideration of rare-earth or alkaline-earth chalcogenides for high-energy-gap lead-chalcogenide alloys, since all of these materials (except Be and Mg chalcogenides) share the rock-salt (face-centered cubic) crystal structure. One thus hopes to obtain completely miscible materials and a very wide range of energy bandgaps (Figs. 2 and 3).

The original work in the area of alloying was done with MBE-grown PbEuTe and PbYbTe (Suryanarayanan and Paparoditis, 1968, 1969; Paparoditis and Suryanarayanan, 1972). Their preliminary conclusion was that

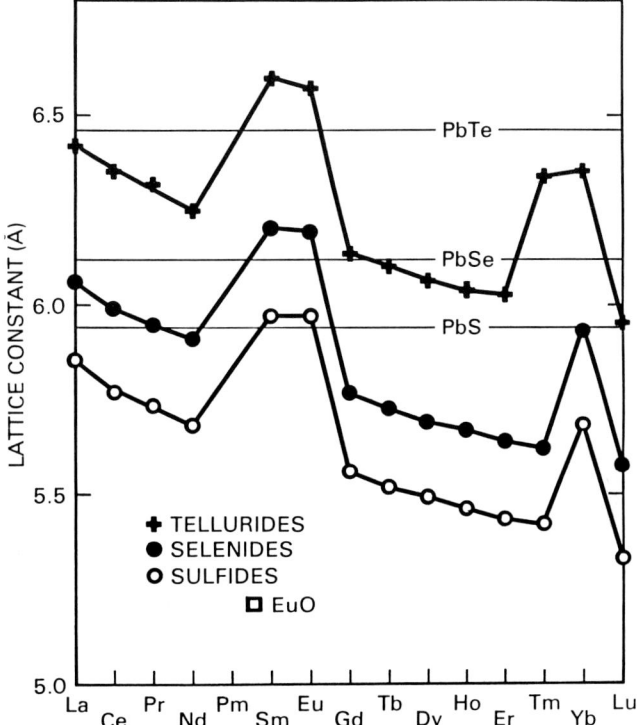

FIG. 2. Lattice constants of the rare-earth monochalcogenide compounds in the face-centered cubic crystal structure.

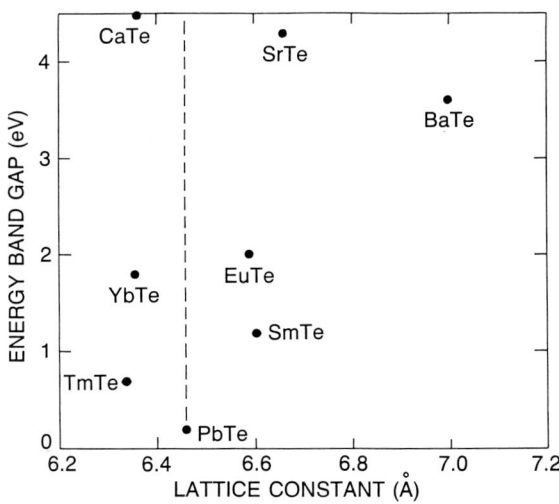

FIG. 3. Energy bandgap versus lattice constant for several face-centered cubic telluride compounds and for CdTe.

complete solid solubility was attained in both of these material systems. More recently, detailed studies of MBE-grown PbYbTe and PbSnYbTe were made (Partin, 1983a,b). It was found that Yb incorporates well into PbTe under Te-rich MBE growth conditions and has a negligible diffusion coefficient. The lattice constant of $Pb_{1-x}Yb_xTe$ does not obey Vegard's law but varies by a nearly negligible amount, being given (in Å) by $6.460 - 0.15x$, for $x \leqslant 0.42$. The energy gap varies as $dE_g/dx = 3.3 \, eV$ for $x < 0.04$. However, Yb has a valence instability, undergoing a transition $Yb^{+2} \to Yb^{+3} + e^-$, thus introducing a deep donor level into the energy bandgap and thereby making it impossible to obtain heavily p-type material, which is required for diode-laser heterostructures. However, this donor level moves deep into the valence band when Sn is added, making it possible to obtain p-type $(Pb_{1-x}Sn_x)_{0.97}Yb_{0.03}Te$ for $x > 0.10$. Thus, this material is suitable for relatively long-wavelength diode lasers. More recently, the energy-band structure of PbYbTe has been explored (Sayadian et al., 1986). Diode lasers have been successfully fabricated in the PbSnYbTe materials system, which required adequate passivation from atmospheric effects (Partin, 1985a).

PbEuTe has recently been studied very extensively. It incorporates well into PbTe and has a small diffusion coefficient under Te-rich MBE growth conditions (Partin, 1984). The lattice constant varies strongly and nonlinearly with composition, which led to the use of $Pb_{1-x}Eu_xSe_yTe_{1-y}$ for lattice-matched heterojunctions with PbTe. The energy gap of $Pb_{1-x}Eu_xTe$ increases strongly with x, varying as $dE_g/dx = 3.5 \, eV$ for $x \leqslant 0.044$. Furthermore, the material could be doped heavily n-type or p-type for a reasonable range of Eu concentrations, and the material was found to be highly stable under atmospheric conditions. These properties indicated that it was very suitable for diode-laser applications, and justified further study. Since this material system has only recently been extensively explored, a brief review of some of its salient properties is given here.

Many of the material properties of PbEuTe were studied in films grown by HWE (Krost et al., 1985). The properties found earlier in MBE-grown PbEuTe (Partin, 1984) were generally confirmed, although the study of lattice properties and energy-band structure was more detailed. More recent transmission electron microscopy (TEM) studies of MBE-grown $Pb_{1-x}Eu_xTe$ have shown that the material is single phase for $0 \leqslant x \leqslant 0.25$ and $0.75 \leqslant x \leqslant 1$, but undergoes a type of spinoidal decomposition at intermediate compositions, resulting in natural superlattices with a periodicity of ~ 18 Å (Salamanca-Young et al., 1988a,b). However, this instability appears to be caused by the strain that exists between the PbEuTe film and the BaF_2 substrate. Current studies indicate that $Pb_{0.5}Eu_{0.5}Te$, grown on more nearly lattice-matched PbTe, is single phase (Salamanca-Young et al., 1989).

Interface abruptness in PbTe/PbEuTe heterojunctions was initially characterized by Auger electron spectroscopy (AES) depth profiling and found to be < 100 Å. More recent TEM studies have shown that the interfaces are no more than three monolayers in width. This has made it possible to make single-quantum-well PbTe/PbEuSeTe diode lasers with well widths as small as 200 Å. The quantum shift in diode-laser-emission wavelength was in reasonable agreement with calculations, indicating minimal interdiffusion. A variety of optical and transport studies in PbTe/PbEuSeTe quantum-well structures with well widths as small as 50 Å are generally consistent with these determinations of extremely abrupt interfaces. In another study, it was found that "spikes" in the Eu-concentration depth profile can occur if the MBE growth process is temporarily interrupted. This effect is observed at the upper part of the usual MBE-substrate temperature range (> 340°C) because of the much higher vapor pressure of PbTe compared with EuTe. However, at 300°C, the PbEuTe surface composition was stable and remained free of background vacuum contaminants (e.g., C, O) for extended periods at $\sim 10^{-8}$ torr (Partin and Thrush, 1984).

The energy bandgap of $Pb_{1-x}Eu_xTe$ has been measured by photoluminescence and absorption techniques (Krost *et al.*, 1985) and from the emission wavelengths of diode lasers with PbEuSeTe active regions. The energy gap increases rapidly up to $x = 0.05$, and then much more slowly. Various studies have given information on minority-carrier properties. From photoluminescence measurements, the presence of deep energy levels has been observed (Goltsos *et al.*, 1985a,b). Double-heterojunction $Pb_{1-x}Eu_xSe_yTe_{1-y}$ diode lasers, with active region compositions up to $x = 0.046$, $y = 0.054$, have been fabricated that lase at wavelengths as short as 2.6 μm, indicating that subgap carrier recombination is not dominant at high carrier-injection levels (Partin and Thrush, 1984).

$Pb_{1-x}Eu_xSe$ has now been grown by MBE and characterized (Norton *et al.*, 1985; Norton and Tacke, 1986; Rosman *et al.*, 1987). Good film morphologies were obtained for $x \leq 0.20$, provided that Se-rich growth conditions were used. The energy bandgap is a very rapidly increasing function of x. Transport properties of the material were studied, and high-performance diode lasers using PbSnEuSe, as well as PbEuSe, were fabricated. Both lattice-matched PbEuSeTe/PbTe and near-lattice-matched PbEuSe/PbSe materials systems are now able to produce diode lasers operating up to ~ 175 K CW (Partin, 1988; Tacke *et al.*, 1989). Recently, the PbEuS material system has also begun to be explored (Ishida *et al.*, 1988).

The alkaline-earth chalcogenides have even higher energy bandgaps than the rare-earth monochalcogenides and still share the rock-salt crystal structure of the lead-salt compounds. Furthermore, the extreme stability of the alkaline-earth metals in the +2 valence suggests that they would be free of the valence instabilities of the Eu and Yb chalcogenides that result in deep energy levels. The exceptions are found in PbMgTe and PbMgSe alloys, which have a limited range of solid solubility caused by a phase transition (Rogers and Crocker, 1971; Rogers, 1971). Whereas bulk crystals of these materials were extensively characterized, MBE growth of these materials has not been explored to our knowledge.

The first attempt to grow lead-alkaline-earth chalcogenide alloys by MBE was in the PbCaS and PbSrS alloy systems (Holloway and Jesion, 1982). Complete immiscibility was found over the entire composition range for both of these alloy systems, with energy bandgaps ranging from 0.4 to 4.6 eV in the case of PbSrS. Furthermore, the instability of alkaline-earth chalcogenides upon exposure to air was not a problem as long as a small amount of PbS (e.g., a few percentages) was incorporated into the lattice (H. Holloway, private communication). The energy-band structure of SrS-rich compositions has been further characterized (Tamor et al., 1983). More recently, the properties of MBE-grown lead-alkaline-earth tellurides have been explored. In the case of $Pb_{1-x}Ca_xTe$, AES depth profiling was used to show good incorporation and minimal interdiffusion at heterojunctions. The energy gap increases rapidly with Ca concentration, and the lattice constant was nearly independent of x for $0 \leqslant x \leqslant 0.15$. The index of refraction decreased monotonically in this composition range, but increased anomalously at $x = 0.26$, possibly indicating multiphase material (Partin et al., 1986). $Pb_{1-x}Sr_xTe$ was also grown, and the lattice constant was found to obey Vegard's law up to $x = 0.15$, beyond which, phase segregation was observed (Partin et al., 1987). $Pb_{1-x}Ba_xTe$ was briefly explored, and it was found that the solid-solubility range was limited to $x < 0.04$. The energy bandgap versus composition for PbEuTe, PbCaTe, and PbSrTe is shown in Fig. 4.

Another general feature of these alloys is that alloy-disorder scattering typically becomes very significant for alloy additions of 10–20% to lead chalcogenides. The carrier mobilities are thus greatly reduced, leading to a series resistance that can significantly affect the performance of high-current-density devices such as diode lasers (Fig. 5). Thus, what has made the PbEuSeTe material system so attractive for these applications is the very rapid increase of bandgap with Eu concentration, so that relatively little Eu is needed for a given change in bandgap. Relatively little Se is then needed for lattice matching to PbTe, which is also important, since Se additions also reduce carrier mobilities (Hohnke and Hurley, 1976). This suggests that PbCaTe might be a competing material system, since little or no other additions are necessary for lattice matching to PbTe. Currently, PbSrSe is under study for diode-laser applications, since PbSrSe has somewhat higher carrier mobility than PbEuSe of the same energy bandgap (M. Tacke, private communication).

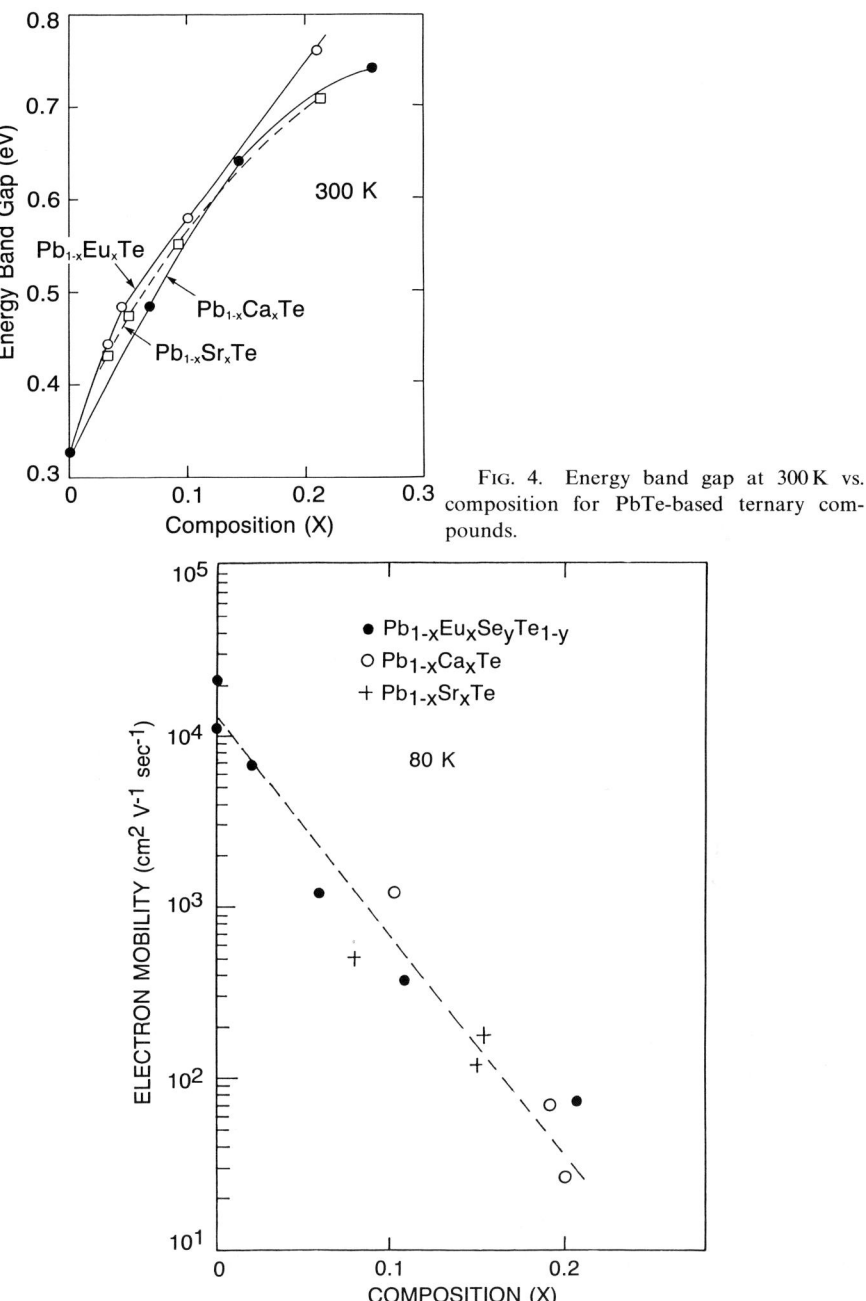

FIG. 4. Energy band gap at 300 K vs. composition for PbTe-based ternary compounds.

FIG. 5. Electron mobility of PbTe-based alloys versus composition at 80 K.

Some general features of MBE growth of lead chalcogenides may now be summarized. In each case, an elemental source must be used for alloying rare-earth or alkaline-earth metal, since the vapor pressures of their chalcogenides are very low. A separate source of elemental chalcogenide is then required to maintain stoichiometry, and a third source oven is used for the lead chalcogenide. The alloy composition is determined by the ratio of lead-chalcogenide flux from one oven to that of the alloying metal from another oven. In the case of PbEuTe, where precise lattice matching to PbTe has been desired, a lattice-matched quaternary, PbEuSeTe, has been grown. In this case, both Se and Te could be supplied from elemental sources. However, this may give rise to uncertainties in controlling the Se-to-Te ratio incorporated into the film, since both of these elements have a very low efficiency of incorporation into the film once the crystal-stoichiometry condition has been satisfied. This has led to the use of an elemental Te source for stoichiometry control and a binary PbSe source for lattice matching, such that PbSe, which almost completely evaporates as a molecule, is incorporated into the film with near-unity efficiency. A slight excess of Te_2 flux compared to Eu flux (typically 10–20% excess on an atomic basis) is then used to assure a stoichiometric crystal.

D. IN SITU EVALUATION UTILIZING RHEED

There have been only a few published reports of the use of reflection high-energy electron diffraction (RHEED) during MBE growth of IV-VI compounds. In the earliest study, the orientation of PbTe films deposited upon $LiNbO_3$ substrates was determined (Semiletov et al., 1980). Secondly, observations of PbTe/CdTe and PbTe/GaAs heterostructures have been made (Yoshino et al., 1987). This permitted a study of the conditions under which high-quality heterojunctions could be grown. More recently, RHEED intensity oscillations were used to show that the growth rate of PbTe and of PbSe is independent of substrate temperature over a wide range (Fuchs et al., 1988). In general, surface reconstruction is not observed on PbTe or PbSe surfaces, decreasing the utility of RHEED. Furthermore, observation of the initial stages of epitaxy of PbTe on BaF_2, a commonly used substrate, is precluded by the fact that a BaF_2 surface rapidly degrades with electron-beam exposure. In general, however, the growth of new heterojunctions and materials (e.g., PbEuSe), often under lattice-mismatched conditions, would benefit greatly from detailed RHEED studies. Also, fundamental issues in the growth of PbTe and PbSe need to be addressed. For example, after a PbTe molecule adsorbs onto a PbTe surface, do the Pb and Te atoms dissociate and migrate separately to the growth front, or do they migrate as a bound pair? In the latter case, the ability to grow high-quality PbTe by MBE at substrate temperatures as low as 260°C implies a significant lateral diffusion coefficient for such an anomalously large atomic species. Also, the diffusion mechanism for PbTe grown from separate PbTe source ovens might be different. Growth of alloys in which a metal monomer (such as Eu) is deposited at significant concentrations (with Te_2 and PbTe) might also have very different surface properties. There is a clear need for more work in this area.

Whereas RHEED is normally used in conjunction with MBE growth to gain information on surface structure, scanning tunneling microscopy (STM) is another technique addressing surface-atom positions. In a very recent paper, the surface of (100)-oriented PbS was studied by STM near the point where an 8 KeV krypton ion struck the surface (Wilson et al., 1988). The unreconstructed surface normally seen with RHEED was confirmed for regions remote from the impact site.

E. CONDUCTIVITY CONTROL VIA INCORPORATION OF IMPURITY SPECIES

A major difference between IV-VI and II-VI compounds is in the degree of difficulty of doping. With IV-VI compounds, lead vacancies can readily dope the material p-type to the 10^{18}–10^{19} cm^{-3} range, and nonmetal vacancies can dope the material n-type to a comparable range. This contrasts with the situation involving II-VI compounds, where it is usually difficult to heavily dope the material with vacancies, and where self-compensating defects often prevent one from doping a given compound both n-type and p-type. Controlling carrier concentrations with deviations from stoichiometry has not been used very much in IV-VI compounds in recent years because of the belief that impurity dopants would diffuse more slowly than vacancies (Walpole and Guldi, 1974). However, it has recently been shown that PbTe-doping superlattices can be grown by using vacancy dopants at temperatures as high as 350°C (Jantsch et al., 1987). This works at low to moderate carrier densities. However, at the high (10^{19} cm^{-3}) carrier densities needed for some regions of high-current-density devices such as diode lasers (Partin, 1988), impurity doping is still preferred. Long-term annealing has been used to prepare PbS bulk crystals with carrier densities as low as 5×10^{15} cm^{-3} (Harman and Strauss, 1976).

Impurity doping is often used to control carrier concentrations of lead-salt semiconductors, since many elements can be used as dopants, especially for n-type doping. However, most recent work has focussed on Tl (thallium) as a p-type dopant and Bi as a n-type dopant, since they were reported to have relatively low diffusion coefficients in PbSnTe and PbSnSe (Smith and Pickhardt, 1978). Also, these dopants are relatively unaffected by changes in crystal stoichiometry compared with others (Cu, Ag, In, As, and Sb) (Strauss, 1973). Smith and Pickhardt (1975) used elemental Tl as an MBE dopant, but it oxidizes very rapidly in air. This led some workers to use Tl_2Te, which is much more stable, and Bi_2Te_3, although Bi is reasonably stable in air (Partin, 1981, 1988). Also, according to Strauss (1973), the doping action of Tl and Bi in PbTe is readily understood from the fact that they both occupy Pb lattice sites. With Tl having one atomic number less than Pb, and Bi having one more, their respective acceptor and donor action is easily understood. Therefore, the use of tellurium compounds rather than elements as sources of these dopants helps maintain crystal stoichiometry and lessens the need for adding a large Te flux from a separate source (Smith and Pickhardt, 1978). The issue of dopant diffusion in PbTe has recently been re-examined for the case of Tl and Bi impurities (Partin et al., 1989). Whereas Bi appears to be very well controlled, Tl diffuses appreciably at high MBE growth temperatures and at high concentrations and surface segregates at very high ($\sim 10^{20}\,cm^{-3}$) concentrations.

Bi tends to work well as a n-type dopant even for high-energy bandgap alloys, whereas Tl generally begins to produce a deep acceptor level as the energy gap increases beyond a certain point, leading to low-temperature freeze-out of holes. This trend is observed for PbSrS (Holloway and Jesion, 1982), PbSrTe (Partin et al., 1987, PbEuSeTe, and to a somewhat lesser extent for PbCaTe (Partin et al., 1986). This problem of low-temperature freeze-out of holes in Tl-doped high-energy-gap alloys has led to the exploration of Ag as an alternative p-type dopant for PbSrS (Holloway and Jesion, 1982) and PbEuSe (Norton and Tacke, 1986), despite its relatively large diffusion coefficient. In both of these cases, low-temperature freeze-out of holes is still observed to some extent. In a recent paper on the MBE growth of PbCdSSe, Tl was used as a p-type dopant (Koguchi et al., 1987). Doping of high-energy-gap lead-salt alloys clearly needs more study.

An alternative donor that has been considered is Sb (Smith and Pickhardt, 1978), which does not incorporate well at high concentrations in PbTe or PbSnSe. Other possible donors are Ho, Dy, and Er (Partin, 1985b). However, these were shown to be ineffective in maintaining PbTe n-type under sufficiently Te-rich conditions. They also have large diffusion coefficients at low concentration. Sodium is an alternative acceptor that has been well characterized in bulk PbTe (Crocker and Dorning, 1968), but has not been used during MBE growth, possibly because of concern about its very high vapor pressure.

Thus, it is relatively easy to dope the Pb-salt high-energy-gap alloys heavily n-type, with Bi being the preferred dopant. Tl and Ag have been the preferred acceptors, with some carrier freeze-out for sufficiently high-energy-gap alloys as a remaining problem. The other significant problem is the difficulty of obtaining low carrier concentrations ($< 10^{17}\,cm^{-3}$) without long-term annealing.

III. Use of Lattice-Mismatched Substrates

A. GaAs and Other III-V Compound Substrates

There has recently been an attempt to grow PbTe on (100)-oriented GaAs (Yoshino et al., 1987). A mixture of (100)- and (111)-oriented PbTe was reported for most growth conditions. However, if a very thin film of PbTe was predeposited at 100°C, it underwent solid-phase (100)-oriented epitaxy while being raised to 300°C, and subsequent growth preserved the (100) orientation. An additional serious problem with PbTe grown on GaAs is the appearance of microcracks in films thicker than 4000 Å upon thermal cycling (Clemens et al., 1988). This is caused by the large difference in thermal expansion coefficients between PbTe and GaAs. It may be that use of intermediate alkaline-earth fluoride layers can alleviate this problem, as discussed in the following.

B. Alkaline-Earth Fluoride-Compound Substrates

The alkaline-earth fluorides (MF_2) have been one of the most commonly used substrates for the growth of IV-VI compounds since the pioneering studies of Holloway (Holloway and Walpole, 1979). Some data are given in Table I, which shows the attractiveness of the MF_2 compounds, especially BaF_2, compared with the alternative alkali-halide substrates (MX). The MF_2 compounds have negligible vapor pressures at typical MBE growth temperatures, whereas the MX compounds have relatively high vapor pressures,

TABLE I

Properties of the Lead Chalcogenides and Some Substrates

Compound	Lattice Const. (Å) at 300 K	Thermal Expn. Coef. ($K^{-1} \times 10^6$) near 300 K	Cleavage	p(torr) at 700 K
PbS	5.94	20	(100)	
PbSe	6.12	19	(100)	
PbTe	6.46	20	(100)	
NaCl	5.64	39	(100)	4.5×10^{-7}
NaBr	5.86	42	(100)	3.7×10^{-6}
NaI	6.46	45	(100)	4.0×10^{-5}
KCl	6.29	37	(100)	2.2×10^{-6}
KBr	6.59	38	(100)	1.1×10^{-5}
KI	7.05	40	(100)	3.9×10^{-5}
CaF_2	5.40	19	(111)	$\sim 6.0 \times 10^{-20}$
SrF_2	5.80	18	(111)	$\sim 1.0 \times 10^{-20}$
BaF_2	6.20	18	(111)	$\sim 3.0 \times 10^{-17}$

which may lead to difficulties with contamination of MBE-grown films with alkali metals, which are p-type dopants in IV-VI compounds. Also, the MF_2 compounds are closely thermal-expansion matched to the lead chalcogenides, which the MX compounds are not. A difficulty with the MF_2 compounds is that their natural cleavage plane is (111), which is not the preferred orientation for the lead-salt compounds. Thus, if contaminants are inadvertently present on the BaF_2 surface, PbTe will nucleate in a (100) orientation. BaF_2 can be cut and polished in the (100) BaF_2 orientation, but obtaining a polished (100) BaF_2 surface with adequate quality for epitaxy is not easily reproduced. Thus, cleaved, (111)-oriented BaF_2 has been the substrate of choice for many investigations, since it is commercially available, electrically insulating, and optically transparent over a wide wavelength range.

The lattice mismatch between BaF_2 and PbSe is about 1% (Table I), and many investigations of PbSe and PbEuSe have been made with this substrate (Norton and Tacke, 1986). PbS/PbSe heterojunctions recently have been grown on BaF_2 (Chu et al., 1987). PbTe has a 4% lattice mismatch with BaF_2; and all of its important alloys have been grown on BaF_2. The strains introduced into PbSnTe grown on BaF_2 have been well characterized (Fantner et al., 1984). Thermal cycling relieves strains in PbTe/BaF_2 and results in reduced carrier mobilities (Restorff et al., 1981). It is possible that annealing will improve the situation (Holloway and Logothetis, 1971).

In the case of $Pb_{1-x}Eu_xTe$, phase segregation was reported for films grown on BaF_2 by hot-wall epitaxy for $x > 0.05$ (Krost et al., 1985). In an initial study of MBE-grown $Pb_{1-x}Eu_xTe$ on BaF_2, only single-phase material was observed up to at least $x = 0.13$. However, detailed TEM studies of MBE-grown $Pb_{1-x}Eu_xSe_yTe_{1-y}$ have shown that the material does not nucleate well on BaF_2 for $x = 0.20$, $y = 0.15$. Since this composition has approximately the same lattice constant as PbTe, the problem is not caused by additional lattice mismatch from the BaF_2 substrate introduced by Eu. Additional TEM studies were done of $Pb_{1-x}Eu_xTe$ with higher x grown on BaF_2 upon which a PbTe layer was nucleated. This again allowed high-quality epitaxy to be obtained until $x > 0.25$, where natural superlattices were observed, as described earlier (Section II.C). As mentioned there, single-phase $Pb_{0.5}Eu_{0.5}Te$ on BaF_2 is obtained if an adequately thick PbEuTe buffer layer is first grown for strain relief. Thus, BaF_2 is a versatile substrate that has found extensive use in the growth of IV-VI compounds, despite a lattice mismatch of typically 4%.

C. GROUP-IV SUBSTRATES

There have not been many attempts to grow IV-VI compounds directly on Si because of the large difference in lattice constants and in thermal expansion coefficients. A much more promising approach makes use of the growth of a graded CaF_2-BaF_2 layer on (111)-oriented Si (Zogg, 1986, Wittmer et al., 1986). Here, a very thin CaF_2 layer is first nucleated on Si followed by a BaF_2 layer, with a graded composition layer sometimes present. CaF_2 has a 2% lattice mismatch from Si at the growth temperature of 700°C. Its presence greatly improves the quality of epitaxy of BaF_2, which is subsequently grown (mismatch 14% to Si). Furthermore, despite this large degree of lattice mismatch and a large thermal-expansion-coefficient mismatch to Si, the BaF_2 is found to be almost completely strain-relieved near room temperature, provided the BaF_2 layer is at least 2500 Å thick. These BaF_2/CaF_2 films on Si have been used as substrates for the growth of IV–VI-compound thin-film growth (Maissen et al., 1988; Zogg et al., 1986). PbSe, PbSnSe, and PbTe were all grown on such surfaces, and very high carrier mobilities were attained. Poor surface morphologies were obtained under Se-rich growth conditions but were absent for metal-rich growth conditions. PbTe and PbSnSe detectors fabricated on these substrates were nearly background-noise limited. Growth of CdTe on similarly graded fluoride layers on Si substrates has also been reported (Zogg and Blunier, 1986).

D. IV-VI SUBSTRATES

For many years it was believed that IV-VI-compound films could be grown on substrates to which they were lattice mismatched to some extent without paying too severe a penalty in materials or device properties. This belief was founded in part upon the observation that the IV-VI compounds have narrow bandgaps, small carrier effective masses, and large static dielectric constants, and therefore it was reasoned that defect energy levels associated with dislocations or point defects would tend to be very near the conduction- or valence-band edges (for donor and acceptor levels, respectively), or even in the bands. In the simplest model of shallow, hydrogenlike impurity levels (Milnes, 1973), the binding energy is proportional to carrier effective mass and inversely proportional to the static dielectric constant squared. This leads to binding energies of the order of μeV for IV-VI compounds. Such calculations are supported by the fact that, unlike the more conventional semiconductors such as Si, there is no carrier freeze-out in IV-VI compounds with conventional dopants (e.g., vacancies) at the lowest temperatures experimentally accessible. Optical absorption studies sometimes have shown that dopants introduce resonant levels within the bands. Thus, it was reasoned, in the absence of "deep" energy levels near the middle of the energy bandgap, minority-carrier properties (relevant to infrared sources and detectors) would not be significantly degraded. Furthermore, early success in making sensitive photoconductive lead-salt detectors from polycrystalline thin films deposited on foreign substrates seemed to support this view, although highly proprietary, poorly characterized techniques for passivating or "activating" grain boundaries in the films are often used to gain significant improvements in detector sensitivity (Harris, 1983). Early studies in natural crystals of PbS (galena) showed that the room-temperature lifetime is inversely proportional to the square of the majority-carrier density (Moss, 1953). This is typical of Auger recombination, which is the expected minority-carrier recombination mechanism for the case of a narrow-energy-gap semiconductor at moderately high temperature and carrier density. More recently, increasingly detailed and sophisticated studies of minority-carrier properties of high-quality lead-salt epitaxial layers have shown that Shockley–Read (deep-level) defect centers are often present (Schlicht *et al.*, 1978; Zogg *et al.*, 1982).

Of special relevance to the growth of IV-VI compounds on lattice-mismatched substrates is the introduction of dislocations and their properties. In early work, dislocations were found to reduce the minority-carrier lifetime in PbS bulk crystals (Scanlon, 1957). More recently, detailed TEM studies of PbS/PbSe, PbTe/PbSe, and SnTe/PbSe heterojunctions have been made (Matthews, 1961, 1971; Yagi *et al.*, 1971; Palatnik and Fedorenko, 1981; Samaras *et al.*, 1984). There is general agreement that there is no initial nucleation stage at the first stages of growth of the heterojunction. The growth proceeds in a pseudomorphic mode as originally described by Frank and Van der Merwe (1949). Beyond a critical thickness of the overlayer, misfit dislocations are introduced, and the lattice constant of the overgrown material relaxes to its bulk value. If contaminants are present at the heterojunction, the overlayer initially grows via a nucleation process (Yagi *et al.*, 1971). Despite the introduction of dislocations, optoelectronic IV-VI-compound devices have frequently been fabricated on lattice-mismatched lead-salt substrates, especially in the PbSSe and PbSnSe systems (Preier, 1979; Norton and Tacke, 1986). For recent MBE-grown diode-laser structures, PbTe and PbSe have been used (Rosman *et al.*, 1987; Partin, 1988; Tacke *et al.*, 1989). The softness of pure PbTe is addressed by adding Tl, a *p*-type dopant, at concentrations of 2×10^{19} cm^{-3}, which hardens the material. In the case of PbSe, adding a small amount of PbTe hardens the material, and makes it lattice match to PbEuSe epitaxial layers (K. H. Bachem, private communication).

E. OTHER SUBSTRATES

Historically, alkali-halide substrates have been widely used for IV-VI-compound epitaxial thin films, since they mainly have the rock-salt crystal structure as do the lead-salt compounds, and they have comparable lattice constants. They also cleave on (100) planes, which is the preferred growth orientation for IV–VI compounds. Some of this data is summarized in Table I.

Sodium chloride (100) has been used as a substrate for PbTe, PbSe, and PbS (Mojejko and Subotowicz, 1981; Yagi et al., 1971; Matthews, 1971). In general, (100)-oriented islands nucleate, which coalesce with increasing deposition. These films sometimes have relatively large dislocation densities. PbS layers grown simultaneously on NaCl and BaF_2 showed much higher carrier mobilities and lower carrier concentrations in the films grown on BaF_2 (Bleicher et al., 1977). Similar results have been obtained for PbTe films on these substrates (Lopez-Otero, 1975). It has been speculated that BaF_2 is a superior substrate for IV-VI-compound growth, since it is nonhygroscopic, whereas alkali-halide substrates tend to be more or less hygroscopic, depending on which one is under consideration (Holloway and Walpole, 1979).

Other substrates have been used for IV-VI-compound growth. $LiNbO_3$ has been evaluated, and high-quality PbTe films were obtained in some cases (Semiletov et al., 1980). Mica has also been used (Dawar et al., 1980). However, most recent work has been done on IV-VI or BaF_2 substrates, and one important future direction of IV-VI-compound research is the growth of $BaF_2/CaF_2/Si$ structures as substrates for IV-VI-compound work. This work potentially opens the way for growth on a high-thermal-conductivity, large-area, and widely available substrate. Infrared detector arrays or other devices made in IV-VI films could conceivable be integrated with control circuitry elsewhere on the Si substrate.

IV. Strained-Layer Superlattices and Heterojunctions

A. ENERGY BAND OFFSETS

Energy band offsets are generally rather difficult to measure accurately but are important for fundamental studies of heterojunctions and have practical device implications. Band offsets have been studied in less detail in IV-VI compounds than in some III-V compounds, and this study is made more difficult by the narrow energy bandgaps of the IV-VI compounds.

In PbTe/PbSnTe heterojunctions, one group maintains that Type-I' offsets occur in which the conduction-band edge of the relatively narrow energy gap PbSnTe is higher in energy than the conduction-band edge of PbTe (Ishida et al., 1985a; Isida et al., 1986). The other group finds that these heterojunctions are the normal Type I, in which both band edges of PbSnTe lie between those of PbTe (Pascher et al., 1986; Kriechbaum et al., 1988). Most recently, interdiffusion effects have been taken into account (Kriechbaum et al., 1989). Band offsets in this material system continue to be studied in order to resolve these disagreements.

In PbEuTe/PbTe or in lattice-matched PbEuSeTe/PbTe heterojunctions, all studies find Type-I band offsets. Most studies indicate that the conduction- and valence-band offsets are roughly equal (Partin, 1988), although nonequal offsets have also been reported (Kim et al., 1987). One experimental observation that was made since this latter investigation was done is that there is a high density of holes at an MBE-grown PbTe/BaF$_2$ interface (Partin et al., 1988). This high hole density may be associated with the 4% lattice mismatch between PbTe ($a = 6.460$ Å) and BaF$_2$ ($a_0 = 6.200$ Å) and with the high dislocation density observed at that interface (Salamanca-Young et al., 1987). These holes probably also exist at interfaces between BaF$_2$ and PbTe-rich PbSnTe or PbEuTe. Thus, they may be interfering not only in PbTe/PbEuTe band-offset determinations for material grown on a BaF$_2$ substrate, but also in similar investigations of the PbTe/PbSnTe band offset. Use of KCl substrates may help with this problem, although this has not been clearly demonstrated (Ishida et al., 1986). It has been also noted that uncertainties in carrier densities and effective masses further complicate these investigations (Dios de Leyva and Rodriguez, 1988). Most recently, band offsets in PbTe/PbEuSeTe (and in PbSe/PbEuSe) heterojunctions have been reported to be strongly temperature dependent (Heinrich et al., 1989).

In PbTe/CdTe heterojunctions grown by ionized cluster-beam epitaxy (Takagi et al., 1985), a Type-I system was reported with a relatively small valence-band offset. We note that the similarity of electron and hole effective masses in PbTe, and the fact that the heavy-hole band edge is deep in the light-hole band normally makes it difficult to distinguish holes from electrons. However, in this case, dopant levels of cadmium probably diffused imto the PbTe, doping it n-type, and perhaps helping to resolve this ambiguity.

In the PbS/PbSe heterojunction system, a Type-I band alignment has been reported, with the valence-band offset being somewhat larger than the conduction-band offset (Chu et al., 1987). This determination was made with a novel photovoltaic response technique that may be useful in the study of band offsets in other narrow-energy-gap semiconductors.

It is clear from the TEM studies mentioned in Section II.C that IV–VI heterojunctions normally have large dislocation densities except in the case of very thin films and relatively small lattice mismatch. These findings have been confirmed with dislocation etches in the case of PbTe/PbGeTe (Partin, 1982), where the critical thickness for the introduction of dislocations was quantitatively related to the Ge concentration as predicted by Matthews (1971). Misfit dislocations in PbTe/PbSnTe heterojunctions have also been studied (Yoshikawa et al., 1979). X-ray studies of PbTe/PbSnTe superlattices have demonstrated that for sufficiently thin layers and small Sn concentrations, the lattice mismatch is almost completely accomodated by misfit strain (Fantner et al., 1984). In another study of PbSnTe/PbSeTe superlattices, the lattice mismatch was systematically varied through zero by varying the Se concentration (Ishida *et al.*, 1985b). It was found that in these hot-wall epitaxially grown films, Sn diffusion was significant even at a substrate temperature of 250°C. It was suggested that thermal radiation from the wall (500°C) or the source (500°C) may have caused this anomalously high diffusion. Thermal radiation from the much higher-temperature heater windings, which were apparently not shielded from the sample by a metal or other infrared-opaque material, was another source of optical energy. PbSeTe/PbSnTe superlattices that were nominally lattice matched were observed to have significant lattice distortions because of Sn diffusion into the PbSeTe layers. However, PbTe/PbSnTe superlattices that were nominally mismatched had their lattice distortion reduced via Sn diffusion, although, of course, at the cost of reduced heterojunction abruptness.

Another way of characterizing lattice-mismatched heterojunctions is to examine the effect of this mismatch on device performance. In the PbSSe material system, double-heterojunction diode lasers have been fabricated with a $PbS_{0.6}Se_{0.4}$ active region and PbS confinement layers (1.4% lattice mismatch) that had low-temperature threshold-current densities of 30 A/cm^2 (Preier *et al.*, 1976a). This low threshold-current density implies very little minority-carrier recombination at dislocations. Comparable PbS/PbSe/PbS devices (3.5% lattice mismatch) had ~ 500 A/cm^2 threshold-current densities at low temperature (Preier *et al.*, 1976b). Although this is much higher, it still illustrates the fact that a relatively large lattice mismatch can be tolerated in the PbSSe material system from a device point of view, and suggests that dislocations may not be so electrically active as an earlier observation on PbS implied (Scanlon, 1957). The rather nonlinear increase in threshold current with lattice mismatch that these data indicate may be caused by interactions between dislocations at very high dislocation densities that bend them out of the plane of the first-grown heterojunction. A large density of threading dislocations in the PbSe-active region would then appear.

In the PbSnTe material system, lattice-matched PbSnTe/PbSeTe photodiodes were shown to have a quantum efficiency of 43%, close to the maximum possible of 50% without an antireflection coating (Rotter *et al.*, 1982). Comparable PbSnTe/PbTe devices (0.4% lattice mismatch) had 28% quantum efficiency. A comparable study of double-heterojunction diode lasers in the same lattice-mismatched material system showed that the interface recombination velocity was increased with increasing lattice mismatch, in a typical case leading to a reduction in the internal quantum efficiency of a device of 80% (~ 0.3% lattice mismatch) (Kasemset and Fonstad, 1979). Thus, lattice mismatch appears to have a much more serious effect on devices in the PbSnTe material system than in the PbSSe system. However, it has been shown that high-quality PbSnTe lasers can be grown on lattice-mismatched PbTe substrates as long as the PbSnTe active region is sandwiched between lattice-matched PbSeTe confinement layers (Shinohara et al., 1984). Threshold current densities as low as 30 A/cm² were achieved, with internal quantum efficiencies as high as 40%. This apparently implies that most dislocations in a lattice-mismatched PbTe/PbSeTe heterojunction remain near the plane of the heterojunction, with relatively few of them being inclined to this plane so that they penetrate the active laser region. In more recent work, the same group has interposed a strained-layer PbTe/PbSeTe superlattice between the PbTe substrate and the PbSnTe-active laser region in an apparent effort to reduce threading dislocations (Ishida et al., 1986). Thus, the picture that has so far emerged is that dislocations are much more deleterious to devices in the PbSnTe material system than in the PbSSe material system, but that very-high-quality PbSnTe can be grown on lattice-mismatched substrates if it is suitably buffered from the interface where the lattice mismatch is accomodated.

Semimetallic PbTe/SnTe superlattices have been grown in which a lattice mismatch between PbTe ($a = 6.460$ Å) and SnTe ($a = 6.323$ Å) of 2.2% exists, using stress-free lattice parameters. In one study, it was reported that a Type-II superlattice formed, with a large density of electrons (in PbTe) coexisting with a nearly equal density of holes (in SnTe) (Ishida et al., 1985b; Takaoka et al., 1986). More recently, such a two-carrier system was not observed except possibly in the case in which Bi, a n-type dopant, was incorporated (Tamor et al., 1988). SnTe and PbTe have the face-centered cubic NaCl structure, which PbSe also shares. However, SnSe has the orthorhombic SnS-type structure, which is a distorted NaCl-type structure. The lattice mismatch between PbSe and SnSe is 3%. Thus, a PbSe/SnSe superlattice is strained not only by virtue of lattice mismatch, but also because of a difference in crystal symmetry. In a recent TEM study of such strained-layer superlattices, it was found that abrupt heterojunctions were formed, permitting the layer thicknesses to be varied over a wide range (Hiroi et al., 1987). For PbSe thicknesses of less than about 30 Å, it adopts the SnS structure. For larger thicknesses, it relaxes back to the NaCl structure, with loss of coherency.

B. MAGNETIC-SEMICONDUCTOR SUPERLATTICES

There is a significant amount of research underway on the growth and characterization of lead-salt semiconductors containing magnetically active impurities such as Mn (Yoshino et al., 1987; Bartkowski et al., 1986) and Eu (Goltsos et al., 1986b; Braunstein et al., 1987). The growth of PbTe/EuTe superlattices allows probing of the details of magnetic ordering in EuTe (Heremans and Partin, 1988). These superlattices have been used in field-effect-transistor structures that are very sensitive to magnetic fields at low temperatures (Partin et al., 1988). The microstructure of PbEuSeTe/PbTe heterojunctions has been characterized by TEM (Salamanca-Young et al.,

1987), as has the structure of high Eu-content PbEuTe films (Salamanca-Young et al., 1988a, 1988b). TEM studies of EuTe/PbTe superlattices are currently underway. Whereas EuTe exhibits antiferromagnetic ordering, ferromagnetic ordering has been observed in the SnTe–MnTe system (Mathur et al., 1970). Recently, the effect of carrier concentration on the temperature at which ferromagnetic ordering occurs in this material system was investigated (Story et al., 1986). The potential of superlattices in this material system has not yet been explored.

V. Conclusions

The interest in IV-VI compounds has been greatly stimulated by the introduction of molecular-beam-epitaxy growth techniques. MBE has made possible the growth of a number of new alloys, such as PbEuSeTe and PbCdSeS. These new alloys can be grown with very well-controlled compositions and thicknesses, and at growth temperatures that are low enough to essentially eliminate solid-state diffusion as an interface-broadening mechanism. This has allowed highly dimensionally defined heterojunctions, quantum wells, and superlattices to be grown with these new materials. These structures are not only interesting in their own right, especially in the area of magnetic-field interactions, but have led to important advances in long-wavelength diode lasers.

Acknowledgments

The work at General Motors Research Laboratories on IV-VI compounds has involved much indispensible collaboration with J. Heremans, C. M. Thrush, and the late W. Lo, whose contributions to the advances made here are gratefully acknowledged. L. Salamanca-Young, currently at the University of Maryland, also made many contributions through her TEM measurements on superlattices.

References

Bartkowski, M., Reddoch, D. F., Williams, D. F., Lamarche, G., and Korczak, Z. (1986). *Solid State Commum.* **57**, 185.
Bleicher, M., Wurzinger, H. D., Maier, H., and Preier, H. (1977). *J. Materials Sci.* **12**, 317.
Braun, F. (1874). *Ann. Physik Chem.* **153**, 556.
Braunstein, G., Dresselhaus, G., Heremans, J., and Partin, D. L. (1987). *Phys. Rev.* **B35**, 1969.
Butler, J. F., Calawa, A. R., Phelen, R. J., Harman, T. C., Strauss, A. J., and Rediker, R. H. (1964). *Appl. Phys. Lett.* **5**, 75.

Chu, T. K., Agassi, D., and Martinez, A. (1987). *Appl. Phys. Lett.* **50**, 419–421.
Clemens, H., Krenn, H., Tranta, B., Ofner, P., and Bauer, G. (1988). *Superlattices and Microstructures* **4**, 591.
Crocker, A. J., and Dorning, B. F. (1968). *J. Phys. Chem. Solids* **29**, 155.
Dalven, R. (1969). *Infrared Physics* **9**, 141.
Dawar, A. L., Taneja, O. P., Krishna, K. V., and Mathur, P. C. (1980). *J. Electrochem. Soc.* **127**, 976–978.
Dios de Leyva, M., and Rodriguez, A. M. (1988). *Solid State Commun.* **66**, 549.
Fantner, E. J., Clemens, H., and Bauer, G. (1984). "Advances in X-ray Analysis," Vol. 27 (J. B. Cohen, J. C. Russ, D. E. Leyden, C. S. Barrett, and P. K. Predeski, eds.). Plenum Press, New York, p. 171.
Farrow, R. F. C. (1985). *Nato Adv. Study Inst. on Molecular Beam Epitaxy and Heterostructures.* Erice, Sicily, p. 227.
Frank, F. C., and Van der Merwe, J. H. (1949). *Proc. Roy Soc. A* **198**, 216.
Fuchs, J., Feit, Z., and Preier, H. (1988). *Appl. Phys. Lett.* **53**, 894–896.
Goltsos, W., Nakahara, J., Nurmikko, A. V., and Partin, D. L. (1986a). *Surface Science* **174**, 288.
Goltsos, W., Nurmikko, A. V., and Partin, D. L. (1986b). *Solid State Commun.* **59**, 183–186.
Harman, T. C., and Melngailis, I. (1974). In "Applied Solid State Science," Vol. **4** (R. Wolfe, ed.). Academic Press, New York, pp. 1–94.
Harman, T. C., and Strauss, A. J. (1976). *J. Electron. Mater.* **5**, 621–644.
Harris, R. E. (1983). "Laser Focus/Electro-Optics," 87–96. Penn Well Publ. Co., Westford, MA.
Heinrich, H., Panhuber, C., Eisenbeiss, A., Preier, H., and Feit, Z. (1989). *Superlattices and Microstructures* **5**, 175–179.
Heremans, J., and Partin, D. L. (1988). *Phys. Rev.* **B37**, 6311–6314.
Hiroi, Z., Nakayama, N., and Bando, Y. (1987). *J. Appl. Phys.* **61**, 206–214.
Hohnke, D. K., and Hurley, M. D. (1976). *J. Appl. Phys.* **47**, 4975–4979.
Hohnke, D. K., Holloway, H., and Kaiser, S. (1972). *J. Phys. Chem. Solids* **33**, 2053.
Holloway, H., and Jesion, G. (1982). *Phys. Rev.* **B26**, 5617–5622.
Holloway, H., and Logothesis, E. M. (1971). *J. Appl. Phys.* **42**, 4522–4525.
Holloway, H., and Walpole, J. N. (1979). *Prog. in Cryst. Growth and Charac.* **2**, 49–73.
Horikoshi, Y. (1985). "Semiconductors and Semimetals," Vol. **22**, Part C (W. T. Tsang, ed.). Academic Press, New York, pp. 93–151.
Ishida, A., Aoki, M., and Fujiyasu, H. (1985a). *J. Appl. Phys.* **58**, 797–801.
Ishida, A., Aoki, M., and Fujiyasu, H. (1985b). *J. Appl. Phys.* **58**, 1901–1903.
Ishida, A., Matsuura, S., Fujiyasu, H., Ebe, H., and Shinohara, K. (1986). *Superlattices and Microstructures* **2**, 575.
Ishida, A., Nakahara, N., Okamura, T., Sase, Y., and Fujiyasu, H. (1988). *Appl. Phys. Lett.* **53**, 274–275.
Isida, A., Fujiyasu, H., Ebe, H., and Shinohara, K. (1986). *J. Appl. Phys.*, **59**, 2032.
Jain, M., Warrier, A. V. R., and Sehgal, H. K. (1984). *Phys. Status Solidi* **82**, K181–K184.
Jantsch, W., Lischka, K., Eisenbeiss, A., Pichler, P., Clemens, H., and Bauer, G. (1987). *Appl. Phys. Lett.* **50**, 1654–1656.
Karczewski, G., Kowalczyk, L., and Szczerbakow, A. (1981). *Solid State Commun.* **38**, 499–501.
Kasemset, D., and Fonstad, C. G. (1979). *Appl. Phys. Lett.* **34**, 432–434.
Katayama, S., and Murase, K. (1980). *Solid State Commun.* **36**, 707.
Kim, L. S., Drew, H. D., Doezema, R. D., Heremans, J. P., and Partin, D. L. (1987). *Phys. Rev.* **B35**, 2521–2523.
Koguchi, N., Kiyosawa, T., and Takahash, S. (1987). *J. Cryst. Growth* **81**, 400.
Kressel, H., and Butler, J. K. (1977). "Semiconductor Lasers and Heterojunction LEDs." Academic Press, New York, pp. 439–454.

Kriechbaum, M., Kocervar, P., Pascher, G., and Bauer, G. (1988). *IEEE J. Quantum Electronics* **24**, 1727–1743.
Kriechbaum, M., Pascher, H., Rothlein, P., Bauer, G., and Clemens, H. (1989). *Superlattices and Microstructures* **5**, 93–98.
Krost, A., Harbecke, B., Faymonville, R., Schlegal, H., Fautner, E. J., Ambrosch, K. E., and Bauer, G. (1985). *J. Phys. C* **18**, 2119–2143.
Lopez-Otero, A. (1975). *Appl. Phys. Lett.* **26**, 470–472.
Maissen, C., Masels, J., Zogg, M., and Blunier, S. (1988). *Appl. Phys. Lett.* **53**, 1608–1610.
Mathur, M. P., Deis, D. W., Jones, C. K., Patterson, A., Carr, Jr., W. J., and Miller, R. C. (1970). *J. Appl. Phys.* **41**, 1005–1007.
Matthews, J. W. (1961). *Phil. Mag.* **6**, 1347–1351.
Matthews, J. W. (1971). *Phil. Mag.* **23**, 1405–1416.
Milnes, A. G. (1973). "Deep Impurities in Semiconductors." Wiley, New York.
Mojejko, K., and Subotowicz, M. (1981). *Thin Solid Films*, **78**, 319–326.
Moss, T. S. (1953). *Proc. Phys. Soc. (London)* **B66**, 993–1002.
Norton, P., and Tacke, M. (1986). *Society of Photo-Optical Instrumentation Engineers* **659**, 195.
Norton, P., Knoll, G., and Bachem, K. H. (1985). *J. Vac. Sci. Technol.* **3**, 782–783.
Palatnik, L. S., and Fedorenko, A. I. (1981). *J. Cryst. Growth* **52**, 917–924.
Paparoditis, C., and Suryanarayanan, R. (1972). *J. Cryst. Growth* **13/14**, 389–392.
Partin, D. L. (1981). *J. Electron. Mater.* **10**, 313–325.
Partin, D. L. (1982). *J. Vac. Sci. Technol.* **21**, 1–5.
Partin, D. L. (1983a). *J. Vac. Sci. Technol.* **B1**, 174–177.
Partin, D. L. (1983b). *J. Electrom. Mater.* **12**, 917–929.
Partin, D. L. (1984). *J. Electron. Mater.* **13**, 493–504.
Partin, D. L. (1985a). *Optical Engineering* **24**, 367–370.
Partin, D. L. (1985b). *J. Appl. Phys.* **57**, 1997–2000.
Partin, D. L. (1988). *IEEE J. Quantum Electronics* **24**, 1716–1726.
Partin, D. L., and Thrush, C. M. (1984). *Appl. Phys. Lett.* **45**, 193–195.
Partin, D. L., Clemens, B. M., Swets, D. E., and Thrush, C. M. (1986). *J. Vac. Sci. Technol.* **B4**, 578–580.
Partin, D. L., Thrush, C. M., and Clemens, B. M. (1987). *J. Vac. Sci. Technol.* **B5**, 686–689.
Partin, D. L., Heremans, J., Thrush, C. M., Green, L., and Olk, C. H. (1988). *Phys. Rev.* **B38**, 3549–3552.
Partin, D. L., Thrush, C. M., Simko, S. J., Gaarenstroom, S. W. (1989). *J. Appl. Phys.*, **66**, 6115–6120.
Pascher, H., Pichler, P., Bauer, G., Clemens, H. J., Faunter, J., and Kriechbaum, M. (1986). *Surf. Sci.* **170**, 657–664.
Preier, H. (1979). *Appl. Phys.* **20**, 189–206.
Preier, H., Bleicher, M., Riedel, W., and Maier, H. (1976a). *Appl. Phys. Lett.* **28**, 669–671.
Preier, H., Bleicher, M., Riedel, W., and Maier, H. (1976b). *J. Appl. Phys.* **47**, 5476–5477.
Restorff, J. B., Allgaier, R. S., and Houston, B. (1981). *J. Appl. Phys.* **52**, 6185–6189.
Rogers, L. M. (1971). *J. Phys. D* **4**, 1025.
Rogers, L. M., and Crocker, A. J. (1971). *J. Phys. D* **4**, 1016.
Rosenberg, A. J., Grierson, R., Wooley, J. C., and Nikolic, P. (1964). *Trans. Met. Soc. AIME* **230**, 342.
Rosman, R., Katzir, A., Norton, P., Bachem, K. H., and Preier, H. M. (1987). *IEEE J. Quantum Electronics* **QE-23**, 94–102.
Rotter, S., Kasemset, D., and Fonstad, C. G. (1982). *IEEE Elec. Dev. Lett.* **EDL-3**, 66–68.
Salamanca-Young, L., Partin, D. L., Heremans, J., and Dresselhaus, G. M. (1987). *Mater. Res. Soc. Proc.* **77**, 199.

Salamanca-Young, L., Wuttig, M., Partin, D. L., and Heremans, J. (1988a). *Mat. Res. Soc. Symp. Proc.* **103**, 133–138.
Salamanca-Young, L., Partin, D. L., and Heremans, J. (1988b). *J. Appl. Phys.* **63**, 1504–1508.
Salamanca-Young, L., Nahm, S., Wuttig, M., Partin, D. L., and Heremans, H. (1989). *Phys. Rev. B*, in press.
Samaras, I., Papadimitriou, L., Stoemenos, J., and Economou, N. A. (1984). *Thin Solid Films* **115**, 141–154.
Sayadian, H. A., Drew, H. D., and Partin, D. L. (1986). *Solid State Commun.* **60**, 745–748.
Scanlon, W. W. (1957). *Phys. Rev.* **106**, 718–720.
Schlicht, B., Dornhaus, R., Nimty, G., Haas, L. D. and Jakohus, T. (1978). *Solid-State Electron.* **21**, 1481–1485.
Semiletov, S. A., Rakova, E. V., and Zheludeva, S. I. (1980). *Thin Solid Films* **66**, 11–24.
Shinohara, K., Nishijima, Y., Fukuda, H., Ebe, H., and Murase, K. (1984). *Proc. Ninth Intl. Conf. on IR and mm Waves*, Takarazuka, p. 71.
Silberg, E., Sternburg, Y., and Yellin, N. (1981). *J. Solid State Chem.* **39**, 100.
Smith, D. L., and Pickhardt, V. Y. (1975). *J. Appl. Phys.* **41**, 2366–2374.
Smith, D. L. and Pickhardt, V. Y. (1978). *J. Electrochem. Soc.* **125**, 2042–2050.
Smorodina, T. A., and Tsuranov, A. P. (1980). *Inorganic Materials* **15**, 1068–1070.
Strauss, A. J. (1973). *J. Electron. Mater.* **2**, 553–569.
Story, T., Galazka, R. R., Frankel, R. B., and Wolff, P. A. (1986). *Phys. Rev. Lett.* **56**, 777–778.
Suryanarayanan, R., and Paparoditis, C. (1968). *J. Phys.* **29**, C4-46–C4-49.
Suryanarayanan, R., and Paparoditis, C. (1969). *Colloq. Int. C.N.R.S.* No. **180**, 149–155.
Tacke, M., Spanger, B., Lambrecht, A., Norton, P. R., and Bottner, H. (1989). *Appl. Phys. Lett.*, in press.
Takagi, T., Takaoka, H., Kuriyama, Y., and Matsubara, K. (1985). *Thin Solid Films* **126**, 149–154.
Takaoka, S., Okumura, T., Murase, K., Ishida, A., and Fujiyasu, H. (1986). *Solid State Commun.* **58**, 637–640.
Tamor, M. A., Davis, L. C., and Holloway, H. (1983). *Phys. Rev.* **B28**, 3320–3323.
Tamor, M. A., Holloway, H., Davis, L. C., Baird, R. J., and Chase, R. E. (1988). *Superlattices and Microstructures* **4**, 493–496.
Vanyarkho, V. G., Zlomanov, V. P., and Novoselova, A. V. (1970). *Inorg. Mat.* **6**, 1352.
Vujatovic, S. S., Nikolic, P. M., and Pavlovic, M. (1982). *J. Cryst. Growth* **58**, 285–287.
Walpole, J. N., and Guldi, R. L. (1974). "Physics of IV–VI Compounds and Alloys" (S. Rabii, ed.). Gordon and Breach, London.
Wilson, I. H., Zeng, N. J., Knipping, U., Tsong, I. S. T. (1988). *Appl. Phys. Lett.* **53**, 2039.
Wittmer, M., Smith, D. A., Segmuller, A., Zogg, H., and Melchior, H. (1986). *Appl. Phys. Lett.* **49**, 898–900.
Yagi, K., Takayanagi, K., Kobayashi, K., and Hongo, G. (1971). *J. Cryst. Growth* **9**, 84–97.
Yamada, I., Takaoka, H., Ulsui, H., and Takagi, T. (1986). *J. Vac. Sci. Technol.* **A4**, 722–727.
Yoshikawa, M., Ito, M., Shinohara, K., and Ueda, R. (1979). *J. Cryst. Growth* **47**, 230.
Yoshino, J. H., Munekata, H., and Chang, L. L. (1987). *J. Vac. Sci. Technol.* **B5**, 683.
Zogg, H. (1986). *Appl. Phys. Lett.* **49**, 933–935.
Zogg, H., and Blunier, S. (1986). *Appl. Phys. Lett.* **49**, 1531–1533.
Zogg, H., Vogt, W., and Baumgartner, W. (1982). *Solid-State Electron.* **25**, 1147–1155.
Zogg, H., Vogt, W., and Melchior, H. (1986). *Proc. Soc. Photo-Optical Instrum. Eng.* **587**, 588.

CHAPTER 6

Molecular-Beam Epitaxy of II-VI Semiconductor Microstructures

Robert L. Gunshor

SCHOOL OF ELECTRICAL ENGINEERING
PURDUE UNIVERSITY
WEST LAFAYETTE, INDIANA

Leslie A. Kolodziejski

DEPARTMENT OF ELECTRICAL ENGINEERING
AND COMPUTER SCIENCE
MASSACHUSETTS INSTITUTE OF TECHNOLOGY
CAMBRIDGE, MASSACHUSETTS

Arto V. Nurmikko

DIVISION OF ENGINEERING AND DEPARTMENT OF PHYSICS
BROWN UNIVERSITY
PROVIDENCE, RHODE ISLAND

Nobuo Otsuka

SCHOOL OF MATERIALS ENGINEERING
PURDUE UNIVERSITY
WEST LAFAYETTE, INDIANA

I.	Introduction	338
II.	Growth by Molecular-Beam Epitaxy	340
	A. Fundamental Material Properties	340
	B. MBE Effusion-Cell Sources.	340
	C. In Situ Evaluation Utilizing RHEED	342
	D. Intentional Substitutional Doping versus Self-Compensation	345
	E. Alloying the II-VI Compounds	353
III.	Use of Lattice-Mismatched Substrates	363
	A. III-V Compound Bulk and Epitaxial-Layer Substrates . .	363
	B. Elemental Group-IV Substrates	375
	C. II-VI Substrates.	375
IV.	Strained-Layer Superlattices and Heterojunctions	376
	A. ZnSe-Based Superlattices	376

    B. CdTe-Based Superlattices 381
    C. II-VI/III-V Heterojunctions and Multiple-Quantum Wells . 386
    D. Metastable Zinc-Blende Magnetic Semiconductors . . . 397
 V. Conclusions 402
    Acknowledgments 403
    References 403

I. Introduction

Major new advances in epitaxial growth techniques have contributed to an improved understanding of the properties of II-VI materials and their alloys; these advances have provided the creation of sophisticated heterojunctions, multiple-quantum wells, and strained-layer superlattices composed of II-VI/II-VI and II-VI/III-V multilayers. The objective of the chapter is to provide an overview of the research activity in the area of II-VI-based structures, with significant recent achievements highlighted.

The renewed interest in the family of II-VI semiconductors has been primarily attributed to their successful epitaxial growth by molecular-beam epitaxy (MBE), atomic-layer epitaxy (ALE), metal-organic chemical vapor deposition (MOCVD), and hot-wall epitaxy (HWE). These advanced growth techniques have resulted in a much improved crystalline quality as well as significant progress toward controlled substitutional doping. The discussion of II-VI-semiconductor microstructures will be restricted to those that have been grown by MBE. Thus far, the properties of *zinc-blende* MBE-grown chalcogenides have been reported; the list includes both binary and alloy compounds involving ZnSe, ZnS, ZnTe, CdTe, MnSe, MnTe, and FeSe. (Hexagonal CdSe has recently been grown by MBE (J. K. Furdyna, private communication).) A study reporting the optical and magnetic properties of the heretofore hypothetical zinc-blende MnSe and zinc-blende MnTe (Gunshor *et al.*, 1989) represents a recent advance. The metastable thin-film MBE growth of FeSe has also been achieved by Jonker *et al.* (1988b). By alloying the II-VI chalcogenides with Mn or Fe, an interesting group of materials referred to as semimagnetic or dilute-magnetic semiconductors (DMS) has been grown as epitaxial layers. Employing the MBE growth technique, a series of wide-gap II-VI superlattice and multiple-quantum-well structures, incorporating barrier layers containing the magnetic element Mn, has been formed and extensively characterized. These new superlattice structures were found to possess very interesting magnetic and optical properties. Exploitation of the magnetic tunability of the bandgap, which is a potentially important characteristic of the DMS materials, provides a unique advantage to scientists to aid in the understanding of the physics of these DMS-containing superlattices, and in superlattice structures in general.

The II–VI compounds currently receiving the most attention involve HgTe, CdTe, ZnTe, and ZnSe. The interest in compounds containing Hg originates from the importance of these materials for the development of infrared detectors; CdTe has been studied primarily for its utilization in HgTe-based configurations. (A representative review of mercury containing II-VI materials can be found in Faurie (1986); the vast amount of research on (Hg,Cd)Te will not be addressed in this discussion of II-VI materials.) In contrast, ZnTe, with a bandgap in the green, and ZnSe, having a bandgap in the blue portion of the spectrum, are the focus of an extensive effort to develop light-emitting devices.

The nonequilibrium growth techniques, such as MBE, have provided a new motivation to readdress doping of the II-VI compounds. From significant research on bulk crystals (Aven and Prener, 1967), it is well known that the II-VI chalcogenide compounds tend to resist substitutional amphoteric doping; the tellurides tend to be p-type, whereas the selenides are usually n-type. An entire field of defect chemistry deals with the tendency of equilibrium-grown II-VI compounds to generate compensating defects in response to the incorporation of substitutional impurities. Recently, there has been speculation that nonequilibrium epitaxial growth techniques can provide the means for avoiding the self-compensation and defect-generation problems associated with prior equilibrium growth techniques and thus provide the opportunity to obtain p-type ZnSe, to cite one example. Employing laser-assisted epitaxy or creative "delta-doping" techniques also provide encouragement that conductivity control will eventually be achieved in the II-VI compound family.

Multiple-quantum wells and superlattices composed of wide-bandgap II-VI compounds were first fabricated in the mid 1980s. The last five years have seen a rapid advancement in the understanding of the microstructural, optical, and electrical properties of these structures as well as an increase in the number of research groups worldwide addressing this field. The MBE technique has also made a significant approach toward achievement of an "ideal" II-VI/III-V heterojunction by providing opportunities for epilayer-on-epilayer interfaces.

From a materials-oriented point of view, much of the recent research activities have centered on epitaxial growth techniques (especially MBE, MOCVD, and HWE), development of new alloys with higher-energy bandgaps, exploration of magnetically active alloys, growth on lattice-mismatched substrates, and the growth of superlattices that are lattice mismatched or have components with different crystal structures. The goal of this chapter is to highlight the significant advances in these exciting new areas of materials science, physics, and engineering.

II. Growth by Molecular-Beam Epitaxy

A. FUNDAMENTAL MATERIAL PROPERTIES

A number of references on the fundamental properties of II-VI compounds are available with extensive reviews found in Aven and Prener (1967), Hartmann et al. (1982), and Yao (1985a). However, some particularly important properties should briefly be discussed here. It is important to note, for example, that all of the compounds listed in the introduction have direct bandgaps (hence the interest in light emission), and are more polar than, for example, the III-V group of semiconducting compounds. The II-VI compounds of interest span a wide range of lattice parameters, from 5.409 Å for ZnS to 6.481 Å for CdTe, with available bandgaps ranging from 3.6 eV for ZnS to the negative bandgap (-0.15 eV) of HgTe (see Fig. 1). The close-packed crystal structure is generally zinc-blende or wurtzite under equilibrium growth. However, it has proven possible to impose a particular crystal structure, for example zinc-blende, in the form of metastable thin films, by means of kinetic (nonequilibrium) growth techniques. The II-VI compounds have similar vapor pressures for cation and anion components, necessitating MBE growth techniques that are quite different than those used for the growth of GaAs, for example. The vapor pressures of the elemental constituents are relatively high, further complicating the MBE process.

B. MBE EFFUSION-CELL SOURCES

The MBE growth of II-VI compounds can utilize both compound and

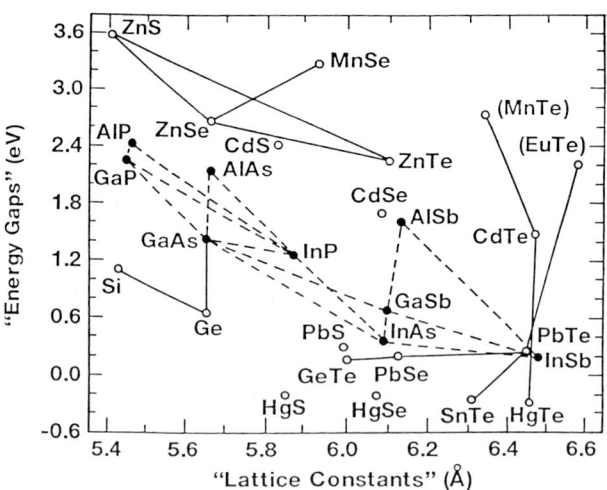

FIG. 1. Energy bandgap versus lattice parameter for a variety of semiconductors (Chang, 1986).

elemental sources. In the case of CdTe, a compound source is primarily used (Farrow *et al.*, 1981) with the addition of a small diameter aperture over the opening of the crucible, to more closely approximate an ideal Knudsen effusion cell. The compound source provides a stoichiometric flux of Cd atoms and Te_2 molecules over the temperature range necessary to result in typical MBE growth rates ($\sim 1\, \mu m/hr$). Substrate temperatures that are relatively low (180–240°C) result in the growth of high-quality, stoichiometric CdTe layers. When forming an alloy of CdTe, however, as in the example of (Cd,Mn)Te, an additional Te source was implemented in some cases (Bicknell *et al.*, 1984a), while not employed in others (Kolodziejski *et al.*, 1984a). The growth of other Cd chalcogenide alloys primarily use a CdTe source with the addition of Zn or Hg sources.

The MBE growth of compound semiconductors is often performed by using elemental sources, and under conditions where the flux ratio is a critical parameter. Often, an ionization gauge is used to determine a beam-equivalent pressure ratio, a quantity that, at least in principle, can be related to the flux ratio. More ideally, the fluxes are monitored directly by using a quartz-crystal monitor (QCM) within the growth chamber, positioned as closely as possible to the growth position of the substrate. Any difference in the geometrical positions of the QCM and substrate results in different flux values at their respective locations. Since the stoichiometry and crystalline quality of the epilayers are greatly influenced by the flux ratio, flux-calibration experiments are essential. Venkatasubramanian *et al.* (1989) performed several different experiments aimed at flux calibration for the MBE growth of ZnSe. Flux levels, deduced by using thickness measurements obtained from the growth of ZnSe under an overflux (10:1) of one component (such that unity sticking is expected for the minority element), were compared to theoretical calculations based on Knudsen's equations and QCM values. The agreement between flux values obtained was within 9%.

At least three research groups (Park *et al.*, 1985a; Prinz *et al.*, 1986; Cammack *et al.*, 1987) have used a compound source for the MBE growth of ZnSe, while Feldman *et al.* (1988) used a compound source for the growth of ZnTe, combined with an additional elemental Zn source. At relatively low growth temperatures ($< 300°C$), very fine control of each anion and cation flux must be maintained for growth of high-electrical- and optical-quality layers. (A discussion of the effects of the flux ratio on the optical and electrical properties of ZnSe will be detailed in Section II.D). The high vapor pressures of the II-VI elemental sources at low source temperatures make precise determination and control of each individual flux difficult. At Purdue, both ZnSe and ZnTe are grown from elemental sources with flux ratios measured by means of a QCM positioned near the substrate position. For the growth of ZnSSe, a ZnSSe compound source was employed (Cammack *et al.*, 1987) to minimize overloading the vacuum system with sulfur, since elemental sulfur tends to have a low sticking coefficient. Recent results involving the metal-organic molecular-beam epitaxy of ZnSe, ZnTe, and ZnS have been reported (Ando et al., 1986, 1987) where gaseous sources of dimethylzinc, diethylzinc, hydrogen selenide, hydrogen sulfide, and diethylselenide are employed as sources of the group-II and VI elements.

C. IN SITU EVALUATION UTILIZING RHEED

Reflection high-energy electron diffraction (RHEED) has been an invaluable tool to aid in the understanding of the molecular-beam epitaxial growth of II-VI materials. The most common use of RHEED in the epitaxy of II-VI compounds was to provide information concerning the surface stoichiometry by indicating either a cation-rich or anion-rich surface. At a particular substrate temperature, the transition between the two surfaces provided a reference point for the adjustment of flux ratio. RHEED intensity oscillations were observed and utilized for the growth of magnetic-semiconductor superlattices composed of ultrathin zinc-blende MnSe layered with ZnSe (Kolodziejski et al., 1986a, 1987a). In these experiments (discussed in Section IV.D.), the thickness of the MnSe layers was controlled to a resolution of one molecular monolayer by using the RHEED intensity oscillations to provide a real-time measure of growth rate. The most recent utilization of RHEED involved studies of the thermally activated electron-stimulated desorption of Se from ZnSe surfaces.

Similar to the growth of GaAs by MBE, a surface phase diagram can be mapped out as a function of flux ratio and substrate temperature for many of the II-VI compounds. Such a phase diagram has been experimentally determined by Menda et al. (1987) for ZnSe surfaces. The various reconstruction patterns that can be observed provide a relative real-time indication of the Se-to-Zn flux ratio impinging onto the surface. For ZnSe surfaces, the typical reconstruction patterns observed are the $c(2 \times 2)$ for a Zn-stabilized surface and the (2×1) reconstructed surface for Se stabilization. During optimum growth of ZnSe, at near one-to-one Zn-to-Se flux ratios and growth temperatures around 300°C, the RHEED pattern commonly observed is the (2×1) Se-stabilized surface. Similar surface reconstructions are observed in the atomic layer epitaxy of the Zn chalcogenides. Yao and Takeda (1986) have reported a (2×1) reconstruction pattern for either a Se- or Te-covered surface (for growth of ZnSe or ZnTe, respectively) and both a $c(2 \times 2)$ or (1×2) reconstructed surface when covered with Zn. Cornelissen et al. (1988) have monitored the growth of Zn(S,Se) with RHEED and have observed both a (5×2) and a (5×1) surface reconstruction with sulphur concentrations below 16%, although the reconstructions were not clearly developed. The same reconstruction was observed during the growth of ZnSe/Zn(S,Se) superlattice structures. In superlattice structures that are heavily strained, the RHEED patterns developed diagonal streaking occurring between spots; such RHEED patterns were believed to arise from the presence of defects, such as stacking faults and twins, in the heavily strained ZnS layers (Cornelissen *et al.*, 1988).

RHEED intensity oscillations have been observed for the II-VI compounds that include ZnSe, MnSe, CdTe, and ZnTe, as well as during the formation of II-VI/III-V heterojunctions (see Sections III.A) and magnetic semiconductor superlattices (see Section IV.D.1). In all cases, one period of intensity oscillation was found to be equivalent to the growth of one molecular monolayer of the compound. In the work reported by Yao *et al.* (1986), RHEED intensity oscillations were reported for growth of ZnSe on ZnSe buffer layers grown on (100) GaAs bulk substrates. In these experiments, the substrates underwent thermal treatment both in the presence of an As_4 flux and without an impinging arsenic flux; the buffer layer of ZnSe was then grown on the GaAs surface. When the substrate was thermally prepared in the presence of As, Yao and coworkers observed that the nucleating ZnSe exhibited a streaked RHEED pattern after a shorter period of time as compared with nucleation when the substrate was prepared without As. Figure 2 shows the RHEED intensity variations observed in the $\langle 100 \rangle$ during growth of ZnSe (at a Zn : Se flux ratio of 1 : 3) on Se-stabilized and Zn-stabilized surfaces maintained at 350°C. The intensity of the reflected RHEED beam is seen to increase for growth on the Se-stabilized surface, whereas it decreases on the Zn-stabilized surface. Yao and coworkers reported that differing flux ratios and lower substrate temperatures did not result in the presence of RHEED intensity oscillations; the absence of the

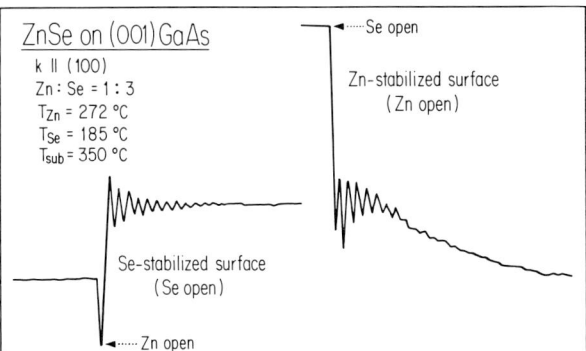

FIG. 2. RHEED intensity variations observed during the MBE growth of ZnSe on Se-stabilized (left trace) and Zn-stabilized (right trace) surfaces of ZnSe epilayers. (Reprinted with permission from the Japan Journal of Applied Physics, T. Yao, Y. Miyoshi, Y. Makita, S. Maekawa (1977). *Japan J. Appl. Phys.* **25**, L952.)

oscillations was attributed to three-dimensional growth via slow surface diffusion of ZnSe molecules on the surface. In contrast to the few oscillations observed during the growth of ZnSe, Yao (1988) observed more than 150 oscillations during the growth of ZnTe on ZnTe (observed in the [100] at a substrate temperature of 350°C).

Mathine et al. (1989) have studied the nucleation of ZnTe on GaSb substrates and epitaxial layers of GaSb and AlSb. (The III–V epilayers were both grown on GaSb substrates in a MBE chamber dedicated to the growth of III-V materials.) After growth, the AlSb (or GaSb) epilayers were transferred via an ultrahigh-vacuum transfer module to the II-VI MBE growth chamber. The epitaxial layers were heated to 320°C for the nucleation of the ZnTe. The early stages of ZnTe growth were studied by using RHEED. Prior to the ZnTe nucleation, both GaSb and AlSb epilayers exhibited (1 × 3) Sb-stabilized reconstructed surfaces. When the ZnTe was nucleated, a (2 × 1) reconstructed surface was observed; the (2 × 1) Te-stabilized surface remained during the subsequent ZnTe growth. RHEED intensity oscillations were also observed during nucleation of ZnTe on the III-V epilayers, which suggested the occurrence of two-dimensional nucleation. In contrast, nucleation on an Sb-stabilized GaSb bulk *substrate* always resulted in spotty RHEED patterns and was characteristic of three-dimensional nucleation Streaked RHEED patterns evolved as the ZnTe film growth continued.

Benson et al. (1986) examined the variation in the surface reconstruction of CdTe epilayers in an attempt to distinguish various surface stoichiometries. Under normal growth conditions using a compound CdTe source, the usual (2 × 1) reconstructed surface was observed. However, for a static surface held at 300°C, the twofold periodicity disappeared after 5–10 minutes. When an additional Cd flux was employed during growth, the twofold reconstruction in the {110} azimuth again disappeared, while at the same time, twofold reconstruction lines were observed along the [100] azimuth. Subsequent to impingement of a Te_2 flux, the surface lifetime of Te was measured by observing the time required for the RHEED pattern to change from (2 × 1) to (1 × 1) on a static surface. In later publications, Benson et al. (1988, 1989) reported the effect of laser illumination on the desorption-activation energies of Cd, Te, and Sb for CdTe surfaces. The photon beams (from He–Ne and Ar-ion lasers) had little effect on the desorption of Cd and Sb, but enhanced the desorption of Te. The enhanced desorption is speculated to result in an increased number of sites available for the substitutional incorporation of Sb and may support the successful doping results obtained by using photo-assisted MBE (see Section II.D).

Although RHEED is typically employed as an in situ characterization tool to monitor the real-time nucleation or growth of a semiconductor or metal layer, RHEED has also been employed as a method to stimulate desorption of an atom from the growing surface (Farrell *et al.*, 1989). In the work of Farrell and coworkers, impingement of a 10 keV electron beam, with a nominal current of 150 mA, thermally activated a process of electron-stimulated desorption of Se atoms from (100) ZnSe surfaces. Following the growth of a Se-stabilized (2 × 1) ZnSe surface, the Se flux was turned off, and the electron beam was allowed to illuminate the surface of the ZnSe. The time required for the twofold reconstruction pattern to disappear was recorded as a function of temperature. The resultant thermally activated desorption process was seen to occur with an activation energy of 0.6 ± 0.1 eV (Farrell *et al.*, 1989). For comparison, Menda *et al.* (1987) measured an activation energy for desorption of Se from a Se-stabilized surface to be 1.02 eV. Such observations (Farrell *et al.*, 1989) of the desorption of Se is believed to be primarily due to an electron-stimulated desorption process, since the twofold reconstruction pattern was observed to remain on parts of the sample not illuminated by the electron beam. In their experiment, Farrell and coworkers found that the desorption of Se was more pronounced at higher substrate temperatures and lower growth rates, and was not insignificant when compared with the growth rates typically employed in MBE growth. The experiments suggest that the surface lifetimes of atoms can be influenced by the presence of an electron beam, with significant effects occurring when the electron-stimulated desorption rate approaches the growth rate.

Monte Carlo simulations of the MBE growth of ZnSe (Venkatasubramanian *et al.*, 1987, 1989) were based on a growth-kinetic model closely following that developed for GaAs growth by Madhukar (1983) and Singh and Bajaj (1984). The surface-kinetic processes considered were (i) incorporation of Zn on Se and Se on Zn, (ii) surface migration of Zn and Se, and (iii) re-evaporation of Zn and Se. The incorporation of Se on Zn consisted of two types of Se atoms, those arriving on Zn itself, and those arriving on Se and migrating over to a Zn-covered surface. The processes were assumed to be Arrhenius type. The lattice model was a rigid lattice gas zinc-blende model. Experimental data on growth rate, sticking coefficients, type of surface coverage (Zn or Se), for various flux ratios (also the behavior of observed RHEED intensity oscillations for unity flux ratio), were compared with predictions from the simulation. The agreement was good, indicating that the kinetic model developed was useful for a description of the MBE growth of ZnSe. Farrell *et al.* (1988a) have also developed a model for the MBE and ALE growth of ZnSe that describes the layer-by-layer growth via incorporation of only electronically stable surface species.

D. INTENTIONAL SUBSTITUTIONAL DOPING VERSUS SELF-COMPENSATION

Semiconductor device applications rely on the ability to control electron and hole concentrations by the selective incorporation of donors and acceptors.

The wide-bandgap II-VI compounds, however, have exhibited a difficulty in achieving amphoteric doping. In general, it was found that the tellurides were readily doped p-type, whereas the selenides were more easily doped n-type. As mentioned above, a traditional problem associated with the use of II-VI compounds in device applications is their tendency to defect generation, together with an associated self-compensation, which results as dopant impurity species are introduced into the material. For example, when indium atoms are incorporated during the MBE growth of CdTe, the photoluminescence (PL) is degraded while little evidence of doping activation is obtained. To overcome this tendency for self-compensation, Bicknell et al. (1986a,b) have introduced a technique of photo-assisted MBE, employed during the doping of CdTe with In. By illuminating the growing film with a low-intensity beam from an argon laser ($150\,\text{mW/cm}^2$), they reported that a high degree of dopant activation was obtained. A similar success has been reported in the case of MBE-grown CdTe doped with Sb and As (Bicknell et al., 1986a; Harper et al., 1988). Comparisons of photoluminescence and transport properties have shown dramatic improvements when laser illumination is employed for both n- and p-type doping. The electrical properties of a heavily doped CdTe:As (grown on a (100) CdTe substrate) grown via the photo-assisted MBE technique (Harper et al., 1989) are found in Fig. 3a. At 290 K, the hole mobility is reported to be $74\,\text{cm}^2/\text{Vs}$ and increases as the temperature is decreased to a value of $157\,\text{cm}^2/\text{Vs}$ at 230 K. The hole concentration of the film at 230 K is greater than $2 \times 10^{18}\,\text{cm}^{-3}$, which is near the degenerate doping level for CdTe. When compared with doping levels obtained in bulk CdTe crystals ($5 \times 10^{16}\,\text{cm}^{-3}$, 300 K), the high p-type carrier concentration that has been achieved represents an increase of over two orders of magnitude. Figure 3b shows the photoluminescence spectrum taken for the same heavily doped CdTe:As film (Harper et al., 1989). The dominant feature was attributed to excitons bound to neutral As acceptors. The feature at 1.5937 eV was associated with unintentionally incorporated donor impurities, and the feature at 1.5967 eV was associated with free excitons. The free-exciton energy was shifted slightly from that observed in undoped CdTe; the shift was attributed to a reduction in the dielectric constant due to the free-carrier concentration in the doped sample. The laser-assisted doping technique has also been used during dopant incorporation in CdTe/(Cd,Mn)Te superlattices where the diluted magnetic-semiconductor barrier layer is doped with In and Sb (Bicknell et al., 1986b; Harper et al., 1988). In the case of relatively wide CdTe wells, the measured mobility values exceeded those obtained single-layer, In-doped CdTe films. As the wells became narrower, however, the mobilities tended to decrease, suggesting that interface roughness may play a role in limiting the mobility for carrier transport parallel to the superlattice layers. Having achieved the

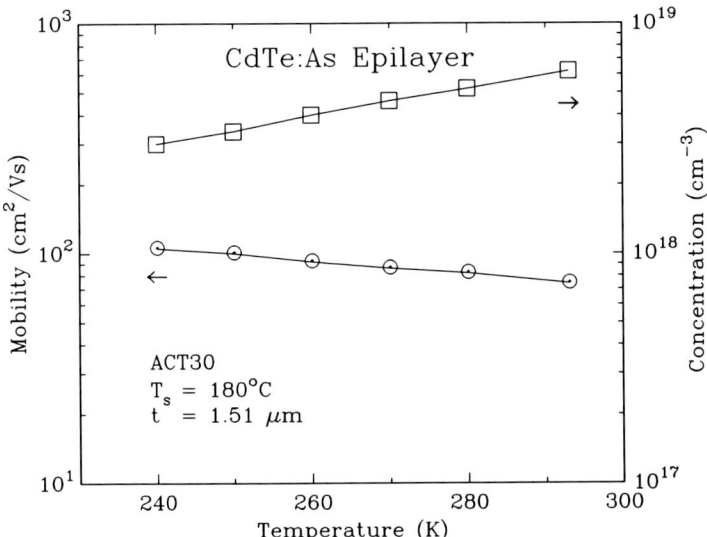

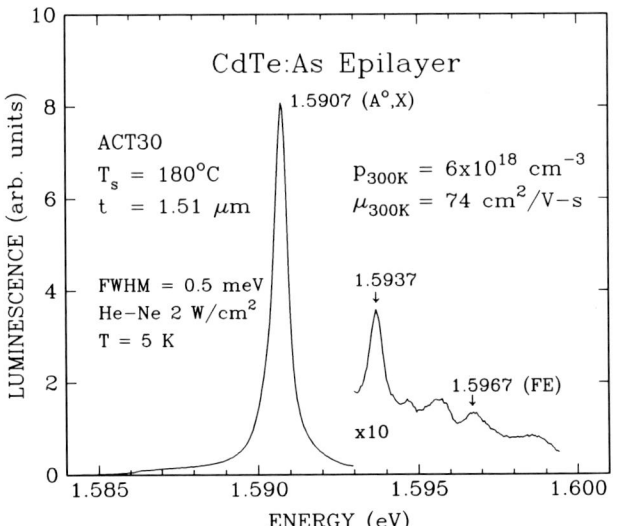

FIG. 3. (a) Mobility and carrier concentration as a function of temperature; (b) low-temperature photoluminescence obtained from heavily doped CdTe:As grown by photo-assisted MBE. Reprinted with permission from J. F. Schetzina (from Harper et al., 1989).

controlled substitutional doping of MBE-grown CdTe, a variety of devices, including $p-n$ diodes and metal-semiconductor field-effect transistors, have been fabricated and studied (Dreifus et al., 1987). Although the mechanism by which the photons interact with the growing surface layer during photoassisted MBE is not yet completely understood, this work has generated a great deal of interest and clearly warrants further investigation. Based on the work of Benson et al. (1988, 1989), it has been speculated that the effect of the laser could be to improve the surface stoichiometry of the growing surface by promoting desorption of excess Te. (This hypothesis is supported by work at Purdue involving the use of laser illumination during the growth of Ga-doped ZnSe and is discussed in the following.)

Reasonable carrier concentrations have been obtained in the n-type doping of MBE-grown ZnSe. In these various experiments, the dopant species were incorporated during the growth process itself; annealing at elevated temperatures (as in bulk crystals) was not required in order to activate the substitutional donors. Attempts to produce p-type ZnSe by MBE have been reported, and photoluminescence has indicated the presence of ionized substitutional acceptors. However, to our knowledge, there have been no transport measurements involving observation of a net hole current.

Studies of dopant incorporation have been complicated by the presence of unintentional impurities found in both the elemental and compound source materials that are typically used in MBE. In most cases, undoped ZnSe grown by MBE, using commercially available source material of 6 nines purity, has been n-type with low ($\sim 1\,\Omega$-cm) resistivity (Yao et al., 1979). The low resistivity of the ZnSe material implied that the ZnSe was of good stoichiometry, because a relatively small deviation from a unity Zn-to-Se flux ratio, toward either Zn-rich or Se-rich conditions, was found to result in high-resistivity material (Yao, 1985a). In Yao's experiments, the defects generated during growth under nonstoichiometric conditions appeared to compensate the nonintentionally incorporated impurities. For example, in doping experiments performed at Purdue for a given Ga oven temperature, the resistivity was found to increase by two orders of magnitude when the flux ratio (Se:Zn) went from unity to 2:1. (The fluxes of the elements were measured by a quartz-crystal monitor placed at the position of the substrate.) Through enhancement of the purity of the source material, by vacuum distillation and/or zone refining, nominally undoped ZnSe grown under similar conditions of substrate temperature and flux ratio exhibited high resistivity ($\sim 10^4\,\Omega$-cm). The enhanced purity of the resultant ZnSe material was confirmed in photoluminescence measurements where free-exciton features were more prominent, having intensities similar to, and sometimes greater than, bound exciton-related transitions. The use of purity-enhanced source material was reported by three groups (Yoneda et al., 1984; Kolodziejski et *al.*, 1984a,b, 1985a; Ohkawa *et al.*, 1986). Yoneda *et al.* (1984) performed a study in which they reported the variation of carrier concentration, resistivity, and relative intensity of free-exciton emission for undoped ZnSe as a function of the number of purification cycles, where the Se source material was vacuum distilled. The undoped carrier concentrations ranged from 1×10^{17} cm^{-3} to less than 7×10^{14} cm^{-3} as the number of purification cycles of the Se were varied from 1 to 9, respectively. At Purdue, we have consistently used vacuum-distilled source material (Zn, Se, Mn, and CdTe) prepared in-house for the growth of (Cd,Mn)Te (Kolodziejski *et al.*, 1984a), ZnSe, and (Zn,Mn)Se (Kolodziejski *et al.*, 1985a). Depending on the conditions of the MBE apparatus and particular charge of source material, we have measured both high-resistivity ($\sim 10^4$ Ω-cm) undoped ZnSe and lower-resistivity ZnSe (on the order of 3 Ω-cm). We have also grown undoped ZnSe using commercially available vacuum-distilled source material obtained from Osaka Asahi Mining Company. Again we found nominally undoped ZnSe to have a resistivity greater than 10^4 Ω-cm. Our observations agreed with the results of Ohkawa *et al.* (1986) where they obtained high-resistivity (10^4 Ω-cm) undoped ZnSe using the purity-enhanced source material purchased from the commercial vendor mentioned.

Two elements (Cl, Ga) have been successfully employed to obtain *n*-type MBE-grown ZnSe. The highest carrier concentration reported (1×10^{19} cm^{-3}) was obtained by the use of Cl (Ohkawa *et al.*, 1986). In this work, ZnCl$_2$ (5 nines purity) was used as the dopant source with oven temperatures varying from 150 to 250°C. The Cl atom behaved as a donor with an ionization energy of 26.2 meV. Incorporation of the dopant species caused the resistivity of the ZnSe to decrease from a value of 10^4 to 3×10^{-3} Ω-cm at the highest carrier concentration. Room-temperature mobilities ranged from 200 to 400 cm^2/Vs.

The majority of experiments reporting the *n*-type doping of MBE-grown ZnSe involved the incorporation of Ga. Earlier work (Niina *et al.*, 1982; Yao and Ogura, 1982) emphasized the effect of Ga incorporation on the photoluminescence properties of ZnSe grown under a variety of conditions. The doping experiments performed at Purdue (Vaziri *et al.*, 1989) agree with more recent reports (Yao, 1985a) where carrier concentrations for Ga-doped ZnSe were found in the range of $2-5 \times 10^{17}$ cm^{-3}, with resistivities as low as 0.05 Ω-cm and having room-temperature mobilities of 200–400 cm^2/Vs. During a series of experiments at Purdue involving the substitutional doping of ZnSe with Ga, a low-power (150 mW/cm^2) laser (with excitation wavelengths in the visible, 5145 Å and 4880 Å) was used to examine the effect of photon illumination on the incorporation of the dopant species. The experiments were performed with Se-to-Zn flux ratios of 2:1 and 1:1. At the higher flux ratio, the samples were found to exhibit high resistivity as a result of compensation. Photoluminescence measurements revealed the presence of a defect band (near 2.1 eV), usually attributed to the presence of Zn vacancies, the amplitude of which was proportional to the Ga source temperature. It was found that the ratio of near-bandgap excitonic emission to defect-band emission was greatly increased for regions of the sample illuminated by the laser, suggesting a reduction in the generation of defects (as seen in Fig. 4). In subsequent experiments, the flux ratio during growth was adjusted to values that were closer to unity. Under these, presumably more stoichiometric flux conditions, the defect-band emissions were greatly reduced for similar Ga flux levels consistent with the observations described earlier for CdTe. As the tendency for defect compensation diminished, the expected carrier concentrations and mobilities were obtained. No apparent effect of the laser was observed when growth proceeded under more stoichiometric growth conditions. One can speculate that the laser-assisted doping experiments in CdTe are in fact related, suggesting that the laser illumination during CdTe doping serves to improve the surface stoichiometry during growth, thereby reducing the degree of compensation and improving the substitutional incorporation of In or Sb. The growth of CdTe by MBE, employing a compound CdTe source, may be improved if a second source of Cd was utilized as a way of fine tuning the impinging fluxes.

MBE growth of undoped ZnSe has also been studied (Ohishi *et al.*, 1989) during irradiation of ultraviolet (325 nm, 3.81 eV) light originating from a

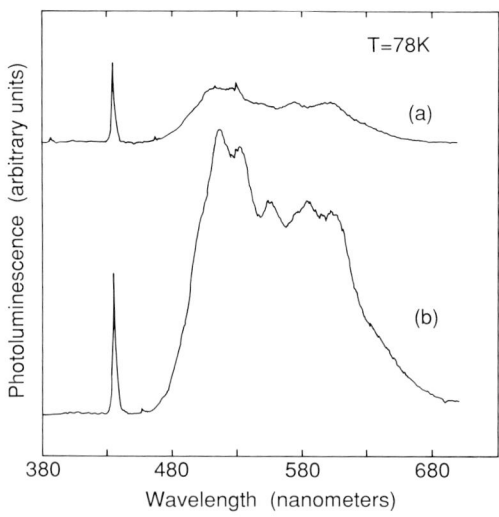

FIG. 4. Photoluminescence intensity as a function of energy for ZnSe:Ga using a 2:1 Se-to-Zn flux ratio (a) under laser illumination and (b) without laser illumination.

He–Cd laser source. The ultraviolet (UV) source was utilized in an attempt to dissociate Se_2 molecules into Se atoms; Se_2 molecules have a dissociation energy of 3.55 eV. The power density was maintained at $\sim 40\,mW/cm^2$ in the work described in the following, with the Se:Zn flux ratio measured with an ion gauge of 4:1. In situ observation of the evolution of growth via RHEED showed definite differences between samples un-irradiated and irradiated with the UV light. For irradiated samples, the streaked pattern of the GaAs substrate continued during subsequent nucleation and growth of the ZnSe layer. The un-irradiated sample exhibited a spotty RHEED pattern upon nucleation. Differences observed in the surface morphology provided additional evidence that the presence of the laser improved the migration of surface atoms and molecules, and confirmed similar suggestions provided by the RHEED observations. The $2-2\frac{1}{2}$ times increase in growth rate that was observed suggested that the Se_2 molecules were dissociated as was anticipated; samples mounted near those samples illuminated with the UV light also exhibited enhanced growth rates. Distinct differences were also noted in photoluminescence spectra. For samples exposed to the UV radiation, the free and bound I_2 and I_X lines increased in intensity, whereas emissions attributed to Zn vacancies were seen to decrease (Ohishi et al., 1989). These photo-assisted effects suggest the possibility of controlling the difficulties associated with native defects in II-VI materials. Farrell et al. (1988b) have suggested possible mechanisms that may be occurring during the MBE growth process, whereby light interacts with the surface species.

de Miguel et al. (1988) have recently reported a series of experiments involving the doping of ZnSe, where Ga atoms were incorporated by using a planar or "delta-doping" technique; in this method, the dopant species are not uniformly distributed, but instead are deposited in discrete planes (in these experiments separated by 35 or 70 Å). The best results were obtained when shutter sequences were arranged such that the Ga and Zn were simultaneously impinged onto a Zn-stabilized surface; the implication is that the number of defects related to Zn vacancy–Ga complexes were minimized. Figure 5 shows the carrier concentration versus Ga oven temperature for ZnSe:Ga grown under a variety of surface-stabilized conditions. The delta-doping procedure resulted in the highest reported carrier concentrations of $1.3 \times 10^{18}\,cm^{-3}$ for ZnSe:Ga. Self-compensation effects are observed if the planar carrier concentration becomes greater than $1 \times 10^{12}\,cm^{-2}$. Subsequent delta-doping experiments performed at Purdue, also using Ga, tend to confirm that a reduction in defect concentration occurs when employing the delta-doping technique; a tenfold reduction in trap density for a given carrier concentration (over that measured for uniform doping) has been observed in deep-level transient spectroscopy (DLTS) measurements (Venkatesan et al., 1989).

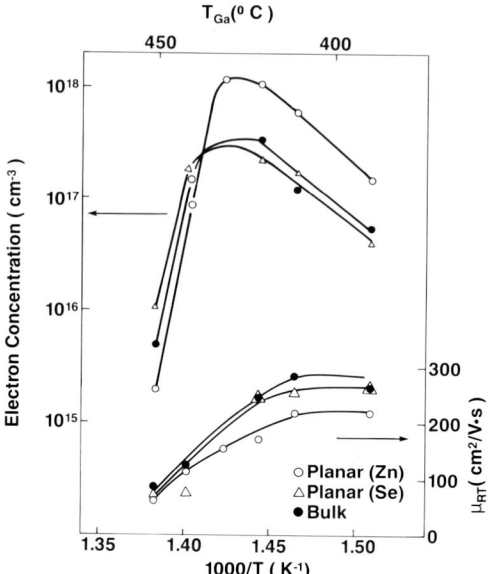

FIG. 5. Electron concentration and room-temperature mobility as a function of inverse Ga cell temperature for a series of planar doped ZnSe:Ga. The open triangles represent doping on Se-terminated surfaces, the open circles represent Zn-terminated surfaces, and the solid circles represent uniformly doped samples. Reprinted with permission from M. C. Tamargo (from de Miguel *et al.*, 1988).

A more difficult challenge is the observation of a net hole current that has not been measured in *p*-type ZnSe grown by MBE. Transport measurements of *p*-type material have thus far eluded the efforts of several groups working on MBE of ZnSe. The problem is twofold: (i) One must first achieve an $N_A - N_D$ greater than zero, and (ii) one must form ohmic contacts to this wide-bandgap material, which, in initial experiments, is expected to have relatively low net carrier concentrations. The early results that have been reported include attempts at *p*-type doping of ZnSe: (i) by using Zn_3P_2 (Yao, 1985b, 1986) as a source of phosphorus, (ii) by using a low-energy ionized beam of NH_3 gas (Ohkawa *et al.*, 1987) or un-ionized gas sources of N_2 or NH_3 (Park *et al.*, 1985b) as a source of N, and (iii) by using Na_2Se (Yao and Taguchi, 1984) as a source of Na. In the reported experiments involving the phosphorus doping of ZnSe, as the phosphorus incorporation increased, the ZnSe material remained *n*-type with the acceptor carrier concentration estimated through decreases in the net electron carrier concentration. (In this study, however, the undoped ZnSe had a resistivity of the order of 1 Ω-cm due to unintentional impurities originating from the Zn and Se source material.) In attempts to use nitrogen as a *p*-type dopant species, photoluminescence measurements indicated the presence of an acceptor-bound transition attributed to the substitutional incorporation of nitrogen.

A number of experiments have been reported by Cheng *et al.* (1988) in which potential *p*-type impurities have been incorporated into MBE-grown ZnSe. In several cases, photoluminescence has indicated the presence of acceptor impurities, whereas carrier type has been identified by apparent success or failure of Au (*p*-type) and In (*n*-type) to form ohmic contacts. There has (at the time of writing) as yet been no report of *p*-conduction in Hall measurements. Recent reported results for ZnSe grown by MOCVD, using Li_3N to introduce *p*-type dopants, include transport measurements demonstrating *p*-type conduction (Yasuda *et al.*, 1988). Although it is not clear which of the *p*-type elements (N or Li) are most effective in these experiments, there are potential problems associated with the practical use of Li as a dopant for device applications, because the small size of the Li atom enhances diffusion. *P*-type conductivity has also been reported for ZnSe layers grown by using vapor-phase transport (Stucheli and Bucher, 1989). A metal/*p*-ZnSe/*n*-GaAs heterojunction exhibited dominant blue 2.68 room-temperature electroluminescence.

There have been few experiments involving the *p*-doping of ZnTe, for which as-grown samples are *p*-type. In one reported series of experiments, both As and Sb were used (Kitagawa *et al.*, 1980). The lack of success in using As was attributed to a zero sticking coefficient for As molecular species at the growth temperature. The experiments with Sb were more successful; however, the antimony flux required was comparable to the host species. Phosphorus was used as a *p*-type dopant for MBE-grown films of ZnTe on GaAs substrates by the group at Sanyo (Hishida *et al.*, 1989); hole concentrations of 1.2×10^{16}–4×10^{17} cm^{-3} were obtained for these films, which were grown by using elemental sources.

E. ALLOYING THE II-VI COMPOUNDS

The II-VI family of semiconductors has been alloyed with a variety of elements, including magnetic transition metals (Mn, Fe, and Co), to provide either (i) a range of lattice parameters for lattice matching, or (ii) a wider- (narrower-) bandgap material for use as barrier (well) layers in multiple-quantum-well structures. The random incorporation of the magnetic transition metals results in a group of materials known as diluted magnetic (DMS) or semimagnetic semiconductors. As in the example of (Zn, Mn)Se, the band structure of the host II-VI semiconductor (ZnSe) is not directly modified by the presence of Mn, since the two *s* electrons of the outer shell replace those of Zn and become part of the band electrons in extended states. The five electrons in the unfilled 3*d* shell of Mn, however, give rise to localized magnetic moments that are partially aligned in an external magnetic field. The resultant magnetic moment interacts with the band electrons, causing a Zeeman splitting that is two orders of magnitude larger at low lattice temperature than for the host II-VI semiconductor. The presence of the magnetic ion (with the associated Zeeman shifts) provides a unique and useful feature for the superlattices and multiple-quantum-well structures in which Mn is incorporated. The magnetic field-induced changes in optical transition energies in superlattices incorporating the DMS material (Hefetz et al., 1985) provide additional insight into excitonic behavior and valence-band offset in strained-layer structures in general. The binary semiconductor ZnSe has been alloyed with Mn, Fe, and Co and represents the only epitaxial thin-film DMS Zn chalcogenides, while alloys of S- and Te-containing ZnSe have also been studied. Epitaxial Cd chalcogenides have been alloyed with Mn, Hg, Zn, and Se.

Although the presence of the magnetic transition metal Mn results in very interesting magnetic and magneto-optical properties, the primary motivation for the thin-film growth of (Zn, Mn)Se originated from the need for a wider-bandgap material to layer with ZnSe. In addition, the MBE growth of the pseudobinary material system ZnSe–MnSe provided a unique opportunity to investigate metastable zinc-blende crystals over a large range of alloy

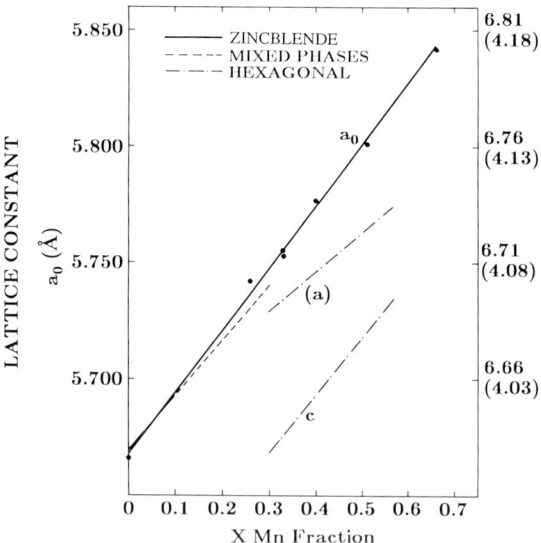

FIG. 6. Experimental lattice constant (a_0) versus Mn mole fraction (x) for MBE-grown zinc-blend $Zn_{1-x}Mn_xSe$ epitaxial layers. Hexagonal and mixed phases apply to bulk crystals. (For the hexagonal lattice parameter (a, c) data, the axis on the right should be used.)

fractions that are unavailable by conventional bulk equilibrium growth techniques. Thick (1–3 μm) epilayers of zinc-blende $Zn_{1-x}Mn_xSe$ have been grown by MBE over the $0 < x \leqslant 0.66$ composition range (see Fig. 6), whereas bulk crystals exist with pure zinc-blende crystal structure only up to $x < 0.10$ (Twardowski et al., 1983). The ability to achieve zinc-blende (Zn, Mn)Se having appreciable Mn content was crucial, because the small variation in bandgap with Mn concentration made it necessary to grow barrier layers with a high Mn fraction to achieve sufficient band offset for carrier confinement. Figure 7 shows the variation in excitonic bandgap (6.5 K and 77 K) versus Mn concentration for the zinc-blende MBE-grown (Zn, Mn)Se epilayers; these data were obtained from photoluminescence and reflectance measurements. The data show an initial bowing to lower energy at small Mn concentrations ($x \leqslant 0.10$) and a monotonic increase in bandgap as more Mn is substituted into the host zinc-blende lattice. In photoluminescence, two dominant competing optical transitions are observed and originate from radiative recombination of excitons near the band edge, as well as due to emission from transitions associated with excited Mn ions. For Mn fractions that are reasonably low ($0 < x \leqslant 0.1$), blue excitonic recombination dominates, whereas very intense yellow luminescence originating from the Mn-ion internal transitions dominates at higher Mn fractions. For samples containing Mn fractions greater than 0.5, the only emission observed is the yellow emission around 2.1 eV; no near-band-edge emission is observed (Hefetz et al., 1986a).

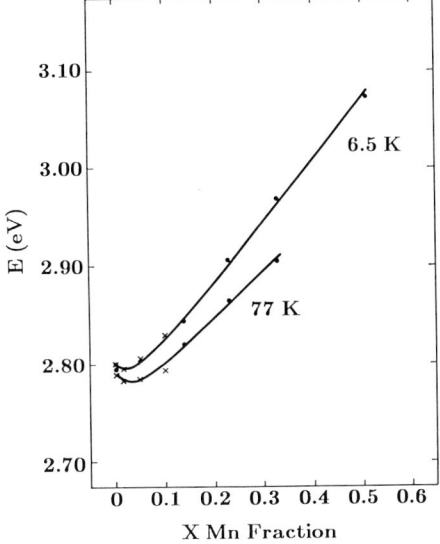

FIG. 7. Energy of Dominant near-band-edge feature versus Mn concentration for MBE-grown zinc-blend $Zn_{1-x}Mn_xSe$ epilayers. (Cross hatch is data obtained on bulk crystals obtained by Twardowski et al. (1983).)

Epitaxial thin-film ZnSe has been alloyed with Fe over the composition range of $0 < x \leq 1.0$ by employing molecular-beam-epitaxy growth techniques (Jonker et al., 1987a, 1988a). Bulk crystals of $Zn_{1-x}Fe_xSe$ have been synthesized by Twardowski et al. (1987) over a much lower composition range ($0 < x \leq 0.22$) with evidence of mixed crystal structures. Pure FeSe thin films have been grown by MBE, although the films exhibit a tetragonal crystal structure rather than a cubic structure above some growth-parameter-dependent critical thickness (Jonker et al., 1988b). Bulk FeSe crystals exist in the hexagonal NiAs crystal structure (Wyckoff, 1963). For the MBE-grown films, both GaAs (100) 2° off-axis and on-axis substrates were used with growth temperatures of 325°C; MBE effusion cells containing Zn, Se, and Fe (790–980°C) were utilized during the growth. Observation of the RHEED patterns during nucleation and growth of the (Zn, Fe)Se layers suggested a two-dimensional (2D) growth behavior as the somewhat spotty RHEED pattern originating from the GaAs substrate quickly changed into a streaky pattern as the (Zn, Fe)Se was nucleated. Following growth of several hundred angstroms, well-defined Kikuchi lines were observed indicating a high degree of bulk order, persisting for the entire growth period (for 1-μm-thick layers). Figure 8 shows a cross-sectional high-resolution electron-microscope

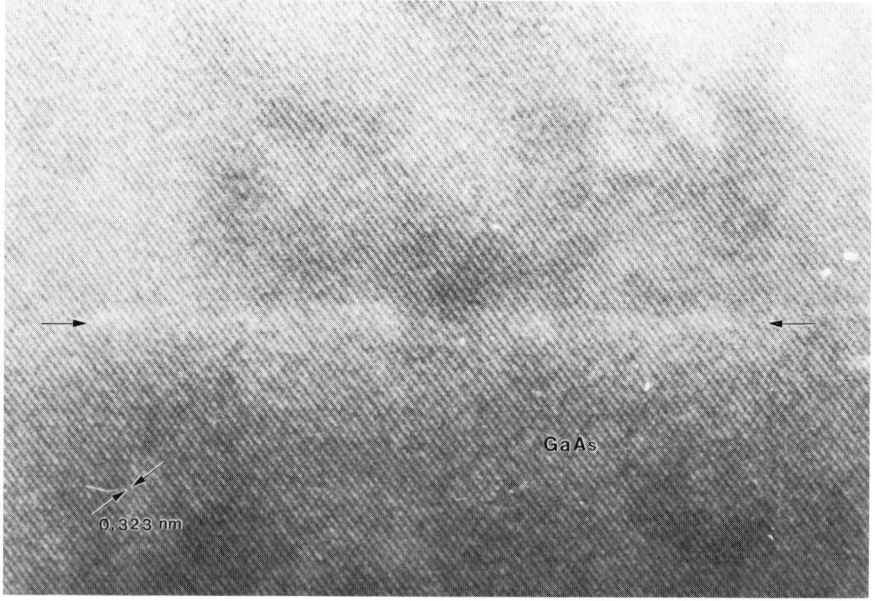

FIG. 8. Cross-sectional HREM image of the interface between $Zn_{0.78}Fe_{0.22}Se$ and a GaAs substrate. Reprinted with permission from L. Salamanca-Young.

(HREM) image of the interface between the GaAs substrate and an epitaxial layer of $Zn_{0.78}Fe_{0.22}Se$ (printed with permission by L. Salamanca-Young). The image indicates the presence of a coherent interface and confirms the ordered epitaxial growth of the (Zn, Fe)Se, as suggested by the RHEED patterns. For reasonably large Fe fractions ranging from $0.30 \leqslant x \leqslant 0.62$, the initial RHEED patterns suggested a well-ordered growth for the deposition of a few thousand angstroms. However, later RHEED patterns suggested a deterioration of crystalline quality as the layer thickness increased beyond ~ 3000 Å. The decrease in crystalline quality is attributed to either the degree of lattice misfit between overlayer and substrate or to a disruption in the growth of the zinc-blende crystal. For 1-μm-thick layers composed of only zinc-blende crystal structure, x-ray θ–2θ measurements indicate an average lattice-parameter variation with Fe content of $\bar{a}(x) = (5.6684 \pm 0.0005) + (0.058 \pm 0.004)x$ Å (Jonker et al., 1987a, 1988a,b,c, and Qadri et al., 1988a).

A variety of optical, magneto-optical, and magnetic measurements have been performed on the (Zn, Fe)Se layers, including photoluminescence and magnetoreflectivity (Liu et al., 1988a,b), and SQUID magnetometry (Jonker et al., 1988b). The room-temperature optical bandgap changes approximately linearly with Fe concentration for $0 < x \leqslant 0.22$ with a net change of only ~ 50 meV, which is an amount of approximately a factor of two smaller than the variation in bandgap for zinc-blende (Zn, Mn)Se films. Magnetoreflectivity measurements (Liu et al., 1988a,b), performed at low temperature in the Faraday geometry, have provided measurements of the conduction- and valence-band exchange integrals (α and β, respectively) and the spin–orbit splitting of the lowest Fe^{2+} states. Figure 9 shows a plot of the energy of the $(-1/2 \rightarrow -3/2)$ interband excitonic feature relative to its energy at zero magnetic field versus $x\langle s \rangle$, where x is the Fe concentration and $\langle s \rangle$ is the average projection of the Fe^{2+} spin along the field, obtained independently from SQUID magnetometry data. The data show a straight line with a slope of $N_0\alpha - N_0\beta = 1560$ meV (N_0 represents the number of unit cells per unit volume). At $B = 8$ T experimental values for the exchange integrals were determined to be $N_0\alpha = 226 \pm 10$ meV and $N_0\beta = -1334 \pm 14$ meV (Liu et al., 1988b). Jonker et al. (1988b) have measured the magnetization of 2-μm-thick epitaxial layers of zinc-blende $Zn_{0.78}Fe_{0.22}Se$ by SQUID magnetometry. Paramagnetic Van Vleck behavior is observed, although the number of magnetically free Fe^{2+} ions decreases as the Fe concentration increases. The behavior suggests an increasing number and/or size of antiferromagnetic clusters occurring as the Fe content is increased.

The most recent diluted magnetic semiconductor to be grown successfully by molecular-beam epitaxy is $Zn_{1-x}Co_xSe$ (Jonker et al., 1988d, 1989; Krebs et al., 1988). Since the Co–Co exchange interaction is larger than that of Mn in many compounds, it can be anticipated that Co-based DMS structures

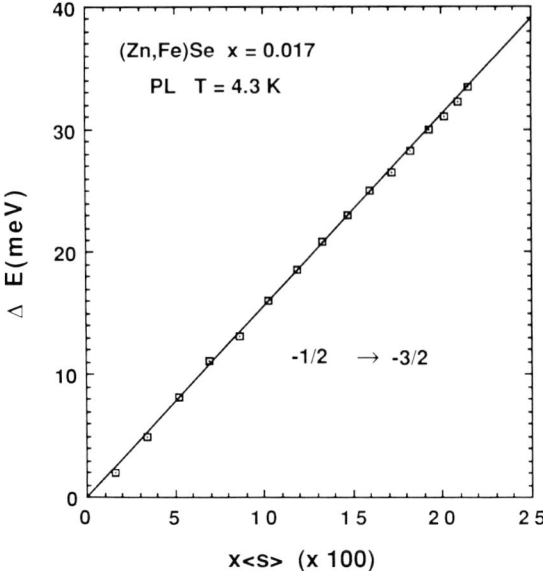

FIG. 9. Experimental measurement of the conduction- and valence-band exchange integrals; the slope provides an indication of $N_0(\alpha - \beta)$ for (Zn, Fe)Se. Reprinted with permission from B. T. Jonker.

may exhibit magnetic ordering behavior at higher temperatures. The first epitaxial structures have been grown on GaAs (100) substrates at growth temperatures of 330°C by using elemental MBE sources. RHEED patterns observed during the growth suggest that well-ordered zinc-blende crystals can be achieved over the composition range of $0 < x \leqslant 0.095$. Figure 10 shows the interfacial region between a $Zn_{0.905}Co_{0.095}Se$ epilayer and the GaAs substrate, as viewed in cross-sectional transmission electron microscopy (TEM) high-resolution imaging (Jonker et al., 1989). A high-quality interface is seen with few defects observed either at the interface or in the epilayer; there is no evidence of twin formation. For this relatively large Co fraction, a well-ordered epitaxial layer can be grown on GaAs. A variety of x-ray measurements have been used to assess the microstructural quality of the MBE-grown epilayers as well as to determine the variation of lattice parameter with increasing Co concentration. The average lattice parameter for samples, having x values between $0 < x \leqslant 0.036$, increases linearly with Co concentration varying as $\bar{a}(x) = 5.6676 + 0.0584$ Å (Jonker et al., 1988d) and is similar to the variation in lattice parameter versus Fe content measured for (Zn, Fe)Se. X-ray double-crystal rocking-curve measurements show an increase in the full width at half maximum (FWHM) as the Co

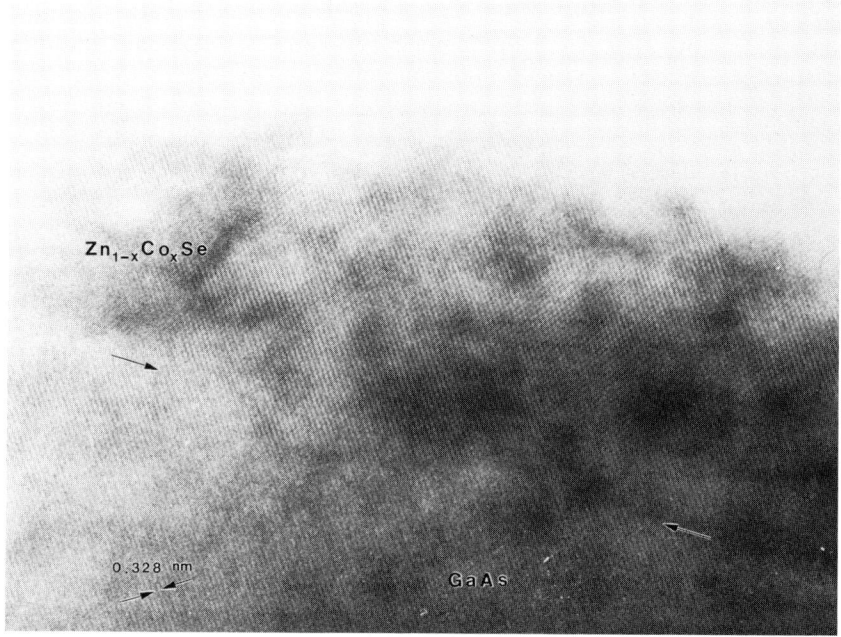

FIG. 10. Cross-sectional HREM image of the interface between zinc-blend (Zn, Co)Se with a Co fraction of 9.5% and GaAs. Reprinted with permission from B. T. Jonker.

concentration increases over a relatively small range, and are greater than the FWHM variation with incorporation of Fe into the ZnSe host lattice. The x-ray data suggest that the incorporation of Co is more disruptive to the ZnSe lattice and that the (Zn, Co)Se layers may prefer to relax into their bulk crystalline structure of hexagonal NiAs. Thin (Zn, Co)Se layers ($\sim 0.3\,\mu$m) exhibit very narrow rocking curves, however, which are indicative of a crystalline quality approaching that of the GaAs substrate (Jonker et al., 1989).

SQUID magnetometry measurements have been performed on the Co-containing epilayers to address the issue of magnetization. The magnitude of the measured moment is a factor of ~ 0.4 smaller than expected for the number of Co^{2+} ions in the sample. Jonker et al. (1988d) attribute such magnetic behavior to the existence of antiferromagnetic-exchange-coupled pairs of Co^{2+} ions or to the occupation of low-symmetry sites in the vicinity of dislocations resulting in a magnetically active ground state (Kribs et al., 1988). If the placement of Co atoms on nearest-neighbor sites gives rise to the large reduction in magnetic moment for (Zn, Co)Se layers having large amounts of Co, then the magnetization data would suggest that clustering is more pronounced in this DMS material than in the Fe-containing DMS. Figure 11 shows the energy difference between the two main interband excitonic transitions as a function of $x\langle s \rangle$; the slope provides a measurement of $N_0(\alpha - \beta)$ whose value is 2530 meV. The $N_0(\alpha - \beta)$ value is much larger than that obtained for DMS (Zn, Fe)Se and (Zn, Mn)Se epitaxial layers and illustrates the much larger exchange interaction present in the Co-containing DMS material.

Non-DMS alloys of ZnSe that have been grown by MBE include Zn(Se, Te) and Zn(S, Se). MBE growth of Zn(S, Se) is extremely difficult due to the problems of having S in an ultrahigh-vacuum environment. Cammack et al. (1987) circumvented some of these problems by utilizing a ZnSSe compound as the source of S in various ZnSe/Zn(S, Se) superlattice structures (see Section IV.A). Alloying ZnSe with Te to form Zn(Se, Te) permits adjustment of the lattice constant, a lowering of the bandgap (roughly from 2.8 to 2.4 eV), and provides opportunities for amphoteric doping. Zn(Se, Te) is of particular interest due to the observation that the photoluminescence yield of the alloy can be significantly enhanced over that of bulk ZnSe crystals and epitaxial layers (Lee et al., 1987a), due to localization of excitons in the random alloy. The exciton self-capture phenomena observed in the bulk crystal, as well as in films and superlattice structures, may be useful in enhancing the quantum efficiency in superlattices at noncryogenic temperatures, where defects might otherwise limit the luminescence from free or weakly bound excitons. (The results concerning the self-trapping in various superlattice structures is discussed in Section IV.A.) The growth of these Zn chalcogenide alloys experiences unique difficulties for the incorporation of Te

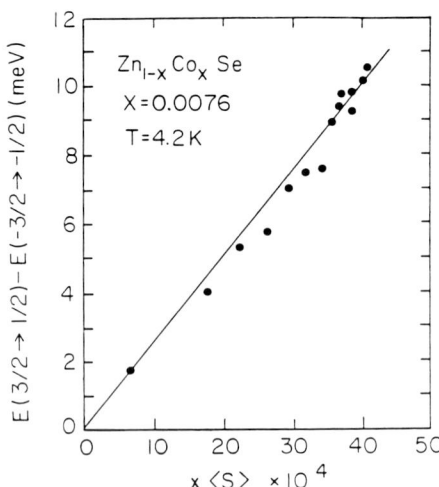

FIG. 11. Experimental measurement of the conduction- and valence-band exchange integrals; the slope provides an indication of $N_0(\alpha - \beta)$ for (Zn, Co)Se. Reprinted with permission from B. T. Jonker.

(as well as for S). In the case of Zn(Se, Te), large overpressures of Te must be employed during the growth over the entire composition range. The alloy composition is determined by the Te-to-Se ratio. Control of the very-high-vapor-pressure ovens, such as Se and Te, presents many problems for controlling the composition. In the work reported by Yao et al. (1978), a Te-to-Se flux ratio of 3 to 10 was required over the entire range of the Te fraction. At Purdue, a number of Zn(Se, Te) epilayers with varying fractions of Te were grown; a particular difficulty was encountered when a small fraction of Te was desired, resulting in widely varying compositions under what appeared to be similar growth conditions. Specially designed ZnSe/ZnTe superlattice structures can be viewed as a means to circumvent the difficulties encountered in the MBE growth of the Zn(Se, Te) alloy (Kolodziejski et al., 1988a).

(Cd, Mn)Te has been grown by MBE with Mn mole fractions between $0 < x \leqslant 0.53$ (Kolodziejski et al., 1984a). Both (111)- and (100)-oriented (Cd, Mn)Te epilayers have been grown on (100) GaAs substrates. The Mn concentration was investigated after film growth by using in situ Auger electron spectroscopy and depth profiling. The concentrations of the Cd and the Mn were uniform throughout the film. X-ray diffraction performed on the (Cd, Mn)Te epilayers to determine the lattice parameter indicated the absence of MnTe and $MnTe_2$ phases. (Cd, Mn)Te has also been grown using atomic-layer epitaxy (Herman et al., 1984).

The alloy (Cd, Zn)Te (Magnea et al., 1986; Tuffigo et al., 1988) is of interest due to the large lattice-constant variation accessible over the composition range bounded by CdTe at 6.481 Å and ZnTe at 6.102 Å. Of major interest are the alloy fractions closer to CdTe, which can be adjusted to match the most important (Hg, Cd)Te lattice parameters. Qadri and Dinan (1985) have reported the properties of the alloy for small Zn fractions, which is the region of interest for focal-plane-array devices. Olego et al. (1985) and Feldman et al. (1986) reported measurements of photoluminescence and x-ray rocking curves, respectively, for a range of alloy fractions extending to both binary limits. The x-ray measurements indicated a dramatic increase in FWHM values as compared with the binary limits for virtually any alloy fraction. The rocking-curve widths seemed much too large to be accounted for by increases in misfit dislocations. On the other hand, the photoluminescence, dominated by bound-exciton emissions, exhibited FWHM values that closely followed the expected alloy broadening resulting from a random statistical distribution of the cations. It was found that theory developed for III-V ternaries provided a good fit to the experimental sata.

1. *Epitaxial Metal-on-Semiconductor Structures*

The aforementioned dilute magnetic semiconductors have attracted a significant amount of attention in recent years due to their unique magneto-optical and magnetotransport properties (see for example Aggarwal et al., 1987). Creative utilization of the magnetic semiconductors in device configurations, however, will require application of magnetic fields in planar thin-film configurations. Attention to this specific issue has provided the motivation of Prinz and coworkers to initiate research on the epitaxial growth of ferromagnetic metal films on compound semiconductor materials. Body-centered cubic (BCC) α-Fe has been successfully grown on both (100) and (110) ZnSe by molecular-beam epitaxy (Prinz et al., 1986; Jonker et al., 1987b), in addition to single-crystal Ni films (Jonker et al., 1988e).

Although bulk ZnSe exists in the zinc-blende crystal structure and α-Fe is BCC, the (001) surface net of Fe forms a $c(1 \times 1)$ construction on the (001) surface of ZnSe (and similarly on the surface of (001) GaAs); with this epitaxial arrangement, the Fe film undergoes a 1.1% compression due to the slight lattice-constant mismatch. For the growth of Fe onto ZnSe epitaxial layers, the ZnSe was grown from a compound source by MBE onto GaAs substrates. Following the growth of ZnSe, the film was cooled to 175°C, and the Fe layer was grown at a growth rate of only 9 Å/min. The nucleation of single-crystal Fe onto epilayers of ZnSe exhibited a three-dimensional (3D) character as evidenced by an initial spotty RHEED pattern. To protect the samples from oxidation following removal from the ultrahigh-vacuum environment, the samples were overgrown with polycrystalline ZnSe. (Polycrystalline ZnSe growth occurred, since the substrate temperature was maintained very low to minimize diffusion of the Fe into the ZnSe layer.)

The iron layers were magnetically characterized by using ferromagnetic resonance and vibrating sample magnetometry (Prinz et al., 1986; Jonker et al., 1987b; Krebs et al., 1987). As an indication of the very high quality of the Fe epilayers, the FWHM was measured to be 45 Oe for a 35 GHz measurement, and is the narrowest ferromagnetic line width measured for single-crystal iron. The FWHM line-width measurement was by a factor of three times narrower than that measured (Krebs et al., 1982) for epitaxial Fe grown on GaAs substrates (which had experienced chemical and thermal treatment). M versus H hysteresis data, obtained on the Fe-on-ZnSe samples and on the Fe-on-GaAs samples, show differences in the easy-direction coercive field H_c. The coercive force is seen to vary for the (001) and (110) ZnSe surface with smaller values of H_c measured on the (001) surface (~ 5 Oe versus 24–80 Oe, respectively). In contrast, the H_c values measured for Fe-on-GaAs samples were always larger (~ 20 Oe for (001) and ~ 80 Oe for (110)) than for the ZnSe epilayers (Prinz et al., 1986). The (001) ZnSe surface would thus be very desirable in device applications requiring easy switching. The origin of the different magnetic properties of the α-Fe films on ZnSe or GaAs have not been completely resolved. In the case of ZnSe, the degree of lattice misfit is reduced, and the presence of the epitaxial surface provides a more optimum surface for nucleation of the Fe. In an effort to incorporate ferromagnetic materials with thin-film layers of diluted magnetic semiconductors, Fe was deposited onto the closely lattice-matched (Zn, Mn)Se (Jonker et al., 1988g). In the structures examined, the magnetic and structural quality of the Fe layers were comparable to that obtained when Fe was deposited onto ZnSe epitaxial films.

For the growth of Ni films onto (001) ZnSe epilayers (Jonker et al., 1988f), a growth temperature of 175°C was again used with low growth rates of 5–10 Å/min. RHEED observation during the growth showed island formation, as evidenced by spotty patterns, whereas Auger electron spectroscopy (AES) suggested some degree of intermixing of the interface. The RHEED indicated that the (001) of the Ni film oriented onto the (001) of the ZnSe epilayer. However, the surface net of the Ni film was rotated 45° about the surface normal of the underlayer. Such a rotation reduced the surface-net mismatch between the Ni and the ZnSe from 24% to ~12%. Magnetization measurements of the single-crystal Ni films provide $4\pi M$ values that are lower than those observed for bulk Ni. Jonker and coworkers attribute such behavior to the formation of nonmagnetic Ni–Se compound formation occurring at the Ni/ZnSe interface, as suggested by the AES data. Also observed for the samples are coercive fields that are large compared with Fe grown on ZnSe.

III. Use of Lattice-Mismatched Substrates

A. III-V Compound Bulk and Epitaxial-Layer Substrates

An ever-present problem in the epitaxial growth of II-VI compounds, which is generally not an issue for the III-V-compound family, is the difficulty in obtaining high-quality substrates for homo- and heteroepitaxy. Bulk crystals of, for example CdTe, exhibiting sufficiently high crystalline quality, have only recently become available and are limited to relatively small areas. The alloying of bulk CdTe material with Zn (Qadri et al., 1985) or Se (Dean and Johnson, 1987) has been implemented as a means of improving the crystalline quality and providing a variable lattice parameter. As suitable substrates become available, however, problems associated with the preparation of the surface for epitaxy remain. Substrates of many of the III-V compounds, such as GaAs, InP, and InSb, can be chemically prepared such that a native oxide is formed and is thermally desorbed in vacuum prior to the start of growth. Because GaAs is readily available at low cost and high quality and is an important optoelectronic material, GaAs is an attractive candidate for use as a substrate for the heteroepitaxy of CdTe. Furthermore, a suitable surface of GaAs can be obtained without impingement of the group-V-element flux during the thermal desorption of the oxide; during the growth of II-VI compounds, the group-V element can potentially provide a source of dopant species. Having noted the many advantages of GaAs as a substrate for CdTe, the presence of a 14.6% lattice-constant mismatch would appear to be a formidable obstacle to achieving high-quality, single-crystalline epitaxial growth. Nevertheless, high-quality CdTe epilayers have been grown on GaAs. A fascinating consequence of the large lattice mismatch is the nucleation of two orientations of CdTe, (111) and (100), on (100)-oriented GaAs substrates. The early activity associated with the growth of CdTe on GaAs was characterized by a number of groups working independently and ostensibly following similar procedures, but obtaining differing results. Some groups reported the observation of consistent nucleation of only one orientation or the other, while others experienced the seemingly random occurrence of (111) CdTe on (100) GaAs or (100) CdTe on (100) GaAs under what appeared to be identical growth conditions. With the potential importance of using CdTe-on-GaAs as an alternative substrate replacing bulk CdTe, a great deal of effort was focussed on determining the factors that contributed to the selective orientation of either (111) or (100) CdTe on (100) GaAs.

During investigations of the nucleation of CdTe on GaAs performed at Purdue, two eutectic phase changes were used routinely to calibrate the substrate temperature for every film growth, ensuring complete GaAs oxide desorption; the procedure was found to result in consistent nucleation of (111) CdTe (and (111) $Cd_{1-x}Mn_xTe$) on (100) GaAs (Kolodziejski et al., 1984a,b). The two eutectic samples were chosen such that one eutectic phase change occurred near the growth temperature (500 Å Au on Ge: 356°C) and one phase change occurred near the GaAs oxide desorption temperature (500 Å Al on Si: 577°C). Prior to insertion into the molecular-beam-epitaxy system, the GaAs substrates were prepared in the standard manner using a $5 H_2SO_4 : 1 H_2O_2 : 1 H_2O$ etch. The etching process removed approximately 10 μm of material in order to eliminate the damaged surface resulting from the mechanical polish. This accepted standard GaAs wafer-preparation technique resulted in the growth of a passivating oxide layer that was subsequently thermally desorbed in situ at around 580°C. For the MBE growth of GaAs or (Ga, Al)As, the oxide desorption occurred in the presence of an arsenic ambient or in the presence of an impinging arsenic flux. For the case of the growth of CdTe, however, oxide desorption occurred without the presence of As. The substrate was held at the oxide-desorption temperature for only two minutes to ensure complete oxide desorption, while minimizing loss of As from the GaAs surface. (Oxide desorption occurred in the analysis chamber while using AES.) The substrate temperature was then reduced to the growth temperature of 325°C, at which time, RHEED revealed a (4 × 6) reconstructed GaAs surface. The RHEED pattern suggested that the surface consisted of a transition structure between a Ga-stabilized and an As-stabilized GaAs surface (Cho, 1976). During inspection of the GaAs with RHEED, the background vacuum-chamber pressure, with the source ovens at their operating temperature, was typically 1×10^{-9} torr. Evaluating the substrate surface with AES both prior to and after RHEED examination revealed a very small Te peak representative of considerably less than one monolayer of Te on the surface. Under these growth conditions, the (111) orientation of CdTe was always nucleated when the CdTe source shutter was opened (also for (Cd, Mn)Te). RHEED observations show that the bulk streaks of the substrate fade and are replaced by streaks originating from the epilayer. The absence of a spotty pattern suggested that the initial growth occurred two-dimensionally. In every (111) CdTe and (111) (Cd, Mn)Te film grown, RHEED observations showed the [$\bar{2}$11] of the (111)-oriented films to be parallel to the [011] of the GaAs substrate (Kolodziejski et al., 1984a). The aforementioned technique, when implemented by means to ensure accurate substrate temperatures, invariably resulted in the (111) CdTe//(100) GaAs epitaxial orientation. Mar et al. (1984a,b) using MBE, and Cheung (1983) using laser-assisted deposition and annealing, have also found (111) CdTe to nucleate on (100) GaAs with the [$\bar{2}$11] of CdTe parallel to the [011] of GaAs.

Several groups (Nishitani et al., 1983; Bicknell et al., 1984b; Faurie et al., 1984), using what appeared to be similar growth techniques as described earlier, reported the observation of (100) CdTe on (100) GaAs, resulting in confusion in recognizing the relevant factors controlling the orientation. The first insight into the details of the CdTe/GaAs interface was provided by Otsuka et al. (1985a,b), who used cross-sectional transmission electron microscopy. In this study, the (111)-oriented CdTe layer was grown at Purdue under the growth conditions described in detail earlier; the (100)-oriented CdTe layer was grown at North Carolina State University. HREM images reveal the presence of a thin (10 Å) residual GaAs oxide at the interface between the (100) CdTe and (100) GaAs materials. The presence of the residual oxide could be explained, since a much lower temperature (500–550°C) was used for the oxide-desorption step.

In the case of the (111)-oriented CdTe, the HREM lattice images show that the epilayer is in direct and intimate contact with the underlying (100) GaAs substrate. Figure 12a and b are HREM images of the interface between a (111) CdTe epitaxial film and a (100) GaAs substrate. In Fig. 12a, the beam direction is parallel to the colinear [011] and [01$\bar{1}$] axes of GaAs and CdTe, respectively. The image shows (111) lattice fringes and weak (200) lattice fringes in both crystals. Microtwins parallel to the interface (Otsuka et al., 1985a) are also seen in the CdTe layer, supporting the implication of a layer-by-layer growth behavior as observed with RHEED in the initial stage of film growth. Orientations of these lattice fringes exhibit the (111) epitaxial

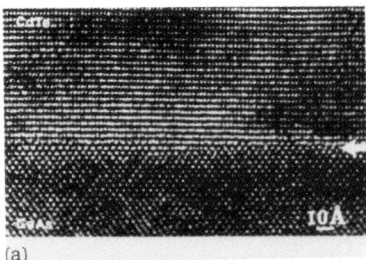

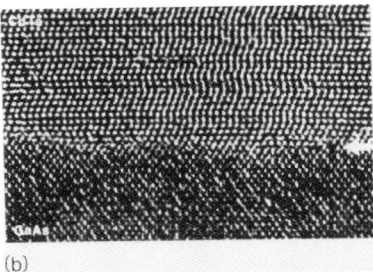

FIG. 12. High-resolution electron micrographs of the (111) CdTe//(100) GaAs interface showing the (a) [11$\bar{2}$] projection and (b) [$\bar{1}$10] projection of the CdTe epilayer using the 1 MV TEM instrument.

relation. A nearly perfect one-to-one correspondence of lattice fringes is seen at the interface. (In this direction, there is a 0.7% lattice mismatch.) It is also seen that the interface is abrupt within the resolution limit of the instrument, with no abnormal interplanar spacings found in the interfacial region. Figure 12b shows the HREM image of the (111) CdTe//(100) GaAs interface in the perpendicular direction ([$\bar{2}$11] of the CdTe epilayer) obtained by using the 1 MV TEM. The image indicates the absence of any misfit dislocations originating from the 14.6% lattice mismatch in this direction; the image shows that the (111) CdTe epilayer forms an incommensurate interface with the (100) GaAs substrate.

Although it is interesting that the presence of a thin oxide layer can affect the orientation of CdTe on GaAs, an oxide layer is not necessary to achieve parallel epitaxy. Two specific growth techniques can be used to nucleate the (100) epitaxial orientation on "clean" (100) GaAs. One technique involves the nucleation of the (100) CdTe at an elevated substrate temperature, whereas the second technique involves the initial nucleation of (Cd, Zn)Te. Faurie *et al.* (1986) have observed a range of Zn mole fractions that nucleate (100) (Cd, Zn)Te on (100) GaAs. The following growth technique (suggested by W. Schaffer, private communication) is found (Kolodziejski *et al.*, 1985b) to nucleate a (100) CdTe epilayer on a (100) GaAs substrate. The calibrated substrate temperature is rapidly raised to 582°C to desorb the GaAs oxide. The oxide desorption is monitored with RHEED. Once the oxide is desorbed, the CdTe source shutter is opened. At this point, the GaAs substrate RHEED pattern changes dramatically as a result of a monolayer of Te bonding to the Ga-rich GaAs surface. Before the CdTe shutter is opened, the GaAs substrate RHEED pattern is confined to the zero-order Laue zone near the shadow edge. However, once the CdTe shutter is opened, the RHEED pattern changes substantially; now the integral-order bulk streaks are observed to increase in intensity, increase in number, and elongate to completely fill the entire area of the fluorescent screen. (Investigating the resultant surface with Auger electron spectroscopy has indicated the presence of approximately a monolayer of Te at the surface.) The substrate heater is then turned off, allowing the substrate temperature to fall rapidly. The CdTe epilayer does not nucleate until the substrate temperature is sufficiently reduced to allow adsorption of the impinging Cd atoms ($\sim 350°C$). As the substrate temperature continues to fall, the (100) CdTe film is nucleated by a three-dimensional growth mechanism. (At a substrate temperature of 325°C, the temperature is maintained constant.) (100) nucleation also occurs if $Cd_{0.60}Zn_{0.40}Te$ is grown on the (100) GaAs rather than CdTe. In this case, the typical growth technique of opening the source shutters with the calibrated substrate temperature stabilized at 325°C has been employed. For all growth techniques of (100) nucleation employed, RHEED observations show the occurrence of an initially spotty pattern (indicating three-dimensional nucleation), which eventually elongates into streaks of uniform intensity; this contrasts with the two-dimensional nucleation observed with the (111) orientation.

By using select growth techniques (as were described in detail earlier) to nucleate the (100) CdTe layer on (100) GaAs, (100) CdTe//(100) GaAs interfaces have been prepared without an interfacial (oxide) layer and have been studied with TEM (Kolodziejski *et al.*, 1985b, 1986b; Ponce *et al.*, 1986). Figure 13 shows a HREM image (obtained with the 1 MV TEM microscope) of a CdTe/GaAs interface, resulting from nucleating the (100) CdTe epilayer at an elevated temperature to induce the occurrence of the (100) epitaxial relationship. The images show the presence of a misfit-dislocation array that exists in the plane of the (100) CdTe//(100) GaAs interface. The presence of additional (111) and (11$\bar{1}$) fringes in the image indicates that these misfit dislocations have a Burgers vector of 1/2[011] and are pure edge dislocations (Kolodziejski *et al.*, 1986b; Ponce *et al.*, 1986). Figure 13 is representative of both perpendicular directions ([011] and [0$\bar{1}$1]). The occurrence of these misfit dislocations tends to generate threading dislocations that propagate perpendicular to the interface. These threading dislocations are not present in HREM imaging, but are readily visible in dark-field imaging. The density of these dislocations are quite high near the interface (10^{12} to 10^{11} cm^{-2}), but are dramatically reduced as the CdTe film thickness is increased. (At the CdTe–GaAs interface, however, the density of threading dislocations is still

FIG. 13. High-resolution electron micrograph of the (100) CdTe//(100) GaAs interface. The arrows indicate the presence of purely edge-type misfit dislocations with Burgers vector of 1/2 [011], originating from the 14.6% lattice mismatch.

two orders of magnitude lower than expected from the 14.6% lattice mismatch.) For a film thickness of $\sim 4\,\mu$m, the dislocation density reduces to 10^6 to 10^5 cm^{-2}. A minimum value of 10^4 cm^{-2} dislocation line density has been reported at the surface of a 6.6 μm MBE-grown CdTe epilayer (Bicknell et al., 1984b).

From the early AES (Mar et al., 1984a,b), RHEED (Kolodziejski et al., 1984a, 1985b; Otsuka et al., 1985a) surface studies, and the HREM interface analysis (Otsuka et al., 1985a,b; Kolodziejski et al., 1986b, and Ponce et al., 1986) of this facinating II-VI/III-V heterointerface, growth techniques were identified to allow the selection of one epitaxial orientation or the other. Many researchers now agree on the conditions required to achieve the nucleation of a (111) CdTe film on a (100) GaAs substrate. More elusive, however, is the identification of factors relevant to the nucleation of (100) CdTe on (100) GaAs. It seems clear that the (100) epitaxial occurrence requires a perturbation of the "state" of the (100) GaAs substrate. This perturbation could be the result of a very thin layer of residual oxide for example, or the presence of a monolayer of adsorbed Te. More recently, several groups have performed experiments to identify different types of perturbation and their effects on the epitaxial orientation. Feldman and Austin (1986) have studied the various surface structures that result from the adsorption of Te on a (100) GaAs surface; it is believed that both the surface structure and the interaction of the group-II element with the surface affects the resultant orientation of the CdTe. The effect of the surface stoichiometry of the GaAs substrate on the epitaxial orientation of the CdTe has also been studied (Srinivasa *et al.*, 1987). In this work, both As-stabilized and Ga-stabilized GaAs surfaces were found to control the orientation such that (100) CdTe nucleated on an As-stabilized surface, while (111) nucleation occurred on the Ga-stabilized surface. Theoretical models (Otsuka *et al.*, 1985a; Cohen-Solal *et al.*, 1986) are also being developed to help identify epitaxial bonding relationships at the II-VI/III-V heterointerface.

Several groups have reported the growth of CdTe on InSb by MBE (Farrow *et al.*, 1981; Wood *et al.*, 1984; Williams *et al.*, 1985; Mackey *et al.*, 1986; and Zahn *et al.*, 1986). The attractiveness of such heterostructures results from the fact that the lattice mismatch is very small, of the order of that for GaAs/AlAs. The result of the close lattice match is an absence of misfit dislocations for relatively thick epitaxial layers. A very high degree of perfection is possible in this substrate/epilayer combination. Although the quality of substrate and epilayer can approach ideal, limitations to the use of InSb as a substrate material include problems associated with requirements for the substrate to be highly insulating, or optically transparant (for use in infrared detector arrays, as one example). The opportunity for obtaining high-quality epilayers can be further enhanced by incorporating a homoepitaxial buffer layer (Wood *et al.*, 1984) on the substrate prior to nucleation of the CdTe layer. Figure 14 shows an x-ray rocking curve (taken by using a four-crystal Si monochromator) for a structure consisting of a 1.1 μm CdTe epilayer grown on a homoepitaxial InSb buffer layer, illustrating the high crystalline quality achievable for these heterostructures. For

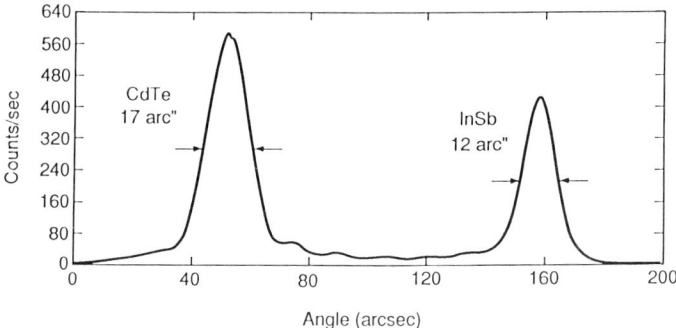

FIG. 14. X-ray rocking curve of a CdTe/InSb heterostructure grown on an InSb substrate, which was prepared for CdTe nucleation by the MBE growth of an InSb buffer layer (0.5 μm in thickness).

various CdTe/InSb heterostructures, examination of the InSb buffer layer/InSb substrate interface with TEM reveals an array of small (~ 30 Å) dotlike features (Wood et al., 1984 and Kolodziejski et al., 1988b). These features, probably representing precipitates of In, Sb, or oxides thereof, when imaged with TEM, are found to be coherent with the InSb, such that threading dislocations or other defects do not propagate into the InSb (and hence the CdTe) buffer layer.

Farrow et al. (1985) describe the effect of growth temperature on the photoluminescence of (100) CdTe films grown on InSb substrates. Subsequent to loading with new material, the CdTe compound source was heated to above 700°C to establish a steady state for congruent sublimation prior to film growth. The CdTe source was constructed with a small orifice to prevent excessive loss of charge during thermal preparation. The CdTe layers were grown at substrate temperatures between 160 and 220°C and evaluated by using several techniques. Double-crystal x-ray rocking curves indicated little difference between films grown at different growth temperatures. However, photoluminescence (Fig. 15) revealed a systematic trend in the reduction of

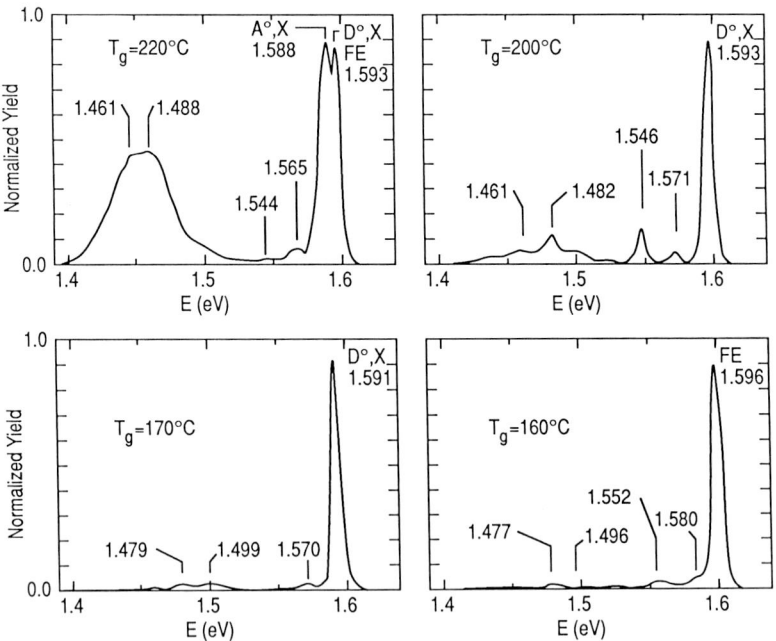

FIG. 15. Photoluminescence at low temperature of MBE-grown CdTe on InSb. Note the progressive dominance of near-band-edge features with decreasing growth temperature. Reprinted with permission from R. F. C. Farrow (from Farrow et al., 1981).

emissions related to native defects as the growth temperature was reduced. As the growth temperature is reduced, the figure also shows the increasing dominance of near-band-edge emissions over donor–acceptor pair and defect-related emissions centered at 1.45 eV. For the films grown at 160°C, the dominant emission feature at 1.596 eV corresponded to free excitons.

Most of the work involving the MBE growth of ZnSe has employed GaAs as the substrate material (Smith and Pickhardt, 1975; Yao et al., 1977, 1983; Kitagawa et al., 1980; Park and Salansky, 1984; Yoneda et al., 1984; Kolodziejski et al., 1985a; Gunshor et al., 1987a; Gunshor and Kolodziejski, 1988; Tamargo et al., 1988a). GaAs is an attractive substrate material for ZnSe primarily because of the relatively small lattice-constant mismatch with GaAs (0.25%). The ZnSe/GaAs heterostructure has potential not only for the use of GaAs as a closely lattice-matched substrate of ZnSe, but also for use of ZnSe as a passivation (Studtmann et al., 1988) or barrier layer for GaAs. For either device application, the structural, compositional, and electrical quality of the ZnSe/GaAs heterointerface can have substantial consequences. A significant amount of effort has been underway to understand this very important II-VI/III-V heterojunction.

Recent work has shown that the stoichiometric condition of the GaAs surface can dramatically affect the subsequent ZnSe layer. Although a variety of desorption times and temperatures are reported, a typical bulk GaAs surface, prepared for MBE growth, shows a reconstructed diffraction pattern indicating arsenic deficiency when examined with RHEED. The surface reconstruction could be altered by impingement of an arsenic beam (Cho, 1976), by use of surface-passivation techniques for preserving an MBE-grown GaAs epitaxial layer (Gunshor et al., 1987a), or by use of modular MBE growth systems where the epitaxial GaAs layer can be maintained in an ultrahigh vacuum via transfer modules (Gunshor et al., 1988, Tamargo et al., 1988b).

For typical MBE growth procedures, subsequent to oxide desorption, the GaAs substrate temperature is reduced to a value ranging between 250–400°C, and nucleation occurs as the Zn and Se source shutters are opened. (In most cases, elemental sources of Zn and Se are used, but a compound ZnSe source has been used by some groups.) Monitoring the RHEED diffraction pattern in the [110], the streaked pattern of the GaAs substrate, showing reconstruction, is rapidly replaced by that originating from the ZnSe. Nucleation occurs in a three-dimensional manner as the streaked GaAs RHEED pattern is replaced by a spotty "fishnet" pattern. Under the aforementioned growth conditions, all research groups observe the three-dimensional nucleation. Three-dimensional nucleation is also observed in the ALE of ZnSe on GaAs substrates (Yao and Takeda, 1986).

The early stages of growth of ZnSe have been studied by using RHEED intensity oscillations (Kolodziejski et al., 1987a; Gunshor et al., 1987a), where both GaAs substrates and MBE-grown GaAs epilayers were employed. Substantial differences were indicated in interfaces formed between epilayer-on-epilayer or epilayer-on-substrate structures. In the nucleation study, two MBE systems were used; in one system, the GaAs epilayers were grown with the resultant as-grown GaAs surface maintained via arsenic-passivation techniques (Price, 1981; Waldrop et al., 1981), whereas the other system was used for the II-VI MBE growth. (The deposition of amorphous Se onto epitaxial layers of ZnSe can also passivate the II-VI semiconductor surface for protection and regrowth (Jonker et al., 1988f,g).) As the GaAs layer was cooled toward room temperature in an arsenic beam, an amorphous As layer was deposited. The As layer was desorbed at 290°C in the analytical chamber of the separate II-VI MBE, following transfer in air from the III-V MBE. Prior to As passivation and subsequent to As desorption, a (2×4) reconstructed GaAs surface was observed. The evolution of the RHEED diffraction pattern (Gunshor et al., 1987a) during nucleation of ZnSe on an MBE-grown GaAs epilayer clearly contrasted the nucleation on a GaAs substrate (which was described earlier). Once the ZnSe was nucleated, the early observation (after 9 seconds) of a strongly streaked RHEED pattern and the early presence of reconstruction lines suggested a more two-dimensional character of the nucleation. The two-dimensional nucleation was confirmed in observations of RHEED intensity oscillations as shown in Fig. 16a (Kolodziejski et al., 1987a; Gunshor et al., 1987a). The RHEED intensity oscillations were observed on the specular spot with an incident angle of less than 1° (off-Bragg conditions). Strong intensity oscillations, characteristic of layer-by-layer growth (Joyce et al., 1984), were observed for nucleation on the GaAs epilayer. RHEED intensity oscillations were not seen when ZnSe was nucleated on a substrate (Fig. 16b); instead, the variation of the specular RHEED intensity on a substrate was similar to observations reported for three-dimensional nucleation of InGaAs on GaAs epilayers (Lewis et al., 1984). It should be noted that the RHEED intensity oscillations just described are unique in that they describe nucleation at a II-VI/III-V interface. In a recent study, employing the modular MBE-system experimental approach (Tamargo et al., 1988b), two-dimensional nucleation was also found to occur when growth commenced on an As-rich GaAs bulk substrate surface, whereas three-dimensional nucleation was observed when growth occurred on a Ga-rich MBE-grown GaAs epilayer. Various GaAs surface stoichiometries were prepared, and Tamargo et al. (1988b) determined that for either substrates or epilayers, two-dimensional growth occurred, as suggested by the evolution of the RHEED pattern during nucleation, on As-stabilized GaAs surfaces. Whether nucleation occurred on Ga-stabilized substrates or epilayers, three-dimensional nucleation resulted.

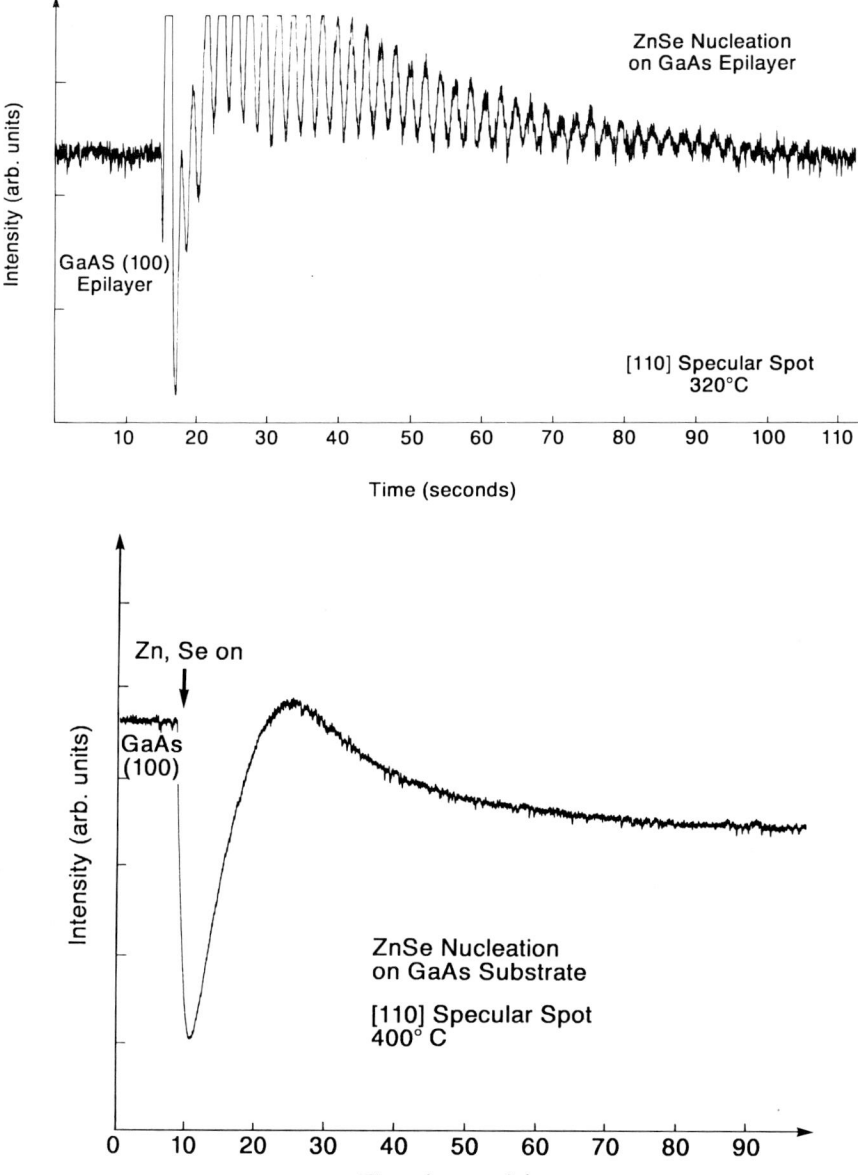

FIG. 16. RHEED intensity oscillations obtained during the nucleation of ZnSe on (a) a GaAs MBE-grown epilayer (at a substrate temperature of 320°C and viewed in the [210] azimuth); and (b) a bulk GaAs substrate (400°C substrate temperature). The higher temperature would be expected to favor two-dimensional nucleation.

By using photoluminescence data (FWHM) as a quality indicator, Tamargo and coworkers provide evidence that the ZnSe epilayers grown on the As-stabilized surfaces are of higher quality than layers nucleated on Ga-stabilized surfaces. From these experimental results, it has been proposed (Tamargo et al., 1988a,b; Farrell et al., 1988a) that an instability exists at the ZnSe/GaAs interface due to an electronic imbalance of differing numbers of Zn–As bonds and Ga–Se bonds. The instability subsequently affects the resultant overlayer by the presence of a "disordered" interface. Their model proposed that a (2 × 4) As-surface has half coverage of As such that a mixed layer of As and Se results upon nucleation. The optimum growth would then occur, since the number of electron-deficient Zn–As bonds would roughly equal the number of electron-rich Se–Ga bonds.

In an effort to provide a closer lattice match to ZnSe than is provided by use of GaAs substrates, ZnSe has been grown on substrates of (In, Ga)As, and epilayers of AlAs and Ga(Al, As) (Skromme et al., 1988). The primary emphasis in this work was a comparison of the photoluminescence obtained under the more lattice-matched growth conditions. It was found that the excitonic features (dominated by bound excitons) were considerably narrower (FWHM) than those observed when ZnSe was nucleated on GaAs directly. It is anticipated that x-ray rocking curves (not yet reported) would also become more narrow when using a substrate/buffer layer of closer lattice match. When x-ray rocking curves are obtained for ZnSe nucleated on GaAs substrates or epilayers, narrow FWHM values for relatively thick ($>1 \mu$m) layers are approximately 126 arc sec (Gunshor and Kolodziejski, 1988); the broadening is a result of the formation of an array of misfit dislocations at the GaAs/ZnSe heterointerface. A detailed study of the crystalline quality of ZnSe grown on GaAs as measured by x-ray double-crystal rocking curves and topography have been performed by Qadri et al. (1988b). When pseudomorphic ZnSe films are grown on GaAs, misfit dislocations do not form; however, the narrowness of x-ray rocking curves for these thin ($\sim$ 1000 to 1500 Å) layers are limited by the thinness of the films. The theoretical FWHM for a 940 Å film of ZnSe is approximately 178 arc sec; experimental rocking-curve measurements using a Ge monochromating crystal indicate a FWHM of 195 arc sec (Gunshor, unpublished).

Pseudomorphic layers of ZnSe have been used to passivate epitaxial GaAs. Using this configuration, a series of depletion-mode transistors were fabricated, and their characteristics were reported (Studtmann et al., 1988). In these transistors, ZnSe formed the pseudoinsulator for GaAs field-effect transistors. It was found that a post-deposition thermal anneal could result in interface-state densities that are comparable (Qian et al., 1989) to those reported for the (Al, Ga)As/GaAs interface, whereas the Fermi level could be swept (without hysteresis or significant frequency dispersion) across the bandgap of GaAs. The low interface-state density is especially interesting when one considers the valence difference across the interface in the case of the II-VI/III-V heterointerfaces.

Another wide-bandgap compound of interest is ZnTe. In contrast to ZnSe, ZnTe tends to be as-grown p-type. There have been efforts to form p–n-junction structures consisting of alternate layers of n-ZnSe and p-ZnTe. A major difficulty with such structures is the large lattice-constant mismatch ($\sim 7.4\%$) between these compounds. ZnTe, a compound that has been successfully grown by MBE, is also attractive because of the potential for forming a closely matched heterointerface with AlSb. The predicted band offsets for this interface suggest the possibility of an n-AlSb/p-ZnTe heterojunction that is suitable for forming a light-emitting device. A discussion of recent results describing the formation of the ZnTe/AlSb heterostructure is found in Section IV.C.3.

B. ELEMENTAL GROUP-IV SUBSTRATES

In addition to the use of GaAs as a substrate, ZnSe has been heteroepitaxially grown on the closely lattice-matched Ge (0.17%) and on lattice-mismatched Si (4.0%) substrates. In the case of ZnSe on Ge, ZnSe has been grown on both Ge substrates (Park and Mar, 1986) and MBE-grown Ge epilayers (grown on Si substrates) (Park *et al.*, 1988). With the presence of a ZnSe/Ge superlattice incorporated in the buffer layer of the Ge epilayer, the quality of the subsequent ZnSe was greatly improved. ZnSe has also been grown on (100) and (111) Si substrates (Mino *et al.*, 1985).

Lo *et al.* (1983) described the MBE growth of CdTe on both (111) and (100) Si substrates; the lattice-constant mismatch for this heterostructure is 16%. It was reported that films grown at substrate temperatures below 300°C exhibited both cubic and hexagonal phases in x-ray diffraction. The best results were obtained for growth temperatures between 330 and 350°C on (111)-oriented substrates for which x-ray indicated cubic single crystals of (111) CdTe. Films grown on (100) Si substrates at the higher temperature range had x-ray diffraction peaks that indicated the formation of twins.

C. II-VI SUBSTRATES

ZnSe has been homoepitaxially grown on ZnSe substrates. However, because of the inability to acquire high-quality ZnSe substrates with useful large areas, only a few initial results of the homoepitaxy have been reported. Park *et al.* (1985c) have reported the MBE growth of ZnSe on both (111) and (100) ZnSe substrates that were prepared by using ion milling and annealing. Under the conditions studied, although the (100) epilayer was of high quality, the (111)-epitaxial-layer quality was significantly inferior, showing no excitonic emission in photoluminescence. Recently, Isshiki *et al.* (1986) have reported the fabrication of very-high-quality bulk substrates using zone-refined Zn as one of the starting materials. These substrates may prove to be more useful for achieving high-quality homoepitaxial ZnSe material.

Due to the large number of misfit dislocations found at the interface between CdTe and GaAs, originating from the 14.6% lattice-constant mismatch, CdTe has also been grown homoepitaxially on commercially available CdTe bulk substrates and on slightly lattice-mismatched bulk (Cd, Zn)Te substrates. (The latter substrates have Zn concentrations that provide a lattice constant to match appropriate (Hg, Cd)Te alloy fractions.) The major difficulty in using the II-VI bulk substrates is the preparation of their surface prior to growth. A substrate-preparation technique that has been found to produce suitable surfaces for nucleation, routinely and repeatably, is as follows (R. N. Bicknell, private communication). The substrate (either CdTe or CdZnTe) is first degreased with TCA, acetone, and methanol, and dried with nitrogen gas. A 60-second etch in a 1% solution of bromine in methanol is performed; and upon completion of the etching period, the etchant is repeatedly diluted with methanol, taking care not to allow the substrate to be exposed to air. The sample is then removed from the methanol and dried with nitrogen gas. Finally, the substrate is mounted onto the Mo sample holder with Ga metal and introduced into the vacuum. Thermal treatment of the CdTe substrates prior to growth consists of elevating the substrate temperature to between 325 and 370°C, followed by a reduction to the desired growth temperature for nucleation of a CdTe buffer layer. In situ observation utilizing RHEED typically indicates streaked patterns complete with reconstruction lines for well-prepared samples.

IV. Strained-Layer Superlattices and Heterojunctions

A. ZnSe-Based Superlattices

ZnSe has been layered with the II-VI semiconductor compounds ZnTe, ZnS, and Zn(S, Se), the semimagnetic semiconductor $Zn_{1-x}Mn_xSe$, and the magnetic semiconductor MnSe to form wide-gap II-VI superlattices and multiple-quantum-well structures. For all of these materials, the lattice constant varies substantially from that of ZnSe, such that layered structures form strained-layer superlattices, provided the layer thicknesses are below the critical thickness where misfit dislocations form. In this section, some of the more recent results involving the wide-bandgap superlattice structures are discussed; however, a complete review can be found in Kolodziejski et al. (1989).

From photoluminescence and reflectance measurements of the zinc-blende

(Zn, Mn)Se epilayers, the bandgap (Fig. 7) has been determined as a function of Mn concentration (Kolodziejski et al., 1986c). For the multiple-quantum-well (MQW) structures reported, the bandgap differences between barrier and well materials were in the range of 100–300 meV. Detailed studies of the magneto-optical behavior of the MQW structures indicated that the magnetic-field-induced band shifts are primarily due to exchange interactions associated with the penetration of the hole wave functions into the (Zn, Mn)Se barrier layers. Concluding that much of the hole envelope wave function resides in the (Zn, Mn)Se barrier layer implied that the strained valence-band offset is small and is likely to be less than 20 meV (Hefetz et al., 1985).

The (Zn, Mn)Se MQWs have been characterized by a variety of techniques including transmission electron microscopy, photoluminescence, and photoluminescence excitation (PLE) spectroscopy, time-resolved luminescence spectroscopy, nonlinear absorption spectroscopy, and observations of stimulated emission and gain spectra. A detailed discussion of these measurements has recently been reviewed by Kolodziejski et al. (1986c).

Laser oscillations have been observed in ZnSe-based structures under both optical excitation and electron-beam pumping. To our knowledge, the first report of lasing in an MBE-grown ZnSe structure was based on experiments with ZnSe/(Zn, Mn)Se multiquantum wells (Bylsma et al., 1985). The experiments were performed on cleaved cavities after removal (by selective etching) of the GaAs substrate. Gain spectra were measured, and thresholds of stimulated emission were determined for various emission wavelengths. Optically pumped lasers were fabricated from these MQWs and found to operate in the blue portion of the visible spectrum. Lasing was observed at temperatures up to 80 K. These first (Zn, Mn)Se MQWs had thresholds for stimulated emission that exhibited an improvement of an order of magnitude over previously reported results (Catalano et al., 1982), using single-crystal ZnSe grown from a melt. Further improvements in the thresholds of the MQWs of this material system are anticipated based on the additions of cladding layers to provide maximal optical confinement and the optimization of growth parameters.

ZnSe epilayers (Potts et al., 1987; Cammack et al., 1987) and ZnSe/ZnSSe superlattices (Cammack et al., 1987) have also shown lasing under electron-beam pumping. Room-temperature lasing of an electron-beam-pumped ZnSe epilayer was obtained with a threshold of 5 A/cm^2, whereas the superlattice structures had a 12 A/cm^2 threshold-current density (Cammack et al., 1987). Electron-beam and optically pumped lasers have been used initially due to the difficulty of achieving both p- and n-type material in wide-gap II-VI compounds.

Compound semiconductors such as (Ga, Al)As and (Zn, Mn)Se can exhibit bleaching of the free-exciton resonance under intense optical illumination; the result is a significant nonlinear optical absorption. Most previous work has involved studies of the (Ga, Al)As material system in which MQWs have had a saturation intensity of 580 W/cm^2 at room temperature (Miller *et al.*, 1982) for a quasi-2D exciton. Measurements of nonlinear excitonic absorption in ZnSe films and ZnSe/(Zn, Mn)Se MQWs at 77 K have been reported (Andersen *et al.*, 1986); experiments have also been performed for ZnSe films at room temperature (Andersen *et al.*, 1986; Peyghambarian *et al.*, 1988). Calculations indicate that both screening and phase space filling of the excitonic states contributed to the nonlinear absorption. Transmission measurements were performed for the ZnSe epilayer and superlattice structures by removing the GaAs substrate, using mechanical lapping followed by a selective chemical etch. Figure 17 shows the change in excitonic absorption as a function of intensity measured at 77 K for a ZnSe epilayer; the measured saturation intensity was 10 kW/cm^2. The cited saturation intensity was approximately an order of magnitude lower than other reported data (Peyghambarian *et al.*, 1988).

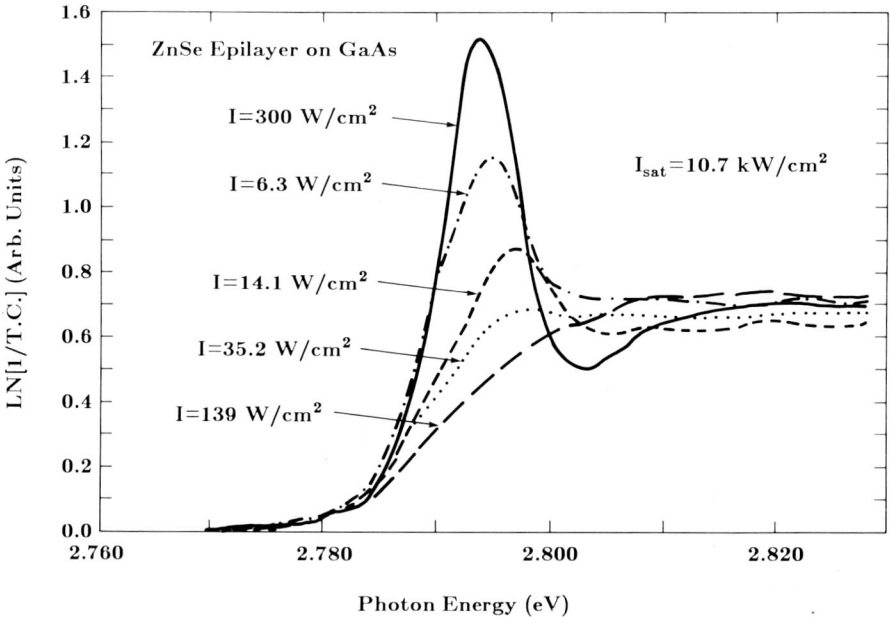

FIG. 17. Change in the excitonic absorption as a function of intensity for a 1.3 μm ZnSe film at $T = 77$ K. The transmission coefficient (T.C.) on the ordinate axis is defined as output power divided by input power.

The quasi-2D electron and hole confinement in a semiconductor quantum well has consequences beyond enhancing the binding energy and oscillator strength of an exciton. For example, enhancement in the stability of an excitonic molecule is also expected. This quasiparticle is a molecularlike entity composed of two electron–hole pairs bound together. Following the calculation by Kleiman (1983), one can estimate that the possible increase in the binding energy of such a molecule (biexciton) in a quantum well may be as much as one order of magnitude. To date, however, there is very scant evidence for biexciton formation in quantum wells (Miller et al., 1982), in part because the binding in a GaAs or similar III-V heterostructure is still weak (less than 1 meV). On the other hand, for bulk ZnSe, there are reports (Nozue et al., 1981) of biexcitons with binding energies $B \sim 3.5$ meV so that in a well-designed heterostructure, the molecular state should be unambiguously observable. Recently, the biexciton state has been detected in ZnSe/(Zn, Mn)Se MQW structures (Fu et al., 1988). The molecular binding energy was found to increase by one order of magnitude over the bulk value in narrow quantum wells.

The difficulty in obtaining p-type ZnSe to serve as an injector of holes motivated the study of ZnSe/ZnTe superlattice structures. However, there are difficulties associated with the use of ZnSe/ZnTe heterostructures. Predicted band offsets suggest a Type-II superlattice, and there is a large lattice-constant mismatch between ZnSe and ZnTe (7.4%). Strained-layer superlattice structures are still possible, provided layer thicknesses are restricted to a few tens of angstroms. Structures containing primarily ZnSe allow a reasonable lattice match to GaAs; whereas structures containing approximately equal amounts of ZnSe and ZnTe provide lattice matching to InP. ZnSe/ZnTe superlattice structures have been grown by a variety of techniques including hot-wall epitaxy (Fujiyasu et al., 1986), molecular-beam epitaxy (Takeda et al., 1986 and Kobayashi et al., 1986a), atomic-layer epitaxy (Takeda et al., 1986), and by a combination of ALE and MBE (Kolodziejski et al., 1988a). Photoluminescence studies (Kobayashi et al., 1986b) of MBE-grown structures indicate wavelength variations as the ratio of ZnSe to ZnTe layer thicknesses are varied. Photoluminescence emission is observed from the red to the green portion of the spectrum.

ZnSe/ZnTe superlattice structures can also be viewed as a means to circumvent the difficulties encountered in the MBE growth of the Zn(Se, Te) alloy (Kolodziejski et al., 1988a). As discussed in Section II.E, Zn(Se, Te) is of interest, since the photoluminescence yield of the alloy can be significantly enhanced over that of epitaxial layers due to the localization of excitons in the alloy. To avoid difficulties in controlling alloy concentrations during MBE growth, we have implemented ZnSe-based structures consisting of ultrathin layers of ZnTe spaced by appropriate dimensions to approximate a

Zn(Se, Te) mixed crystal with low or moderate Te composition. As an illustration, such a "pseudoalloy" was used to modify the ZnSe quantum well in a ZnSe/(Zn, Mn)Se heterostructure. In these MQWs, a ZnTe monolayer sheet was placed in the center of each ZnSe well; the well thicknesses ranged from 44 to 130 Å. In an effort to optimize the interface abruptness of the ZnTe monolayers, the ALE of ZnTe was performed on a recovered ZnSe surface, which made up, for example, the first half of a quantum well, whereas the remainder of the structure was grown by MBE. Although the architecture of these structures was substantially different from a bulk alloy, their optical transitions, as viewed in photoluminescence, were dominated by features that were quite similar to those found in the bulk alloy crystals at low to moderate Te composition. Figure 18 illustrates exciton trapping at the ZnTe monolayer sheets present in the ZnSe quantum well. For comparison, Fig. 18a shows the low-temperature photoluminescence spectrum of a more conventional ZnSe/(Zn, Mn)Se MQW structure (Kolodziejski et al., 1986c) for which the luminescence was dominated by sharp (FWHM < 5 meV), bright, blue exciton recombination at the $n = 1$ (light-hole) quantum-well transition.

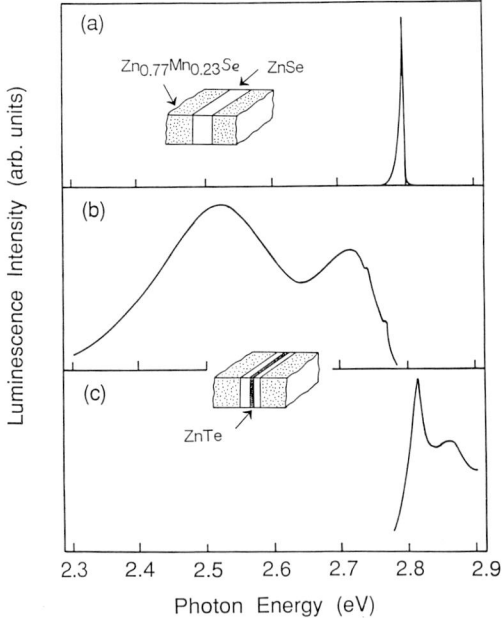

FIG. 18. Comparison of photoluminescence spectra at $T = 2$ K: (a) ZnSe/(Zn, Mn)Se MQW sample, (b) structure similar to that of (a) but with the insertion of a monolayer sheet of ZnTe in the middle of the quantum well, and (c) photoluminescence excitation spectrum of sample in (b). The amplitude of emission in (a) has been reduced to bring the peak to scale.

In contrast to Fig. 18a, the photoluminescence from a ZnSe/(Zn, Mn)Se MQW, which incorporates a ZnTe "sheet" in each quantum well, is shown in Fig. 18b. The broad luminescence features at lower energy were the result of exciton localization at the ZnTe/ZnSe heterointerfaces and were similar to those seen from bulk Zn(Se, Te) mixed crystals (Lee et al., 1987a). Figure 18c, showing a PLE spectrum, indicates that the position of the lowest energy-exciton transition was not significantly shifted by the presence of the ZnTe sheets. Experiments designed to take advantage of the presence of the Mn ions in the diluted magnetic-semiconductor microstructures containing the Te isoelectronic traps have allowed direct microscopic probing of the strongly binding center for excitons. Fu et al. (1989) have performed various magneto-optical measurements and determined that the hole wave function is strongly localized at Te sites with an orbit size comparable to that of a unit cell.

The occurrence of exciton self-trapping in these structures results in luminescence over a range of wavelengths. Also, the quantum efficiency for temperatures above LHe temperatures is enhanced by the exciton localization. The enhancement results, since the localized excitons are more likely to recombine radiatively as they have less opportunity to encounter a nonradiative center. Such phenomena could prove important for the realization of practical devices in situations where the localization persists at room temperature. It should be noted that the predicted Type-II nature of the ZnSe/ZnTe interface would result in the holes residing in ZnTe. Caution, however, must be exercised when considering band offsets for layers that are one or two monolayers thick. It is interesting to consider experiments such as those described here as a means for the study of the transition between localization at Te sites within ZnSe, and hole transfer across a Type-II ZnSe/ZnTe heterointerface.

B. CdTe-Based Superlattices

CdTe is an important II-VI-compound semiconductor, which apart from potential optoelectronic applications, can be used as a substrate for fabrication of focal-plane arrays based on (Hg, Cd)Te alloys and superlattices. Alloying CdTe with Mn results in epitaxial layers of the dilute magnetic semiconductor; the growth of (Cd, Mn)Te by MBE presents an opportunity for developing integrated optical isolators, modulators, and switches in the visible and near-infrared portions of the spectrum. Recent work has established $Hg_{0.08}Cd_{0.67}Mn_{0.25}Te$ as an attractive material for isolator applications in the 820 nm range, whereas $Cd_{0.55}Mn_{0.45}Te$ performs well at the somewhat shorter wavelengths of interest to optical-disc-storage applications.

The first superlattices in the CdTe/(Cd, Mn)Te material system were grown by MBE with superlattice interfaces parallel to (111) planes (Bicknell et al., 1984a and Kolodziejski et al., 1984a). Unusual optical and magneto-optical properties exhibited by these (111) strained-layer superlattices were attributed to the existence of interface-localized excitons associated with the presence of (111) interfacial planes (Zhang et al., 1985). The ability to grow either (111) or (100) CdTe on (100) GaAs provided a unique opportunity to compare directly the effect of orientation on the optical properties of the superlattices. The microstructural, optical, and magneto-optical properties of the many CdTe/(Cd, Mn)Te superlattice structures have been more thoroughly summarized in Kolodziejski et al. (1989).

A number of superlattice configurations have been fabricated with both (100) and (111) orientations. For the (111) orientation, both CdTe and $Cd_{1-x}Mn_xTe$ have been used as the quantum-well material, with a range of Mn mole fractions in the $Cd_{1-x}Mn_xTe$ barrier material (Bicknell et al., 1984a and Kolodziejski et al., 1984b). For superlattices having the (100) orientation, only CdTe has so far been used as the well material. These superlattice structures were at first grown on $1-2\ \mu m$ (Cd, Mn)Te or CdTe buffer layers on (100) GaAs substrates (Kolodziejski et al., 1985b) and were subsequently grown on CdTe substrates (Bickness et al., 1987).

In the (Cd, Mn)Te material system, the sense of variation of lattice constant versus Mn mole fraction is opposite to that for the (Zn, Mn)Se system. In the case of (Cd, Mn)Te, the lattice constant decreases as the Mn fraction increases (Kolodziejski et al., 1984a). As a result, the strain subjects the well material to expansive uniaxial strain normal to the interface, and compressive strain parallel to the interface. The hydrostatic component of the strain increases the optical gap, whereas the uniaxial component acts to remove the valence-band degeneracy at $k = 0$, so that the heavy-hole band (in the direction of the superlattice axis) now moves up in energy relative to the light-hole band.

TEM observations have been made for both (100)- and (111)-oriented superlattice structures. High-resolution images of the superlattice interfaces were observed, and no discontinuity in lattice fringes at the interfaces were found. The implication of the interface coherence is that the lattice mismatch between superlattice layers is accomodated by elastic strain rather than by misfit-dislocation networks, resulting in strained-layer superlattices (Kolodziejski et al., 1984b). Figure 19 is a HREM image of an interface between a 120 Å CdTe quantum well and a 120 Å $Cd_{0.6}Mn_{0.4}Te$ barrier layer in a (111)-oriented multiquantum-well structure. Where superlattice thicknesses exceed the critical layer thickness for elastic accomodation of the strain, misfit-dislocation networks were observed (Choi et al., 1986). All samples have shown highly regular superlattice structures in both images and diffraction patterns. The sharpness of the interfaces between (100) superlattice layers appears similar to (or better than) that of (111)-oriented superlattices.

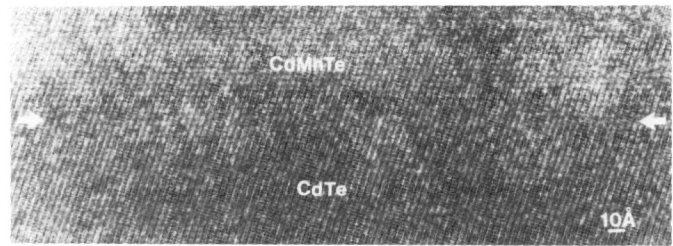

FIG. 19. HREM image of the interface between a (111)-oriented CdTe well and a $Cd_{0.60}Mn_{0.40}Te$ barrier in a multiple-quantum well structure, with each layer having dimensions of 120 Å.

By studying the low-temperature photoluminescence (Fig. 20) from a number of (111)-oriented MQW structures as a function of well thickness, it was discovered that their intense low-temperature recombination originated from localized excitons, particularly in narrower well (< 100 Å) samples. Such localized exciton-emission luminescence spectra were found readily to redshift in external magnetic fields (Zhang *et al.*, 1985b). Whereas spectral shifts were expected from considering the effects by spin exchange for the finite penetration of the exciton wave function into the (Cd, Mn)Te barriers, they were unexpectedly large and also showed pronounced anisotropy regarding the direction of the applied field with respect to the superlattice axis. The

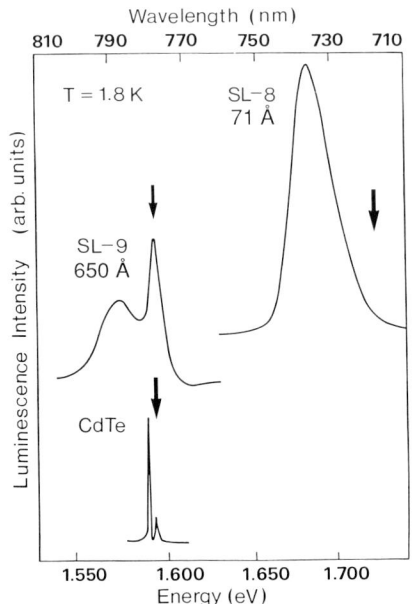

FIG. 20. Photoluminescence spectra at $T = 1.8$ K for two (111)-oriented CdTe/$Cd_{0.74}Mn_{0.26}$Te MQW samples (well widths 71 Å and 650 Å) in comparison with high-quality bulk CdTe (bottom trace). The arrows denote the assigned free-exciton energies for the (111)-oriented MQW and the bulk sample as determined from reflectance spectra.

measured low field shifts for the narrow-well MQWs show that as much as half of the hole wave function lies within the (Cd, Mn)Te layers. This is consistent with a small valence-band offset. The anisotropy has been used to argue that the heavy-hole wave function is substantially two-dimensional (Zhang et al., 1985a,b; Nurmikko et al., 1985). Furthermore, magneto-optical data at very high magnetic fields (> 200 KG), following the evident paramagnetic saturation of the contributing Mn-ion spins, shows little change in the exciton ground-state energy, thereby supporting the notion that the exciton is rather two-dimensional as well (Zhang et al., 1985b). The combination of large penetration of the hole wave function into the Mn-containing region of the MQWs, together with the substantially two-dimensional character of the hole and the exciton, has provided an empirical insight that strongly suggests that the recombining excitons are physically localized at the (111) CdTe/(Cd, Mn)Te heterointerface (Zhang et al., 1985a and Petrou et al., 1985).

The ability to grow the (Cd, Mn)Te superlattices in both (111) and (100) crystalline orientations (Kolodziejski et al., 1985b) provided an opportunity to study possible differences in associated electronic characteristics in these strained-layer structures for the two polar interfaces. The contrast between the two growth orientations was pronounced in luminescence excitation spectra (Chang et al., 1986). Whereas a strong ground-state exciton absorption peak was evident for the (100) case (which also showed strain-split valence-band and excited-state structure), this resonance was almost completely broadened out in the (111) structures. It should be noted that taking excitation spectra on (111) structures at considerably lower Mn concentrations ($x = 0.06$) has shown how well-defined features do indeed emerge even for this orientation (Warnock et al., 1985).

Raman scattering has provided clear evidence for acoustic phonon zone folding in a narrow well/barrier (111)-oriented MQW structure, in good agreement with an elastic continuum model, and consistent with a high degree of crystalline quality (Venugopalan et al., 1984). Similar Raman measurements show an unexpected contrast between (111)- and (100)-oriented CdTe/(Cd, Mn)Te MQWs in studies of the longitudinal-optical (LO) phonon dispersion in samples comparable to those discussed earlier. These Raman data show LO phonon confinement occurring in the separate layers of the (100) structures, but is not evident in the (111) case. It is possible that the differences seen in the Raman experiments (Suh et al., 1987) have a common underlying physical origin to that which produces the large difference in the excitation spectra between the two orientations.

Further evidence of the optical quality of the MBE-grown structures is supported by the demonstration of stimulated emission in both CdTe/(Cd, Mn)Te (Bicknell et al., 1985a and Bonsett et al., 1987) and

$Cd_{1-x}Mn_xTe/Cd_{1-y}Mn_yTe$ quantum-well structures (Bicknell et al., 1985b). Laser samples were prepared by using a selective chemical etch to remove the GaAs substrate; the resultant free-standing epilayers were cleaved and mounted on a copper heat sink (Holonyak and Scifres, 1971) for optical-pumping experiments. For lasers structured with (111) CdTe as the active material, lasing was obtained at wavelengths of 763 to 766 nm at 25 K with a threshold-power density of 1.35×10^4 W/cm^2 (Bicknell et al., 1985a). When (111) $Cd_{1-x}Mn_xTe$ ($x = 0.19$) was the active quantum-well material, lasing occurred in the red spectral region at 665–670 nm at 15 K; the threshold for laser action was 2.0×10^4 W/cm^2 (Bicknell et al., 1985b). The exchange interaction occurring in the DMS material allows a shift in the energy of the quantum-well states such that the application of a magnetic field would allow for tuning of the output energy of the DMS laser (Fig. 21). For the DMS lasers studied, a magnetic-field tuning rate of 3.4 meV/tesla was obtained at 1.9 K (Isaacs et al., 1986). This magnetic-field-induced shift was approximately one-fifth that obtained from bulk $Cd_{1-x}Mn_xTe$ with a comparable x value. (100)-oriented quantum wells, containing CdTe as the active well material, have also exhibited stimulated emission up to 119 K (Bonsett et al., 1987). The

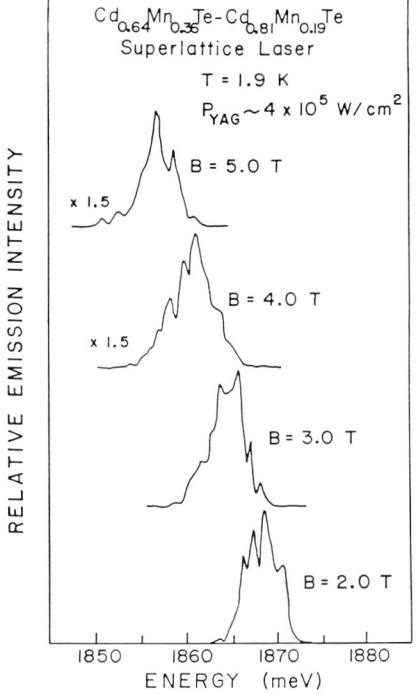

FIG. 21. Stimulated emission from a CdMnTe/CdMnTe superlattice structure as a function of applied magnetic field illustrating magnetic field tunability due to the presence of the Mn ion in the active laser region. Reprinted with permission from J. F. Schetzina (from Isaacs et al., 1986).

effects of strain in these strained-layer superlattices results in TE-polarized stimulated emission from the sample edge, and has been compared with the TM-polarized edge emission from oppositely strained (Zn, Mn)Se MQW structures (Bonsett et al., 1987).

The binary superlattice CdTe/ZnTe (Monfroy et al., 1986) has also been a subject of optical spectroscopy as a possibly useful pseudomorphic structure. The situation involves a very large lattice mismatch (approximately 6.4%). Initial optical studies had suggested a reasonable agreement between experimentally derived superlattice bandgap values and calculations in the "free-standing" superlattice limit (assuming also the absence of a finite valence-band offset (Miles et al., 1986). A small valence-band offset would follow the situation already established with CdTe/(Cd, Mn)Te and ZnSe/(Zn, Mn)Se, in that the finite lattice-mismatch strain (through hydrostatic and uniaxial components) may be the main factor determining the actual valence-band offsets, thus making them dependent on individual sample parameters. Apart from the band-offset issue, subsequent resonance Raman experiments, however, have cast doubts on the arguments on attaining the free-standing superlattice limit in the CdTe/ZnTe system (Menendez et al., 1987).

One important consequence of a small band offset in a real, highly strained system subject to small but finite structural irregularities, however, has been established by using the lowest interband exciton resonances as an indicator in photoluminescence experiments. In particular, time-resolved and resonantly excited spectroscopies have shown that excitons exhibit unusual localization in the CdTe/ZnTe system (Hefetz et al., 1986b). Qualitatively, variations in the layer thickness on a monolayer scale in a highly strained structure is sufficient to produce significant fluctuations in the local strain about some mean value. In the absence of strong confinement (i.e., small offset), the associated random potential fluctuations may be quite efficient in capturing electronic quasiparticles at low or moderate lattice temperatures.

Low-threshold, optically pumped lasers, emitting in the yellow-orange portion of the visible spectrum and fabricated with $Cd_{0.25}Zn_{0.75}Te/ZnTe$ superlattice structures, have been reported by Glass et al. (1988). The binary CdTe/ZnTe superlattices are heavily strained; thus, the laser structures incorporated alloys of CdZnTe as the well layer to reduce strain. In the superlattice structures studied, the lasing wavelength increased from 575 nm at 8 K to 602 nm at 310 K. At low temperatures, the threshold pump intensity was found to be quite low at $7\,kW/cm^2$. At room temperature, the threshold pump intensity increased to $\sim 55\,kw/cm^2$ and represented the first report of room-temperature, optically pumped lasing in a II-VI superlattice.

C. II-VI/III-V Heterojunctions and Multiple-Quantum Wells

There are several II-VI/III-V heterostructures that possess potential technological importance. Three examples that have received attention are ZnSe/GaAs, ZnTe/(Al, Ga)Sb, and CdTe/InSb. The common property shared by all three is that each pair, although composed of compounds giving considerably different bandgap energies, are closely lattice matched. The implication of the close lattice match is the opportunity for a dislocation-free interface by requiring one of the pairs to be strained (pseudomorphically), such that the lattice spacings match in the plane of the interface.

The primary motivating factors in consideration of materials for a heterostructure configuration are usually the bandgaps of the two materials. The next consideration is the issue of lattice-constant match. The latter consideration is especially interesting where MBE is to be used, since the nonequilibrium growth technique presents the possibility of employing materials that, in the form of bulk crystals, do not share the same crystal structure. Finally one must consider the band lineup in order to estimate the energy-band diagram for the heterostructure. The question of band offset is the most difficult consideration to deal with, since there exist a wide range of theoretical estimates for heterojunctions of interest. Furthermore, although various experimental techniques have been used to measure the band-offset values, the results are controversial for virtually all configurations except (Al, Ga)As.

The motivation for considering the CdTe/InSb heterostructure is somewhat different than that for the other two. Simply stated, the interest is to obtain InSb quantum-well structures; the problem to be solved, by using a II-VI/III-V heterojunction, is the absence of suitable III-V compounds to serve as the barrier layers in conjunction with an InSb quantum well. There are no available III-V compounds having lattice constants compatible with InSb.

The interest in the ZnSe/GaAs system has two origins. One motivating consideration is the difficulty of obtaining p-ZnSe for incorporation in injection-light-emitting devices operating in the blue/blue-green portion of the spectrum. The proposal is to inject holes from p-GaAs into n-ZnSe. The inherent problem is that of injecting charge from a narrow-bandgap semiconductor into a wide-bandgap material. Furthermore, the evidence is for much of the energy-band discontinuity to occur between valence bands, which is unfavorable for the desired hole injection. The second motivation for the study of the ZnSe/GaAs heterointerface is the opportunity for the passivation of GaAs with a charge barrier(s) exceeding that of the other successful epitaxial "insulator" such as (Ga, Al)As.

Interest in the ZnTe/(Ga, Al)Sb system is the most recent and arises from the problem of obtaining amphoteric doping with ZnTe, which again is necessary to realize a widegap- (green) injection-emitting device. In this section, we present recent results of MBE epitaxy and evaluation of the three heterostructures.

1. InSb/CdTe Multilayered Structures

One of the more interesting quantum-well structures that have been proposed involves InSb as the well material with CdTe as the barrier layer. Theoretical predictions (Tersoff, 1986) of band offsets agree fairly well with experimental measurements (Mackey et al., 1986) and suggest that these quantum wells will be of Type I with substantial conduction- and valence-band confinement. Minimal strain effects are expected because these two materials are very closely lattice matched ($\sim 0.05\%$), whereas a perfect lattice match can be achieved by incorporating a few percentages of either Zn or Mn into the CdTe barrier layer. Large quantum shifts in the bandgap energy are predicted for relatively wide quantum wells as a result of the small effective mass of electrons and light holes. For example, a 75 Å quantum well has a ground-state transition energy twice that of bulk InSb (Van Welzenis and Ridley, 1984). Structures involving reasonable well dimensions allow a wavelength range of $2-5.5\ \mu m$ to be accessed. The high carrier mobilities and the large de-Broglie wavelength provide the possibility for a wide variety of interesting devices. The realization of proposed device structures, however, has been hampered by the significant materials problems associated with this II-VI/III-V materials system.

Several research groups have reported the MBE growth of InSb on CdTe substrates (Sugiyama, 1982), and CdTe on both InSb substrates (Mackey et al., 1986; Farrow et al., 1981; Wood et al., 1984; Williams et al., 1985; and Zahn et al., 1987) and InSb epilayers (Wood et al., 1984). These studies have primarily focussed on interfaces between epitaxial layers and substrates, whereas recent work has been reported that involves epilayer/epilayer interfaces (Kolodziejski et al., 1988b; Glenn et al., 1989). The majority of previous studies employed InSb and CdTe bulk substrates that were ion etched and thermally annealed prior to epitaxy. The resultant epitaxial layers were of very high quality; however, close examination of the interfacial region revealed a variety of problems. These difficulties must be eliminated because in quantum-well structures the interfaces can completely dominate the electronic and optical behavior. Many problems arise due to the widely differing optimum growth temperatures for the two materials; high-quality CdTe on InSb is grown at temperatures as low as 160°C (Farrow et al., 1984), whereas InSb with superior electrical properties is grown at temperatures at or above 400°C. Interfacial problems occurring at the CdTe/InSb interface include interdiffusion (Sugiyama, 1982; Mackey et al., 1986; Zahn et al., 1987), precipiate formation (metallic indium or Sb segregation) (Wood et al., 1984; Williams et al., 1985; Zahn et al., 1987), and intermediate layer formation of In_2Te_3 (Mackey et al., 1986; Zahn et al., 1987). It is unclear which of these problems will prove important for the case of epilayer/epilayer interfaces. In addition to the interface problem, a fundamental difficulty associated with the

CdTe/InSb system is the tendency for autodoping. A recent study (Williams et al., 1986) indicated that when both compounds are grown in the same MBE chamber, Te seriously contaminated the Sb source so that it was difficult to control the carrier concentration of the InSb material.

In order to circumvent some of these problems, recent work (Kolodziejski et al., 1988b; Glenn et al., 1989) involved (i) employment of two separate growth chambers, connected by an ultrahigh-vacuum transfer module, to eliminate autodoping problems; (ii) use of an antimony cracker as a source of Sb_2 in an effort to improve the low-temperature growth of InSb; and (iii) study of the effects on the heterointerface when a large Cd overpressure is used during multilayered growth (Golding et al., 1988). To achieve high-quality InSb material at substrate temperatures near 300°C, the use of Sb_2 was employed in anticipation of achieving similar results as obtained for the low-temperature growth of GaAs using As_2 (Neave et al., 1980; Missous and Singer, 1987) in place of As_4. Very low growth rates (0.18 Å/s) were also used for the growth of the InSb in various heterostructure configurations; in addition, an epitaxial buffer layer of CdTe was grown when a CdTe substrate was used, and an epitaxial layer of InSb was grown on the InSb substrate. (Growths employing the Sb cracker were performed in a single growth chamber.)

Golding et al. (1988) have studied the effect of Cd:Te flux ratio and InSb growth rate on the interfacial properties of multilayered structures using AES and depth profiling. It was reported that the tendency for In_2Te_3 formation at the interface during nucleation of CdTe on InSb was associated with a deficiency of Cd at the growth surface. It was found that a considerable reduction in the tendency for interfacial compound formation resulted when an overpressure of Cd was used during nucleation of the CdTe layer. For the growth of InSb on CdTe at 300°C, the evolution of RHEED patterns, from spotty to streaked, indicated three-dimensional nucleation. For growth rates of InSb that were less than 0.15 μm/hr, however, the spotty pattern remained unchanged during the growth period. At the lower growth rates, AES analysis and depth profiling revealed a complete degradation of the 400 Å CdTe layer lying below the interface, and severe intermixing throughout.

When two growth chambers (Kolodziejski et al., 1988b; Gunshor et al., 1988) were used for the growth of InSb/CdTe single-quantum-well structures, the InSb epilayer was transferred from the InSb growth chamber via an ultrahigh-vacuum transfer module (4×10^{-10} torr) to a separate growth chamber for the CdTe epitaxy. Following the growth of CdTe, the structure was returned to the InSb chamber for formation of the quantum well, after which the second CdTe barrier or cap layer was formed. Single-quantum wells and multiple-quantum wells (20 periods) of InSb/CdTe have also been grown by using a single growth chamber; the Sb cracker was employed for the growth of the InSb layers with substrate temperatures of 280°C. (A similar epitaxial growth approach has been employed for the growth of ZnSe/GaAs/ZnSe heterostructures, as described in Section IV.C.2.)

Figure 22 shows a dark-field TEM micrograph of a 20-period CdTe/InSb multiple-quantum well structure. The dark contrast represents the 163 Å InSb well, while the light contrast is the 168 Å CdTe barrier layer. The presence of an In_2Te_3 layer at CdTe/InSb interfaces has been reported by Zahn et al. (1987) but is not confirmed in these TEM investigations. In one growth sequence, MQW structures were grown under similar growth conditions with the exception that differing Ta cracking-tube temperatures were used. For one 20-period MQW structure, the cracking zone was kept at 850°C, whereas a second structure was grown by using a temperature of 1040°C for the cracking zone. TEM investigations revealed a high density of dislocations generated in the region of the InSb/CdTe multiple-quantum well for the former structure. For approximately the same growth rate, the second structure, grown with the higher cracking-zone temperature, exhibited the reduction of an order of magnitude in the number of dislocations occurring in the superlattice region. The generation of dislocations in the multiple-quantum-well region is unexpected since these two materials are very closely lattice matched. We speculate that the reduction in the number of dislocations was related to a decrease in the number of Sb precipitates that may form at such low growth temperatures for InSb.

Structural characterization of the InSb/CdTe MQWs has also been performed with x-ray rocking-curve diffraction. Figure 23 shows an x-ray rocking curve obtained from a 15-period superlattice structure having a periodic spacing of approximately 833 ± 10 Å. Satellite peaks are present with spacing of 208 arc sec, and indicate a superlattice periodicity of 870 Å agreeing well with the TEM. The higher-angle feature (FWHM = 22 arc sec) is attributed to the zero-order diffraction peak of the multilayer structure, whereas the other high-intensity feature (FWHM = 11 arc sec) in the spectrum is the (004) reflection of the InSb buffer layer/substrate. Assignment of

FIG. 22. Dark-field TEM micrograph of an InSb/CdTe multiple-quantum-well structure, having 20 periods, with layer dimensions of 163 Å and 168 Å, respectively.

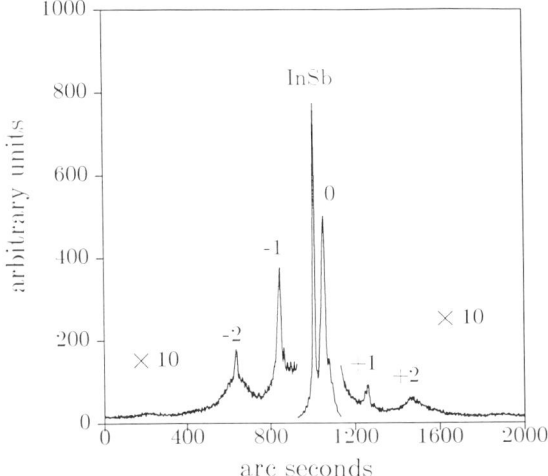

FIG. 23. X-ray rocking curve of a 15-period superlattice structure of InSb/CdTe.

the high-angle peak, as corresponding to the average lattice spacing of the multilayer in the growth direction, assumes that each interface of the MQW structure contains ultrathin interfacial layers of In_2Te_3. Since the lattice spacing in the growth direction for the InSb buffer/substrate should be smaller than the average lattice-plane spacing associated with the periodic structure, the zero-order diffraction peak from the superlattice is expected to lie at a lower angle than the peak corresponding to InSb. A simple calculation, involving minimization of the strain energy in the multilayer, will predict the measured angular positions of the diffraction peaks, provided several (~ 5) monolayers of the assumed interfacial In_2Te_3 are present per period of the superlattice.

Infrared photoluminescence has been used to examine the optical properties of the InSb epilayers (Kolodziejski et al., 1988b), double heterostructures (Gunshor et al., 1988), and multiple-quantum wells. The quantum efficiency of the double heterostructure is lower than that obtained from an epilayer or substrate, but is still reasonable. The double heterostructure consisted of a 0.42 μm InSb buffer layer grown on an InSb substrate, followed by a 1.63 μm CdTe buffer layer, the active 160 Å InSb layer, and a 2200 Å CdTe cap. (In this case, the structure was fabricated by the interrupted-growth approach using two separate growth chambers.) The spectrum contained two features whose higher energy peak was assigned to band-to-band recombination. In comparison with bulk InSb, the spectrum was approximately five times broader, however, measuring approximately 20 meV at $T = 10$ K.

2. ZnSe/GaAs Heterojunctions

The development of a metal-insulator-semiconductor (MIS) technology based on the important compound semiconductor GaAs has met with considerable difficulties due to the inadequate electrical characteristics of interfaces formed by using native oxides or by deposition of various insulators onto the GaAs surface. In the "nonepitaxial" insulator/GaAs interfaces, the existence of a large number of interface states (Wieder, 1978; Meiners, 1978; Kohn and Hartnagel, 1978) limits the extent to which the bands can be bent. An alternative technology has been developed, however, in which the wider-bandgap semiconductor (Al, Ga)As is used as a semi-insulator in a variety of field-effect-transistor (FET) structures where the (Al, Ga)As layer is either doped (Mimura et al., 1980; Solomon and Morkoç, 1984) or undoped (Matsumoto et al., 1984; Solomon et al., 1984; Cirillo et al., 1985). An "epitaxial" heterojunction is utilized in these transistor structures due to the nearly identical lattice match between (Al, Ga)As and GaAs, and due to the ability to fabricate such a heterointerface without growth interruption at the electrical junction. A very different "epitaxial" II-VI/III-V heterojunction has recently been employed in a FET structure where pseudomorphic ZnSe (Gunshor et al., 1987a; Kolodziejski et al., 1987b) forms a pseudoinsulator on doped epitaxial layers of GaAs. Typical transistor characteristics were exhibited, and channel modulation indicated that the Fermi level could be varied over a large portion of the GaAs bandgap (Studtmann et al., 1988).

In comparison to GaAs, ZnSe has a bandgap twice that of GaAs; ZnSe possesses a close lattice constant (0.25% mismatch); and the two semiconductors have highly compatible thermal-expansion coefficients. The wider bandgap of ZnSe, compared to that of (Al, Ga)As (2.0 eV for an Al mole fraction of 0.5), suggests a variety of device applications where ZnSe may present an alternative to (Al, Ga)As for passivation of GaAs. Photoluminescence (Gunshor et al., 1987a; Kolodziejski et al., 1987b), transmission electron microscopy (Gunshor et al., 1987a; Kolodziejski et al., 1987b; Studtmann et al., 1988), and piezo-modulated reflectance spectroscopy (Lee et al., 1988) indicate the high quality of pseudomorphic ZnSe grown by molecular-beam epitaxy onto epitaxial layers of GaAs. The heterointerface is interrupted during growth for the transfer from the III-V to the II-VI growth chamber. In the earlier experiments performed at Purdue, the as-grown GaAs was protected by the use of an amorphous arsenic layer (Waldrop et al., 1981; Price, 1981). The samples were transferred in air between MBE machines; later work employed a modular MBE with the transfer between III-V and II-

VI chambers occurring under ultrahigh-vacuum (UHV) conditions. The high-resolution TEM images of the structures suggest that the ZnSe/GaAs interface possesses a similar degree of structural perfection as the (Al, Ga)As/GaAs interface. An electrical characterization of the pseudomorphic ZnSe/GaAs heterojunction, consisting of capacitance-voltage (C-V) and current-voltage (I-V) measurements of metal/ZnSe/GaAs capacitors, illustrated successful passivation of the GaAs epitaxial surface. In situ or ex situ thermal processing of the ZnSe/GaAs epitaxial structures have a beneficial effect on the electrical properties of the II-VI/III-V heterointerface.

A number of prototype device structures have been fabricated in which pseudomorphic ZnSe forms the "insulator" in GaAs depletion-mode (FETs) (Studtmann et al., 1988). The I_D–V_{DS} curves for a 45 μm gate prototype device is shown in Fig. 24. The FET curves show good depletion-mode characteristics with complete pinch-off and current saturation. The modulation of the channel carrier concentration indicates that the Fermi level positioning at the ZnSe/n-GaAs interface can be varied by at least 0.6 eV. Although the transconductance (g_m) appears to be low (3.5 mS/mm), when the effect of series resistance is included, a value of 5.1 mS/mm is obtained, which agrees fairly well with a theoretical maximum prediction of 8.5 mS/mm.

A series of MIS capacitors have been fabricated for which the (C-V and the I-V characteristics have been investigated (Qian et al., 1989). Following growth of the ZnSe/GaAs heterostructures, capacitors (125 μm in diameter) were formed by using a lift-off process for the metal gate. For each epilayer grown, two different sets of capacitors were analyzed; in the first case, capacitors were fabricated on "as-grown" heterostructures, whereas the second set had capacitors fabricated after a post-deposition (ex situ) anneal of

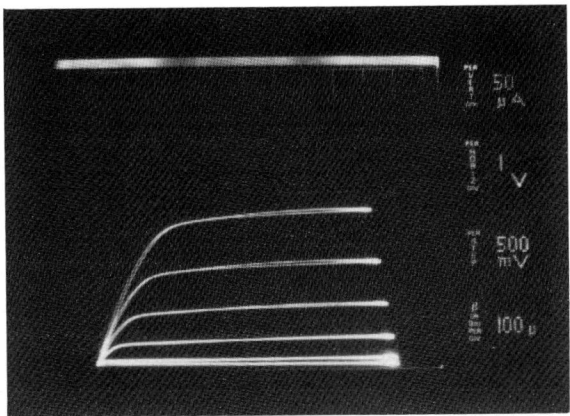

FIG. 24. Room-temperature I-V characteristic of the metal/ZnSe/n-GaAs FET with a gate width and length of 45 μm. The vertical scale is 50 μA/div, and the horizontal scale is 1 V/div. The gate bias is decreased by 0.5 V/div starting at 0 volts.

the ZnSe/GaAs interface. The annealing was performed at 600°C in flowing nitrogen gas for a duration of 30 minutes, with a GaAs wafer as a cap. The data described here were taken in the dark at room temperature. Figure 25 shows the high-frequency (1 MHz) C-V characteristics for a ZnSe/p-type GaAs structure that was subjected to the post-deposition anneal. The doping density, as determined from a plot of $1/C^2$ versus voltage, is unchanged by the thermal process. It is seen than an accumulation condition exists, whereas for more positive voltages (~ 4 V), the capacitor is depleted. For further voltage increases, the device goes into deep depletion. Also apparent for this particular sample is the presence of a negative fixed charge. The C-V measurements were performed for a range of frequencies between 10 kHz and 1 MHz; very little (approximately 2.5%) frequency dispersion of the accumulation capacitance was observed. The comparison between the behavior of annealed capacitors and the theoretical curve indicates the nearly ideal C-V characteristics. From the high-frequency C-V curve, the surface-state density for the heterointerface has been estimated. The best samples have interface-state densities that average below 10^{11} cm^{-2} eV^{-1} over the bandgap; these values are comparable to interface-state densities reported for the (Al, Ga)As/GaAs interface (Chung et al., 1987).

The C-V characteristics exhibited by as-grown capacitors differ greatly from those that have been thermally processed. For "as-grown" capacitors,

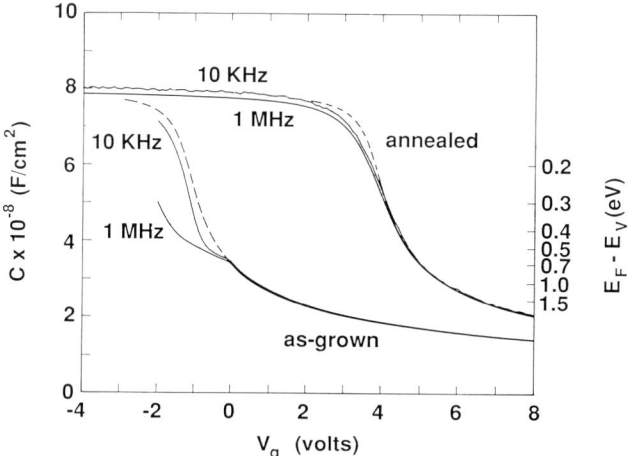

FIG. 25. Capacitance versus voltage characteristics of "as-grown" and annealed metal/ZnSe/p-GaAs capacitors. The sweep rate was 8 mV/s for all experimental curves. The theoretical C-V curve (dashed line) has been shifted to facilitate comparison with the experimental curves. (Parameters used in the calculation include 7.9×10^{-8} F/cm^2 for the ZnSe capacitance and 2.1×10^{16} cm^{-3} for the doping density of the p-GaAs.)

accumulation cannot be achieved in the structures as current begins to flow. At voltages necessary to achieve the accumulation region, a variation can be observed in the capacitance for various frequencies. Measurements of current versus voltage for the capacitors also exhibited differences between structures that were thermally processed and as-grown structures. For the unannealed sample, a very rapid rise in current was observed with increased negative bias on the gate. After the anneal process, however, the current-voltage data show very little conduction current (10^{-7} to 10^{-8} A/cm^2, a value corresponding to the background-noise level of the picoammeter used) for virtually the entire range of accumulation-biasing conditions. These data indicate that the annealed structure now contains a barrier to holes. Structures fabricated with n-GaAs do not exhibit accumulation, but do form an inversion layer. Although the band discontinuities for the ZnSe/GaAs heterojunction have been predicted (Harrison, 1977) and measured by x-ray photoelectron spectroscopy (Grant et al., 1983), discussion is still ongoing.

The thermal processing of the interface could be expected to substantially alter the chemical nature and structural properties of the II-VI/III-V interface. Tu and Kahn (1984, 1985) have suggested that when ZnSe is evaporated on (100) GaAs at similar temperatures reported here, a GaSe layer can form at the interface. When they performed a high-temperature anneal at 500–540°C subsequent to growth, a layer of Ga_2Se_3 was found by using Auger electron spectroscopy. In their case, the thick ZnSe was strain relaxed through formation of misfit dislocations; the presence of dislocations is known to enhance diffusion. Additional experiments are required to provide an understanding of the mechanisms giving rise to the improved electrical characteristics of the epitaxial ZnSe/GaAs interface. Experiments in progress in our laboratory indicate that modifications during the actual formation of the interface can result in the high-quality electrical properties of the heterojunction; variations to the temperature of nucleation of the ZnSe on GaAs, or stoichiometries of the two epitaxial layers, appear to provide alternate techniques to the ex situ anneal. Recent theoretical predictions (Farrell et al., 1988a) suggest that optimum interfacial conditions between II-VI and III-V semiconductors result when a layer of mixed cations is produced at the interface. It is conceivable that the steps (described above) leading to improved electrical properties of the interface are related to these theoretical predictions.

Preliminary experiments addressing the possibility of growth by MBE of multilayered ZnSe/GaAs structures have been reported by Tamargo et al. (1988c), where the ZnSe represented the barrier material. In their work, GaAs was deposited onto epitaxial MBE-grown ZnSe surfaces at low growth temperatures ($\sim 300°$C). These first results suggested the existence of an instability due to an electronic imbalance (Harrison et al., 1978; Harbison and Farrell, 1988; Farrell et al., 1988a) at the mixed II-VI/III-V interface. In situ RHEED, x-ray photoelectron spectroscopy, and ex situ TEM and PL have been used to characterize the heterostructures; these various measurements indicated the presence of a degree of interface disorder, and they have suggested that control of the ZnSe surface stoichiometry may alleviate some of the problems.

3. *ZnTe/AlSb/GaSb Heterojunctions*

As described above, the ZnSe-on-GaAs epilayer structures can exhibit interface-state densities that are comparable to those obtained in (Al, Ga)As/GaAs heterostructures. Measurement of hole accumulation (p-GaAs) and inversion layers (n-GaAs) suggest a substantial valence-band offset, with a small conduction-band offset, resulting in a band lineup that is probably not optimal for hole injection into the n-ZnSe layer. Although the ZnSe/AlAs heterostructure should be investigated for application to blue-light-emitting devices, another heterojunction that seems to be suitable for wide-bandgap emission consists of p-ZnTe/n-AlSb, which promises to have a favorable band lineup for carrier injection (McCaldin and McGill, 1988). The first epilayer/epilayer ZnTe/AlSb/GaSb heterojunctions grown by MBE on GaSb substrates were recently fabricated (Mathine et al., 1989); the structures were evaluated using RHEED, TEM, x-ray rocking curves, PL, and modulated-reflectance spectroscopy.

To avoid autodoping, two isolated II-VI and III-V growth chambers, connected by a UHV transfer tube, were used during the growth of the heterostructures. ZnTe was grown by using elemental sources, and flux ratios were measured by using a crystal monitor placed at the substrate position. Thick ($\sim 2 \mu$m) ZnTe layers were grown on GaSb substrates with various flux ratios. Pseudomorphic ZnTe layers were nucleated on pseudomorphic AlSb layers; the AlSb was grown on homoepitaxial GaSb buffer layers.

Microstructural quality was examined by using TEM and x-ray rocking curves. For the pseudomorphic samples, cross-sectional TEM images showed planar, atomically flat interfaces and an absence of dislocations and stacking faults in the interface areas. Despite the presence of a low density of misfit dislocations, the x-ray rocking curve FWHM of a nearly pseudomorphic sample with 2200 Å of ZnTe is 80 arc sec, which is close to the theoretically expected broadening (73 arc sec) due to the finite layer thickness. The measured spacings between (004) diffraction peaks of GaSb and AlSb, and GaSb and ZnTe correspond to the predicted tetragonal distortion.

Optical properties of the ZnTe layers were studied by photoluminescence and modulated-reflectance spectroscopy. Free-exciton-related features were observed in photoluminescence (and confirmed in modulated reflectance) for both thick (strain-relaxed) and pseudomorphic ZnTe epilayers. Dominant features of the luminescence corresponded to near-band-edge transitions, and only weak deep-level features were observed. Transition energies of pseudomorphic samples are shifted to higher values when compared to thick ZnTe epilayers, providing an opportunity to obtain the deformation potentials (Mathine et al., 1989).

D. METASTABLE ZINC-BLENDE MAGNETIC SEMICONDUCTORS

1. *Superlattices Containing Ultrathin MnSe*

The first growth of the hypothetical zinc-blende magnetic semiconductor MnSe (Kolodziejski et al., 1986a) resulted from efforts to increase the fractional incorporation of Mn in (Zn, Mn)Se. The existence of zinc-blende MnSe is a consequence of the kinetic (nonequilibrium) nature of the MBE growth method employed; bulk crystals of MnSe have the NaCl crystal structure (Pajaczkowska, 1978). Extrapolations of data on lattice constant and bandgap, obtained for zinc-blende (Zn, Mn)Se epilayers with up to 66% Mn, predict a bandgap and lattice constant for zinc-blende MnSe of 3.4 eV (6.5 K) and 5.93 Å, respectively. A variety of superlattice structures composed of ZnSe layered with ultrathin MnSe (3 to 32 Å) have been grown by MBE; the layer thickness of the MnSe (with monolayer resolution) was controlled by the use of RHEED intensity oscillations (Kolodziejski et al., 1987a,c; Gunshor et al., 1987b).

RHEED provided the first observation of the zinc-blende crystal structure for MnSe. During MBE growth, as the Zn shutter was closed and the Mn shutter was opened to begin the growth of MnSe on ZnSe, a virtually instantaneous increase in the intensity of the twofold reconstruction streaks was observed. In addition, the RHEED pattern revealed a high degree of ordering as demonstrated by the presence of Kikuchi lines and, with the exception of somewhat more intense reconstruction lines, was quite similar to the pattern observed from ZnSe. In all cases, the zinc-blende crystal structure was confirmed by the TEM examination of cross-sectional specimens. The observed electron-diffusion patterns corresponded only to zinc-blende phases with no indication of the wurtzite or rock-salt crystal structures. The metastable zinc-blende MnSe, with thicknesses ranging from 3 to 32 Å, was incorporated in strained-layer superlattice structures. When grown as a "thick" epilayer (400 Å), the zinc-blende crystal structure remained in the presence of strain-relieving misfit dislocations.

The zinc-blende MnSe served as the barrier when combined with ZnSe in superlattice structures. By controlling the thickness of, and spacing between, layers of the magnetic semiconductor, an opportunity was provided to study magnetic ordering as the dimensionality varied from 3D to quasi-2D. It was crucial in a study of this type for the MnSe layer thickness to be controlled to one monolayer. Although RHEED intensity oscillations had not been previously reported for growth of II-VI compounds, they were employed during the superlattice fabrication to provide the requisite one-monolayer resolution.

RHEED intensity oscillations were observed during the homoepitaxial growth of ZnSe on ZnSe and during the lattice-mismatched heteroepitaxial growth of ZnSe on MnSe (4.7% mismatch). One period of oscillation was found to be equivalent to one monolayer of growth as confirmed by TEM. It was interesting to compare the homoepitaxy of ZnSe with the growth of ZnSe on MnSe under identical growth conditions. For homoepitaxy, six or seven oscillation periods were observed after interruption and re-initiation of growth; in contrast, the nucleation of ZnSe on MnSe resulted in up to 30 oscillation periods of enhanced amplitude (Kolodziejski et al., 1986c). Figure 26 shows the effect of alternating the cation species without growth interruption during the fabrication of a binary superlattice composed of MnSe/ZnSe. For this particular binary superlattice, each period consisted of

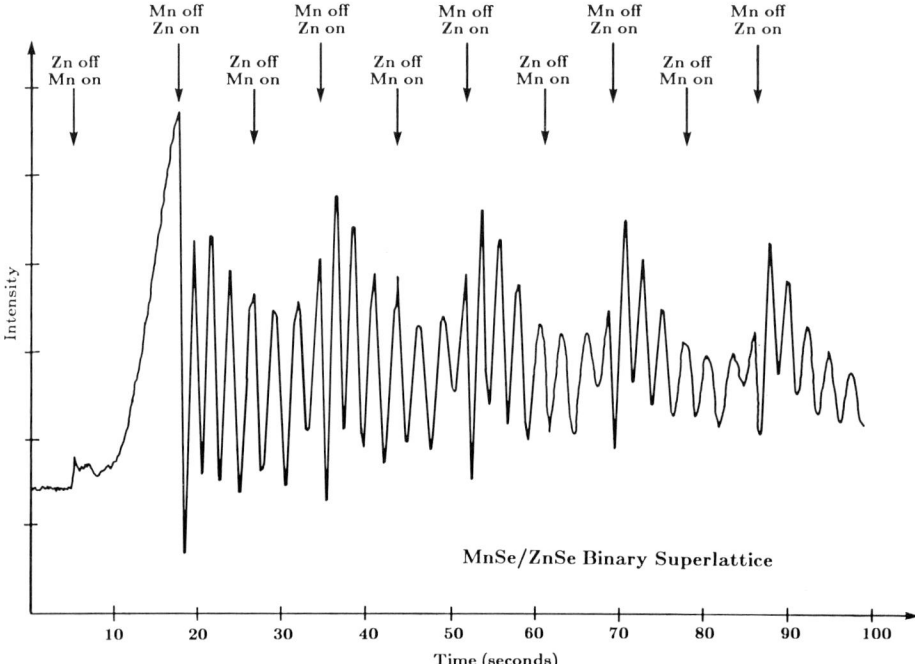

FIG. 26. RHEED intensity oscillation observed during growth of a ZnSe/MnSe binary superlattice. During the growth, the Se source shutter remained open throughout. (The first MnSe layer is grown on a thick ZnSe buffer layer.)

four monolayers of ZnSe separated by three monolayers of MnSe. The layer thicknesses were controlled by counting the number of oscillation periods and confirmed by HREM imaging of cross-sectional specimens. Throughout the entire growth of the five-period superlattice, intensity oscillations were recorded in both the ZnSe and MnSe layers. The observation of RHEED intensity oscillations during the growth provided the means to control the layer thicknesses with one-monolayer resolution.

A series of "comblike" superlattices consisting of 30 to 100 periods were grown with MnSe layer thicknesses of one, three, and four *monolayers*; the MnSe layers were separated by approximately 45 Å of ZnSe. For comparison, the superlattices with one- and three-monolayer MnSe thicknesses were grown with and without growth-interruption techniques at each interface. A fourth related superlattice structure consisted of 30 periods containing ten monolayers of MnSe alternated with 24 Å of ZnSe.

Photoluminescence measurements performed in the presence of an external magnetic field (up to 5 tesla) were initially used as an indicator of the magnetic behavior in these magnetic superlattices; complementary information was subsequently acquired through magnetization measurements using a SQUID magnetometer (Lee *et al.*, 1987b,c; Chang *et al.*, 1987). In photoluminescence, the observed magneto-optical shifts of the ground-state excitonic transition originate from the exchange interaction between electron-hole states of the superlattice and the magnetic moments of the Mn ions in the thin MnSe layers. In the absence of a magnetic field, the observed optical transition energies are in agreement with predictions of a Kronig–Penney model for these superlattices with ultrathin MnSe barriers. These calculations were performed by using a bandgap energy for zinc-blend MnSe of 3.4 eV, a value predicted by extrapolation from optical data obtained on zinc-blende (Zn, Mn)Se epilayers.

The rock-salt crystal structure of MnSe is known to exhibit antiferromagnetic ordering at low lattice temperatures. Recent susceptibility measurements (Furdyna *et al.*, unpublished) performed on Bridgman-grown bulk crystals of $Zn_{1-x}Mn_xSe$ existing in the wurtzite phase (the maximum x value obtained was 0.57) indicated an increased antiferromagnetic ordering (at low temperature) as the Mn mole fraction was increased. Magneto-optic measurements on zinc-blend MBE-grown epilayers showed a reduction of Zeeman shifts with increasing Mn mole fraction; the decreasing red shift was attributed to the tendency for antiferromagnetic ordering. The similarity in magnetic behavior between zinc-blend and wurtzite (Zn, Mn)Se crystals was not surprising, because nearest-neighbor distances are the same for the two crystal structures. Since due to the presence of an external magnetic field, no spectral red shift was anticipated for a superlattice structure containing antiferromagnetic MnSe, the degree of paramagnetic behavior observed was significant (Chang et al., 1987). The largest Zeeman shifts occurred for superlattices containing MnSe layer thicknesses approaching the monolayer limit; no Zeeman shift was observed for the superlattice containing ten monolayers of MnSe.

Direct magnetization measurements (Lee et al., 1987b,c; Chang et al., 1987) were made by using a SQUID magnetometer. Subtracting the diamagnetic (negative) GaAs substrate contribution, the striking result exhibited by the superlattice samples was the large positive contribution at low temperature. In strong contrast, the 400 Å MnSe epilayer behaved qualitatively as a normal-bulk antiferromagnetic insulator at lower temperatures.

Although further studies are necessary to rigorously identify the origin of the frustrated antiferromagnetism, the present interpretation is that the origin of the tendency for spins to align in an external magnetic field arises from "loose" spins at the MnSe/ZnSe heterointerfaces. The thinner the MnSe layer, the greater is the influence of the heterointerface, whereas for thicker MnSe layers, the antiferromagnetic "inner core" begins to dominate the magnetic behavior (Chang et al., 1987).

2. *Quantum-Well Structures with MnTe Barriers*

Recently, the first epitaxial layers of the zinc-blend phase of MnTe have also been grown by MBE (Gunshor et al., 1989); bulk crystals of MnTe have the hexagonal NiAs crystal structure. The variation of lattice parameter and excitonic bandgap with Mn concentration for zinc-blend (Cd, Mn)Te epilayers extrapolates to values of 6.328 Å and 3.18 eV (10 K), respectively (Lee and Ramdas, 1988), for zinc-blend MnTe. The difference in the bandgap energy for the two crystal structures is especially dramatic because the NiAs phase has a bandgap of 1.3 eV (Allen et al., 1977). The MnTe growth studies were undertaken in response to a considerable amount of speculation in recent years concerning the expected physical properties of the hypothetical zinc-blend MnTe (Wei and Zunger, 1986; Tersoff, 1986; Furdyna and Kossut, 1986; Ehrenreich, (1987); McCaldin and McGill, 1988). A series of double-barrier heterostructures were fabricated with MnTe forming the barrier layers for quantum wells of either CdTe or ZnTe; also relatively thick epilayers of zinc-blend MnTe have been grown to thicknesses of 0.5 μm. Optical evaluation of the structures provides information with regard to the bandgap of the zinc-blend MnTe as well as insight into the band offsets when layered with either CdTe or ZnTe. The microstructure of the MnTe epilayers was characterized by using TEM and x-ray diffraction; optical properties were studied by using reflectance, photoluminescence, Raman, and resonant Raman spectroscopies.

Three types of structures were fabricated in which the MnTe was incorporated: (i) thick epitaxial layers (up to 0.5 μm), (ii) double-barrier quantum-well structures, where MnTe forms the barrier when layered with CdTe, and (iii) double-barrier quantum-well structures, where MnTe is the barrier when layered with ZnTe. CdTe substrates having a CdTe buffer layer were used for the structures involving relatively thick epilayers of MnTe. In the series of MnTe/CdTe double-barrier structures, the CdTe layers were grown (1.2 Å/sec) using a compound source, while the MnTe layers were grown at 300°C at a growth rate of 1.1 Å/sec using elemental sources. The Te flux used during the growth of MnTe had the same intensity as the Te flux used during the growth of CdTe; the Mn flux was adjusted to approximate a unity cation/anion flux ratio. The ZnTe epilayers were grown at a substrate temperature of 320°C with a growth rate of 1.5 Å/sec.

The crystal structure was monitored during growth using RHEED. The electron-diffraction patterns provided evidence of the growth of metastable zinc-blend MnTe. During the sequential growth of barrier and well layers for the MnTe/CdTe structures, the (2 × 1) RHEED patterns appeared to be virtually identical, in contrast to the case of MnSe/ZnSe, where the intensity of the reconstruction lines sharply varied between alternate layers. The zinc-blend phase of MnTe was also confirmed in x-ray diffraction ($\theta-2\theta$) scans. The diffraction peaks obtained from x-ray diffraction measurements performed on samples containing relatively thick epilayers of MnTe could be identified as corresponding only to the zinc-blend phases of CdTe and MnTe.

The microstructure of MnTe epilayers and MnTe double-barrier structures have been examined by cross-sectional transmission electron microscopy (Gunshor et al., 1989). The MnTe layers have the zinc-blend structure and have formed epitaxially on CdTe or ZnTe epilayers. From the diffraction patterns, the MnTe epilayer appears to have formed a perfect epitaxial relationship with the CdTe crystal. In the diffraction pattern, each diffraction spot of the MnTe crystal forms a pair with a spot of the CdTe crystal, as expected from the zinc-blend structure of the MnTe crystal. Nearly complete relaxation of the lattice mismatch at the interface, due to formation of misfit dislocations, is suggested by the degree of separation of the 220-type diffraction spots.

In four MnTe/CdTe/MnTe double-barrier structures, for which the CdTe well thickness varies from approximately 56 to 28 Å (the MnTe barrier layer thickness is 35 Å), strong photoluminescence, originating from the well, was detected in all samples, even under excitation at energies below the MnTe barrier-layer absorption edge. Strong evidence of carrier confinement is obtained from the systematic shift to higher photon energy of the photoluminescence emission with decreasing CdTe quantum-well thickness (Fig. 27). For the narrowest well sample (28 Å), an overall blue shift of nearly 400 meV was realized. (Optical-reflectance measurements of (bulk) thin

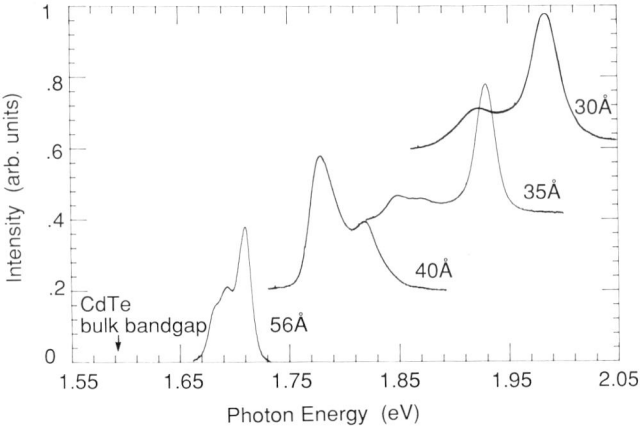

FIG. 27. The photoluminescence energy shift relative to the excitonic band edge of CdTe based on the average energy of the emission lines originating from the quantum well. A value of 200 meV is used for the valence-band offset in the calculations; the effects of strain are neglected..

MnTe films have yielded a value of 3.2 eV for the s–p bandgap.) Comparisons between the experimentally determined excitonic energies of the quantum-shifted excitonic features and predictions based on calculated transmission (Cahay et al., 1987) through the double-barrier structure are in good agreement. Since the MnTe is lattice mismatched to CdTe by 2.4%, the valence-band offset is expected to be significantly influenced by the strain. The strain appears primarily in the MnTe layers.

V. Conclusions

The research summarized in the chapter describes the advances that have been achieved by the use of molecular-beam epitaxy for the growth of II-VI materials and associated microstructures. The first stages necessary in the contemporary development of II–VI epitaxy are now completed. The generally optimistic attitude of researchers in the field suggests the beginning of a new era addressing II-VI-device development. There is increasing evidence that the nonequilibrium growth property of MBE (and MOCVD) will provide the means to surmount the many obstacles of the past that have held back the development of the II-VI-compound family. Significant understanding of the epitaxy process and the optical behavior of quantum-well and superlattice structures incorporating wide-bandgap II-VI compounds has been gained, with new results concerning the transport properties of the II-VI compounds under intense investigation.

Acknowledgments

The authors have attempted to review the latest progress made in the research of II-VI semiconductor microstructures. However, some relevant works may unfortunately have been overlooked. We would like to thank the authors who supplied figures, reprints, and preprints describing their work. We wish to thank the many graduate students and colleagues who contributed to the work at Purdue and Brown. The research was supported by the Office of Naval Research, the Air Force Office of Scientific Research, the Defence Advanced Research Projects Agency, the National Science Foundation, and the Naval Research Laboratory.

References

Aggarwal, R. L., Furdyna, J. K., and Molnar, S. von (1987). "Diluted Magnetic (Semimagnetic Semiconductors," *Proceedings of the Materials Research Society*, Vol. **89**.
Allen, J. W., Lucovsky, G., and Mikkelsen, J. C. (1977). *Solid State Commun.* **24**, 367.
Andersen, D. R., Gunshor, R. L., Kolodziejski, L. A., Datta, S., Kaplan, A. E., and Nurmikko, A. V. (1986). *Appl. Phys. Lett.* **48**, 1559.
Ando, H., Taike, A., Kimura, R., Konagai, M., and Takahashi, K. (1986). *Japan J. Appl. Phys.* **25**(4), L279–281.
Ando, H., Taike, A., Konagai, M., and Takahashi, K. (1987). *Japan J. Appl. Phys.* **62**(4), 1251–1256.
Aven, M., and Prener, J. S. (1967). "Physics and Chemistry of II–VI Compounds." North-Holland Publishing Company, Amsterdam.
Benson, J. D., and Summers, C. J. (1988). *J. Crystal Growth* **86**, 354.
Benson, J. D., Wagner, B. K., Torabi, A., and Summers, C. J. (1986). *Appl. Phys. Lett.* **49**, 1034.
Benson, J. D., Rajavel, D., Wagner, B. K., Benz II, R., and Summers, C. J. (1989). *J. Crystal Growth* **95**, 543.
Bicknell, R. N., Yanka, R. W., Giles-Taylor, N. C., Blanks, D. K., Buckland, E. L., and Schetzina, J. F. (1984a). *Appl. Phys. Lett.* **45**, 92.
Bicknell, R. N., Yanka, R. W., Giles, N. C., Schetzina, J. F., Magee, T. J., Leung, C., and Kawayoshi, H. (1984b). *Appl. Phys. Lett.* **44**, 313.
Bicknell, R. N., Giles-Taylor, N. C., Schetzina, J. F., Anderson, N. G., and Laidig, W. D. (1985a). *Appl. Phys. Lett.* **46**, 238.
Bicknell, R. N., Giles-Taylor, N. C., Blanks, D. K., Schetzina, J. F., Anderson, N. G., and Laidig, W. D. (1985b). *Appl. Phys. Lett.* **46**, 1122.
Bicknell, R. N., Giles, N. C., and Schetzina, J. F. (1986a). *Appl. Phys. Lett.* **49**, 1095.
Bicknell, R. N., Giles, N. C., and Schetzina, J. F. (1986b). *Appl. Phys. Lett.* **49**, 1735.
Bicknell, R. N., Giles, N. C., and Schetzina, J. F. (1987). *Appl. Phys. Lett.* **50**, 691.
Bonsett, T. C., Yamanishi, M., Gunshor, R. L., Datta, S., and Kolodziejski, L. A. (1987). *Appl. Phys. Lett.* **51**, 499.
Bylsma, R. B., Becker, W. M., Bonsett, T. C., Kolodziejski, L. A., Gunshor, R. L., Yamanishi, M., and Datta, S. (1985). *Appl. Phys. Lett.* **47**, 1039.
Cahay, M., McLennan, M., Datta, S., and Lundstrom, M. (1987). *Appl. Phys. Lett.* **50**, 612.
Cammack, D. A., Dalby, R. J., Cornelissen, H. J., and Khurgin, J. (1987). *J. Appl. Phys.* **62**, 3071.
Catalano, I. M., Cingolani, A., Ferrara, M., and Lugara, M. (1982). *Solid State Commun.* **43**, 371.

Chang, L. L. (1986). "Layered Structures and Epitaxy," Vol. **56**, *Proceedings of the Materials Research Society* (J. M. Gibson, G. C. Osbourn, and R. M. Tromp, eds.), p. 267.
Chang, S.-K., Nurmikko, A. V., Kolodziejski, L. A., and Gunshor, R. L. (1986). *Phys. Rev.* **B33**, 2589.
Chang, S.-K., Lee, D., Nakata, H., Nurmikko, A. V., Kolodziejski, L. A., and Gunshor, R. L. (1987). *J. App. Phys.* **62**, 4835–4838.
Cheng, H., DePuydt, J. M., Potts, J. E., and Smith, T. L. (1988). *Appl. Phys. Lett.* **52**, 147.
Cheung, J. T. (1983). *Appl. Phys. Lett.* **43**, 462.
Cho, A. Y. (1976). *J. Appl. Phys.* **47**, 2841.
Choi, C., Otsuka, N., Kolodziejski, L. A., Gunshor, R. L. (1986). *Mat. Res. Soc. Symp. Proc.* **56**, 235.
Chung, Sung-koo, Wu, Y., Wang, K. L., Sheng, N. H., Lee, C. P., and Miller, D. L. (1987). *IEEE Trans. Electron Devices* **ED-34**, 149.
Cirillo, Jr., N., Shur, M. S., Vold, P. J., Abrokwah, J. K., and Tufte, O. N. (1985). *IEEE Elec. Dev. Lett.* **EDL-6**, 645.
Cohen-Solal, G., Bailly, F., and Barbe, M. (1986). *Appl. Phys. Lett.* **49**, 1519–1521.
Cornelissen, H. J., Cammack, D. A., and Dalby, R. J. (1988). *J. Vac. Sci. Technol.* **B6**, 769.
Dean, B. E., and Johnson, C. J. (1987). *Proc. of the Third Intern. Conf. on II–VI Compounds*, Monterey, July 12–17.
de Miguel, J. L., Shibli, S. M., Tamargo, M. C., and Skromme, B. J. (1988). *Appl. Phys. Lett.* **53**, 2065–2067.
Dreifus, D. L., Kolbas, R. M., Harris, K. A., Bicknell, R. N., Harper, R. L., and Schetzina, J. F. (1987). *Appl. Phys. Lett.* **51**, 931.
Ehrenreich, H. (1987). *Science* **235**, 1029.
Farrel, H. H., Tamargo, M. C., and de Miguel, J. L. (1988a). *J. Vac. Sci. Technol.* **B6**, 767.
Farrell, H. H., Nahory, R. E., and Harbison, J. P. (1988b). *J. Vac. Sci. Technol.* **B6**, 779.
Farrell, H. H., de Miguel, J. L., and Tamargo, M. C. (1989). *J. Appl. Phys.* **65**, 4084.
Farrow, R. F. C., Jones, G. R., Williams, G. M., and Young, I. M. (1981). *Appl. Phys. Lett.* **39**, 954.
Farrow, R. F. C., Noreika, A. J., Shirland, F. A., Takei, W. J., Wood, S., Greggi, Jr., J., and Francombe, M. H. (1984). *J. Vac. Sci. Technol.* **A2**, 527.
Farrow, R. F. C., Wood, S., Greggi, Jr., J. C., Takei, W. J., Shirland, F. A., and Furneaux, J. (1985). *J. Vac. Sci. Technol.* **B3**, 681.
Faurie, J. P. (1986). *IEEE J. Quantum Electronics* **QE-22**, 1656.
Faurie, J. P., Sivananthan, S., Boukerche, M., and Reno, J. (1984). *Appl. Phys. Lett.* **45**, 1307.
Faurie, J. P., Shu, C., Sivananthan, S., and Chu, X. (1986). *Surface Science* **168**, 473.
Feldman, R. D., and Austin, R. F. (1986). *Appl. Phys. Lett.* **49**, 954.
Feldman, R. D., Austin, R. F., Dayem, A. H., and Westerwick, E. H., (1986). *Appl. Phys. Lett.* **49**, 797.
Feldman, R. D., Austin, R. F., Bridenbaugh, P. M., Johnson, A. M., and Simpson, W. M. (1988). *J. Appl. Phys.* **64**, 1191.
Fu, Q., Lee, D., Mysyrowicz, A., Nurmikko, A. V., Gunshor, R. L., and Kolodziejski, L. A. (1988). *Phys. Rev.* **B37**, 8791–8794.
Fu, Q., Lee, D., Nurmikko, A. V., Kolodziejski, L. A., and Gunshor, R. L. (1989). *Phys. Rev.* **B39**, 3173.
Fujiyasu, H., Mochizuki, K., Yamazaki, Y., Aoki, M., Kuwabara, H., Nakanishi, Y., Shimaoka, G. (1986). *Surface Science* **174**, 543.
Furdyna, J. K., and Kossut, J. (1986). *Superlattices and Microstructures* **2**, 89.
Furdyna, J. K., Frankel, R. B., and Debska, U., unpublished.
Glass, A. M., Tai, K., Bylsma, R. B., Feldman, R. D., Olson, D. H., and Austin, R. F. (1988). *Appl. Phys. Lett.* **53**, 834.

Glenn, J. L., O, Shunki, Kolodziejski, L. A., Gunshor, R. L., Kobayashi, M., Li, D., Otsuka, N., Haggerott, M., Pelekanos, N., and Nurmikko, A. V. (1989). *J. Vac. Sci. Technol.* **B7**, 249–252.
Golding, T. D., Martinka, M., and Dinan, J. H. (1988). *J. App. Phys.* **64**, 1873.
Grant, E., Kraut, E. A., Kowalczyk, S. P., and Waldrop, J. R. (1983). *J. Vac. Sci. Technol.* **B1**, 320.
Gunshor, R. L., and Kolodziejski, L. A. (1988). *IEEE J. Quantum Electronics* **24**, 1744.
Gunshor, R. L., Kolodziejski, L. A., Melloch, M. R., Vaziri, M., Choi, C., and Otsuka, N. (1987a). *Appl. Phys. Lett.* **50**, 200–202.
Gunshor, R. L., Kolodziejski, L. A., Otsuka, N., Gu, B. P., Lee, D., Hefetz, Y., and Nurmikko, A. V. (1987b). *Superlattices and Microstructures* **3**, 5–8.
Gunshor, R. L., Kolodziejski, L. A., Melloch, M. R., Otsuka, N., and Nurmikko, A. V. (1988). "Growth and Optical Properties of Wide-Gap II–VI Low Dimensional Structures," *NATO ASI Series B*, Vol. **200** (T. C. McGill, C. M. Sotomayor Torres, and W. Gebhardt, eds.). Plenum Press, New York, pp. 229–238.
Gunshor, R. L., Kolodziejski, L. A., Kobayashi, M., Nurmikko, A. V., and Otsuka, N. (1989). "Ultrathin Magnetic Films and Multilayers," *Proceedings of the Materials Research Society Symposium* (B. T. Jonker, J. P. Heremans, and E. E. Marinero, eds.).
Harbison, J. P., and Farrell, H. H. (1988). *J. Vac. Sci. Technol.* **B6**, 733.
Harper, R. L., Hwang, S., Giles, N. C., Bicknell, R. N., Schetzina, J. F., Lee, Y. R., and Ramdas, A. K. (1988). *J. Vac. Sci. Technol.* **A6**, 2627.
Harper, R. L., Hwang, S., Giles, N. C., Schetzina, J. F., Dreifus, D. L., and Myers, T. H. (1989). *Appl. Phys. Lett.* **54**, 170.
Harrison, W. A. (1977). *J. Vac. Sci. Technol.* **14**, 1016.
Harrison, W. A., Kraut, E. A., Waldrop, J. R., Grant, R. W. (1978). *Phys. Rev.* **B18**, 4402.
Hartman, H., Mach, R., and Selle, B. (1982). In "Current Topics in Materials," Vol. **9** (E. Kaldis, ed.). North-Holland Publishing Company, Amsterdam, pp. 1–414.
Hefetz, Y., Nakahara, J., Nurmikko, A. V., Kolodziejski, L. A., Gunshor, R. L., and Datta, S. (1985). *Appl. Phys. Lett.* **47**, 989–991.
Hefetz, Y., Goltsos, W. C., Nurmikko, A. V., Kolodziejski, L. A., and Gunshor, R. L. (1986a). *Appl. Phys. Lett.* **48**, 372–374.
Hefetz, Y., Lee, D., Nurmikko, A. V., Sivananthan, S., Chu, Z., and Faurie, J. P. (1986b). *Phys. Rev.* **B34**, 4423.
Herman, M. A., Jylha, O., and Pessa, M. (1984). *J. Crystal Growth* **66**, 480.
Hishida, Y., Ishii, H., Toda, T., Nina, T. (1989). *J. Crystal Growth* **95**, 517.
Holonyak, N. and Scifres, D. R. (1971). *Rev. Sci. Instrum.* **12**, 1885–1886.
Isaacs, E. D., Heiman, D., Zayhowski, J. J., Bicknell, R. N., and Schetzina, J. F. (1986). *Appl. Phys. Lett.* **48**, 275.
Isshiki, M., Yoshida, T., Igaki, K., Uchida, W., and Suto, S. (1986). *J. Crystal Growth*, **72**, 162.
Jonker, B. T., Krebs, J. J., Qadri, S. B., and Prinz, G. A. (1987a). *Appl. Phys. Lett.* **50**, 848.
Jonker, B. T., Krebs, J. J., Prinz, G. A., and Qadri, S. B. (1987b). *J. Crystal Growth* **81**, 524.
Jonker, B. T., Qadri, S. B., Krebs, J. J., and Prinz, G. A. (1988a). *J. Vac. Sci. Technol.* **A6**, 1946.
Jonder, B. T., Krebs, J. J., Qadri, S. B., Prinz, G. A., Volkening, F., and Koon, N. C. (1988b). *J. Appl. Phys.* **63**, 3303.
Jonker, B. T., Qadri, S. B., Krebs, J. J., and Prinz, G. A. (1988c). *J. Vac. Sci. Technol.* **A6**, 1946.
Jonker, B. T., Krebs, J. J., and Prinz, G. A. (1988d). *Appl. Phys. Lett.* **53**, 450.
Jonker, B. T., Krebs, J. J., and Prinz, G. A. (1988e). *J. Appl. Phys.* **64**, 5340.
Jonker, B. T., Krebs, J. J., and Prinz, G. A. (1988f). "Deposition and Growth: Limits for Microelectronics" (G. W. Rubloff, ed.). American Institute of Physics, New York, p. 347.
Jonker, B. T., Krebs, J. J., and Prinz, G. A. (1988g). *J. Appl. Phys.* **63**, 5885.
Jonker, B. T., Qadri, S. B., Krebs, J. J., and Prinz, G. A. (1989). *J. Vac. Sci. Technol.* **A7**, 1360.

Joyce, B. A., Neave, J. H., Dobson, P. J., and Larson, P. K. (1984). *Phys. Rev.* **B29**, 814–819.
Kitagawa, Fumitaka, Mishima, Tomoyoshi, and Takahashi, Kiyoshi (1980). *J. Electrochem. Soc.* **127**, 937–943.
Kleinman, D. A. (1983). *Phys. Rev.* **B28**, 871.
Kobayashi, M., Mino, N., Konagai, M., and Takahashi, K. (1986a). *Surface Science* **174**, 550.
Kobayashi, M., Mino, N., Katagiri, H., Kimura, R., Konagai, M., and Takahashi, K. (1986b). *J. Appl. Phys.* **60**, 773.
Kohn, E., and Hartnagel, H. L. (1978). *Solid-State Electronics* **21**, 409.
Kolodziejski, L. A., Sakamoto, T., Gunshor, R. L., and Datta, S. (1984a). *Appl. Phys. Lett.* **44**, 799–800.
Kolodziejski, L. A., Bonsett, T. C., Gunshor, R. L., Datta, S., Bylsma, R. B., Becker, W. M., and Otsuka, N. (1984b). *Appl. Phys. Lett.* **45**, 440.
Kolodziejski, L. A., Gunshor, R. L., Bonsett, T. C., Venkatasubramanian, R., Datta, S., Bylsma, R. B., Becker, W. M., and Otsuka, N. (1985a). *Appl. Phys. Lett.* **47**, 169–171.
Kolodziejski, L. A., Gunshor, R. L., Otsuka, N., Zhang, X.-C., Chang, S.-K., and Nurmikko, A. V. (1985b). *Appl. Phys. Lett.* **47**, 882–884.
Kolodziejski, L. A., Gunshor, R. L., Otsuka, N., Gu, B. P., Hefetz, Y., and Nurmikko, A. V. (1986a). *Appl. Phys. Lett.* **48**, 1482–1484.
Kolodziejski, L. A., Gunshor, R. L., Otsuka, N., and Choi, C. (1986b). *J. Vac. Sci. Technol.* **A4**, 2150–2151.
Kolodziejski, L. A., Gunshor, R. L., Otsuka, N., Datta, S., Becker, W. M., and Nurmikko, A. V. (1986b). *IEEE J. Quantum Electronics* **QE-22**, 1666–1676.
Kolodziejski, L. A., Gunshor, R. L., Nurmikko, A. V., and Otsuka, N. (1987a). "Thin Film Growth Techniques for Low Dimensional Structures," Series B, Physics Vol. 163 (R. F. C. Farrow, S. S. P. Parkin, P. J. Dobson, J. H. Neave, and A. S. Arnott, eds.). Plenum Press, New York, pp. 247–260.
Kolodziejski, L. A., Gunshor, R. L., Melloch, M. R., Vaziri, M., Choi, C., and Otsuka, N. (1987b). "Growth of Compound Semiconductors," SPIE Vol. 796 (R. L. Gunshor and H. Morkoç, eds.), pp. 98–103.
Kolodziejski, L. A., Gunshor, R. L., Otsuka, N., Gu, B. P., Hefetz, Y., and Nurmikko, A. V. (1987c). *J. Crystal Growth* **81**, 492–494.
Kolodziejski, L. A., Gunshor, R. L., Fu, Q., Lee, D., Nurmikko, A. V., Gonsalves, J. M., and Otsuka, N. (1988a). *Appl. Phys. Lett.* **52**, 1080.
Kolodziejski, L. A., Gunshor, R. L., Otsuka, N., and Nurmikko, A. V. (1988b). *Mat. Res. Soc. Symp.* **102**, 113.
Kolodziejski, L. A., Gunshor, R. L., and Nurmikko, A. V. (1989). "Strained-Layer Superlattices" (R. M. Biefeld, ed.). Trans Tech Publications, Switzerland.
Krebs, J. J., Rachford, F. J., Lubitz, P., and Prinz, G. A. (1982). *J. Appl. Phys.* **53**, 8058.
Krebs, J. J., Jonker, B. T., and Prinz, G. A. (1987). *J. Appl. Phys.* **61**, 3744.
Krebs, J. J., Jonker, B. T., and Prinz, G. A. (1988). *IEEE Trans. on Magnetics*.
Lee, D., Mysyrowicz, A., Nurmikko, A. V., and Fitzpatrick, B. J. (1987a). *Phys. Rev. Lett.* **58**, 1475.
Lee, D., Chang, S.-K., Nakata, H., Nurmikko, A. V., Kolodziejski, L. A., and Gunshor, R. L. (1987b). *Mat. Res. Soc. Symp. Proc.* **77**, 252–258.
Lee, D., Chang, S.-K., Nakata, H., Nurmikko, A. V., Kolodziejski, L. A., and Gunshor, R. L. (1987c). *J. Phys.* **C5**, 311.
Lee, Y. and Ramdas, A. K. (1988). *Phys. Rev.* **B38**, 10600.
Lee, Y., Ramdas, A. K., Kolodziejski, L. A., and Gunshor, R. L. (1988). *Phys. Rev.* **B38**, 143–149.
Lewis, B. F., Lee, T. C., Grunthaner, F. J., Madhukar, A., Fernanadez, R., and Maserjian, J. (1984). *J. Vac. Sci. Technol.* **B2**, 419–424.

Liu, X., Petrou, A., Jonker, B. T., Prinz, G. A., Krebs, J. J., and Warnock, J. (1988a). *J. Vac. Sci. Technol.* **A6**, 1508.
Liu, X., Petrou, A., Jonker, B. T., Prinz, G. A., Krebs, J. J., and Warnock, J. (1988b). *Appl. Phys. Lett.* **53**, 476.
Lo, Yawcheng, Bicknell, R. N., Myers, T. H., Schetzina, J. F., and Stadelmaier, H. H. (1983). *J. Appl. Phys.* **54**, 4238.
Mackey, K. J., Allen, P. M. G., Herrenden-Harker, W. G., Williams, R. H., Whitehouse, C. R., and Williams, G. M. (1986). *Appl. Phys. Lett.* **49**, 354.
Madjukar, A. (1983). *Surface Science* **132**, 344.
Magnea, M., Dal'Bo, F., Pautrat, J. L., Million, A., DeCioccio, L., and Feuillet, G. (1986). "Materials for Infrared Detectors and Sources," *Proceedings of the Materials Research Society* (R. F. C. Farrow, J. F. Schetzina, and J. T. Cheung, eds.).
Mar, H. A., Chee, K. T., and Salansky, N. (1984a). *Appl. Phys. Lett.* **44**, 237–239.
Mar, H. A., Salansky, N., and Chee, K. T. (1984b). *Appl. Phys. Lett.* **44**, 898–900.
Mathine, D. L., Durbin, S. M., Gunshor, R. L., Kobayashi, M., Menke, D. R., Pei, Z., Gonsalves, J., Otsuka, N., Fu, Q., Haggerott, M., and Nurmikko, A. V. (1989). *Appl. Phys. Lett.* **54**, xxx.
Matsumoto, K., Ogura, M., Wada, T., Hashizume, N., Yao, T., and Hayashi, Y. (1984). *Electron. Lett.* **20**, 462.
McCaldin, J. O., and McGill, T. C. (1988). *J. Vac. Sci. Technol.* **B6**, 1360.
Meiners, L. G. (1978). *J. Vac. Sci. Technol.* **15**, 1402.
Menda, K., Takayasu, I., Minato, T., and Kawashima, M. (1987). *Japan J. Appl. Phys.* **26**, L1326.
Menendez, J., Pinczuk, A., Valladares, J. P., Feldman, R. D., and Austin, R. F. (1987). *Appl. Phys. Lett.* **50**, 1101.
Miles, R., Wu, G., Johnson, M., McGill, T., Faurie, J. P., and Sivananthan, S. (1986). *Appl. Phys. Lett.* **48**, 1383.
Miller, D. A. B., Smith, S. D., and Wherrett, B. S. (1980). *Opt. Commun.* **35**, 221.
Miller, R. C., Kleinman, D. A., Gossard, A. C., and Munteanu, O. (1982). *Phys. Rev.* **B25**, 6545.
Mimura, T., Hiyamizu, S., Fujii, T., and Nanbu, K. (1980). *Japan J. Appl. Phys.* **19**, L225.
Mino, N., Kobayashi, M., Konagai, M., and Takahashi, K. (1985). *J. Appl. Phys.* **58**, 793.
Missous, M., and Singer, K. E. (1987). *Appl. Phys. Lett.* **50**, 694.
Monfroy, G., Sivananthan, S., Chu, X., Faurie, J. P., Knox, R. D., and Staudenmann, J. L. (1986). *Appl. Phys. Lett.* **49**, 152.
Neave, J. H., Blood, P., and Joyce, B. A. (1980). *Appl. Phys. Lett.* **36**, 311.
Niina, T., Yoneda, K., Toda, T., Minato, T., and Hishida, Y. (1982). *Collected Paper of MBE-CST-2*, Tokyo.
Nishitani, K., Ohkata, R., and Murotani, T. (1983). *J. Elec. Mat.* **12**, 619–623.
Nozue, Y., Itoh, M., and Cho, K. (1981). *J. Phys. Soc. Japan* **50**, 889.
Nurmikko, A. V., Zhang, X.-C., Chang, S.-K., Kolodziejski, L. A., Gunshor, R. L., and Datta, S. (1985). *J. Lumin.* **34**, 89.
Ohishi, M., Saito, H., Okano, H., Ohmori, K. (1989). *J. Crystal Growth* **95**, 538.
Ohkawa, K., Mitsuyu, T., and Yamazaki, O. (1986). *Extended Abstracts of the 18th Conference on Solid State Sevices and Materials*, Tokyo, pp. 635–638.
Ohkawa, K., Mitsuyu, T., and Yamazaki, O. (1987). *J. Crystal Growth* **86**, 329–334.
Olego, D. J., Faurie, J. P., Sivananthan, S., and Raccah, P. M. (1985). *Appl. Phys. Lett.* **47**, 1172.
Otsuka, N., Kolodziejski, L. A., Gunshor, R. L., Datta, S., Bicknell, R. N., and Schetzina, J. F. (1985a). *Appl. Phys. Lett.* **46**, 860–862.
Otsuka, N., Kolodziejski, L. A., Gunshor, R. L., Datta, S., Bicknell, R. N., and Schetzina, J. F. (1985b). *Mater. Res. Soc. Symp. Proc.* **37**, 449–454.
Pajaczkowska, A. (1978). *Prog. Crystal Growth Charact.* **1**, 289–326.
Park, R. M., and Mar, H. A. (1986). *J. Mater. Res.* **1**, 543.

Park, R. M., and Salansky, N. M. (1984). *Appl. Phys. Lett.* **44**, 249–251.
Park, R. M., Mar, H. A., and Salansky, N. M. (1985a). *J. Vac. Sci. Technol.* **B3**, 676–680.
Park, R. M., Mar, H. A., and Salansky, N. M. (1985b). *J. Appl. Phys.* **58**, 1047–1049.
Park, R. M., Mar, H. A., and Salansky, N. M. (1985c). *J. Vac. Sci. Technol.* **B3**, 1637.
Park, R. M., Mar, H. A., and Kleiman, J. (1988). *J. Crystal Growth*, **86**, 335.
Petrou, A., Warnock, J., Bicknell, R. N., Giles-Taylor, N. C., and Schetzina, J. F. (1985). *Appl. Phys. Lett.* **46**, 692.
Peyghambarian, N., Park, S. H., Koch, S. W., Jeffery, A., Potts, J. E., and Cheng, H. (1988). *Appl. Phys. Lett.* **52**, 182–184.
Ponce, F. A., Anderson, G. B., and Ballingall, J. M. (1986). *Surface Science* **168**, 564–570.
Potts, J. E., Smith, T. L., and Cheng, H. (1987). *Appl. Phys. Lett.* **50**, 7.
Price, G. L. (1981). "Collected Papers of the 2nd International Symposium on Molecular Beam Epitaxy and Related Clean Surface Techniques," Japan Soc. Appl. Phys., Tokyo, p. 259.
Prinz, G. A., Jonker, B. T., Krebs, J. J., Ferrari, J. M., and Kovanic, F. (1986). *Appl. Phys. Lett.* **48**, 1756.
Qadri, S. B., and Dinan, J. H. (1985). *Appl. Phys. Lett.* **47**, 1066.
Qadri, S. B., Skelton, E. F., Webb, A. W., and Kennedy, J. (1985). *Appl. Phys. Lett.* **46**, 257–259.
Qadri, S. B., Jonker, B. T., Prinz, G. A., and Krebs, J. J. (1988a). *Thin Solid Films* **164**, 111.
Qadri, S. B., Jonker, B. T., Prinz, G. A., and Krebs, J. J. (1988b). *J. Vac. Sci. Technol.* **A6**, 1526.
Qian, Q.-D., Qiu, J., Melloch, M. R., Cooper, Jr., J. A., Kolodziejski, L. A., and Gunshor, R. L. (1989). *Appl. Phys. Lett.* **54**, 1359.
Singh, J., and Bajaj, K. K. (1984). *J. Vac. Sci. Technol.* **132**, 276.
Skromme, B. J., Tamargo, M. C., Turco, F. S., Shibli, S. M., Nahory, R. E., and Bonners, W. A. (1988). Paper presented at the Electrochemical Society Meeting, Chicago, October.
Smith, D. L., and Pickhardt, V. Y. (1975). *J. Appl. Phys.* **41**, 2366–2374.
Solomon, P. M., and Morkoç, H. (1984). *IEEE Trans. Electron Devices* **ED-31**, 1015.
Solomon, P. M., Knoedler, C. M., and Wright, S. L. (1984). *IEEE Elec. Dev. Lett.* **EDL-5**, 379.
Srinivasa, R., Panish, M. B., and Temkin, H. (1987). *Appl. Phys. Lett.* **50**, 1441–1443.
Stucheli, N., and Bucher, E. (1989). *J. Electronic Materials* **18**, 105.
Studtman, G. D., Gunshor, R. L., Kolodziejski, L. A., Melloch, M. R., Cooper, Jr., J. A., Pierret, R. F., Munich, D. P., Choi, C., and Otsuka, N. (1988). *Appl. Phys. Lett.* **52**, 1249.
Sugiyama, K. (1982). *J. Crystal Growth* **60**, 450.
Suh, E.-K., Bartholomew, D. U., Ramdas, A. K., Venugopalan, S., Kolodziejski, L. A., and Gunshor, R. L. (1987). *Phys. Rev.* **B36**, 4316.
Takeda, T., Kurosu, T., Lida, M., and Yao, T. (1986). *Surface Science* **174**, 548
Tamargo, M. C., de Miguel, J. L., Hwang, D. M., and Farrell, H. H. (1988a). *J. Vac. Sci. Technol.* **B6**, 784–787.
Tamargo, M. C., de Miguel, J. L., Turco, F. S., Skromme, B. J., Hwang, D. M., Nahory, R. E., and Farrell, H. H. (1988b). "Growth and Optical Properties of Wide-Gap II-VI Low Dimensional Structures," *NATO ASI Series* **B**, Vol. **200** (T. C. McGill, C. M. Sotomayer Torres, and W. Gebhardt, eds.). Plenum Press, New York, pp. 239–243.
Tamargo, M. C., de Miguel, J. L., Hwang, D. M., Skromme, B. J., Meynadier, M. W., Nahory, R. E., and Farrell, H. H. (1988c). *Mat. Res. Soc. Symp. Proc.* **102**, 125.
Tersoff, J. (1986). *Phys. Rev. Lett.* **56**, 2755.
Tu, D.-W., and Kahn, A. (1984). *J. Vac. Sci. Technol.* **A2**, 511.
Tu, D.-W., and Kahn, A. (1985). *J. Vac. Sci. Technol.* **A3**, 922.
Tuffigo, H., Cox, R. T., Magnea, N., d'Aubigne, Y. M., Million, A. (1988). *Phys. Rev.* **B37**, 4310.
Twardowski, A., Dietl, T., and Demianuk, M. (1983). *Solid State Commun.* **48**, 845–848.
Twardowski, A., Ortenberg, M von, and Demianiuk, M. (1987). *Solid State Commun.* **64**, 63.
Van Welzenis, R. G., and Ridley, B. K. (1984). *Solid-State Electron.* **27**, 113.

Vaziri, M., Reifenberger, R., Gunshor, R. L., Kolodziejski, L. A., Venkatesan, S., and Pierret, R. F. (1989). *J. Vac. Sci. Technol.* **B7**, 253–258.
Venkatasubramanian, R., Otsuka, N., Datta, S., Kolodziejski, L. A., and Gunshor, R. L. (1987) "Growth of Compound Semiconductors," *SPIE Proceedings*, Vol. **796** (R. L. Gunshor and H. Morkoc, eds.). p. 121.
Venkatasubramanian, R., Otsuka, N., Qiu, J., Kolodziejski, L. A., and Gunshor, R. L. (1989). *J. Crystal Growth* **95**, 533.
Venkatesan, S., Pierret, R. F., Gunshor, R. L., Qiu, J., Kobayashi, M., and Kolodziejski, L. A. (1989). Submitted to *J. Appl. Phys.*
Venugopalan, S., Kolodziejski, L. A., Gunshor, R. L., and Ramdas, A. K. (1984). *Appl. Phys. Lett.* **45**, 974.
Waldrop, J. R., Kowalczyk, S. P., Grant, R. W., Kraut, E. A., and Miller, D. L. (1981). *J. Vac. Sci. Technol.* **19**, 573.
Warnock, J., Petrou, A., Bicknell, R. N., Giles-Taylor, N. C., Blanks, D. K., and Schetzina, J. F. (1985). *Phys. Rev.* **B32**, 8116.
Wei, S. H., and Zunger, A. (1986). *Phys. Rev. Lett.* **56**, 2391.
Wieder, H. H. (1978). *J. Vac. Sci. Technol.* **15**, 1498.
Williams, G. M., Whitehouse, C. R., Chew, N. G., Blackmore, G. W., and Cullis, A. G. (1985). *J. Vac. Sci. Technol.* **B3**, 704.
Wood, S., Greggi, Jr., J., Farrow, R. F. C., Takei, W. J., Shirland, F. A., and Noreika, A. J. (1984). *J. Appl. Phys.* **55**, 4225.
Wyckoff, R. W. G. (1963). "Crystal Structures." Wiley, New York, p. 124.
Yao, T. (1985a). "The Technology and Physics of Molecular Beam Epitaxy" (E. H. C. Parker, ed.). Plenum Press, New York, p. 313.
Yao, T. (1985b). *J. Crystal Growth* **72**, 31–40.
Yao, T. (1986). *Japan J. Appl. Phys.* **25**, 821–827.
Yao, T. (1988). "Growth and Optical Properties of Wide-Gap II–VI Low Dimensional Structures," *NATO ASI Series* B, Vol. **200** (T. C. McGill, C. M. Sotomayor Torres, and W. Gebhardt, eds.). Plenum Press, New York, pp. 209–218.
Yao, T., and Ogura, M. (1982). *Collected Paper of MBE-CST-2*, Tokyo.
Yao, T., and Taguchi, T. (1984). *Proceedings of the 13th International Conference on Defects in Semiconductors*, Coronado, California, August 12–17.
Yao, T., and Takeda, T. (1986). *Appl. Phys. Lett.* **48**, 160–162.
Yao, T., Miyoshi, Y., Makita, Y., and Maekawa, S. (1977). *Japan J. Appl. Phys.* **16**, 369–370.
Yao, T., Makita, Y., and Maekawa, S. (1978). *J. Crystal Growth* **45**, 309.
Yao, T., Makita, Y., and Maekawa, S. (1979). *Appl. Phys. Lett.* **35**, 97–98.
Yao, T., Ogura, M., Matsuoka, S., and Morishita, T. (1983). *Appl. Phys. Lett.* **43**, 499–501.
Yao, T., Taneda, H., and Funaki, M. (1986). *Japan J. Phys.* **25**, L952.
Yasuda, T., Mitsuishi, I., Kukimoto, H. (1988). *Appl. Phys. Lett.* **52**, 57.
Yoneda, K., Hishida, Y., Toda, T., Ishii, H., and Niina, T. (1984). *Appl. Phys. Lett.* **45**, 1300–1302.
Zahn, D. R. T., Mackey, K. J., Williams, R. H., Munder, H., Geurts, J., and Richter, W. (1986). Presented at the 4th International Conference on Molecular Beam Epitaxy, York, England, September 7–10.
Zahn, D. R. T., Mackey, K. J., Williams, R. H., Munder, H., Geurts, J., and Richter, W. (1987). *Appl. Phys. Lett.* **50**, 742.
Zhang, X.-C., Chang, S.-K., Nurmikko, A. V., Kolodziejski, L. A., Gunshor, R. L., and Datta, S. (1985a). *Phys. Rev.* **B31**, 4056.
Zhang, X.-C., Chang, S.-K., Nurmikko, A. V., Heiman, D., Kolodziejski, L. A., Gunshor, R. L., and Datta, S. (1985b). *Solid State Commun.* **56**, 255.

Index

A

Activated transport, 273
Activation barrier, 254
Al(Ga)As/GaAs, 19–20, 60, 61, 62–63
Alkaline earth alloy, IV-VI, 320
Alloy
  clustering, 63–64
  disorder, 168
  ordering, 62–63
  II-VI compounds, 353–361
    dilute magnetic semiconductors, 353–354
      MBE growth of
        (Cd,Mn)Te, 361
        (Cd,Zn)Te, 361
        Zn(S,Se), 360
        (Zn,Co)Se, 357-260
        Zn(Se,Te), 360–361
          HREM image, 356
          lattice parameter, 358
          magnetization measurements, 359
        (Zn,Fe)Se, 356–358
          HREM image, 356
          lattice parameter, 357
          magnetization measurements, 357
          optical bandgap, 357
        (Zn,Mn)Se, 353–355
          energy bandgap, 355
          lattice constant, 354
Annealing, 104
Antimony, 252
Atomic layer epitaxy, 7
Atomic number, 225
Auger electron spectroscopy, sputter depth profiles, 92–93
Avalanche photodetection, 288
Azimuth (RHEED), 249

B

Band alignment, 74, 88, 257–273
  conduction band, 88–90, 103, 108
  table, 88, 273
  IV-VI, 329
  type I, 258, 268, 273
  type II, 263, 268, 273
  valence band, 88
Band dispersion, 268, 278, 280
Bandgap
  direct, 283
  InAs mole fraction dependence, 87
  indirect, 281
  narrowing, 288
  quasi-direct, 278, 281
Band offsets, *see* Band alignment
Band structure, 85–89, 274–287
Barium fluoride
  epitaxy, 327
  interface defects, 330
  substrate, 325
Bond length, 285
Boron, 256
  compounds, 257
  nitride, pyrolytic, 247
Boundary, 205
Brillouin mode, 277
Brillouin zone, 274
Buffer
  Burgers vector, 234
  design, 242, 245
  layer, 175, 212
  relaxed, 262
  strain adjusting, 242
  symmetrizing, 241, 273, 285

C

Carbon impurities, 5–6
Carriers, majority, 299

Cathodo-luminescence (CL), 79-80
Cadmium telluride,
  CdTe/GaAs, 16, 17
  growth, 341-344
    compound source, 341
    flux ratio, 341
    MBE, 341
  superlattices, 381-386
    (Cd,Mn)Te multiple
      quantum wells, 381-386
        HREM image, 383
        photoluminescence, 383
        stimulated emission, 384-386
        CdTe/(Cd,Zn)Te, 386
Channeling, 148, 156
Charged particles, 248
Chemical vapor deposition, 8, 55
Cladding layer, 259, 298
Clustering, 99
Coherence, epitaxial, 143
Coherent structures, 212
Commensurate epitaxy, 141
Conduction band, 110, 116
  diagrams, 116, 117
  discontinuity, 88-90, 110
  dispersion, 278, 280
  effective mass, 85
  offsets, 88-90, 105-106, 110, 271
  splitting, 266, 270
Confinement
  carriers, 281, 284
  phonons, 277
Continuum elastic, 276
$CoSi_2$/Si, 7
Covalent radius, 226
Critical layer thickness (CLT), 29-33, 43-47, 77-78, 90-91, 98, 237, 262, 283
Critical temperature, 254
Crystal quality, 155, 166, 254
Current gain, 300
C-V profiling, 88-89, 94

D

De Broglie wavelength, 257
Debye temperature, 227
Dechanneling, 160-167
  catastropic, 164
  cross section, 167
  dislocations, 167

Deep level transient spectroscopy (DLTS), 94-96
Defects, interface, 94-98
Deformation potential, 269
Density functional, 267, 271
Desorption
  dopant, 256
  electron stimulated, 344-345
  laser illumination, 344
  sputter, 256
  thermal, 256
Diamond, 226
Diamond lattice, 172
Dielectric constant, 229
Diffraction, electron
  LEED, 61, 64
  RHEED, 58-59, 64, 91-92, 190, 322, 342-345, 351, 356-357, 364-365, 372-373, 397-399,
  TEM, 55-58, 62, 64, 79, 128-129
  X-ray, 61-64, 170, 178
    analysis, 186, 276
    diffractometer, 174, 180, 182
    position-sensitive detector, 184
    rocking curve, 79, 276, 369, 374, 391, 396-397
    satellites, 176, 187
    strain, 174, 187
    topography, 195
Diffractometer, 180
  double-crystal diffraction, 174
  triple-axis, 182
Dilute magnetic semiconductors, 353-354
Dimer row, 249
Diode
  avalanche photodetector, 288
  detectors, IV-VI, 312
  impact avalanche transit time (IMPATT), 296
  lasers
    IV-VI, 312, 318, 319
    III-V, 120-127
  mixed tunneling avalanche transit time (MITATT), 296
  photodetector, 288
  quantum well injection (QWITT), 299
  resonant tunneling, 296
Dislocations
  Burgers vector, 234
  dechanneling, 167

distance, 236
edge, 147
energy, 237, 240
filtering (by strained-layer superlattices), 50–51
imaging, 195
misfit, 78–84, 147, 200, 235
nucleation, 208, 239
properties, 27–29
Doping
adlayer, 253
antimony (Sb), 252
boron (B), 256
boron compounds, 257
coevaporational, 252, 254
delta-doping, 351–352
desorption, 256
electrical activation, 252, 254
flash-off, 256
IV-VI, 315, 323
gallium (Ga), 256
incorporation, 254
inhomogeneity, 254
ion implantation, 252
low temperature (LTD), 254
modulation, 259
phosphorus (P), 262
potential enhanced (PED), 253
pre-build-up, 253, 256
recoil implantation, 253
secondary implantation (DSI), 253, 264
segregation, 253, 256
selective, 246
sources, 247, 253
spontaneous incorporation, 256
unintentional, 252
vapor pressure, 252, 256
Double positioning, 16, 61
DX-center, 102, 103, 110

E

Effective mass, 85, 229, 260, 270
Effusion cell, 247, 253
Eigenvalue, 257
Elasticity, modulus of, 145
Electron beam induced imaging (EBIC), 196, 217
Electroreflectance, 283

Electron diffraction
LEED, 61, 64
RHEED, 58–59, 64, 91–92, 190, 322, 342–345, 351, 356–357, 364–365 372–373, 397–399
TEM, 55–58, 62, 64, 79, 128–129
Elemental group IV substrates, 375
Energy band alignment, 74, 88, 257, 266
conduction band, 88–90, 103, 108
IV-VI, 329
type I, 258, 268, 273
type II, 263, 268, 273
valence band, 88
Energy dispersion, conduction band, 280
Epitaxy
atomic layer, 7
barium flouride, 327
CdTe, 341–344
CdTe/GaAs, 16–17
CdTe/InSb, 388–392
(Cd,Mn)Te, 361
(Cd,Zn)Te, 361
chemical vapor deposition (CVD), 246
coherent, 143
differential, 301
GaAs/Si, 11–15, 21–22, 43–44
GeSi/Si, 18–19, 32, 36–42, 46, 47, 56, 57
InAsSb/GaAs, 22–25
InGaAs/GaAs, 18, 31–35, 46, 52–53, 62
layer, 143
limited reaction processing (LRP), 246, 302
metal-on-semiconductor structure
  Fe on ZnSe, 361–363
  Magnetization, 361–363
  Ni on ZnSe, 361–363
MnSe, 397–400
MnTe, 400–402
photoassisted MBE, 346–351
Si/SiGe, 18–19, 32, 36–42, 46, 47, 56, 57, 226, 244
ZnSe/GaAs, 371–375, 392–396
Zn(S,Se), 360
ZnTe/(Ga,Al)Sb, 396–397
Etch-contrast microscopy, 197
Europium alloy, 318
Evaporation source for molecular beam epitaxy, 246–247
Ewald construction, 178
Excess stress, 199, 206

Exchange integrals
   (Zn,Co)Se, 360
   (Zn,Fe)Se, 357–358

F

Fermi level, 261
Field effect transistor (FET)
   high electron mobility (HEMT), 291
   metal-oxide-semi-conductor (MOSFET), 257
   modulation doped (MODFET), 103, 106, 290
      charge-control, 113
      complimentary, 295
      current-gain cut-off frequency, 102, 103, 106–115
      effective electron velocity, 107–116
      electron supplying layer, 113, 116
      knee voltage, 106
      modulation doping, 102
      modulation efficiency, 108, 113–116
      multiple quantum well, 295
      output power, 102–103, 109
         n-channel, 106
         p-channel, 104
      noise figure, 102–103
      real-space transfer, 110
      saturated-velocity, 113–116
      scattering parameters (S-parameters), 106–107
      sheet density, 103, 110–16
      spacer-layer thickness, 101, 111–113
      structure, 100–102
      transconductance, 102, 106–107, 112
      undoped spacer, 112
         n-type, 295
         p-type, 292
      two-dimensional electron gas (TEGFET), 291
      ZnSe/GaAs field effect transistors, 374–375
Flux control
   electron impact emission (EIES), 248
   quadrupole mass spectrometer (QMS), 248
   quartz crystal microbalance, 249
Flux ratio
   ZnSe, 341, 348–350
   ZnTe, 341
IV elements, properties of, table, 225

IV-VI alloy semiconductors
   doping, 323
   energy band gap, 317
   energy band offsets, 329
   heterojunction, 219, 329
   lattice constant, 317
   magnetic, 332
   material property, 313, 318
   RHEED, 322
   sources, MBE, 314
   substrate material, 325
Frequency
   FET current-gain cut-off 102, 103, 106–115
   laser microwave response, 120, 125–126

G

GaAs/Si, 11–15, 21–22, 43–44, 52
Gallium (Ga), 256
Gas source molecular beam epitaxy (GSMBE), 7–8
Germanium (Ge), 226
GeSi/Si, 18–19, 32, 36–42, 46–47, 56–57, 226, 244
GeSn, 226
Graphite, 247, 256
Graphite, pyrolytic, 256
Growth
   conditions for MBE, 89
   II-VI compounds by MBE, 340–345
      compound source, 341
      energy bandgap, 340
      flux ratio, 341
      lattice parameter, 340
      pseudomorphic, 259, 261
      RHEED, 342

H

Hall effect
   measurements, 98, 99
   mobility
      CdTe:As, 347
      electron, 86–87, 98–100 229, 263
      IV-VI, 320
      hole, 85–86, 229
      2DEG, 97, 99, 260–265
   scattering
      alloy, 262

INDEX   **415**

    Coulomb, 261
    ionized impurity, 259, 261
Heterojunction
    bipolar transistor (HBT), 97–98, 118, 299–303
    IV-VI compound, 219, 318, 329, 332
    TEM study, 318, 332
Heterostructure
    CdTe/InSb, 388–392
    coherency, 145
    interfacial layer, 388, 391, 395
    interface state density, 394–395
    laser, buried heterostructure, 125
    lattice-mismatched substrates, 363–375
        CdTe on (100) GaAs, 363–369
        CdTe on InSb, 369–371
        MnSe/ZnSe, 376–381, 397
        MnTe/CdTe, 401–402
        ZnSe on GaAs, 371–375
    II-VI/III-V heterostructures, 386–397
        CdTe/InSb, 388–392
        ZnSe/GaAs, 392–396
        ZnTe/(Ga,Al)Sb, 396–397
High resolution transmission electron microscopy
    (HREM), 55–58, 62, 64, 366–368
Hole
    heavy, 85, 104, 260, 298
    light, 85, 104, 298
    non-parabolicity, 85
    trapping, 289

I

IMPATT, 296
Implantation
    damage, 252, 254
    secondary, 253, 264
Incoherent strained layer, 198
Infrared detectors, 287
In situ electron microscopy, 36–42, 61, 65–66
Interdiffusion, 256
Interface
    diffuseness, 54–61
    quality, 251, 261
    rectification, 96–97
    roughness, 19–20, 54–61, 99
    scattering, 94–98
Ion-beam analysis, 147
Ion channeling, 155, 164–66
Ionized cluster beam epitaxy, IV-VI, 330

Ionized impurity scattering, 111, 259, 261
Iso-relaxation contours, 207

K

Kinematic theory, 187
Kinetical limitation, 254

L

Laser
    buried heterostructure, 125
    confinement factor, 122
    degradation, 124
    microwave frequency response, 120, 125–126
    operating lifetime, 124
    oscillations, 377
    quantum-well, 120–125
    surface-emitting, 120, 127
    threshold currents, 120–124
        table, 122
    wavelength, 120–122
Lattice mismatch, 131, 158, 228, 230, 237, 259, 325, 331
Lattice parameter, 140, 145, 155–156, 227, 317, 340, 354, 358, 400
    table, 140, 159, 227, 325, 340
Layer
    composition, 149
    electron supplying, 113, 116
    nucleation, 16–25, 65
    spacer thickness, 101, 111–113
    strained, 143, 198–202, 240
    thickness, 149
Leakage current, 290
Local-density approximation, 280
Low energy electron diffraction (LEED), 61, 64
Luminescence, 286

M

Magnetic semiconductors, 353–354
Magnetization measurements, 357, 399–400
Magnetoresistance, 260, 265
Many-body effects, 271
Mass, effective, 260, 279
Matthews–Blakeslee criterion, 200
Melting point, 227

Mercury (Hg) alloys of IV-VI compounds, 315
Metastable compounds, 354–355, 397–402
MnSe, 354, 397–400
    magnetization measurements, 399–400
    MBE growth, 397–400
    MnSe/ZnSe superlattices, 397
    RHEED intensity oscillations, 397–399
MnTe, 400–402
    energy bandgap, 400
    lattice parameter, 400
    MBE growth, 400–401
    MnTe/CdTe quantum well structures, 401–402
Microwave frequency response, laser, 120, 125–126
Misfit dislocations, 26–27, 29–52, 65, 78–84, 147, 200, 235
    interactions of, 34, 40–42, 65
    metastable, 141
    nucleation of, 33, 34, 35–36, 65, 208, 239
    propagation of, 33, 34, 37–40, 65
    table, 159, 229
Mobility
    electron, 97–103, 229, 263, 321, 352
    IV-VI, 320
    Hall, 259, 263
    hole, 229, 259, 262
Modulation doped field effect transistors (MODFET)
    charge-control, 113
    complementary, 295
    current-gain cut-off frequency, 102, 103, 106–115
    effective electron velocity, 107–116
    electron supplying layer, 113, 116
    modulation doping, 102
    modulation efficiency, 108, 113–116
    multiple quantum well, 295
    n-type, 295
    output power, 102–103, 109
        p-channel, 104
        n-channel, 106
        noise figure, 102–103
    p-type, 292
    real-space transfer, 110
    saturated-velocity, 113, 114, 116
    scattering parameters (S-parameters), 106–107
    sheet density, 103, 110–116
    spacer-layer thickness, 101, 111–113
    structure, 100–102
    transconductance, 102, 106, 107, 112
    undoped spacer, 112
Modulators, 127
    blue-shift from strain, 127
Modulus of elasticity table, 232
Molecular beam epitaxy (MBE), 3–7, 55, 89, 246–257, 340–345
    congruent sublimation temperature, 90–91
    doping, 253–256, 345–353
    effusion cell sources, 314, 340–342
    evaporation sources, 248–249
    growth conditions, 89, 340–345
    reflection high-energy electron diffraction, 58–59, 64, 91–92, 249–251, 322, 342–343
    substrate temperatures, 90
    surface segregation, 92
Molecular dynamics simulations, 65, 345
Monte Carlo simulation,
    RHEED, 345
    Rutherford backscattering, 152–154
Mosaic broadening of the reciprocal lattice, 177–178

N

Natural superlattice, 318
Ni overlayer on ZnSe, 361–363
Non-parabolicity, 85
Nucleation,
    epitaxial layers, 16–25, 65, 250
    misfit dislocations, 33–36, 65

O

Offset, valence band, 74, 88, 257–273, 329
Optical
    absorption, 289
    detectors, 127, 130–131
    fibre, 290
    transition, first-order, 283
Optimized lattice constant epitaxy (OLE), 127–131
Ordering, spontaneous alloy, 62–63
Oscillator strength, 281, 284
Output power (MODFET), 102–103, 109
    n-channel, 106

noise figure, 102–103
p-channel, 104
Oxygen as an impurity, 5–6

P

Particulates, 6
Patterned
    area substrates, 80–81, 99
    growth, 52, 66, 80–81, 99
Peak-to-valley-ratio, 298
Phonon
    acoustic, 274
    confined mode, 277
    dispersion, 276
    folded mode, 274, 278, 285
    optical, 265, 275
Phosphorus (P), 262
Photocurrent, 284
Photodiode, 288
Photoemission spectroscopy, 61–62
Photoluminescence
    microscopy, 79
    spectroscopy, 19–20, 59–60, 79, 83, 346–351
    topography, 198
Pinch-off, 292
Poisson
    equation, 257
    number, 232
    ratio, 145
Position-sensitive X-ray detector, 184
Potential well, 258, 297
Pseudomorphic, 259, 261
Pseudopotential, 267

Q

Quadrupole mass spectrometer (QMS), 248
Quantum efficiency, 288
Quantum well
    (Cd,Mn)Te multiple, 381–386
    confinement
        carriers, 281, 284
        phonons, 277
    diodes
        quantum well injection (QWITT), 299
        resonant tunneling, 296
    energy band alignment, 74, 88, 257, 266

laser, 120–125
MODFET, 295
phonons
    confined mode, 277
    dispersion, 276
    folded mode, 285
    optical, 265, 275
    zone folding, 274, 278
Quartz crystal microbalance, 249
Quasi-direct energy gap, 278, 281

R

Raman scattering, 61, 64, 192, 264, 266, 274
Rare earth alloy, IV-VI, 316
Real-space transfer, 110
Reciprocal
    lattice vector, 274
    space, 170
Recoil implantation, 253
Reflection high energy electron diffraction (RHEED)
    Monte Carlo simulations, 345
    nucleation, 343–344
    pattern, 58–59, 64, 91–92, 190, 249, 322, 342–345, 351, 356–357, 364–365, 372–373, 397–399
    intensity oscillations, 343, 351
    Se-stabilized, 342–343, 345, 351
    surface reconstruction, 342, 344
    Te-stabilized, 344
    Zn-stabilized, 342–343,
Relaxation, 199, 202, 204, 207
Resonant tunneling, 119, 296
Rocking curves, X-ray
    analysis, 276
    spectra, 79, 217, 276, 369, 374, 391, 396–397, 181
Rutherford backscattering spectroscopy (RBS), 147–159

S

Satellites, X-ray, 187
Saturated velocity, FET, 113, 114, 116
Scanning transmission electron microscopy (STEM), 58
Scattering
    alloy, 262

Coulomb, 261
  ionized impurity, 259, 261
  parameters (S-parameters), 106–107
Schrödinger equation, 257
Self-consistent calculation, 257, 261, 267, 270
Shear modulus, 146
Sheet density of carriers, 103, 110–116
Shubnikov–deHaas oscillations, 260, 265
SiGe/Si, 18–19, 32, 36–42, 46–47, 56–57, 226, 244
Silicon (Si), 226
Silicon carbide (SiC), 226
Solid-phase recrystallization, 256
Sound velocity, 276
Space charge, 258, 261
Spacer layer
  electron supplying, 113, 116, 260
  thickness, 101, 111–113
  undoped, 112
Spectroscopy
  Auger electron spectroscopy (AES), 92–93
  deep level transient spectroscopy (DLTS), 94–96
  optical absorption, 289
  photoemission, 61–62
  photoluminescence, 19–20, 59–60, 79, 83, 346–351
  quadrupole mass spectrometry (QMS), 248
  Raman, 61, 64, 192, 264, 266, 274
  Rutherford backscattering (RBS), 147–159
Stability
  diagram, 203
  meta-stability, 238
  strained layer superlattices, 240
Stable-metastable boundary, 210
Strain
  analysis
    angular scans, 156
    catastrophic dechanneling, 164
    dechanneling, 160
  asymmetric, 259
  biaxial, 266
  compressive, 143–145, 263, 267
  distribution, 155, 241
  energy, 233
  hydrostatic, 266
  in-plane, 232
  nonperiodic, 214
  relaxation, 205

relief, 83, 201
  double-kink, 83
  single-kink, 83
  splitting, 85, 266
  subband, 257, 270
  symmetrization, 241, 263
  tensile, 143, 263, 267
  tetragonal distortion, 140, 232
  uniaxial, 266
  X-ray, 174, 187
Strained layer
  characterization, 143
  critical layer thickness (CLT), 29–33, 43–47, 77–78, 90–91, 98, 237, 262, 283
  incoherent, 143, 198
  perfect coherence, 198
  relaxation, 199, 202
  strain relieved, 198
  superlattice, 241, 274
Substitutional doping of II-VI compounds
  CdTe:As, 346, 347
    mobility, 347
    photoluminescence, 347
  CdTe:In, 346
  CdTe:Sb, 346
  DLTS, 351
  high resistivity undoped ZnSe, 348–349
  photoassisted MBE, 346–351
  photoluminescence, 346–347, 350–351
  p-type ZnSe, 352–353
  p-type ZnTe, 353
  ZnSe:Cl, 349
  ZnSe:Ga, 348–352
    delta-doping, 351–352
  Zn vacancies, 350–351
Substrate cleaning, 8–16
Superlattice,
  buffer, 243–245
  (Cd,Mn)Te, 381–386
    HREM image, 383
    photoluminescence, 383
    stimulated emission, 384–386
  CdTe/(Cd,Zn)Te, 386
  composition, 332
  doping, 312
  laser oscillations, 377
  natural, 318
  nonlinear excitonic absorption, 378
  strained-layer, 241, 274

symmetry, 229, 280
TEM, 55–58, 62, 64, 79, 128–129, 318, 332, 366–368, 383
ZnSe, 376–386
   laser oscillations, 377
   nonlinear excitonic absorption, 378
   (Zn,Mn)Se quantum wells, 376–381
   ZnSe/ZnTe, 379–381
      atomic layer epitaxy, 380
      optical properties, 380–381
      pseudoalloy, 380–381
      self-trapping of excitons, 380–381
Surface
   morphology, 191, 251
   reconstruction, 249
Surface-emitting laser, 120, 127

T

Ternary clustering, 87
Terracing, 177
Temperature, critical, 254
Tetragonal distortion, 140, 232
Thermal
   annealing, 49
   conductivity, 229
   expansion, 227–230
Thickness, critical, 29–33, 43–47, 77–78, 90–91, 98, 237, 262, 283
Threading dislocations, 30–34, 47–52, 65–66
Threshold currents, 120–124
   table, 122
Tin (Sn), 226
Tilt angle, 158
Topography
   photoluminescence, 79, 198
   X-ray, 195–196, 216
Transconductance, 292–293
Transfer
   carrier, 259, 271
   charge, 261
Transistor
   field-effect, 100–116, 290, 374–375
   heterojunction bipolar transistor (HBT), 97–98, 118, 219, 299–303, 318, 329, 332
   high electron mobility (HEMT), 100–116, 291
   metal-oxide-semi-conductor (MOSFET), 257
   modulation doped (MODFET), 103, 106, 290
   charge-control, 113
   complimentary, 295
   current-gain cut-off frequency, 102, 103, 106–115
   effective electron velocity, 107–116
   electron supplying layer, 113, 116
   knee voltage, 106
   modulation doping, 102
   modulation efficiency, 108, 113–116
   multiple quantum well, 295
   noise figure, 102–103
   output power, 102–103, 109
      p-channel, 104
      n-channel, 106
   real-space transfer, 110
   saturated-velocity, 113–116
   scattering parameters (S-parameters), 106–107
   sheet density, 103, 110–116
   spacer-layer thickness, 101, 111–113
   structure, 100–102
   transconductance, 102, 106–107, 112
   undoped spacer, 112
      n-type, 295
      p-type, 292
   two-dimensional electron gas (TEGFET), 291
   ZnSe/GaAs field effect transistors, 374–375
Transmission electron microscopy (TEM)
   cross-section, 23, 46, 125, 129
   dark field, 390
   diffraction, 55–58, 62–64, 79, 128–129
   HBT study, 318–322
   high-resolution electron microscopy (HREM), 12, 17, 356, 359, 366, 368, 383
   plan view, 13, 24, 38, 81–82, 195–196, 318–322
Tunneling, resonant, 119, 296
Two-dimensional electron gas (2DEG)
   carriers, 257
   charge-control, 113
   effective electron velocity, 107–116
   electron gas, 263–265
   electron supplying layer, 113, 116
   hole gas (2DHG), 260
   modulation doping, 102

modulation efficiency, 108, 113–116
multiple quantum well, 295
real-space transfer, 110
saturated-velocity, 113–116
sheet density, 103, 110–116
spacer-layer thickness, 101, 111–113
structure, 100–102
undoped spacer, 112
    n-type, 295
    p-type, 292
II-VI compounds
  energy bandgap, 340
  growth by molecular beam epitaxy, 340
    compound source, 341
    flux ratio, 341
    lattice parameter
      wurtzite, 340
      zincblende, 340
    RHEED, 342
  substrates, 375–376
    CdTe, 376
    ZnSe, 375–376
II-VI on lattice-mismatched substrates
  CdTe-on-InSb, 369–371
    photoluminescence, 370
    X-ray rocking curves, 369
  CdTe on (100) GaAs substrates, 363–369
    HREM interface imaging, 366–368
    (100)-oriented CdTe, 363–369
    (111)-oriented CdTe, 363–369
  ZnSe-on-GaAs, 371–375
    bulk versus epilayer, 371–374
    nucleation, 371–373
    RHEED intensity oscillations, 372–373
    X-ray rocking curves, 374–375
II-VI/III-V heterostructures
  CdTe/InSb, 388–392
    interfacial layers, 388, 391
    MBE growth, 388–390
    photoluminescence, 391–392
    TEM of multiple quantum well structures, 390
    X-ray rocking curves, 391
  ZnSe/GaAs, 392–396
    field effect transistors, 392–393
    interface state density, 394–395
    interfacial layers, 395
    MIS capacitors, 393–395

ZnTe/(Ga,Al)Sb, 396–397
  photoluminescence, 396–397
  X-ray rocking curves, 396–397

U

Ultra-high vacuum, 245
Umklapp process, 275
Uniaxial strain, 266

V

Vacancies, IV-VI, 323
Valence band
  degeneracy, 84–85
  effective mass, 84, 120
  exchange integral, 358–360
  heavy-hole, 85, 104
  light-hole, 85, 104
  non-parabolicity, 85
  offset, 74, 88, 257–273, 379, 387
  splitting, 85, 268
  table, 273
Valley splitting, 266, 270
Vapor pressure, dopant, 252, 256
Vegard's law deviation, table, 230

W

Waveguide, 288
Wurtzite, lattice parameter, 340

X

X-ray, 61–64, 170, 178
  analysis, 186, 276
  diffraction, 61–64, 170
  diffractometer, 180
    double-crystal, 174
    triple-axis, 182
  position-sensitive detector, 184
  rocking curves
    analysis, 276
    spectra, 79, 217, 276, 369, 374, 391, 396–397, 181
  satellites, 176, 187
  strain, 174, 187
  topography, 195

Y

Ytterbium, alloy, 318

Z

Zinc-blende lattice, 173, 340
(Zn,Co)Se
  exchange integrals, 360
  HREM image, 356
  lattice parameter, 358
  magnetization measurements, 359
(Zn,Fe)Se
  exchange integrals, 357–358
  HREM image, 356
  lattice parameter, 357
  magnetization measurements, 357
  optical bandgap, 357
(Zn,Mn)Se
  energy bandgap, 355
  lattice constant, 354
  multiple quantum wells, 376–381
ZnSe
  compound source, 341
  doping
    delta-doping, 351–352
    p-type ZnSe, 352–353
    ZnSe:Cl, 349
    ZnSe:Ga, 348–352
  elemental source, 341
  field effect transistors, 374–375
  flux ratio, 341, 348–350
  high resistivity, 348–349
  MBE growth, 341
  photoassisted MBE, 346–351
  photoluminescence, 346–347, 350–351
  substrate, 375
  superlattice, 376–386
    laser oscillations, 377
    nonlinear excitonic absorption, 378
  Zn vacancies, 350–351
Zn(Se,Te)
  compound source, 341
  elemental source, 341
  flux ratio, 341
  MBE growth, 341
Zn(S,Se), 360
  MBE growth, 341–342
ZnSe/ZnTe, 379–381
  atomic layer epitaxy, 380
  optical properties, 380–381
  pseudoalloy, 380–381
  self-trapping of excitons, 380–381
ZnTe, 341, 344
  p-type ZnTe, 353
  substrates, 344
Zone folding, 274, 278–287

Contents of Previous Volumes

Volume 1 Physics of III–V Compounds

C. Hilsum, Some Key Features of III–V Compounds
Franco Bassani, Methods of Band Calculations Applicable to III–V Compounds
E. O. Kane, The $k \cdot p$ Method
V. L. Bonch-Bruevich, Effect of Heavy Doping on the Semiconductor Band Structure
Donald Long, Energy Band Structures of Mixed Crystals of III–V Compounds
Laura M. Roth and Petros N. Argyres, Magnetic Quantum Effects
S. M. Puri and T. H. Geballe, Thermomagnetic Effects in the Quantum Region
W. M. Becker, Band Characteristics near Principal Minima from Magnetoresistance
E. H. Putley, Freeze-Out Effects, Hot Electron Effects, and Submillimeter Photoconductivity in InSb
H. Weiss, Magnetoresistance
Betsy Ancker-Johnson, Plasmas in Semiconductors and Semimetals

Volume 2 Physics of III–V Compounds

M. G. Holland, Thermal Conductivity
S. I. Novkova, Thermal Expansion
U. Piesbergen, Heat Capacity and Debye Temperatures
G. Giesecke, Lattice Constants
J. R. Drabble, Elastic Properties
A. U. Mac Rae and G. W. Gobeli, Low-Energy Electron Diffraction Studies
Robert Lee Mieher, Nuclear Magnetic Resonance
Bernard Goldstein, Electron Paramagnetic Resonance
T. S. Moss, Photoconduction in III–V Compounds
E. Antončik and J. Tauc, Quantum Efficiency of the Internal Photoelectric Effect in InSb
G. W. Gobeli and F. G. Allen, Photoelectric Threshold and Work Function
P. S. Pershan, Nonlinear Optics in III–V Compounds
M. Gershenzon, Radiative Recombination in the III–V Compounds
Frank Stern, Stimulated Emission in Semiconductors

Volume 3 Optical of Properties III–V Compounds

Marvin Hass, Lattice Reflection
William G. Spitzer, Multiphonon Lattice Absorption
D. L. Stierwalt and R. F. Potter, Emittance Studies
H. R. Philipp and H. Ehrenreich, Ultraviolet Optical Properties
Manuel Cardona, Optical Absorption above the Fundamental Edge
Earnest J. Johnson, Absorption near the Fundamental Edge
John O. Dimmock, Introduction to the Theory of Exciton States in Semiconductors
B. Lax and J. G. Mavroides, Interband Magnetooptical Effects

H. Y. Fan, Effects of Free Carriers on Optical Properties
Edward D. Palik and George B. Wright, Free-Carrier Magnetooptical Effects
Richard H. Bube, Photoelectronic Analysis
B. O. Seraphin and H. E. Bennett, Optical Constants

Volume 4 Physics of III–V Compounds

N. A. Goryunova, A. S. Borschevskii, and D. N. Tretiakov, Hardness
N. N. Sirota, Heats of Formation and Temperatures and Heats of Fusion of Compounds $A^{III}B^V$
Don L. Kendall, Diffusion
A. G. Chynoweth, Charge Multiplication Phenomena
Robert W. Keyes, The Effects of Hydrostatic Pressure on the Properties of III–V Semiconductors
L. W. Aukerman, Radiation Effects
N. A. Goryunova, F. P. Kesamanly, and D. N. Nasledov, Phenomena in Solid Solutions
R. T. Bate, Electrical Properties of Nonuniform Crystals

Volume 5 Infrared Detectors

Henry Levinstein, Characterization of Infrared Detectors
Paul W. Kruse, Indium Antimonide Photoconductive and Photoelectromagnetic Detectors
M. B. Prince, Narrowband Self-Filtering Detectors
Ivars Melngailis and T. C. Harman, Single-Crystal Lead–Tin Chalcogenides
Donald Long and Joseph L. Schmit, Mercury–Cadmium Telluride and Closely Related Alloys
E. H. Putley, The Pyroelectric Detector
Norman B. Stevens, Radiation Thermopiles
R. J. Keyes and T. M. Quist, Low Level Coherent and Incoherent Detection in the Infrared
M. C. Teich, Coherent Detection in the Infrared
F. R. Arams, E. W. Sard, B. J. Peyton, and F. P. Pace, Infrared Heterodyne Detection with Gigahertz IF Response
H. S. Sommers, Jr., Microwave-Based Photoconductive Detector
Robert Sehr and Rainer Zuleeg, Imaging and Display

Volume 6 Injection Phenomena

Murray A. Lampert and Ronald B. Schilling, Current Injection in Solids: The Regional Approximation Method
Richard Williams, Injection by Internal Photoemission
Allen M. Barnett, Current Filament Formation
R. Baron and J. W. Mayer, Double Injection in Semiconductors
W. Ruppel, The Photoconductor–Metal Contact

Volume 7 Application and Devices
Part A

John A. Copeland and Stephen Knight, Applications Utilizing Bulk Negative Resistance
F. A. Padovani, The Voltage–Current Characteristics of Metal–Semiconductor Contacts
P. L. Hower, W. W. Hooper, B. R. Cairns, R. D. Fairman, and D. A. Tremere, The GaAs Field-Effect Transistor
Marvin H. White, MOS Transistors
G. R. Antell, Gallium Arsenide Transistors
T. L. Tansley, Heterojunction Properties

Part B

T. Misawa, IMPATT Diodes
H. C. Okean, Tunnel Diodes
Robert B. Campbell and Hung-Chi Chang, Silicon Carbide Junction Devices
R. E. Enstrom, H. Kressel, and L. Krassner, High-Temperature Power Rectifiers of $GaAs_{1-x}P_x$

Volume 8   Transport and Optical Phenomena

Richard J. Stirn, Band Structure and Galvanomagnetic Effects in III–V Compounds with Indirect Band Gaps
Roland W. Ure, Jr., Thermoelectric Effects in III–V Compounds
Herbert Piller, Faraday Rotation
H. Barry Bebb and E. W. Williams, Photoluminescence I: Theory
E. W. Williams and H. Barry Bebb, Photoluminescence II: Gallium Arsenide

Volume 9   Modulation Techniques

B. O. Seraphin, Electroreflectance
R. L. Aggarwal, Modulated Interband Magnetooptics
Daniel F. Blossey and Paul Handler, Electroabsorption
Bruno Batz, Thermal and Wavelength Modulation Spectroscopy
Ivar Balslev, Piezooptical Effects
D. E. Aspnes and N. Bottka, Electric-Field Effects on the Dielectric Function of Semiconductors and Insulators

Volume 10   Transport Phenomena

R. L. Rode, Low-Field Electron Transport
J. D. Wiley, Mobility of Holes in III–V Compounds
C. M. Wolfe and G. E. Stillman, Apparent Mobility Enhancement in Inhomogeneous Crystals
Robert L. Peterson, The Magnetophonon Effect

Volume 11   Solar Cells

Harold J. Hovel, Introduction; Carrier Collection, Spectral Response, and Photocurrent; Solar Cell Electrical Characteristics; Efficiency; Thickness; Other Solar Cell Devices; Radiation Effects; Temperature and Intensity; Solar Cell Technology

Volume 12   Infrared Detectors (II)

W. L. Eiseman, J. D. Merriam, and R. F. Potter, Operational Characteristics of Infrared Photodetectors
Peter R. Bratt, Impurity Germanium and Silicon Infrared Detectors
E. H. Putley, InSb Submillimeter Photoconductive Detectors
G. E. Stillman, C. M. Wolfe, and J. O. Dimmock, Far-Infrared Photoconductivity in High Purity GaAs
G. E. Stillman and C. M. Wolfe, Avalanche Photodiodes
P. L. Richards, The Josephson Junction as a Detector of Microwave and Far-Infrared Radiation
E. H. Putley, The Pyroelectric Detector—An Update

Volume 13 Cadmium Telluride

Kenneth Zanio, Materials Preparation; Physics; Defects; Applications

Volume 14 Lasers, Junctions, Transport

N. Holonyak, Jr. and M. H. Lee, Photopumped III–V Semiconductor Lasers
Henry Kressel and Jerome K. Butler, Heterojunction Laser Diodes
A. Van der Ziel, Space-Charge-Limited Solid-State Diodes
Peter J. Price, Monte Carlo Calculation of Electron Transport in Solids

Volume 15 Contacts, Junctions, Emitters

B. L. Sharma, Ohmic Contacts to III–V Compound Semiconductors
Allen Nussbaum, The Theory of Semiconducting Junctions
John S. Escher, NEA Semiconductor Photoemitters

Volume 16 Defects, (HgCd)Se, (HgCd)Te

Henry Kressel, The Effect of Crystal Defects on Optoelectronic Devices
C. R. Whitsett, J. G. Broerman, and C. J. Summers, Crystal Growth and Properties of $Hg_{1-x}Cd_xSe$ Alloys
M. H. Weiler, Magnetooptical Properties of $Hg_{1-x}Cd_xTe$ Alloys
Paul W. Kruse and John G. Ready, Nonlinear Optical Effects in $Hg_{1-x}Cd_xTe$

Volume 17 CW Processing of Silicon and Other Semiconductors

James F. Gibbons, Beam Processing of Silicon
Arto Lietoila, Richard B. Gold, James F. Gibbons, and Lee A. Christel, Temperature Distributions and Solid Phase Reaction Rates Produced by Scanning CW Beams
Arto Lietoila and James F. Gibbons, Applications of CW Beam Processing to Ion Implanted Crystalline Silicon
N. M. Johnson, Electronic Defects in CW Transient Thermal Processed Silicon
K. F. Lee, T. J. Stultz, and James F. Gibbons, Beam Recrystallized Polycrystalline Silicon: Properties, Applications, and Techniques
T. Shibata, A. Wakita, T. W. Sigmon, and James F. Gibbons, Metal–Silicon Reactions and Silicide
Yves I. Nissim and James F. Gibbons, CW Beam Processing of Gallium Arsenide

Volume 18 Mercury Cadmium Telluride

Paul W. Kruse, The Emergence of $(Hg_{1-x}Cd_x)Te$ as a Modern Infrared Sensitive Material
H. E. Hirsch, S. C. Liang, and A. G. White, Preparation of High-Purity Cadmium, Mercury, and Tellurium
W. F. H. Micklethwaite, The Crystal Growth of Cadmium Mercury Telluride
Paul E. Petersen, Auger Recombination in Mercury Cadmium Telluride
R. M. Broudy and V. J. Mazurczyck, (HgCd)Te Photoconductive Detectors
M. B. Reine, A. K. Sood, and T. J. Tredwell, Photovoltaic Infrared Detectors
M. A. Kinch, Metal-Insulator-Semiconductor Infrared Detectors

Volume 19 Deep Levels, GaAs, Alloys, Photochemistry

G. F. Neumark and K. Kosai, Deep Levels in Wide Band-Gap III–V Semiconductors
David C. Look, The Electrical and Photoelectronic Properties of Semi-Insulating GaAs
R. F. Brebrick, Ching-Hua Su, and Pok-Kai Liao, Associated Solution Model for Ga–In–Sb and Hg–Cd–Te
Yu. Ya. Gurevich and Yu. V. Pleskov, Photoelectrochemistry of Semiconductors

Volume 20 Semi-Insulating GaAs

R. N. Thomas, H. M. Hobgood, G. W. Eldridge, D. L. Barrett, T. T. Braggins, L. B. Ta, and S. K. Wang, High-Purity LEC Growth and Direct Implantation of GaAs for Monolithic Microwave Circuits
C. A. Stolte, Ion Implantation and Materials for GaAs Integrated Circuits
C. G. Kirkpatrick, R. T. Chen, D. E. Holmes, P. M. Asbeck, K. R. Elliott, R. D. Fairman, and J. R. Oliver, LEC GaAs for Integrated Circuit Applications
J. S. Blakemore and S. Rahimi, Models for Mid-Gap Centers in Gallium Arsenide

Volume 21 Hydrogenated Amorphous Silicon
Part A

Jacques I. Pankove, Introduction
Masataka Hirose, Glow Discharge; Chemical Vapor Deposition
Yoshiyuki Uchida, dc Glow Discharge
T. D. Moustakas, Sputtering
Isao Yamada, Ionized-Cluster Beam Deposition
Bruce A. Scott, Homogeneous Chemical Vapor Deposition
Frank J. Kampas, Chemical Reactions in Plasma Deposition
Paul A. Longeway, Plasma Kinetics
Herbert A. Weakliem, Diagnostics of Silane Glow Discharges Using Probes and Mass Spectroscopy
Lester Guttman, Relation between the Atomic and the Electronic Structures
A. Chenevas-Paule, Experimental Determination of Structure
S. Minomura, Pressure Effects on the Local Atomic Structure
David Adler, Defects and Density of Localized States

Part B

Jacques I. Pankove, Introduction
G. D. Cody, The Optical Absorption Edge of a-Si:H
Nabil M. Amer and Warren B. Jackson, Optical Properties of Defect States in a-Si:H
P. J. Zanzucchi, The Vibrational Spectra of a-Si:H
Yoshihiro Hamakawa, Electroreflectance and Electroabsorption
Jeffrey S. Lannin, Raman Scattering of Amorphous Si, Ge, and Their Alloys
R. A. Street, Luminescence in a-Si:H
Richard S. Crandall, Photoconductivity
J. Tauc, Time-Resolved Spectroscopy of Electronic Relaxation Processes
P. E. Vanier, IR-Induced Quenching and Enhancement of Photoconductivity and Photoluminescence
H. Schade, Irradiation-Induced Metastable Effects
L. Ley, Photoelectron Emission Studies

Part C

Jacques I. Pankove, Introduction
J. David Cohen, Density of States from Junction Measurements in Hydrogenated Amorphous Silicon
P. C. Taylor, Magnetic Resonance Measurements in a-Si:H
K. Morigaki, Optically Detected Magnetic Resonance
J. Dresner, Carrier Mobility in a-Si:H
T. Tiedje, Information about Band-Tail States from Time-of-Flight Experiments
Arnold R. Moore, Diffusion Length in Undoped a-Si:H
W. Beyer and H. Overhof, Doping Effects in a-Si:H
H. Fritzsche, Electronic Properties of Surfaces in a-Si:H
C. R. Wronski, The Staebler–Wronski Effect
R. J. Nemanich, Schottky Barrier on a-Si:H
B. Abeles and T. Tiedje, Amorphous Semiconductor Superlattices

Part D

Jacques I. Pankove, Introduction
D. E. Carlson, Solar Cells
G. A. Swartz, Closed-Form Solution of I–V Characteristic for a-Si:H Solar Cells
Isamu Shimizu, Electrophotography
Sachio Ishioka, Image Pickup Tubes
P. G. LeComber and W. E. Spear, The Development of the a-Si:H Field-Effect Transistor and Its Possible Applications
D. G. Ast, a-Si:H FET-Addressed LCD Panel
S. Kaneko, Solid-State Image Sensor
Masakiyo Matsumura, Charge-Coupled Devices
M. A. Bosch, Optical Recording
A. D'Amico and G. Fortunato, Ambient Sensors
Hiroshi Kukimoto, Amorphous Light-Emitting Devices
Robert J. Phelan, Jr., Fast Detectors and Modulators
Jacques I. Pankove, Hybrid Structures
P. G. LeComber, A. E. Owen, W. E. Spear, J. Hajto, and W. K. Choi, Electronic Switching in Amorphous Silicon Junction Devices

Volume 22 Lightwave Communication Technology
Part A

Kazuo Nakajima, The Liquid-Phase Epitaxial Growth of InGaAsP
W. T. Tsang, Molecular Beam Epitaxy for III–V Compound Semiconductors
G. B. Stringfellow, Organometallic Vapor-Phase Epitaxial Growth of III–V Semiconductors
G. Beuchet, Halide and Chloride Transport Vapor-Phase Deposition of InGaAsP and GaAs
Manijeh Razeghi, Low-Pressure Metallo-Organic Chemical Vapor Deposition of $Ga_xIn_{1-x}As_yP_{1-y}$ Alloys
P. M. Petroff, Defects in III–V Compound Semiconductors

Part B

J. P. van der Ziel, Mode Locking of Semiconductor Lasers
Kam Y. Lau and Amnon Yariv, High-Frequency Current Modulation of Semiconductor Injection Lasers
Charles H. Henry, Spectral Properties of Semiconductor Lasers
Yasuharu Suematsu, Katsumi Kishino, Shigehisa Arai, and Fumio Koyama, Dynamic Single-Mode Semiconductor Lasers with a Distributed Reflector
W. T. Tsang, The Cleaved-Coupled-Cavity (C^3) Laser

Part C

R. J. Nelson and N. K. Dutta, Review of InGaAsP/InP Laser-Structures and Comparison of Their Performance
N. Chinone and M. Nakamura, Mode-Stabilized Semiconductor Lasers for 0.7–0.8- and 1.1–1.6-μm Regions
Yoshiji Horikoshi, Semiconductor Lasers with Wavelengths Exceeding 2 μm
B. A. Dean and M. Dixon, The functional Reliability of Semiconductor Lasers as Optical Transmitters
R. H. Saul, T. P. Lee, and C. A. Burrus, Light-Emitting Device Design
C. L. Zipfel, Light-Emitting Diode Reliability
Tien Pei Lee and Tingye Li, LED-Based Multimode Lightwave Systems
Kinichiro Ogawa, Semiconductor Noise-Mode Partition Noise

Part D

Federico Capasso, The Physics of Avalanche Photodiodes
T. P. Pearsall and M. A. Pollack, Compound Semiconductor Photodiodes
Takao Kaneda, Silicon and Germanium Avalanche Photodiodes
S. R. Forrest, Sensitivity of Avalanche Photodetector Receivers for High-Bit-Rate Long-Wavelength Optical Communication Systems
J. C. Campbell, Phototransistors for Lightwave Communications

Part E

Shyh Wang, Principles and Characteristics of Integratable Active and Passive Optical Devices
Shlomo Margalit and Amnon Yariv, Integrated Electronic and Photonic Devices
Takaaki Mukai, Yoshihisa Yamamoto, and Tatsuya Kimura, Optical Amplification by Semiconductor Lasers

Volume 23 Pulsed Laser Processing of Semiconductors

R. F. Wood, C. W. White, and R. T. Young, Laser Processing of Semiconductors: An Overview
C. W. White, Segregation, Solute Trapping, and Supersaturated Alloys
G. E. Jellison, Jr., Optical and Electrical Properties of Pulsed Laser-Annealed Silicon
R. F. Wood and G. E. Jellison, Jr., Melting Model of Pulsed Laser Processing
R. F. Wood and F. W. Young, Jr., Nonequilibrium Solidification Following Pulsed Laser Melting
D. H. Lowndes and G. E. Jellison, Jr., Time-Resolved Measurements During Pulsed Laser Irradiation of Silicon
D. M. Zehner, Surface Studies of Pulsed Laser Irradiated Semiconductors
D. H. Lowndes, Pulsed Beam Processing of Gallium Arsenide
R. B. James, Pulsed CO_2 Laser Annealing of Semiconductors
R. T. Young and R. F. Wood, Applications of Pulsed Laser Processing

Volume 24 Applications of Multiquantum Wells, Selective Doping, and Superlattices

C. Weisbuch, Fundamental Properties of III–V Semiconductor Two-Dimensional Quantized Structures: The Basis for Optical and Electronic Device Applications
H. Morkoç and H. Unlu, Factors Affecting the Performance of (Al, Ga)As/GaAs and (Al, Ga)As/InGaAs Modulation-Doped Field-Effect Transistors: Microwave and Digital Applications
N. T. Linh, Two-Dimensional Electron Gas FETs: Microwave Applications
M. Abe et al., Ultra-High-Speed HEMT Integrated Circuits
D. S. Chemla, D. A. B. Miller, and P. W. Smith, Nonlinear Optical Properties of Multiple Quantum Well Structures for Optical Signal Processing
F. Capasso, Graded-Gap and Superlattice Devices by Band-gap Engineering
W. T. Tsang, Quantum Confinement Heterostructure Semiconductor Lasers
G. C. Osbourn et al., Principles and Applications of Semiconductor Strained-Layer Superlattices

Volume 25 Diluted Magnetic Semiconductors

W. Giriat and J. K. Furdyna, Crystal Structure, Composition, and Materials Preparation of Diluted Magnetic Semiconductors
W. M. Becker, Band Structure and Optical Properties of Wide-Gap $A^{II}_{1-x}Mn_xB^{VI}$ Alloys at Zero Magnetic Field
Saul Oseroff and Pieter H. Keesom, Magnetic Properties: Macroscopic Studies
T. Giebultowicz and T. M. Holden, Neutron Scattering Studies of the Magnetic Structure and Dynamics of Diluted Magnetic Semiconductors
J. Kossut, Band Structure and Quantum Transport Phenomena in Narrow-Gap Diluted Magnetic Semiconductors
  C. Riqaux, Magnetooptics in Narrow Gap Diluted Magnetic Semiconductors
J. A. Gaj, Magnetooptical Properties of Large-Gap Diluted Magnetic Semiconductors
J. Mycielski, Shallow Acceptors in Diluted Magnetic Semiconductors: Splitting, Boil-off, Giant Negative Magnetoresistance
A. K. Ramdas and S. Rodriquez, Raman Scattering in Diluted Magnetic Semiconductors
P. A. Wolff, Theory of Bound Magnetic Polarons in Semimagnetic Semiconductors

Volume 26 III–V Compound Semiconductors and Semiconductor Properties of Superionic Materials

Zou Yuanxi, III–V Compounds
H. V. Winston, A. T. Hunter, H. Kimura, and R. E. Lee, InAs-Alloyed GaAs Substrates for Direct Implantation
P. K. Bhattacharya and S. Dhar, Deep Levels in III–V Compound Semiconductors Grown by MBE
Yu. Ya. Gurevich and A. K. Ivanov-Shits, Semiconductor Properties of Superionic Materials

Volume 27 Highly Conducting Quasi-One-Dimensional Organic Crystals

E. M. Conwell, Introduction to Highly Conducting Quasi-One-Dimensional Organic Crystals
I. A. Howard, A Reference Guide to the Conducting Quasi-One-Dimensional Organic Molecular Crystals
J. P. Pouget, Structural Instabilities
E. M. Conwell, Transport Properties
C. S. Jacobsen, Optical Properties
J. C. Scott, Magnetic Properties
L. Zuppiroli, Irradiation Effects: Perfect Crystals and Real Crystals

Volume 28   **Measurement of High-Speed Signals in Solid State Devices**

J. Frey and D. Ioannou, Materials and Devices for High-Speed and Optoelectronic Applications
H. Schumacher and E. Strid, Electronic Wafer Probing Techniques
D. Auston, Picosecond Photoconductivity
J. Valdmanis, Electrooptic Measurement Techniques
R. Jain and J. Wiesenfeld, Direct Optical Probing of Integrated Circuits and High-Speed Devices
G. Plows, Electron Beam Probing
A. M. Weiner and R. B. Marcus, Photoemissive Probing

Volume 29   **Very High Speed Integrated Circuits: Gallium Arsenide LSI**

M. Kuzuhara, T. Nozaki, and H. Hashimoto, Active Layer Formation by Ion Implantation
H. Hashimoto, Focused Ion Beam Implantation
T. Nozaki and A. Higashisaka, Device Fabrication Process Technology
M. Ino and T. Takada, GaAs LSI Circuit Design
M. Hirayama, M. Ohmori, and K. Yamasaki, GaAs LSI Fabrication and Performance

Volume 30   **Very High Speed Integrated Circuits: Heterostructure**

H. Watanabe, T. Mizutani, and A. Usui, Fundamentals of Epitaxial Growth and Atomic Layer Epitaxy
S. Hiyamizu, Molecular Beam Epitaxy for High Quality Active Layers
T. Nakanisi, Metal Organic Vapor Phase Epitaxy for High Quality Active Layers
T. Mimura, High Electron Mobility Transistor and LSI Applications
T. Sugeta and T. Ishibashi, Hetero-Bipolar Transistor and LSI Applications
H. Matsueda, T. Tanaka, and M. Nakamura, Opto-Electronic Integrated Circuits

Volume 31   **Indium Phosphide: Crystal Growth and Characterization**

J. P. Farges, Growth of Dislocation-Free InP
M. J. McCollum and G. E. Stillman, High Purity InP Grown by Hybride Vapor Phase Epitaxy
T. Inada and T. Fukuda, Direct Synthesis and Growth of Indium Phosphide by the Liquid Phosphorous Encapsulated Czochralski Method
O. Oda, K. Katagiri, K. Shinohara, S. Katsura, U. Takahashi, K. Kainosho, K. Kohiro, and R. Hirano, InP Crystal Growth, Substrate Preparation and Evaluation
K. Tada, M. Tatsumi, M. Morioka, T. Araki, and T. Kawase, InP Substrates: Production and Quality Control
M. Razeghi, LP-MOCVD Growth, Characterization, and Application of InP Material
T. A. Kennedy and P. J. Lin-Chung, Stoichiometric Defects in InP

Volume 32   **Strained-Layer Superlattices: Physics**

T. P. Pearsall, Strained-Layer Superlattices
Fred H. Pollak, Effects of Homogeneous Strain on the Electronic and Vibrational Levels in Semiconductors
J. Y. Marzin, J. M. Gérard, P. Voisin, and J. A. Brum, Optical Studies of Strained III-V Heterolayers
R. People and S. A. Jackson, Structurally Induced States from Strain and Confinement
M. Jaros, Microscopic Phenomena in Ordered Superlattices